Quantitative

Electron Microprobe Analysis

by Roger Theisen

*With 7 Figures
and 6 Annexed Tables*

Springer-Verlag Berlin Heidelberg GmbH 1965

Dipl.-Ing. Dr. rer. nat. ROGER THEISEN

Euratom Brussels
delegated to the Karlsruhe Fast Breeder Project
Institute for Applied Reactor Physics

Additional material to this book can be downloaded from http://extras.springer.com

ISBN 978-3-662-23130-2   ISBN 978-3-662-25106-5 (eBook)
DOI 10.1007/978-3-662-25106-5

Library of Congress Catalog Card Number 65-25013

Title No. 1317

Preface

The Electron Microprobe X-Ray Analyser conceived by R. CASTAING and A. GUINIER in 1949 has been developed as an extremely powerful tool in spectrochemical analysis for a wide range of applications, ranging from qualitative elementary distribution studies, to highly localised quantitative analysis on a one micron scale.

With the increasing number of versatile instruments, commercially available, the domain of applications — in metallurgy, solid state physics, mineralogy and geology, biology and medicine, arts and archeology — is rapidly expanding, particularly because reliable quantitative analyses can be achieved.

It is well established that in multicomponent specimens, the relative x-ray intensity generated by the electron bombardment — i.e. the intensity ratio of the characteristic x-ray radiation emitted under identical experimental conditions by the specimen and a calibration standard — is not directly correlated to the elementary mass concentration. The use of a wide scale of carefully prepared homogeneous calibration standards is generally very tedious and restricted to binary systems. For more complex specimens, the conversion of recorded x-ray intensity ratios to elementary mass concentration requires, besides carefule selection of experimental conditions, an adequate correction calculation to take account of the various physical phenomenas occurring in the target — electron retardation, electron backscattering, x-ray excitation efficiency, fluorescence enhancement by characteristic and continuous radiation and x-ray mass absorption.

When only an order of magnitude correction is required, an empirical and simplified correction formulae may be used. However any treatment which will yield quantitative results of significant reliability requires the use of an elaborate mathematical treatment. While some of the data — fluorescence yields, x-ray mass absorption coefficients, electron backscatter effect — require improvement, the success achieved with theoretical approaches, when used as performance controls for numerous test preparations from different laboratories, has demonstrated that quantitative analyses may be achieved in any system using only pure elements as calibration standards.

It is hoped that the following computed tables and the annexed data compilations, without requiring a detailed study of the quantitative theory or the use of a well programmed high speed digital computer, are of some help to the average user, in calculating the necessary corrections.

For the more directly interested critical reader, a survey of the physical basis of present approximation methods and a guideline for the development of the proposed approach is included.

Table of Contents

§ 1 General Features of Electron Microprobe Analysis

1.1 Introduction

In his thesis (Paris 1951) R. CASTAING(1) published not only the conception of a new instrument for x-ray emission spectrochemical microanalysis, but also a survey of physical principles involved in this promising, non destructive analytical method.

The principle of the electron microprobe method is rather simple and is a logical extension of the pioneering experiments of MOSELY(2) on the frequencies of characteristic x-ray radiation excited by electron bombardment. The microvolume of a particular point at the surface of a metallographically polished sample is excited by a focused electron beam. The beam about 1 micron in diameter is accelerated by a voltage of 3 to 50 kilovolts. The excited volume emits a complex x-ray spectrum which is selectively analysed by one or several x-ray spectrometers. Comparison between the intensities of characteristix x-rays, emitted under identical experimental conditions by the sample $I_{(A)}$ and by a calibration standard I_A, permits the determination of the respective elementary mass concentrations C_A, to a first approximation. For reliable quantitative information a refined correction method taking account of the different physical phenomena occuring in the anticathode is absolutely necessary. The unique features of microprobe analysis deliberately aspired to by CASTAING are precise localisation under an optical microscope coaxial to the electron beam and microanalysis of an irradiated volume of about 1 cubic micron. Furthermore the absolute character of microprobe analysis permits complete ranges of concentrations to be explored with only one calibration standard — the pure element.

A comparison with x-ray fluorescence spectrochemical analysis underlines this last characteristic. The monochromatic x-ray radiation used as primary excitation for fluorescence emission penetrates at least two orders of magnitude deeper than the electron beam into the surface of the sample. The microvolume of interest therfore surpasses the electron-probe excited volume by an order of 10^6. This volume of a sample is generally inhomogeneous and consequently the primary excitation beam, as well as the secondary fluorescence radiation, are differently absorbed with local variations in mean atomic mass. The ionisation function and the mass absorption of the radiations cannot be evaluated directly. To achieve quantitative results, the non destructive nature and the possibility of direct localisation must be abandonned. Homogenisation is then achieved by a dilution method (aqueous solutions, borax bead). Another possibility consists in the use of a complete set of internal calibration standards with known and neighboring chemical composition, specificly prepared for every series of determinations. By comparison, in electron probe microanalysis, it is possible to consider that the specimen is homogeneous within the region excited by electron bombardment.

As in fluorescence analysis, considerable confusion exists about the "interelement" effect. The interelement effect, generally designated in microprobe analysis as atomic number effect, consists of two partial components — the electron stopping power due to deceleration and diffuse penetration of electrons in solids and the electron backscattering effect. In the first published oversimplified theoretical correction methods (3,4) one of them by the author, an approximate formula was derived for calculating the distribution in depth of the direct emission involving the first factor of the atomic number effect. Further research has proved that:

a) The derivation of a theoretical solution for evaluation of the distribution in depth of direct emission is indirectly (through a modified LENARD coefficient) and directly dependent on the difference $(V - V_c)$, on the electron acceleration voltage (V) and the critical excitation potential of the primary radiation (V_c). This effect is specially important at low accelerating voltages. At initial electron energies higher than 25 kilovolts a compensation of the $(V - V_c)$ dependent factor according to the combination of the LENARD and BOTHE laws occurs, as will be seen below.

b) The intensity ratio of backscattered electrons with energies higher than $V - V_c$ (loss of x-ray excitation) emitted by the reference element and the specimen, deviates strongly from unity, especially for analysis of alloys with components of widely differing atomic number and the use of high electron accelerations.

Following the development of POOLE and THOMAS (5) many microprobe users have tried to improve the correction formulae of PHILIBERT (3) or THEISEN (4) by introducing a supplementary atomic number correction for electron stopping power of the anticathode and for electron backscatter. However, this approach does not give complete satisfaction as it introduces a twofold electron deceleration factor.

For unresolved problems, and to test the application and extension of empirical laws, valuable assistance could be given by electronic computers. The method described by ARCHARD and MULVEY (6) for numerical computations of the electron path and the emission of x-rays from a given specimen is likely to gain in importance if a better model for electron trajectories is found.

1.2 Constructional Elements of the Electron Microanalyser

It is out of the scope of this volume to detail extensively the general structure of the microprobe for which reference may be made to publications, by CASTAING, BIRKS, DUNCUMB (7—9). The important elements of an electron microprobe are the following: (Fig. 1)

a) Electron optical system

The electron beam is generated by a hot filament electron gun working in a vacuum of the order of 10^{-5} torr and an electron acceleration field of generally 2 to 50 kilovolts. The reduced image of the electron cross-over is focalised by one or several electromagnetic lenses to produce on the specimen surface an electron probe from 0,2 µm up to several µm in diameter. The geometrical resolution of microanalysis is however limited to the electron range in the specimen, from V, the incident electron energy up to the critical excitation potential of the measured characteristic radiation. From LANGMUIRS theory (10) for the brightness of an image and ZWORYKINS (11) equation for spherical aberration of magnetic lenses, the incident theoretical electron density as a function of beam diameter is given by

$$i = \frac{d^{8/3}}{c_s^{2/3}} \cdot i_0 \left(\frac{eV}{kT}\right) \cdot \frac{9 \pi^2}{64} \tag{1}$$

where d is the probe diameter, c_s the spherical aberration coefficient, i_o the electron emission of the filament. Unless the spherical aberration coefficient can be decreased by new techniques, the produced x-ray intensity in the sample will be proportional to $d^{8/3}$.

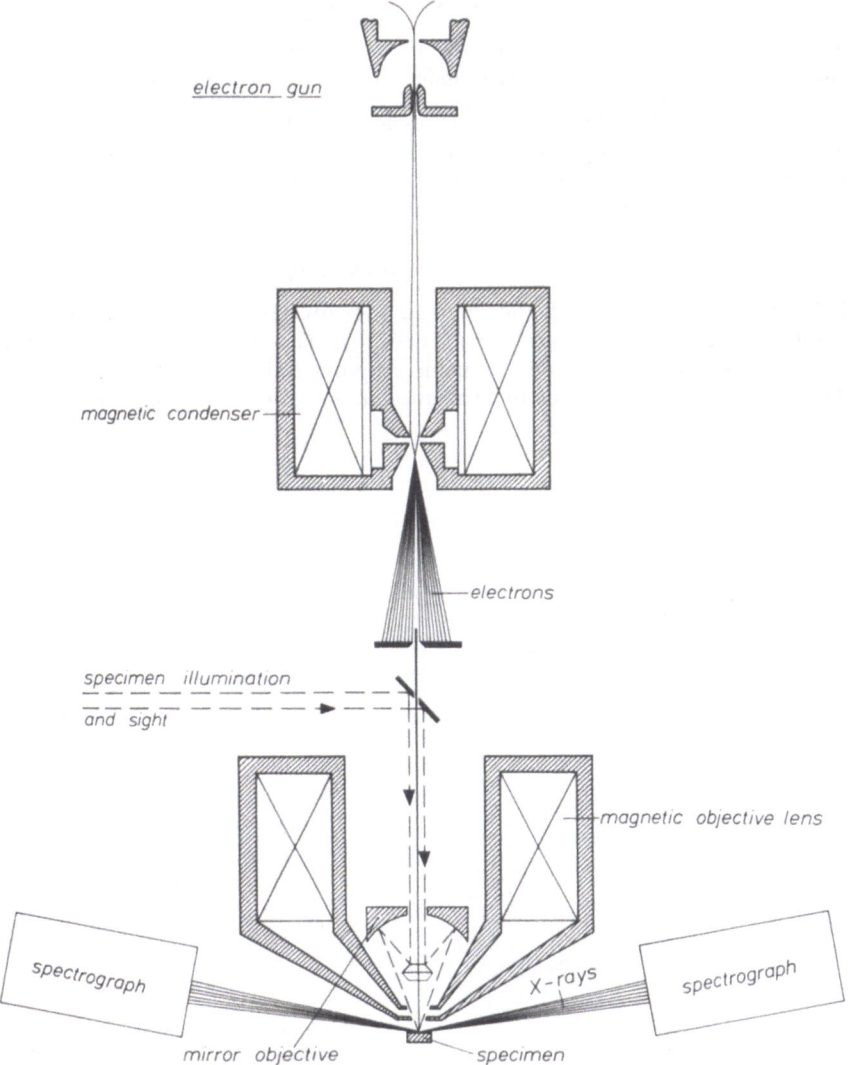

Fig. 1. Schematic diagram of the electronic Micro-Probe. (Cameca)

b) Specimen viewing device

The specimen viewing device is necessary for both the correct selection of the point to be analysed and the precise determination of the focalisation plane of the anticathode surface in respect to the focussing circle of the spectrograph. The most satisfactory solution consists of an optical microscope, coaxial with the electron beam, as already used by CASTAING in 1954. The choice of an objective lens with a short focal distance allows a correct positioning of the specimen within ± 0,5 µm. The uncertainty of the emergent take-off angle of the characteristic radiation is therefore minimised.

For instruments with high take-off angles, or semi-focussing x-ray optics, the correct positioning in the axis of the electron beam is less critical.

c) Specimen stage

The specimen stage consists of an assembly which allows the reference standard and the sample to be brought successively under the impact of the electron probe with a precision at least equal to the geometrical resolution of the electron beam ($< 1\ \mu m$). Since the specimen plane can also be the source for x-ray projection microscopy (12), KOSSEL-line microdiffraction techniques (13), analysis of thin films (14), the specimen stage should be designed to accomodate these accessories (15).

d) X-ray spectrograph

If high peak-to-background counting ratios for precise quantitative analysis is needed, a fully focalising x-ray spectrometer is desirable. The spectrometer consists of a curved monochromating crystal and a suitable proportional, scintillation or G. M. detector. The fully focusing condition can be obtained in one of three possible ways:

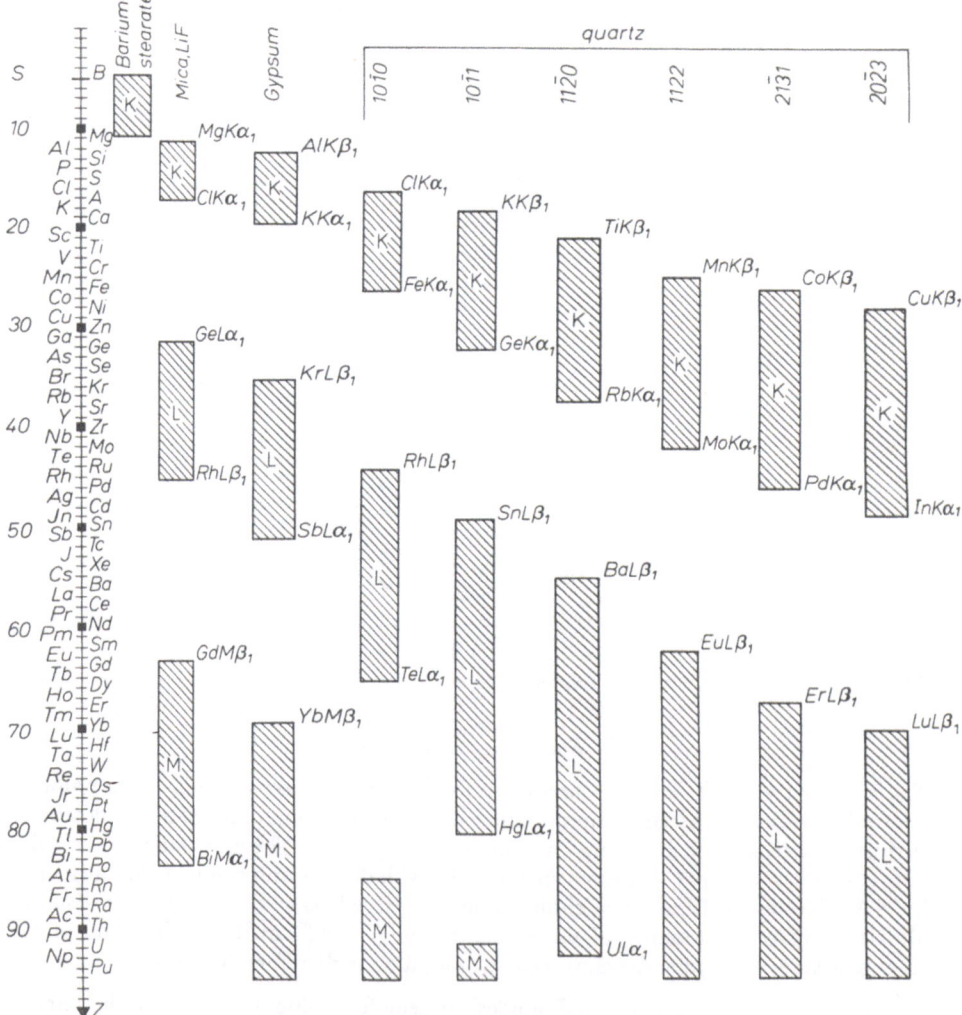

Fig. 2. Optimum range of monochromaters.

α. The crystal is of a fixed curvature and moves on the same ROWLAND circle as the detector (JOHANNSON or JOHANN geometry).

β. The crystal with fixed curvature moves along a straight line and rotates about its own center while the detector follows a non circular line (position on different focussing circles determined by the crystal position).

γ. The source to crystal distance may be kept constant and the radius of the crystal changed according to the diameter of the focussing circle (variable bent crystal spectrometer).

The domain of application of the different possible monochromating crystals as Quartz, Gypsum, Mica, Barium Stearate Decanoate are indicated in (Fig. 2).

1.3 The Characteristic X-ray Emission Spectrum

The major part of the bombarding electrons reaching the target will dissipate their energy in heat; another part will be backscattered with little loss of energy. Incident electrons with sufficient kinetic energy will penetrate the outer shells of an atom in the target and eject a photoelectron from the inner K-shell, leaving the atom in an excited state. The knocked-off electron is replaced by an electron from an outer $L, M,$ or N shell, untill the excited atom is restored to the normal lower potential, (Fig. 3).

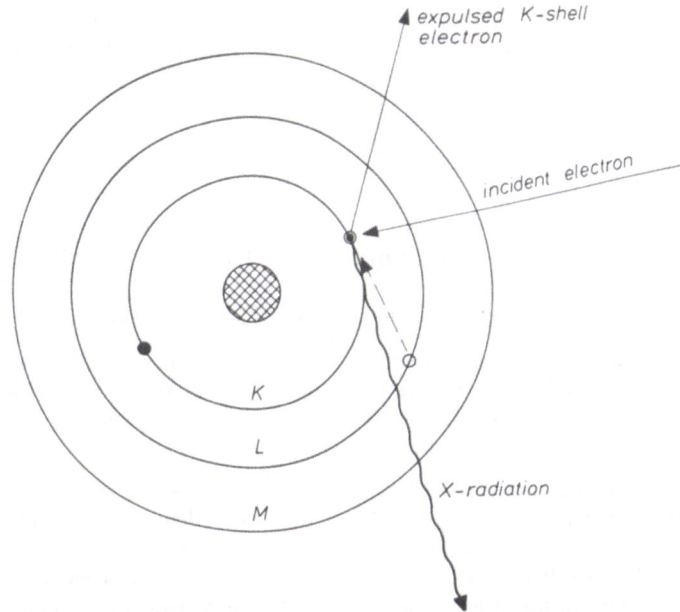

expulsed K-shell electron

incident electron

K

L

M

X-radiation

Fig. 3. Schematic representation of X-ray exitation by electron bombardment.

This release of energy is accompanied by emission of x-radiation with defined wavelength and characteristic of the concerned atom. The heavier the atom, the harder the emitted radiation. According to the experiments of MOSELEY (2) the wavelength of the characteristic x-ray radiation is

$$\lambda = \frac{c}{(Z - \sigma)^2}$$

Z is the atomic number, c and σ are constants dependent on the line series (K, L, M, N).

The possible transitions are illustrated in Fig. 4.

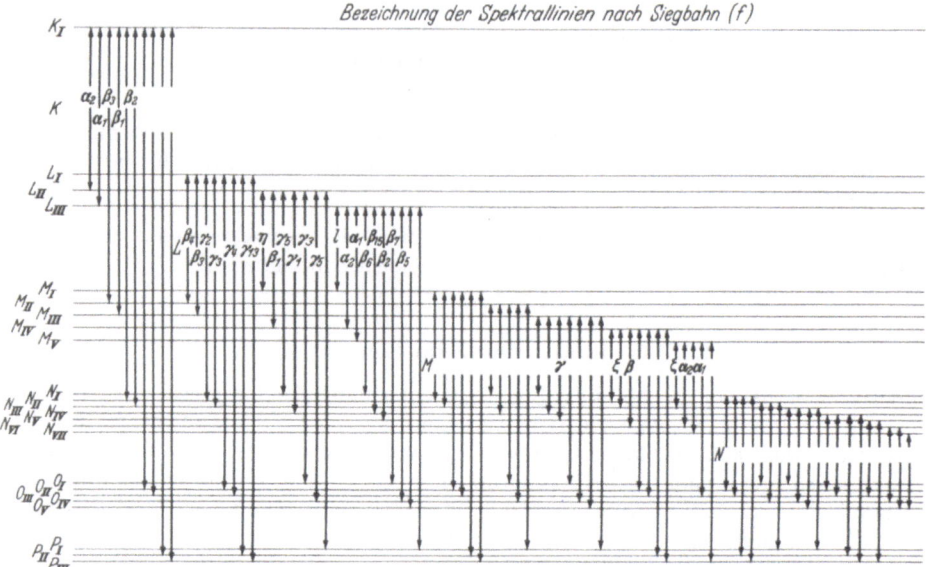

Fig. 4. Thermschema der Röntgenniveaus von Uran. Die vertikalen Pfeile stellen die tatsächlich beobachteten Übergänge dar.
(Ent.: LANDOLDT-BÖRNSTEIN: 1. Teil, 6. Aufl. Berlin/Göttingen/Heidelberg: Springer 1950).
X-ray Transition Series.

§ 2 Fundamentals of Quantitative Electron Microprobe Analysis

2.1 Necessary Corrections of the X-Ray Intensity Ratio

Emission concentration proportionality is, as has been stressed in the introduction, one of the main features of electron microprobe analysis. The correction required to obtain true primary x-ray intensities and to convert them to mass concentration must take account of the following effects in the indicated order:

1. Deceleration of the impinging electrons by the induced target contamination; 2. Shift in wavelength of characteristic x-ray lines due to valence state; 3. Dead-time and background counts; 4. Secondary fluorescence enhancement by the continuous spectrum or the characteristic radiation; 5. X-ray excitation efficiency as a function of electron penetration in the anticathode; 6. X-ray mass absorption in the anticathode; 7. Electron backscattering ratio.

All theoretical correction methods are by neccessity based on extrapolation of electron absorption and diffusion laws theories, such as LENARD and BOTHE's law. A further difficulty, which should not be underestimated, is the insufficient available knowledge concerning the absorption coefficients of soft x-rays and the electron backscatter ratios. The computed correction tables of the present volume, will not be affected by these last two insufficiencies, as their relative contributions are rapidly evaluated in a separate step.

By correct choice of experimental conditions, the departure from the linear relationship between primary x-ray intensities and mass concentration can be generally reduced to less than 25% and, even an error of 20% on the total correction value, would yield a precision better than 5%.

The relative importance of the different sources of correction will be shortly reviewed, before the detailed study of the different correction theories.

1. Deceleration of electrons by target contamination

During microanalysis, hydrocarbon molecules from diffusion pump oil are present at the specimen surface as a partial pressure vapor. These are cracked by the electron beam and a layer of contamination accumulates. The contamination rate depends principally on the interval of analysis, the electron beam current, the residual organic vapor pressure and the nature of the specimen surface. The contamination tends to accumulate at the precise location being analysed. However, it must be inhibited if a) carbon or other ultralight elements are to be analysed as spurious carbon radiation is generated in the layer; b) low electron accelerating voltages are used for high geometric resolution or for minimising the necessary corrections; c) microprobe applications in trace analysis are being performed where long counting intervals would produce a relatively thick contamination layers.

As indicated by CASTAING (16) and CAMPBELL, GIBBONS (17) different methods for the inhibition of contamination are possible: Heating the specimen up to about 250 °C; surrounding it by a cool surface, placed near the point of impact; by burning away the contamination by directing a low-pressure air flow on the point of impact of the probe. The last technique offers the advantage of both simplicity and versatility. All it requires is a capillary of non magnetic material, to prevent any deflection of the electron beam and a high pressure valve to regulate the air admission. With a flow rate of 0,025 mg per second, the contamination spot is avoided perfectly for arbitrary long time intervals. If desired, the air admission valve can be closed immediately after the analysis and during a short supplementary interval, the contamination spot may be allowed to build up for location of the analysis.

The main handicap of this method is, that even with small air flow admission, the life time of the electron emitting filament is greatly reduced, requiring sucessive realignements of the electron optical system. This drawback can be eliminated, as shown recently by the author (18) by replacing the air flow with an inert gas, preferably argon. This method will be described in the chapter on the detection limit of microprobe analysis.

2. Shift in wavelength of characteristic radiation

As has been stressed by FAESSLER (19) the wavelengths of characteristic x-ray lines increase with increasing negative charge of the ion. MENSHIKOV and NEMNONOV's (20) experiences demonstrate that this wavelength change is not to be seen in compounds with metallic or covalent-metallic bond, but only for clearly ionic bonding. According to WITTRY (21) for a change from metallic to ionic bonding, the shift of the energy level might be of the order of an electron volt. Hence, since $dE/E = d\lambda/\lambda$ the effect is most pronounced in the characteristic x-ray lines of long wavelength. For example, the wavelength for maximum intensity of aluminium K radiation is shifted about 0,0023 A in going from pure Al to Al_2O_3. This is of the order of the natural breath of the AlK_α line and about 1/5 of the half — width of the line obtained in fully focusing microanalyzers. Thus, if the spectrometer is not set for the maximum of the shifted line, the intensity of light elements may be lower than the peak by as much as 10—20 %. To take account of this effect WITTRY suggests two methods. The spectrometer may either be reset to the peak of the shifted line, or the factor by which the shift changes the relative intensities may be determined.

For deliberate studies on the chemical state, even for medium elements, the shift of the $K\beta_1(M_{II,III} \rightarrow K)$ line or the long wave satellite $K_\beta I$ should be considered.

3. Dead time of the detecting system and background counts

Fortunately the techniques used in impulse technique and especially fluorescence analysis can be extended completely to microprobe analysis for this effect. As proportional counters are nearly exclusively used, the resolving time of the associated electronic circuitry is higher than the largest dead-time of the detector and the relative loss in the number of registered impulses is negligible.

The background radiation consists of the electron density-dependent x-ray continuum and the time dependent background due to cosmic rays and spurious impulses. The background is generally measured on both sides of line peak (the shift is generally 6 times the half width of the line) for identical counting intervals on the specimen and the standard, and substracted from the peak intensity. For complex samples a complete spectrometer scanning is highly advisable, in order to avoid perturbations by multiple order radiations.

4. Secondary fluorescence enhancement by characteristic radiation or the continuous spectrum

Emission enhancement by fluorescence excited by characteristic radiation of the continuous spectrum, necessitates a rather complicated treatment, since it is required to relate the fluorescence radiation to the measured primary radiation and not to the exciting spectrum.

Fortunately the calculation of fluorescence enhancement is necessary only in relatively few cases, where the radiation of a complex alloy is strongly and specificly absorbed by an element A, resulting in an apparent increase of the concentration C_A. The necessary correction will reach a maximum, if the major characteristic lines of the spectrum are neighbouring the absorption edge of the analysed element. It is important to note, that the fluorescence enhancement is not limited to the microvolume excited by the electrons, but is generated over a region about 10^6 times larger corresponding to the penetration power of the x-rays. For this reason, existing theories can only be applied to samples homogenous over the whole range of x-ray penetration. In most applications, where an important fluorescence enhancement has to be expected, maximum precision can only be achieved with calibration curves.

For homogeneous samples, HENOC (22), KIRIANENKO, HENOC et aliis (23), have shown, that fluorescence uncertainty due to the continuous spectrum is small, mainly because it is generated in deeper layers within the anticathode than the primary radiation and it is absorbed to a greater extent on its way out, specially when the x-ray take-off angle is low.

Secondary fluorescence, excited by the continuum are however not negligible in extreme cases, for instance, in regions of steep concentration gradients or in elementary analysis of a small precipitate in a pure matrix of the analysed element. An empirical method for this correction has been suggested by DILS and ZEITZ (24).

The four theoretical expressions, derived from work by CASTAING (1), WITTRY (25) and BIRKS (8) for calculating the intensity of K or L fluorescent radiation excited by K or L radiation and the evaluation of the fluorescence enhancement, generated by the continuum, after work from HÉNOC and KIRIANENKO, will be exposed in the second volume on quantitative microprobe analysis.

5. X-ray excitation efficiency of the anticathode, absorption correction and electron backscatter

Considerable confusion exists over the quantitative treatment of the most important corrections, the primary x-ray excitation efficiency, the x-ray mass absorption in the anticathode and the electron backscatter. It will be the object of the following chapters to correlate the different approaches, which have been proposed specifically for the quantitative microprobe theory.

Qualitatively the primary x-ray excitation efficiency consists of two oppositely acting effects, the electron stopping power in the anticathode (including electron absorption and small angle scattering) and electron backscatter phenomenon due to RUTHERFORD scattering with little loss of energy.

As the backscattered fraction will cause a loss of possible ionisations, only the average penetration of the efficient electron current, with energies between the initial acceleration and the critical excitation voltage must be known. Then it will be possible to relate the x-ray intensity generated within the target, to the measured radiation intensity (which has undergone mass absorption along its path in the target along the emergent take-off angle) in the spectrometer.

Qualitatively the dependence of electron penetration within the anticathode on the initial electron acceleration and atomic number of the target has been illustrated schematically by DUNCUMB (Fig. 5)

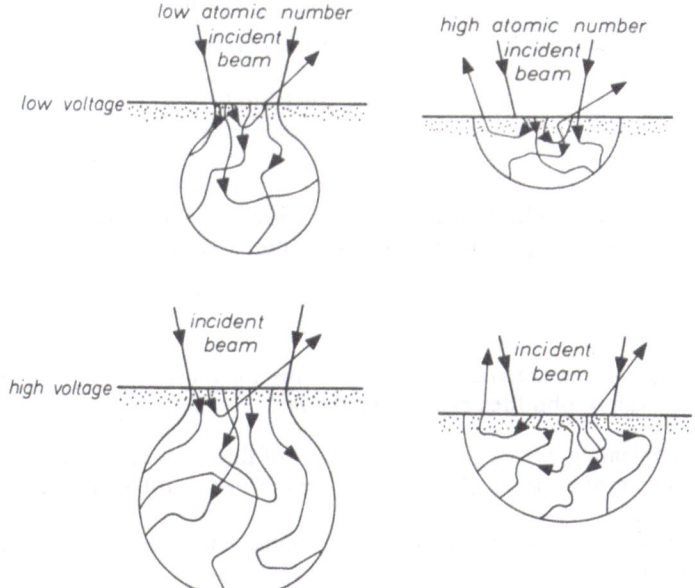

Fig. 5. Section through specimen surface illustrating variation of electron scattering with voltage and atomic number (DUNCUMB, 1963).

For targets with low atomic number, electrons penetrate deeply before changing direction; consequently rather few electrons with sufficient energy to excite the characteristic radiation are backscattered. For targets with high atomic number, penetration occurs nearer the surface and an important backscatter, even multiple scattering, occurs.

The primary x-ray radiation generated in the target is dependent on the energy loss due to the retardation of high energy electrons along its path in the initial direction of the impinging electrons and on the multiple scattering.

In a target of the pure element A, a layer of mass thickness $d(\varrho z)$ at a depth z below the surface will generate the primary emission

$$dI = \varphi_A(\varrho z)\, d(\varrho z)$$

and the total intensity generated within the target will be

$$I_t = \int_0^\infty \varphi_A(\varrho z)\, d(\varrho z) = F(0)$$

The intensity I, recorded by the spectrometer after having undergone x-ray mass absorption in the take-off direction, will be:

$$I = \int_0^\infty \varphi(\varrho z) \exp\left(-\mu/\varrho\, \varrho z\, \text{cosec } \theta\right) d\varrho z$$

Replacing μ/ϱ cosec θ by χ and considering the general case of a complex target with a mass concentration C_A, the integral can be expressed by a LAPLACE transform of a function $f(x)$

$$F(t) = \int_0^\infty f(x)\, e^{-tx}\, dx$$

The quotient $f(\chi) = \dfrac{F(\chi)}{F(0)}$ of the generated primary x-ray intensities within the specimen and the intensities recorded by the spectrometer are given by

$$f(\chi) = \frac{\int_0^\infty \varphi(\varrho z) \exp\left(-\chi \varrho z\right) d(\varrho z)}{\int_0^\infty \varphi(\varrho z)\, d(\varrho z)} = \frac{F(\chi)}{F(0)}$$

It is possible to relate the characteristic radiation intensity emitted by a complex sample I_{AB} and the pure element I_A to the mass concentration C_A, by

$$\frac{I_{AB}}{I_A} = C_A \frac{f(\chi_{AB})}{f(\chi_A)}$$

The function $f(\chi_A)$ for pure elements, which combines the function $\varphi(\varrho z)$ and the absorption correction, can be obtained experimentally by:

1. Direct determination of $\varphi(\varrho z)$ by the method of thin layers (7) with the absorption correction obtained by the LAPLACE transform of this distribution function.

2. Variation of the x-ray emergent angle using a cylindrical sample (1) or by tilting the spectrometer (27).

However, for practical microanalysis it is necessary to extend these determinations to complex samples and for operation at different electron accelerations. It will be shown, that this can be achieved either by combining an electron absorption law (LENARD), an average electron scattering model, and an estimation of the electron backscatter, or by numerical computations considering individual electron trajectories.

2.2 Angle of Incidence of the Electron Beam

In most of the electron microprobes the electron beam is perpendicularly incident to the target surface. For microanalysis with an inclined beam, the mean primary x-ray intensity generation occurs nearer the surface as the penetration diminishes with the oblique electron trajectory. The same function of distribution in depth of the primary x-ray emission can be used as for normal incidence, provided allowance is made for the oblique electron penetration in the calculation of the depth at which electron diffusion occurs.

As multiple scattering occurs nearer the surface of the sample, the angle of electron incidence must be introduced in the calculation of the depth at which complete diffusion begins.

As multiple scattering occurs nearer the surface of the specimen, allowance can be made for the increased backscatter coefficient by a similar treatment, if the CASTAING method is applied, or by empirical determination.

§ 3 Procedures for Correction Calculation

3.1 General Remarks:

In the following development it is assumed that the effects of target contamination, shift in wavelength of the radiation, background counts, resolution time of the electronic circuitry and fluorescence enhancement are corrected previously by the known methods or avoided by the choice of suitable experimental conditions.

Electron penetration, primary x-ray emission efficiency, x-ray mass-absorption and electron backscatter may then be calculated by means of the following:

1. An analytical theory, based on an average scattering model including the interelement effect.

2. An average scattering model and an atomic number theory.

3. Experimental determination by variation of the x-ray emergence.

4. Experimental method, based on a tracer technique.

5. Computational calibration.

Extreme solutions are sometimes required to evaluate simultaneously fluorescence enhancement, These are:

6. Direct determination by calibration.

7. High-precision thin-film analysis.

3.2 Analytical Correction Procedure

a) Correlation between the relative emitted x-ray intensities and the mass concentration

For a first approximation it is assumed that efficiency of the x-ray emission remains constant, regardless of the depth z in the anticathode. According to LENARD's law (28), the electron intensity in the bombarded target varies as follows:

$$I = I_0 \cdot e^{-\sigma \varrho z}$$

where V is the incident electron intensity, ϱ the density and z the depth in the anti-cathode. In LENARD's experiments the coefficient σ is independent of the composition of the alloy and depends only on the initial electron acceleration,

$$\sigma = \frac{c}{V^2}$$

At 30 kilovolt, the experimental σ value is 1750; therefore if V is expressed in kilovolts, we obtain

$$\sigma = 1750 \cdot \left(\frac{30}{V}\right)^2$$

However, in electron microprobe analysis it is not the integral intensity of the electron beam irradiating the target which contributes to the excitation of characteristic radiation, but only the high energy fraction ($V - V_c$) with energies higher than the critical excitation voltage of the measured radiation. Recently DUNCUMB (29) has suggested an empirical σ coefficient $\sigma = (V - V_c)^{1,5}$ but without taking into consideration these values for the determination of the function $\varphi(\varrho z)$ and the interelement parameter h. Effective LENARD coefficients σ_E may be obtained by empirical extension of the LENARD law to the electrons effectively exciting characteristic radiation:

$$\sigma_E = \frac{c}{(V - V_c)^2} = \frac{8,9 \cdot 10^5}{(V - V_c)^2}$$

The application of these effective LENARD coefficients implicitly requires an electron acceleration and an A/Z^2 dependent atomic number parameter h, when calculating the x-ray efficiency of the anticathode.

A further advantage of the use of the "effective LENARD coefficients" is the possible extension to low electron acceleration analysis of the proposed procedure, down to the critical excitation potentiel, where it correctly predicts zero absorption.

In order to simplify further the calculations, let us consider, in this first step, the case of a binary alloy

$$\varrho = m_A + m_C$$

where m_A is the weight of the analyzed element per volume unit and m_c is the weight of the complementary element per volume unit, A and Z being the atomic weight and atomic number of the element (A).

The intensity I_A of the characteristic line (A) in the direction of the emergence of the x-rays can then be expressed as follows:

$$I_A = K \frac{m_A}{A} N I_0 \int_0^\infty e^{-\sigma \varrho z} \, e^{-\mu/\varrho \, \mathrm{cosec}\, \theta} \, dz$$

where K is a constant dependent on the ration V/V_c, $N\frac{m_A}{A}$ is the number of atoms A per unit of volume and μ/ϱ is the mass absorption coefficient.

Let us assume that μ/ϱ cosec $\Theta = \chi$; χ being a linear combination of χ_A (standard) and χ_c (complementary element).

$$\chi_{\text{alloy}} = c_A \cdot \chi_A + (1 - c_A)\chi_c = \chi_{AB}$$

By substituting and integrating the second member of the foregoing equation we obtain

$$I_A = \frac{KI_0 N}{A} \cdot \frac{1}{\sigma + \chi} \cdot \frac{m_A}{\varrho}$$

It should be noted that m_A/ϱ is the mass concentration of element A in the alloy, hence

$$I_A = \frac{KI_0 N}{A} \cdot \frac{c_A}{\sigma + \chi}; \text{ and}$$

$$I_{(A)} = \frac{KI_0 N}{A} \cdot \frac{1}{\sigma + \chi}$$

Then
$$\frac{I_A}{I_{(A)}} = k_A = c_A \frac{\sigma + \chi_A}{\sigma + \chi_{AB}}$$

As the correction term $\quad \Delta c = c_A - k_A$

$$\Delta c = k_A \left(\frac{\sigma + \chi_{AB}}{\sigma + \chi_A} - 1\right)$$

Allowing for the relation

$$\chi_{AB} = c_A \cdot \chi + (1 - c_A)\chi_c$$

we obtain

$$\boxed{\Delta c = k_A(1 - k_A)\frac{\chi_c - \chi_A}{\sigma + \chi_A}}$$

Thus for the first approximation, we obtain by replacing $k_A \cong c_A$

$$\Delta c = k_A(1 - k_A)\frac{\chi_c - \chi_A}{\sigma + \chi_A}.$$

b) Calculation of the x-ray Emission Efficiency $\varphi(\varrho z)$ as a function of Penetration Depth

Contrary to the original hypothesis, allowance is now made for the electron diffusion as a function of the penetration depth since the emission efficiency of a dz layer increases as a function of penetration.

As the electrons penetrate into the anticathode, they deviate from their original course by multiple scattering. Let α be the angle which they form with the normal to the test sample (an electron incidence perpendicular to the sample is assumed).

The ionisation caused by an electron in a dz layer increases by a factor of $\frac{1}{\cos \alpha}$ since it is assumed that the number of ionisation occurrences increase with the path length and that the path length dx is equal to $\dfrac{dz}{\cos \alpha}$ in the dz layer. This relation does not apply however, when α tends towards $\pi/2$ since it is physically inconceivable that the path length should become infinite, when $\alpha = \dfrac{\pi}{2}$.

The calculation of the efficiency function $R(\varrho z)$ amounts to the calculation of a mean value of $\dfrac{1}{\cos \alpha}$ for all electrons. The distribution of the deviation α is provided by BOTHE's (30) law:

$$f(\alpha) = \frac{d\dfrac{N}{N_0}}{d\Omega} = \frac{1}{2\,\Pi\,\Delta^2}\,e^{\frac{-\alpha}{2\,\Delta^2}}$$

where $f(\alpha)$ is the intensity distribution for a solid angle $d\Omega$

and $d\dfrac{N}{N_0}$ is the part of N_0 electrons scattered in the solid angle $d\Omega$

α is the scattering angle

Δ is the most probable scattering angle

The most probable scattering angle is given by

$$\Delta = \frac{800}{V} \cdot \frac{V + 511}{V + 1022} \cdot Z \cdot \sqrt{\frac{\varrho z}{A}} \qquad \text{where}$$

$V =$ Energy of the incident electrons in keV

$Z =$ Atomic number

$\varrho =$ Density in g/cm^3

$A =$ Atomic weight

$z =$ The thickness of the layer in cm

Since V is always relatively low in microprobe analysis (0—50 kilovolt), the following simplified formula can be used:

$$\Delta = \frac{400}{V} Z \sqrt{\frac{\varrho z}{A}}$$

The mean value of $\dfrac{1}{\cos \alpha}$ is given by

$$\varphi(\varrho z) = \frac{1}{2\,\pi\,\Delta^2} \int\limits_0^\infty e^{-\frac{\alpha^2}{2\,\Delta^2}} \cdot \frac{1}{\cos \alpha} \cdot 2\,\pi \sin \alpha\,d\alpha$$

using calculations similiar to those by TONG (unpublished 1960) and THEISEN (4), the efficiency of the anticathode layers can be determined theoretically and the less complicated, substitutions for $\varphi(\varrho z)$ defined. This treatment differs sensibly from that of PHILIBERT (34) who empirically determined $\varphi(\varrho z)$ by comparison with the experimental data of CASTAING.

Thus by introducing $\operatorname{tg} \alpha = \alpha + \dfrac{\alpha^2}{3}$

when $0 < \varrho z < \varrho z_1 = \dfrac{\pi^2}{16} \cdot \dfrac{A}{Z^2} \left(\dfrac{V}{400}\right)^2$ we obtain (4)

$$\varphi(\varrho z) = \lambda + \frac{2\varDelta^2}{3} + \frac{Z^2}{A}\varrho z$$

and when $\varrho z_1 > \varrho z$, $\varDelta^2 = \left(\frac{400}{V}\right)^2$

In the deepest layers, the function $\varphi(\varrho\zeta)$ tends towards the value $\varphi\infty$ as the further distribution of electrons is completely diffuse. For the calculation it is necessary only to consider a deepest layer, where $\varDelta = \pi/4$

so that $\qquad \varrho z_1 = \frac{\pi^2}{16} \cdot \frac{A}{Z^2} \cdot \left(\frac{V}{400}\right)^2 = 4 \cdot 10^{-6} \cdot V^2 \cdot A/Z^2$

By combining BOTHE's distribution and LENARD's penetration law, we obtain

$$\varrho z_1 = 1{,}72 \cdot 10^{-6} \cdot V^2 \cdot A/Z^2 = \frac{h}{\sigma_E}$$

$$\sigma_E = \frac{8{,}9 \cdot 10^5}{(V - V_c)^2}$$

$$\boxed{h = 1{,}72 \cdot 10^{-6} \cdot \sigma_E \cdot V^2 \cdot A/Z^2}$$

c) Relation between mass concentration C_A, $F(\chi)$ and $f(\chi)$

Using the substitution $\qquad \varphi(\varrho\zeta) = \varphi\infty\left(1 - \exp\right)\frac{\varrho\zeta}{\varrho\zeta_1}$

$\varphi(\varrho\zeta)$ becomes $\qquad \varphi\infty\, 1 - \exp\left(\frac{\sigma\varrho\zeta}{h}\right)$

and $F(\chi)$

$$F(\chi) = \int_{\varrho\zeta=0}^{\infty} \varphi(\varrho\zeta)\exp\left[-(\sigma + \chi)\varrho\zeta\right]d\varrho\zeta$$

$$F(\chi) = \frac{\varphi\infty}{\sigma}\left(\frac{1}{\chi/\sigma} - \frac{1}{1 + \frac{1}{h} + \chi/\sigma}\right)$$

$$\frac{1}{F(\chi)} = (1 + \chi/\sigma)\left[1 + h(1 + \chi/\sigma)\right]$$

The total correction amount $\varDelta c$ will be, if $k_A = \frac{I_A}{I(A)}$

$$\boxed{\varDelta c = C_A(1 - C_A)\frac{F(\chi_A) - F(\chi_{AB})}{F(\chi_A)}}$$

As $f(\chi) = \frac{F(\chi)}{F(0)}$ (see chapter 2,5) and

$$\frac{I_{AB}}{I_A} = k_A = C_A \cdot \frac{f(\chi_{AB})}{f(\chi_A)}$$

a simpler expression relating the emitted x-ray ratios to the mass concentration may be found:

$$f(\chi) = \frac{F(\chi)}{F(0)} = \frac{1 + h}{(1 + \chi/\sigma)\,[(1 + h)\,(1 + \chi/\sigma)]}$$

$$\boxed{\frac{1}{f(\chi)} = (1 + \chi/\sigma)\left[\left(1 + \frac{h}{1 + h}\right)\chi/\sigma\right]}$$

This expression will be found directly in the annexed computed tables. The correction will be

$$\boxed{\frac{I_{AB}}{I(A)} = k_A = c_A \cdot \frac{1/f(\chi)\ \text{standard}}{1/f(\chi)\ \text{alloy}}}$$

d) Electron Backscatter Effect

In microprobe analysis, especially for high electron accelerations, only a fraction of the electrons with energies higher than the analyzed characteristic radiation contribute to the ionisation of the anticathode. By RUTHERFORD- and multiple scattering near the surface some high energy electrons will be backscattered out of the target and become lost for the ionisation process. In our first quantitative approach (1961) — and also in similar treatments by PHILIBERT and TONG —the assumption that the ratio of backscattered electrons in the specimen to those in the pure element was close to unity, and could be neglected, was a gross approximation. D. CALAIS and A. KIRIANENKO (26) had selected specimens where all the other corrections were negligible and proved quantitatively the necessity of the electrons backscatter correction.

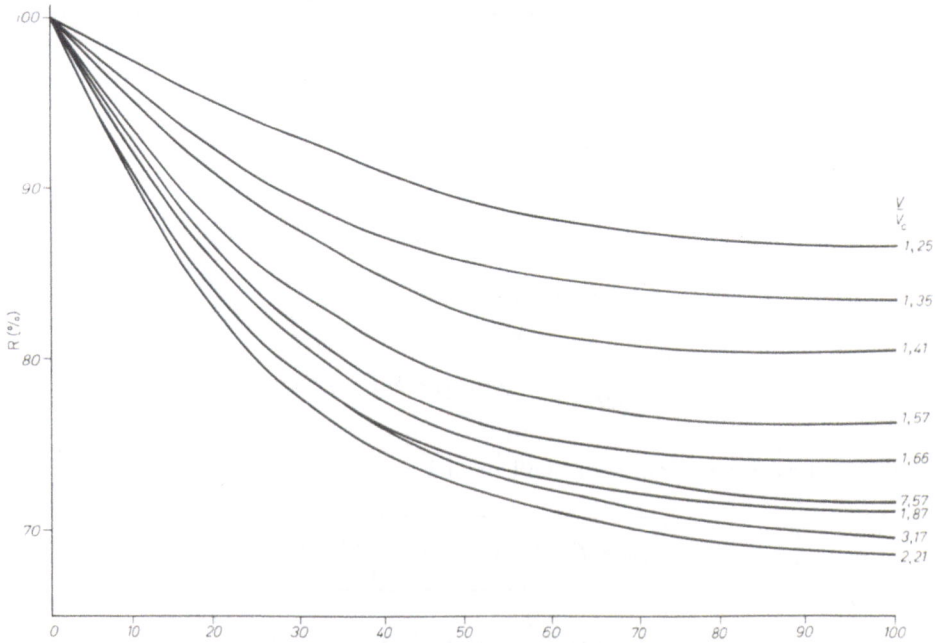

Fig. 6. Variation of effective current factor R with atomic number z-derived from BRANDS' data (3) A. E. R. E. R 4593 (1964) P. M. THOMAS.

The part played by the high energy electrons has been calculated by P. M. THOMAS (31) from data on the energy distribution of electrons backscattered from pure elements published by BRAND (32) and KULENKAMPF and SPYRA (33). The resulting curves relating effective current factor R to mean atomic number $\bar{Z}$ are shown in Fig.(6).

It can be noticed, that the effective current factor R decreases with increasing mean atomic number for a given ratio of initial electron acceleration to a critical excitation potential of the primary analyzed radiation V/V_c and that the variation of effective current with atomic number is to a great extent independent of V/V_c, if this ratio varies between 2 and 7, 6.

It should be emphasized that the effective current factor is not identical with the back-scattering factor λ defined by CASTAING (1).

As an alternate ZIEBOLD (38) proposes to compute the backscatter factor, defined as the ratio of total number of photons actually generated in the sample-to the total number of photons generated, if there were no backscatter. The actual number of ionizations can be calculated from an appropriate model such as those proposed by ARCHARD and BROWN. The same model, but modified so that all the electrons go straight into the sample, is used to compute the denominator of the backscatter factor. Since we are only interested in only the total number of ionizations, this calculation is independent of path shapes and distribution in depth. These detailed numerical computations will probably replace in next future advantageously the effective current values given in table E.

The final correction to convert the x-ray intensity ratio into mass concentration, taking simultaneous account of electron penetration, x-ray emission efficiency, mass absorption and electron backscatter is

$$\boxed{\frac{I_{AB}}{I_{(A)}} = k_A = \frac{R_{AB}}{R_{(A)}} \cdot c_A \cdot \frac{f(\chi_{AB})}{f(\chi_A)}}$$

$$\boxed{k_A = \frac{R_{AB}}{R_A} \cdot c_A \cdot \frac{(1 + \chi_{AB}/\sigma_E)\left[\left(1 + \frac{h_{AB}}{1 + h_{AB}}\right)\frac{\chi_{AB}}{\sigma_E}\right]}{(1 + \chi_A/\sigma_E)\left[\left(1 + \frac{h_A}{1 + h_A}\right)\frac{\chi_A}{\sigma_E}\right]}}$$

For practical application of these correction formulas, it must be noticed, that R alloy, χ alloy and h alloy are related to the unknown mass concentration C_A by:

$$h \text{ alloy} = c_A \cdot h_A + c_B \cdot h_B + c_c \cdot h_c + - \cdots$$

$$\chi \text{ alloy} = c_A \cdot \chi_A + c_B \cdot \chi_B + c_c \cdot \chi_c + \cdots$$

$$\bar{Z} = c_A \cdot Z_A + c_B \cdot Z_B + c_c \cdot Z_c + \cdots$$

Two procedures are possible to overcome this difficulty:

α. Calculation of a theoretical correction curve, plotting C_A versus k. For this purpose mass concentrations, c_A on both sides of the expected value are inserted in the correction equation and the corresponding theoretical x-ray intensity is calculated.

β. Initially the x-ray intensities k_A, k_B, k_C are used as a first approximation for c_A, c_B, c_C in the determination of the R, χ, and h parameters of the alloy. The correct value will be obtained by successive iterations.

e) Application of the correction formula and use of the annexed tables

1) For a given experimental electron acceleration voltage V and from the analysis of the characteristic line A, determine in Table A

$$V - V_c \; ; \; V/V_c$$

2) From $V - V_C$ take σ_E from Table B

3) With Table B, Table C and the relation

$$h = 1{,}72 \cdot 10^{-6} \cdot \sigma_E \cdot V^2 \cdot \overline{A/Z^2}$$

calculate h_A and h alloy.

4) The x-ray mass absorption coefficients are annexed in Table D and permit the calculation of

$$\chi_A = \mu/\varrho \; \text{cosec} \; \Theta$$

$$\text{and} \; \chi_{\text{alloy}} = \sum c_i \, \mu/\varrho \; \text{cosec} \; \Theta$$

5) Insert the χ and h values to obtain the $1/f(\chi)$ values in the computed tables

6) Determine $\overline{Z}$ by $\overline{Z} = \sum c_i Z = C_A \cdot Z_A + C_B \cdot Z_B + \cdots$ and obtain R_A and R_{AB} from Table E.

Finally

$$\frac{I \, \text{alloy}}{I \, \text{standard}} = C_A \cdot \frac{R \, \text{alloy}}{R \, \text{standard}} \cdot \frac{1/f(\chi) \, \text{standard}}{1/f(\chi) \, \text{alloy}}$$

Conclusions

The analytical correction procedure developed above possesses the advantage of simplicity and versatility. Indeed, the complete correction procedure can be passed through in a few minutes with the aid of the annexed tables. The procedure can further be used even for multicomponent systems without any transformation at *arbitrary electron acceleration voltages*. The electron — target interaction effects — electron absorption, scattering and backscattering as a function of the mean atomic number, the electron acceleration and the electron energy range $V - V_C$ (variations with the ionisations cross section), are clearly separated. The results of practical applications to experimental data from more than 100 test samples from different laboratories definitely show an increased precision over previous methods, even for unfavourable experimental conditions (high $V - V_C$). The prediction of zero mass absorption correction for $V = V_C$ is a further confirmation of this analytical procedure.

However it must be noticed, that several arbitrary extensions of physical laws and approximations have been made in deriving the analytical procedure:

a) The LENARD law has been adapted to take account of the effective electron fraction with energies higher than $V - V_C$, and the resulting effective LENARD constant

$k = 8,9 \cdot 10^5$ has been calculated to fit into a limited number of test specimen calculations only.

b) The x-ray excitation function $\varphi(\varrho z)$ has been calculated by the approximation $\operatorname{tg} \alpha = \alpha + \frac{\alpha^3}{3}$, which is only strictly applicable for low α values. The exact integral diverges, when $\alpha \to \pi/2$ because the efficiency, $dz/\cos \alpha$, diverges; the use of this approximation for larger angles sets a limit to the electron path length, but this limit remains arbitrary.

c) The choice of the limit $\pi/4$ is also arbitrary.

d) BOTHE's law applies strictly for small angles only. A more refined method should take account of a supplementary LAMBERT distribution for RUTHERFORD scattering.

e) The calculation of the electron backscattering ratio, from data by BRAND or KULENKAMPF and SPYRA, is only approximate.

3.3 Variation of the Electron Interaction and Atomic Mass Absorption with Mean Atomic Number (POOLE and THOMAS, PHILIBERT)

For the calculation of the x-ray efficiency and mass absorption of x-rays of elementary anticathodes P. H. THOMAS (31) uses the PHILIBERT analytical procedure derived by a similar but simpler method as the above referred procedure. To take account of the change in depth of the x-ray generation with mean atomic number, THOMAS proposes an empirical method for the correction of the absorption term by calculating for the alloy

$$\chi_{\text{alloy}} = \frac{\sum c_i (\mu/\varrho) \operatorname{cosec} \theta}{\sum c_i \dfrac{\alpha_{iA}}{\alpha_{iB}}}$$

The parameter α is equal to the product of S (electron stopping power) and R (backscatter and should take account of the difference of electron absorption and backscatter with mean atomic number of the target, using the same distribution function with depth for both the standard and the specimen.

Two basic errors are inherent in this method.

a) As with the new complete correction proposed by PHILIBERT (36) the atomic number effect is corrected twice. A first correction for the atomic number is contained in the calculation of the distribution function (BOTHE – LENARD) and yet a supplementary independent integral atomic number correction is made.

b) No physical reason can be seen why this new absorption correction should account for the atomic number effect more correctly than the mean atomic number calculation and the theoretically and seperately derived $f(\chi)$ curves for the standard and the specimen.

3.4 Computational Calibration

A completely different approach, originally proposed by ARCHARD and MULVEY (35), is based on the numerical computations of a limited number of electron trajectories and the excitation of x-rays. These authors use a simplified theoretical electron-target interaction model, which according to BOTHE's law accounts simultaneously for electron retardation and backscattering.

The general nature of the predicted effects certainly contribute to the understanding of the basic quantitative principles. The experimental discrepancies for analyses of alloys with components of widely different atomic numbers demonstrate that a method, based on the interaction of only a few, individual, virtual electrons with matter, requires further investigation using a better model.

The use of the fastest available computers — a programme for the IBM 7090 has been devised in 1963 by DUNCUMB and SHIELDS,—would make it possible to apply a rigorous "Monte Carlo" calculation for a large number of electron trajectories. However these calculations for multicomponent specimen are not justified economically and have to be performed individually for every analysis.

The method has been used by GREEN (1963) to calculate the electron backscatter from pure elements as a function of the reduced electron acceleration, V/V_C.

3.5 Direct Determination by Empirical Calibration

For complex systems with uncertain fluorescence enhancement (oxides, complex minerals or other non conducting samples) it is sometimes necessary to apply a rigorous empirical calibration technique.

The use of an individual calibration curve for every considered system, avoids the necessity of correction calculations. However the careful preparation of homogeneous calibration standards is always time consuming and often not possible. Therefore, the only possibility which remains is calibration by the thin-film technique.

3.6 High Precision Microprobe Analysis by Thin Metallic Film Calibration

For high precision analysis of alloys (no limitation in number of components) in the absence of calibration standards homogeneous on a $1/\mu$ scale, the use of thin films, 0.01 to $0.1\,\mu$, obtained by ultramicrotomy permit high resolution analysis with negligible absorption and fluorescence corrections.

Theory

Suppose, an electron beam of intensity I_0 impinges perpendicularly on a metallic film of thickness z cm. The intensity of the characteristic radiation I_t collected from a constant area at an emergent direction θ will be[6]:

$$I_t = K \frac{M_t}{A_t} I_0 \int_{z=0}^{z} e^{-\alpha \varrho z}\, e^{-(\mu/\varrho)\,\cosec\,\theta}\, dz, \tag{1}$$

where

K = constant depending on the voltage ratio V/V_k,

M_t/A_t = number of atoms T per unit volume,

$\alpha = K/V^2 = (V$ being the accelerating voltage),

$\varrho = \sum M_i$ = density,

M_t = mass of element T per volume unit,

A_t = atomic weight of element T,

μ/ϱ = mass absorption for characteristic radiation in analysed alloy.

A considerable simplification is obtained when z and α become so small that the exponential term approaches unity. The physical significance of this simplification is that absorption effects are negligible.

By using sufficiently thin samples $(0.01\,\mu)$ and a highly accelerated electron beam (35 KV and more if possible), it is possible to assume a direct proportionality between the characteristic X-ray emission and the masses per unit area of the analysed elements.

$$\frac{I_A}{I_A} = M_A. \tag{2}$$

$I_A/I_{(A)}$ is the ratio of X-ray intensities emitted respectively in the $K\alpha$ (A) line, under the same experimental conditions by the pure element A and the analysed alloy. Similarly

$$\frac{I_B}{I_B} = M_B, \tag{2'}$$

$$\frac{I_C}{I_C} = M_C. \tag{2''}$$

Considering

$$\varrho = \sum M_i = M_A + M_B + M_C, \tag{3}$$

$$C_A = \frac{M_A}{\sum M_i} = \frac{M_A}{M_A + M_B + M_C + \cdots}, \tag{4}$$

$$C_B = \frac{M_B}{\sum M_i}. \tag{4'}$$

Using equation (4) the chemical composition of a sample is obtained by the determination of the superficial masses of all the elements present in the sample.

Experimental Procedure

The thin films $0.01\,\mu$ thick (samples and pure elements) should be of identical thickness, a condition which can be most easily fulfilled by the automatic advance of an ultramicrotome with a diamond knife.

Films of about 2500 μ^2 are collected on standard electron microscope grids and inserted into a special probe holder (Figure 7). (15)

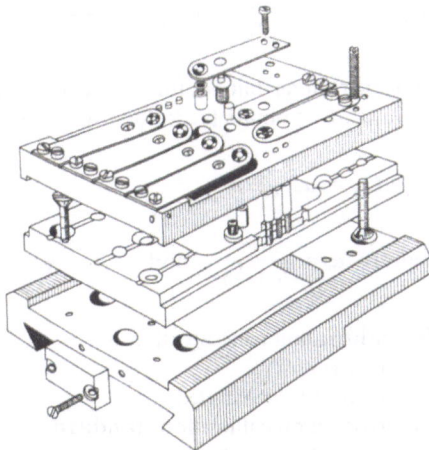

Fig. 7. Special sample holder for thin film Microanalysis (French Microprobe).

The highest available accelerating voltage should be used in order to avoid a notable deceleration of electrons in the anticathodes, to diminish the absorption effect of superficial oxide or contamination layers, as well as to increase the peak-to-background ratio of the analysed radiation.

General use of this method is limited by the delicate and time consuming preparation of samples and standards.

Application includes microanalysis of small inclusions and setting up of calibration curves for complex alloys. Furthermore, projection x-ray metallography combined with electron probe microanalysis may lead to valuable information in physical metallurgy.

§ 4 Detection Limit, Detection Threshold and Microprobe Trace Analysis

Based on theoretical considerations and experimental confirmations, it is possible to demonstrate that the method of X-ray emission microanalysis may be extended to trace analysis. It therefore becomes a powerful evaluation method in a number of metallurgical areas (and in other domains as well) at present inaccessible, if appropriate experimental conditions are chosen.

4.1 Theoretical Basis of Trace Determinations

Quantitative microprobe analysis is based on the comparison of the emission, I_A, of characteristic radiation of element A from a sample with the emission, $I(A)$, of the same characteristic line under the same experimental conditions from a reference target, consisting of either the pure element A or, as in the proposed approach, a sample with low but known concentration of A. The mass concentration C_X of element A in the region analyzed is supplied to a first approximation by CASTAING (1):

$$\frac{I_A}{I(A)} = c_x \cdot c_c$$

If we consider that, for the determination of low concentrations, X-ray absorption in the sample and secondary fluorescence emission effects are negligible and assume that we can performe the low concentration calibration, C_c, without error, we obtain:

$$C_c \cdot C_X = \frac{\bar{N}_x - \bar{N}_{Bx}}{\bar{N}_c - \bar{N}_{Bc}} = \bar{N}_X \frac{1}{\bar{N}_c - \bar{N}_{Bc}} - \frac{\bar{N}_{Bx}}{\bar{N}_c - \bar{N}_{Bc}}$$

For the range of low concentrations, but not for calibration with a pure element, $\mathcal{N}_{Bx}$ is not very different from $\mathcal{N}_{Bc}$ and we obtain the following linear relationship (Fig. 1).

$$C_c \cdot C_X = \alpha \bar{N}_X - \beta$$

$$\text{where } \alpha = \frac{1}{\bar{N}_c - \bar{N}_B} \quad \text{and} \quad \beta = \frac{\bar{N}_B}{\bar{N}_c - \bar{N}_{Bc}}$$

$\bar{N}_c$ = impulse counts for calibration concentration
$\bar{N}_x$ = counts for trace concentration
C_x = unknown concentration to be calculated
$\bar{N}_{Bc}$ = counts for the background on calibration standard
$\bar{N}_{Bx}$ = counts for the background on specimen

The calculation of the concentration C_X from relation (2) if C_X tends to zero is accomplished probability theory and the laws of error (34).

$$\lim_{C_X \to 0} (\Delta C) = \sqrt{2}\; c_c \cdot (\bar{N}_c - \bar{N}_B)^{-1}\; (\Delta \bar{N}_B) \tag{4}$$

if $\bar{r}^x = \dfrac{\bar{N}_x}{\Delta t}$ 　　$\bar{r}^x$ = mean counting rate for random sample
　　　　　　　Δt = time interval for one measurement

$T = n \cdot \Delta t$ 　　T = time interval for the whole random sample
　　　　　　n = number of measurements on the random sample

We get a detection threshold C_{DT} (concentration of an element in a matrix which still can be detected by the analytical method) of

$$C_{DT} = \frac{\sqrt{2} \cdot t_s \cdot \sqrt{\bar{r}^x_B} \cdot c_c}{\sqrt{T}(\bar{r}^x_c - \bar{r}^x_B)} \tag{5}$$

t_s = factor from "Student" distribution, depending only on number of measurements on the random sample (tabulated)

As a detection limit, C_{DL}, we define that concentration of an element, which still can be determined with a given accuracy, i. e. the relative error of the concentration C shall not pass a value "a" in percent.

then 　　$$C_{DL} = \frac{100}{a} \cdot \frac{t_s \cdot \sqrt{2} \cdot \sqrt{\bar{r}^x_B} \cdot C_c}{\sqrt{T}(\bar{r}^x_C - \bar{r}^x_B)} \tag{6}$$

The fundamentally important parameter–the peak to background ratio of the analysed characteristic radiation–is defined by the particular spectrometer set-up.

For statistical reasons the detection limit decreases with increasing total time interval "T" and incrasing number of measurements within the same time interval T for a single point analysis.

4.2 Experimental Procedure

The formation rate of the well known contamination spot on the sample at the point of impact of the electron beam limits the counting intervals on one point in conventional microanalysis to about 50 to 100 seconds per measurement. As the detection limit decreases with longer time intervals, the formation of contamination can be prevented by introducing flowing gas, through a high precision valve and a capillary of non magnetic material into the specimen chamber and directing it on the electron bombarded region in a manner similar to that of CASTAING (16) when he was involved in studies on low accelerated electron beams.

However even extremely small flow rates of air through the capillary 0,025 mg/sec. cause considerable reductions in life-time of the electron-emitting filament. By this fact, the reproducibility of the initial measurements is seriously restricted, as the replacement of this filament necessitates a complete refocussing of the electron optical system. This drawback can be eliminated by the use of an inert gas flow. The prevention of the contamination spot by the presence of a localised gaseous atmosphere at the point of impact of the electron beam is not due to an oxidation phenomenon, but almost certainly to an ion-beam etching effect as demonstrated through the use of several inert gas atmospheres. The etching effect increases regularly with increasing atomic weight of the gas probe. Helium is rather ineffective in preventing contamination, whereas argon-bombarded metallic specimens exhibit perfect surfaces, even after exposures of 10^4 sec. As the life-time of filaments is not influenced by these small quantities of argon, its use as a gaseous probe is suggested.

The upper time limit for the measurement of one concentration is fixed by economical considerations to about 9 000 seconds. Then the gas valve can be closed and the ana-lysed point located by vacuum electron bombardement for about 100 sec. The depen-dence of the Student distribution factor from the number of measurements shows that for more than 15 measurements, there is no further significant decrease in the detection threshold. It has been shown that the instrument stability is high even for these long time intervals and that the fluctuations of a random sample are part of a Gaussian population.

Random parasitic impulses coming from the electric mains may be substracted by an independent pulse height analyzer.

References

(1) CASTAING, R.: Thesis University of Paris (1951) Publ. O. N. E. R. A. No. 55 Adv. Electr., Electron Physics, XIII (1960) Acad. Press.
(2) MOSELY, H. G.: Phil. Mag. *26*, 1024 (1913).
(3) PHILIBERT, J.: J. Int. Metals *90*, 241 (1962).
(4) THEISEN, R.: Euratom Report EUR-I-1 (1961).
(5) POOLE, D. M., and P. M. THOMAS: J. Inst. Metals *90*, 228 (1962).
(6) ARCHARD, G. D., and T. MULVEY: J. Appl. Physics *4*, 626 (1963).
(8) BIRKS, L. S.: Electron Probe Microanalysis Interscience Publishers (1963).
(9) DUNCUMB, P., and G. V. P. LONG; MELFORD, D. A. Electron Probe Microanalysis Hilger and Watts Ltd (1965).
(10) LANGMUIR, D. B.: Proc. I. R. E. *25*, 977 (1937).
(11) ZWORYKIN, V. K., et al.: Electron Optics and Electron Microscope Wiley and Sons (1945).
(12) THEISEN, R., and H. W. SCHLEICHER: Dragon Fuel Element Symposium, Bournemouth 1963, D. P. Report *143* (1963).
(13) KIRIANENKO, A.: J. Microscopie *3*, 354 (1964).
(14) THEISEN, R., and J. LEMAITRE: Proc. *10*, Colloq. Spectr. Int. Maryland, Spartan Books, Washington (1963).
(15) LEMAITRE J., and R. THEISEN: Euratom Report EUR 109 b, (1962).
(16) CASTAING, R., and J. DESCAMPS: Compt. Rend. Acad. Sci. *238*, 1506 (1954).
(17) CAMPBELL, A., and. R. GIBBONS: Manchester private communication (1964).
(18) THEISEN, R.: Proc. Int. Conf. Electr. Beam, Ion Beam Techn., Toronto (1964) Wiley and Sons.
(19) FAESSLER, A.: Proc. X. Colloq. Spectr. Intern. Spartan Books, Washington, *307* (1964).
(20) MENSHIKOV, A. Z., and S. A. NEMNONOV: Inst. Met. Phys. Acad. Sci. U. S.S.R. (1962).
(21) WITTRY, D. B.: A. S. T. M. *x*-ray Symposicum (1963).
(22) HÉNOC, J.: Thesis, University of Paris 1962. Publ. C. N. E. T. No. 635.
(23) KIRIANENKO, A., and J. HÉNOC: Proc. Int. Conf. Toronto (1964) Wiley & Sons.
(24) DILS, R., and L. ZEITZ: Symp. *x*-ray Optics and Microanalysis Stanford (1962).
(25) WITTRY, D. B.: Thesis Cal. Tech. (1957).
(26) KIRIANENKO et al.: 3. Int. Symp. Stanford (1962).
(27) GREEN, M.: ibidem.
(28) LENARD, P.: Quantitatives über Kathodenstr., C. Winter Heidelberg (1918).
(29) DUNCUMB, P.: Private communication (1964).
(30) BOTHE, W.: Handbuch der Physik *24*.
(31) THOMAS, P. M.: A. E. R. E. Harwell, R 4539 (1964).
(32) BRAND, J. D.: Annal. Phys. Leipzig *26*, 609 (1936).
(33) KULENKAMPF, H.: Spyra, W., Z. f. Phys. 137 (1954).
(34) PHILIBERT, J.: Irsid Report B No. 51 (1965).
(35) ARCHARD, G. D. and T. MULVEY: J. Appl. Phys. 14, (1963).
(36) TÄFFNER, K., and R. THEISEN: Int. Symp. E. M. P. Ispra (1963) EUR 1819 e (1964).
(37) THEISEN, R.: Proc. Int. Conf. Electr. Beam Appl., Toronto (1964) Wiley & Sons.
(38) ZIEBOLD, T. O.: M.I. T. private communication (1965).

Annexed Tables

Table A. *Characteristic Wavelengths and Excitation Potentials for K; L; M Series* $\lambda K\alpha_1/\alpha_2 = (2\lambda K\alpha_1 + \lambda K\alpha_2)/3$

Element	Atomic No.	Line	Wavelength λ	Excitation kev	Element	Atomic No.	Line	Wavelength λ	Excitation kev
Na	11	$K\alpha_1/\alpha_2$	11,9 A	1,1	Cu	29	$K\alpha_1/\alpha_2$	1,54	9
		$K\beta_1/\beta_2$	11,6				$K\beta_1/\beta_2$	1,39	
Mg	12	$K\alpha_1/\alpha_2$	9,9	1,3			$L\alpha_1$	13,4	1,1
		$K\beta_1/\beta_2$	9,5				$L\beta_1$	13,1	
Al	13	$K\alpha_1/\alpha_2$	8,3	1,6	Zn	30	$K\alpha_1/\alpha_2$	1,43	9,7
		$K\beta_1/\beta_2$	8				$K\beta_1/\beta_2$	1,29	
Si	14	$K\alpha_1/\alpha_2$	7,1	1,8			$L\alpha_1$	12,2	1,2
		$K\beta_1/\beta_2$	6,8				$L\beta_1$	12	
P	15	$K\alpha_1/\alpha_2$	6,2	2,1	Ga	31	$K\alpha_1/\alpha_2$	1,34	10,4
		$K\beta_1/\beta_2$	5,8				$K\beta_1/\beta_2$	1,2	
S	16	$K\alpha_1/\alpha_2$	5,4	2,5			$L\alpha_1$	11,3	1,3
		$K\beta_1/\beta_2$	5				$L\beta_1$	11	
Cl	17	$K\alpha_1/\alpha_2$	4,7	2,8	Ge	32	$K\alpha_1/\alpha_2$	1,26	11,1
		$K\beta_1/\beta_2$	4,4				$K\beta_1/\beta_2$	1,13	
K	19	$K\alpha_1/\alpha_2$	3,7	3,6			$L\alpha_1$	20,4	1,4
		$K\beta_1/\beta_2$	3,5				$L\beta_1$	10,1	
Ca	20	$K\alpha_1/\alpha_2$	3,4	4	As	33	$K\alpha_1/\alpha_2$	1,18	11,8
		$K\beta_1/\beta_2$	3,1				$K\beta_1/\beta_2$	1,06	
Sc	21	$K\alpha_1/\alpha_2$	3	4,5			$L\alpha_1$	9,7	8,5
		$K\beta_1/\beta_2$	2,8				$L\beta_1$	9,4	
Ti	22	$K\alpha_1/\alpha_2$	2,75	5	Se	34	$K\alpha_1/\alpha_2$	1,1	12,7
		$K\beta_1/\beta_2$	2,51				$K\beta_1/\beta_2$	0,99	
V	23	$K\alpha_1/\alpha_2$	2,5	5,5			$L\alpha_1$	9	1,7
		$K\beta_1/\beta_2$	2,28				$L\beta_1$	8,7	
		$L\alpha_1$	24,3	0,6	Br	35	$K\alpha_1/\alpha_2$	1,04	13,4
Cr	24	$K\alpha_1/\alpha_2$	2,29	6			$K\beta_1/\beta_2$	0,93	
		$K\beta_1/\beta_2$	2,08				$L\alpha_1$	8,4	1,8
		$L\alpha_1$	21,5	0,68			$L\beta_1$	8,1	
		$L\beta_1$	21,2		Rb	37	$K\alpha_1/\alpha_2$	0,93	15,2
Mn	25	$K\alpha_1/\alpha_2$	2,1	6,5			$K\beta_1/\beta_2$	0,83	
		$K\beta_1/\beta_2$	1,9				$L\alpha_1$	7,5	2,1
		$L\alpha_1$	19,4	0,76			$L\beta_1$	7,3	
		$L\beta_1$	19		Sr	38	$K\alpha_1/\alpha_2$	0,88	16,1
Fe	26	$K\alpha_1/\alpha_2$	1,94	7,1			$K\beta_1/\beta_2$	0,78	
		$K\beta_1/\beta_2$	1,75				$L\alpha_1$	6,8	2,2
		$L\alpha_1$	17,6	0,85			$L\beta_1$	6,6	
		$L\beta_1$	17,2		Y	39	$K\alpha_1/\alpha_2$	0,83	17,1
Co	27	$K\alpha_1/\alpha_2$	1,70	7,7			$K\beta_1/\beta_2$	0,74	
		$K\beta_1\ \beta_2$	1,62				$L\alpha_1$	6,4	2,4
		$L\alpha_1$	15,9	0,93			$L\beta_1$	6,2	
		$L\beta_1$	15,6		Zr	40	$K\alpha_1/\alpha_2$	0,79	18,0
Ni	28	$K\alpha_1/\alpha_2$	1,66	8,3			$K\beta_1/\beta_2$	0,7	
		$K\beta_1/\beta_2$	1,5				$L\alpha_1$	6,1	2,5
		$L\alpha_1$	14,5	1			$L\beta_1$	5,8	
		$L\beta_1$	14,2		Nb	41	$K\alpha_1/\alpha_2$	0,75	19,0
							$K\beta_1/\beta_2$	0,67	
							$L\alpha_1$	5,7	2,7
							$L\beta_1$	5,5	

Table A (Continuation)

Element	Atomic No.	Line	Wavelength λ	Excitation *kev*
Mo	42	$K\alpha_1/\alpha_2$	0,71	20,0
		$K\beta_1/\beta_2$	0,63	
		$L\alpha_1$	5,4	2,9
		$L\beta_1$	5,2	
Ru	44	$K\alpha_1/\alpha_2$	0,64	22,1
		$K\beta_1/\beta_2$	0,57	
		$L\alpha_1$	4,8	3,2
		$L\beta_1$	4,6	
Rh	45	$K\alpha_1/\alpha_2$	0,61	
		$K\beta_1/\beta_2$	0,54	
		$L\alpha_1$	4,6	
		$L\beta_1$	4,4	
Pd	46	$K\alpha_1/\alpha_2$	0,59	24,4
		$K\beta_1/\beta_2$	0,52	
		$L\alpha_1$	4,4	3,6
		$L\beta_1$	4,1	
Ag	47	$K\alpha_1/\alpha_2$	0,56	25,5
		$K\beta_1/\beta_2$	0,50	
		$L\alpha_1$	4,2	3,8
		$L\beta_1$	3,9	
Cd	48	$K\alpha_1/\alpha_2$	0,53	26,7
		$K\beta_1/\beta_2$	0,47	
		$L\alpha_1$	3,5	4
		$L\beta_1$	3,7	
In	49	$K\alpha_1/\alpha_2$	0,51	27,9
		$K\beta_1/\beta_2$	0,45	
		$L\alpha_1$	3,9	4,2
		$L\beta_1$	3,7	
Sn	50	$K\alpha_1/\alpha_2$	0,49	29,2
		$K\beta_1/\beta_2$	0,43	
		$L\alpha_1$	3,6	4,4
		$L\beta_1$	3,4	
Sb	51	$K\alpha_1/\alpha_2$	0,47	30,5
		$K\beta_1/\beta_2$	0,41	
		$L\alpha_1$	3,4	4,7
		$L\beta_1$	3,2	
Te	52	$L\alpha_1$	3,28	5
		$L\beta_1$	3,07	
I	53	$L\alpha_1$	3,14	5,2
		$L\beta_1$	2,93	
Cs	54	$L\alpha_1$	2,89	5,7
		$L\beta_1$	2,68	
Ba	56	$L\alpha_1$	2,77	6,0
		$L\beta_1$	2,56	
La	57	$L\alpha_1$	2,66	6,3
		$L\beta_1$	2,45	
Ce	58	$L\alpha_1$	2,56	6,6
		$L\beta_1$	2,35	

Element	Atomic No.	Line	Wavelength λ	Excitation *kev*
Pr	59	$L\alpha_1$	2,46	6,8
		$L\beta_1$	2,25	
Nd	60	$L\alpha_1$	2,37	7,1
		$L\beta_1$	2,16	
Sm	62	$L\alpha_1$	2,2	7,7
		$L\beta_1$	1,99	
		$M\alpha_1/\alpha_2$	11,4	1,7
Eu	63	$L\alpha_1$	2,12	8,1
		$L\beta_1$	1,92	
		$M\alpha_1/\alpha_2$	10,9	1,8
Gd	64	$L\alpha_1$	2,04	8,4
		$L\beta_1$	1,88	
		$M\alpha_1/\alpha_2$	10,4	1,9
Tb	65	$L\alpha_1$	1,97	8,7
		$L\beta_1$	1,77	
		$M\alpha_1/\alpha_2$	10,0	2
		$M\beta$	9,8	
Dy	66	$L\alpha_1$	1,9	9,1
		$L\beta_1$	1,71	
		$M\alpha_1/\alpha_2$	9,5	2
		$M\beta$	9,3	
Ho	67	$L\alpha_1$	1,84	9,4
		$L\beta_1$	1,64	
		$M\alpha_1/\alpha_2$	9,1	2,1
		$M\beta$	8,9	
Er	68	$L\alpha_1$	1,78	9,8
		$L\beta_1$	1,58	
		$M\alpha_1/\alpha_2$	8,8	2,2
		$M\beta$	8,6	
Tm	69	$L\alpha_1$	1,72	10,1
		$L\beta_1$	1,53	
Yb	70	$L\alpha_1$	1,67	10,5
		$L\beta_1$	1,47	
		$M\alpha_1/\alpha_2$	8,1	2,4
		$M\beta$	7,9	
Lu	71	$L\alpha_1$	1,62	10,9
		$L\beta_1$	1,42	
		$M\alpha_1/\alpha_2$	7,8	2,5
		$M\beta$	7,6	
Hf	72	$L\alpha_1$	1,57	11,3
		$L\beta_1$	1,37	
		$M\alpha_1/\alpha_2$	7,5	2,6
		$M\beta$	7,3	
Ta	73	$L\alpha_1$	1,52	11,7
		$L\beta_1$	1,32	
		$M\alpha_1/\alpha_2$	7,2	2,7
		$M\beta$	7	

Table A (Continuation)

Element	Atomic No.	Line	Wavelength λ	Excitation *kev*	Element	Atomic No.	Line	Wavelength λ	Excitation *kev*
W	74	$L\alpha_1$	1,48	12,1	Th	90	$L\alpha_1$	0,96	16,9
		$L\beta_1$	1,28				$L\beta_1$	0,76	
		$M\alpha_1/\alpha_2$	7,0	2,8			$M\alpha_1/\alpha_2$	4,1	4,2
		$M\beta$	6,7				$M\beta$		
Re	75	$L\alpha_1$	1,43	12,5	U	92	$L\alpha_1$	0,91	21,8
		$L\beta_1$	1,24				$L\beta_1$	0,72	
		$M\alpha_1/\alpha_2$	6,7	2,9			$M\alpha_1/\alpha_2$	3,9	5,5
		$M\beta$	6,5				$M\beta$	3,7	
Os	76	$L\alpha_1$	1,39	13,0	Np	93	$L\alpha_1$	0,89	22,4
		$L\beta_1$	1,19				$L\beta_1$	0,70	
		$M\alpha_1/\alpha_2$	6,5	3,0			$M\alpha_1/\alpha_2$	3,7	5,7
		$M\beta$	6,3				$M\beta$	3,5	
Ir	77	$L\alpha_1$	1,35	13,4	Pu	94	$L\alpha_1$	0,87	23,1
		$L\beta_1$	1,16				$L\beta_1$	0,68	
		$M\alpha_1/\alpha_2$	6,2	3,2			$M\alpha_1/\alpha_2$	3,5	5,9
		$M\beta$	6				$M\beta$	3,4	
Pt	78	$L\alpha_1$	1,31	13,9	Am	95	$L\alpha_1$	0,85	23,8
		$L\alpha_2$	1,23				$L\beta_1$	0,66	
		$M\alpha_1/\alpha_2$	6,0	3,3			$M\alpha_1/\alpha_2$	3,4	6,1
		$M\beta$	5,8				$M\beta$	2,4	
Au	79	$L\alpha_1$	1,28	14,4	Cm	96	$L\alpha_1$	0,83	24,5
		$L\beta_1$	1,08				$L\beta_1$	0,64	
		$M\alpha_1/\alpha_2$	5,8	3,4			$M\alpha_1/\alpha_2$	3,3	6,3
		$M\beta$	5,6				$M\beta$		
Hg	80	$L\alpha_1$	1,24	14,8	Be	97	$L\alpha_1$	0,81	25,2
		$L\beta_1$	1,05				$L\beta_1$	0,62	
		$M\alpha_1/\alpha_2$	5,6	3,6			$M\alpha_1/\alpha_2$	3,2	6,6
		$M\beta$	5,4				$M\beta$	3,0	
Tl	81	$L\alpha_1$	1,2	15,3	Cf	98	$L\alpha_1$	0,79	26,0
		$L\beta_1$	1,01				$L\beta_1$	0,60	
		$M\alpha_1/\alpha_2$	5,5	3,7			$M\alpha_1/\alpha_2$	3,	6,8
		$M\beta$	5,2				$M\beta$	2,9	
Pb	82	$L\alpha_1$	1,18	15,9	E	99	$L\alpha_1$	0,78	26,8
		$L\beta_1$	0,98				$L\beta_1$	0,59	
		$M\alpha_1/\alpha_2$	5,3	3,9			$M\alpha_1/\alpha_2$	2,9	7,
		$M\beta$	5,1				$M\beta$	2,8	
Bi	83	$L\alpha_1$	1,14	16,4	Fm	100	$L\alpha_1$	0,76	27,1
		$L\beta_1$	0,95				$L\beta_1$	0,57	
		$M\alpha_1/\alpha_2$	5,1	4			$M\alpha_1/\alpha_2$	2,8	7,3
		$M\beta$	4,9				$M\beta$	2,6	

Table B. Determination of effective LENARD coefficients

$V-V_K$	$(V-V_K)^2$	σ_E	$1{,}72 \cdot 10^{-6} \cdot \sigma_E \cdot$
1	1	890000	1,5
2	4	222500	0,3827
3	9	98890	0,1701
4	16	55630	0,0957
5	25	35600	0,0612
6	36	24720	0,0425
7	49	18160	0,0312
8	64	13910	0,0239
9	81	10990	0,018
10	100	8900	0.0153
11	121	7355	0,0126
12	144	6181	0,0106
13	169	5266	0,00906
14	196	4541	0,00781
15	225	3956	0,00680
16	256	3477	0,00598
17	289	3080	0,00530
18	324	2747	0,00472
19	361	2225	0,00383
20	400	2018	0,00347

*Table C. DETERMINATION of the Penetration Factor $\overline{A/Z^2}$ **

Z	A	element	A/Z^2	Z	A	element	A/Z^2
1	1,008	H	1,008	51	121,76	Sb	0,0466
2	4,003	He	1	52	127,61	Te	0,0466
3	6,94	Li	0,766	53	126,92	I	0,045
4	9,02	Be	0,558	54	131,3	Xe	0,045
5	10.82	B	0,433	55	132,91	Cs	0,044
6	12,01	C	0,33	56	137,36	Ba	0,043
7	14,008	N	0,283	57	138,92	La	0,0425
8	16	O	0,25	58	140,13	Ce	0,042
9	19	F	0,233	59	140,92	Pr	0,040
10	20,183	Ne	0,20	60	144,27	Nd	0,040
11	2,997	Na	0,191	61	147	Pm	0,039
12	24,32	Mo	0,16	62	150,43	Sm	0,039
13	26,98	Al	0,158	63	152	Eu	0,038
14	28,09	Si	0,150	64	156,9	Gd	0,038
15	30,975	P	0,133	65	159,2	Tb	0,0375
16	32,066	S	0,125	66	162,46	Dy	0,0375
17	35,457	Cl	0,125	67	164,94	Ho	0,037
18	39,944	A	0,125	68	167,2	Er	0,036
19	39,1	K	0,108	69	160,4	Tm	0,036
20	40,08	Ca	0,1	70	173,04	Yb	0,035
21	44,96	Sc	0,1	71	174,99	Lu	0,034
22	47,9	Ti	0,1	72	178,6	Hf	0,034
23	50,95	V	0,096	73	180,88	Ta	0,033
24	52,01	Cr	0,092	74	183,92	W	0,033
25	54,93	Mn	0,087	75	186,31	Re	0,0325
26	55,85	Fe	0,08	76	190,2	Os	0,0325
27	58,94	Co	0,081	77	193,1	Ir	0,0325
28	58,69	Ni	0,075	78	195,23	Pt	0,032
29	63,54	Cu	0,0760	79	197,2	Au	0,032
30	65,38	Zn	0,073	80	200,61	Hg	0,031
31	69,72	Ga	0,073	81	204,39	Tl	0,031
32	72,6	Ge	0,071	82	207,21	Pb	0,031
33	74,91	As	0,068	83	209,0	Bi	0,030
34	78,96	Se	0,068	84	210	Po	0,030
35	79,916	Br	0,065	85	211	At	0,029
36	83,8	Kr	0,064	86	222	Rn	0,030
37	85,48	Rb	0,063	87	223	Fr	0,029
38	87,63	Sr	0,061	88	226,05	Ra	0,029
39	88,92	Y	0,058	89	227	Ac	0,028
40	91,22	Zr	0,057	90	232,12	Th	0,028
41	92,91	Nb	0,055	91	231	Pa	0,0275
42	95,95	Mo	0,054	92	238,07	U	0,028
43	98,91	Tc	0,053	93	237	Np	0,0275
44	101,7	Ru	0,052	94	242	Pu	0,0275
45	102,91	Rh	0,051	95	243	Am	0,0266
46	106,7	Pd	0,050	96	243	Cm	0,026
47	107,88	Ag	0,048	97	245	Bk	0,026
48	112,41	Cd	0,0483	98	246	Cf	0,026
49	114,76	In	0,0475	99	252	E	0,026
50	118,70	Sn	0,0475	100	256	Fm	0,026

* The mean penetration factor is given by: $\overline{A/Z^2} = \sum c_i \cdot A_i/Z^2$.

Table D. Table of X-Ray Mass-Absorption Coefficients
compiled by E. Patrassi

The X-ray Mass Absorption Coefficients have been obtained by graphical extrapolation of experimental results, including microprobe determinations by Hughes and coworkers, Descamps, Henoc, Philibert and the author.

Additional information of this book

(Quantitative electron microprobe analysis; 978-3-662-23130-2) is provided:

http://Extras.Springer.com

Table E. Effective Electron Current R in [%] *

V/V_c	1,25	1,35	1,41	1,57	1,66	1,87	2,21	3,17	7,57
z									
	100	100	100	100	100	100	100	100	100
5	98,8	97,5	97,5	96,3	96,3	95,0	95,0	95,0	96,3
10	97,5	95,8	95,0	93,4	92,5	90,4	90,0	90,4	91,7
15	96,3	94,6	93,8	90,4	89,2	86,7	86,7	86,7	88,3
20	95,0	92,1	90,5	87,9	86,3	83,8	82,5	83,7	85,4
25	93,8	90,4	88,7	85,4	83,8	81,2	79,7	81,1	82,9
30	92,5	89,2	87,1	83,8	81,7	78,9	77,5	78,9	80,8
35	91,7	87,9	85,4	82,1	80,0	77,2	75,8	77,1	78,8
40	90,8	86,7	84,2	80,8	78,3	76,1	74,5	75,8	77,2
45	89,7	85,9	83,3	79,6	77,2	74,7	73,3	74,5	76,1
50	88,9	85,4	82,5	78,8	76,3	73,9	72,2	73,6	75,9
55	88,3	84,7	81,7	77,9	75,6	73,1	71,4	72,8	74,5
60	87,8	84,2	81,3	77,1	75,0	72,5	70,8	72,1	73,6
65	87,3	83,9	80,8	76,7	74,7	72,1	70,4	71,6	73,1
70	86,9	83,6	80,4	76,3	74,5	71,7	70,0	71,1	72,5
75	86,7	83,4	80,3	76,1	74,2	71,3	69,6	70,4	72,1
80	86,4	83,1	80,0	75,8	73,9	71,1	69,4	70,3	71,9
85	86,3	83,1	80,0	75,8	73,8	70,8	69,2	70,0	71,4
90	86,3	82,8	80,0	75,6	73,6	70,8	68,9	69,7	71,3
95	86,3	82,8	80,0	75,6	73,6	70,6	68,8	69,6	71,1
100	86,3	82,8	80,0	75,6	73,6	70,6	68,8	69,4	70,8

* Values of R for intermediate atomic numbers are obtained by interpolation.

Table F. Determination of the Efficiency Function $1/f(\chi)$

The Efficiency Function $1/f(\chi)$ has been calculated using an I.B.M. 7070 Model and the basic relation of the analytical correction procedure:

$$1/f(\chi) = (1 + \chi/\sigma)\left(1 + \frac{h}{1+h} \cdot \chi/\sigma\right);$$

an experimentally determined effective Lenard coefficient $\sigma_E = \dfrac{8{,}9 \cdot 10^5}{(V-V_c)^2}$ and an atomic number factor

$$h = 1{,}5308 \cdot \frac{V^2}{(V-V_c)^2} \cdot A/Z^2.$$

SIGMA = 890000.0

V−V_c = 1

H	3.4	3.2	3.0	2.8	2.6	2.4	2.2	2.0	1.8	1.6	1.4	1.2	1.0	0.8	0.6	0.4	0.2	0.10	0.09	0.08
0	1.000	1.000	1.000	1.000	1.000	1.000	1.000	1.000	1.000	1.000	1.000	1.000	1.000	1.000	1.000	1.000	1.000	1.000	1.000	1.000
100	1.000	1.000	1.000	1.000	1.000	1.000	1.000	1.000	1.000	1.000	1.000	1.000	1.000	1.000	1.000	1.000	1.000	1.000	1.000	1.000
200	1.000	1.000	1.000	1.000	1.001	1.001	1.000	1.001	1.000	1.001	1.000	1.000	1.000	1.000	1.000	1.000	1.000	1.000	1.000	1.000
300	1.001	1.001	1.001	1.001	1.001	1.001	1.001	1.001	1.001	1.001	1.001	1.001	1.001	1.000	1.000	1.000	1.001	1.000	1.000	1.000
400	1.001	1.001	1.001	1.001	1.001	1.001	1.001	1.001	1.001	1.001	1.001	1.001	1.001	1.001	1.001	1.001	1.001	1.001	1.000	1.001
500	1.001	1.001	1.001	1.001	1.001	1.001	1.001	1.001	1.001	1.001	1.001	1.001	1.001	1.001	1.001	1.001	1.001	1.001	1.001	1.001
600	1.001	1.001	1.001	1.001	1.001	1.001	1.001	1.001	1.001	1.001	1.001	1.001	1.001	1.001	1.001	1.001	1.001	1.001	1.001	1.001
700	1.001	1.001	1.001	1.001	1.001	1.001	1.001	1.001	1.001	1.001	1.001	1.001	1.001	1.001	1.001	1.001	1.001	1.001	1.001	1.001
800	1.002	1.002	1.002	1.002	1.002	1.002	1.002	1.001	1.001	1.002	1.002	1.002	1.002	1.002	1.001	1.001	1.001	1.001	1.001	1.001
900	1.002	1.002	1.002	1.002	1.002	1.002	1.002	1.002	1.002	1.002	1.002	1.002	1.002	1.002	1.002	1.002	1.001	1.001	1.001	1.001
1000	1.002	1.002	1.002	1.002	1.002	1.002	1.002	1.002	1.002	1.002	1.002	1.002	1.002	1.002	1.002	1.002	1.002	1.001	1.001	1.001
1100	1.002	1.002	1.002	1.002	1.002	1.002	1.002	1.002	1.002	1.002	1.002	1.002	1.002	1.002	1.002	1.002	1.001	1.001	1.001	1.001
1200	1.002	1.002	1.002	1.002	1.002	1.002	1.002	1.002	1.002	1.002	1.002	1.002	1.002	1.002	1.002	1.002	1.002	1.002	1.001	1.002
1300	1.003	1.003	1.003	1.003	1.003	1.003	1.003	1.003	1.003	1.003	1.002	1.003	1.002	1.002	1.002	1.002	1.002	1.002	1.002	1.002
1400	1.003	1.003	1.003	1.003	1.003	1.003	1.003	1.003	1.003	1.003	1.003	1.003	1.002	1.002	1.002	1.002	1.002	1.002	1.002	1.002
1500	1.003	1.003	1.003	1.003	1.003	1.003	1.003	1.003	1.003	1.003	1.003	1.003	1.003	1.003	1.002	1.003	1.002	1.002	1.002	1.002
1600	1.003	1.003	1.003	1.004	1.003	1.003	1.003	1.003	1.003	1.003	1.003	1.003	1.003	1.003	1.003	1.003	1.003	1.002	1.002	1.002
1700	1.003	1.003	1.003	1.003	1.003	1.003	1.003	1.003	1.003	1.003	1.003	1.003	1.003	1.003	1.003	1.002	1.002	1.002	1.002	1.002
1800	1.004	1.004	1.004	1.004	1.003	1.003	1.003	1.003	1.003	1.003	1.003	1.003	1.003	1.003	1.003	1.003	1.003	1.002	1.002	1.002
1900	1.004	1.004	1.004	1.004	1.004	1.004	1.004	1.004	1.004	1.004	1.003	1.003	1.004	1.003	1.003	1.003	1.002	1.002	1.002	1.002
2000	1.004	1.004	1.004	1.004	1.004	1.004	1.004	1.004	1.004	1.004	1.004	1.003	1.004	1.003	1.003	1.003	1.003	1.003	1.002	1.002
2100	1.004	1.004	1.004	1.004	1.004	1.004	1.004	1.004	1.004	1.004	1.004	1.004	1.004	1.003	1.003	1.003	1.003	1.003	1.003	1.003
2200	1.004	1.004	1.004	1.004	1.004	1.004	1.004	1.004	1.004	1.004	1.004	1.004	1.004	1.004	1.003	1.003	1.003	1.003	1.003	1.003
2300	1.005	1.005	1.005	1.004	1.005	1.004	1.004	1.004	1.004	1.004	1.004	1.004	1.004	1.004	1.003	1.003	1.003	1.003	1.003	1.003
2400	1.005	1.005	1.005	1.005	1.005	1.005	1.005	1.005	1.004	1.004	1.004	1.004	1.004	1.004	1.004	1.003	1.003	1.003	1.003	1.003
2500	1.005	1.005	1.005	1.005	1.005	1.005	1.005	1.005	1.005	1.005	1.004	1.004	1.004	1.004	1.004	1.004	1.003	1.003	1.003	1.003
2600	1.005	1.005	1.005	1.005	1.005	1.005	1.005	1.005	1.005	1.005	1.005	1.005	1.004	1.004	1.004	1.004	1.003	1.003	1.003	1.003
2700	1.005	1.005	1.005	1.005	1.005	1.005	1.005	1.005	1.005	1.005	1.005	1.005	1.005	1.004	1.004	1.004	1.004	1.003	1.003	1.003
2800	1.006	1.005	1.006	1.005	1.006	1.005	1.006	1.005	1.005	1.005	1.005	1.005	1.005	1.005	1.004	1.004	1.004	1.004	1.003	1.003
2900	1.006	1.006	1.006	1.006	1.006	1.006	1.005	1.005	1.005	1.005	1.005	1.005	1.005	1.005	1.004	1.004	1.004	1.004	1.004	1.004
3000	1.006	1.006	1.006	1.006	1.006	1.006	1.006	1.006	1.006	1.005	1.005	1.005	1.005	1.005	1.005	1.004	1.004	1.004	1.004	1.004

x

SIGMA = 890000.0

V–V$_c$ = 1

x	3.4	3.2	3.0	2.8	2.6	2.4	2.2	2.0	1.8	1.6	1.4	1.2	1.0	0.8	0.6	0.4	0.2	0.10	0.09	0.08
3100	1.006	1.006	1.006	1.006	1.006	1.006	1.006	1.006	1.006	1.006	1.006	1.005	1.005	1.005	1.005	1.004	1.004	1.004	1.004	1.004
3200	1.006	1.006	1.006	1.006	1.006	1.006	1.006	1.006	1.006	1.006	1.006	1.006	1.005	1.005	1.005	1.005	1.004	1.004	1.004	1.004
3300	1.007	1.007	1.006	1.006	1.006	1.006	1.006	1.006	1.006	1.006	1.006	1.006	1.006	1.006	1.005	1.005	1.004	1.004	1.004	1.004
3400	1.007	1.007	1.007	1.007	1.007	1.007	1.007	1.007	1.006	1.006	1.006	1.006	1.006	1.006	1.005	1.005	1.004	1.004	1.004	1.004
3500	1.007	1.007	1.007	1.007	1.007	1.007	1.007	1.007	1.007	1.006	1.006	1.006	1.007	1.006	1.006	1.005	1.005	1.005	1.004	1.004
3600	1.007	1.007	1.007	1.007	1.007	1.007	1.007	1.007	1.007	1.007	1.006	1.006	1.006	1.006	1.006	1.005	1.005	1.005	1.005	1.004
3700	1.007	1.007	1.007	1.007	1.007	1.007	1.007	1.007	1.007	1.007	1.007	1.007	1.006	1.006	1.006	1.005	1.005	1.005	1.005	1.004
3800	1.007	1.008	1.007	1.007	1.007	1.007	1.007	1.007	1.007	1.007	1.007	1.007	1.007	1.007	1.006	1.006	1.005	1.005	1.005	1.005
3900	1.008	1.008	1.008	1.008	1.008	1.007	1.007	1.007	1.007	1.007	1.007	1.007	1.007	1.007	1.006	1.006	1.005	1.005	1.005	1.005
4000	1.008	1.008	1.008	1.008	1.008	1.008	1.008	1.008	1.007	1.007	1.007	1.007	1.007	1.007	1.006	1.006	1.005	1.005	1.005	1.005
4100	1.008	1.008	1.008	1.008	1.008	1.008	1.008	1.008	1.007	1.007	1.007	1.007	1.007	1.007	1.006	1.006	1.005	1.005	1.005	1.005
4200	1.008	1.008	1.008	1.008	1.008	1.008	1.008	1.008	1.008	1.008	1.007	1.007	1.007	1.007	1.006	1.006	1.006	1.005	1.005	1.005
4300	1.009	1.008	1.008	1.008	1.008	1.008	1.008	1.008	1.008	1.008	1.008	1.007	1.007	1.007	1.007	1.006	1.006	1.005	1.005	1.005
4400	1.009	1.009	1.009	1.009	1.008	1.008	1.008	1.008	1.008	1.008	1.008	1.008	1.007	1.007	1.007	1.006	1.006	1.006	1.006	1.005
4500	1.009	1.009	1.009	1.009	1.009	1.009	1.009	1.008	1.008	1.008	1.008	1.008	1.008	1.007	1.007	1.006	1.006	1.006	1.006	1.005
4600	1.009	1.009	1.009	1.009	1.009	1.009	1.009	1.009	1.009	1.008	1.008	1.008	1.008	1.007	1.007	1.007	1.006	1.006	1.006	1.006
4700	1.009	1.009	1.009	1.009	1.009	1.009	1.009	1.009	1.009	1.009	1.008	1.008	1.008	1.008	1.007	1.007	1.006	1.006	1.006	1.006
4800	1.010	1.010	1.009	1.009	1.009	1.009	1.009	1.009	1.009	1.009	1.009	1.008	1.008	1.008	1.007	1.007	1.006	1.006	1.006	1.006
4900	1.010	1.010	1.010	1.010	1.010	1.009	1.009	1.009	1.009	1.009	1.009	1.009	1.008	1.008	1.008	1.007	1.006	1.006	1.006	1.006
5000	1.010	1.010	1.010	1.010	1.010	1.010	1.009	1.009	1.009	1.009	1.009	1.009	1.009	1.008	1.008	1.007	1.007	1.006	1.006	1.006
5100	1.010	1.010	1.010	1.010	1.010	1.010	1.010	1.010	1.010	1.009	1.009	1.009	1.009	1.008	1.008	1.007	1.007	1.006	1.006	1.006
5200	1.010	1.010	1.010	1.011	1.010	1.010	1.010	1.010	1.010	1.010	1.009	1.009	1.009	1.008	1.008	1.007	1.007	1.006	1.006	1.006
5300	1.011	1.011	1.010	1.011	1.010	1.010	1.010	1.010	1.010	1.010	1.009	1.009	1.009	1.009	1.008	1.007	1.007	1.006	1.006	1.006
5400	1.011	1.011	1.011	1.011	1.011	1.011	1.011	1.010	1.010	1.010	1.010	1.009	1.009	1.009	1.008	1.008	1.007	1.007	1.007	1.007
5500	1.011	1.011	1.011	1.011	1.011	1.011	1.011	1.011	1.011	1.010	1.010	1.010	1.009	1.009	1.008	1.008	1.007	1.007	1.007	1.007
5600	1.011	1.011	1.011	1.011	1.011	1.011	1.011	1.011	1.011	1.011	1.010	1.010	1.010	1.009	1.009	1.008	1.007	1.007	1.007	1.007
5700	1.012	1.011	1.011	1.011	1.011	1.011	1.011	1.011	1.011	1.011	1.010	1.010	1.010	1.009	1.009	1.008	1.008	1.007	1.007	1.007
5800	1.012	1.012	1.012	1.012	1.011	1.012	1.012	1.011	1.011	1.011	1.011	1.010	1.010	1.009	1.009	1.008	1.008	1.007	1.007	1.007
5900	1.012	1.012	1.012	1.012	1.012	1.012	1.012	1.012	1.011	1.011	1.011	1.010	1.010	1.009	1.009	1.008	1.008	1.007	1.007	1.007
6000	1.012	1.012	1.012	1.012	1.012	1.012	1.012	1.012	1.012	1.011	1.011	1.011	1.010	1.010	1.009	1.009	1.008	1.007	1.007	1.007
6100	1.012	1.012	1.012	1.012	1.012	1.012	1.012	1.012	1.012	1.012	1.011	1.011	1.011	1.010	1.009	1.009	1.008	1.007	1.007	1.007
6200	1.012	1.012	1.012	1.012	1.012	1.012	1.012	1.012	1.012	1.012	1.011	1.011	1.011	1.010	1.009	1.009	1.008	1.007	1.007	1.007
6300	1.013	1.013	1.013	1.013	1.012	1.012	1.012	1.012	1.012	1.012	1.012	1.011	1.011	1.010	1.010	1.009	1.008	1.008	1.008	1.008
6400	1.013	1.013	1.013	1.013	1.013	1.013	1.012	1.012	1.012	1.012	1.012	1.011	1.011	1.011	1.010	1.009	1.008	1.008	1.008	1.008
6500	1.013	1.013	1.013	1.013	1.013	1.013	1.013	1.013	1.012	1.012	1.012	1.012	1.011	1.011	1.010	1.009	1.009	1.008	1.008	1.008
6600	1.013	1.013	1.013	1.013	1.013	1.013	1.013	1.013	1.013	1.012	1.012	1.012	1.011	1.011	1.010	1.009	1.009	1.008	1.008	1.008
6700	1.013	1.013	1.013	1.013	1.014	1.013	1.013	1.013	1.013	1.013	1.012	1.012	1.011	1.011	1.010	1.009	1.009	1.008	1.008	1.008
6800	1.014	1.014	1.013	1.013	1.014	1.013	1.013	1.013	1.013	1.013	1.012	1.012	1.012	1.011	1.010	1.010	1.009	1.008	1.008	1.008
6900	1.014	1.014	1.014	1.014	1.014	1.014	1.013	1.013	1.013	1.013	1.012	1.012	1.012	1.011	1.010	1.010	1.009	1.008	1.008	1.008
7000	1.014	1.014	1.014	1.014	1.014	1.014	1.014	1.014	1.013	1.013	1.013	1.012	1.012	1.011	1.011	1.010	1.009	1.008	1.008	1.008
7100	1.014	1.014	1.014	1.014	1.014	1.014	1.014	1.014	1.013	1.013	1.013	1.012	1.012	1.012	1.011	1.010	1.009	1.009	1.009	1.008
7200	1.014	1.014	1.014	1.014	1.014	1.014	1.014	1.014	1.013	1.013	1.013	1.013	1.012	1.012	1.011	1.010	1.009	1.009	1.009	1.009

H

SIGMA = 890000.0

$V-V_c = 1$

H x	0.08	0.09	0.10	0.2	0.4	0.6	0.8	1.0	1.2	1.4	1.6	1.8	2.0	2.2	2.4	2.6	2.8	3.0	3.2	3.4
7300	1.009	1.009	1.009	1.010	1.011	1.011	1.012	1.012	1.013	1.013	1.013	1.014	1.014	1.014	1.014	1.014	1.014	1.014	1.015	1.015
7400	1.009	1.009	1.009	1.010	1.011	1.011	1.012	1.013	1.013	1.013	1.014	1.014	1.014	1.014	1.014	1.015	1.015	1.015	1.015	1.015
7500	1.009	1.009	1.009	1.010	1.011	1.012	1.012	1.013	1.013	1.014	1.014	1.014	1.014	1.014	1.015	1.015	1.015	1.015	1.015	1.015
7600	1.009	1.009	1.009	1.010	1.011	1.012	1.012	1.013	1.013	1.014	1.014	1.014	1.014	1.015	1.015	1.015	1.015	1.015	1.016	1.016
7700	1.009	1.009	1.010	1.010	1.011	1.012	1.013	1.013	1.013	1.014	1.014	1.014	1.015	1.015	1.015	1.016	1.016	1.016	1.016	1.016
7800	1.009	1.009	1.010	1.010	1.011	1.012	1.013	1.013	1.014	1.014	1.015	1.015	1.015	1.015	1.016	1.016	1.016	1.016	1.017	1.017
7900	1.010	1.010	1.010	1.011	1.012	1.012	1.013	1.014	1.014	1.014	1.015	1.015	1.015	1.015	1.016	1.016	1.017	1.017	1.017	1.018
8000	1.010	1.010	1.010	1.011	1.012	1.013	1.013	1.014	1.014	1.015	1.015	1.015	1.016	1.016	1.016	1.017	1.017	1.017	1.018	1.018
8100	1.010	1.010	1.010	1.011	1.012	1.013	1.013	1.014	1.014	1.015	1.015	1.015	1.016	1.016	1.016	1.017	1.017	1.018	1.018	1.019
8200	1.010	1.010	1.010	1.011	1.012	1.013	1.014	1.014	1.014	1.015	1.015	1.016	1.016	1.016	1.017	1.017	1.018	1.018	1.019	1.019
8300	1.010	1.010	1.010	1.011	1.012	1.013	1.014	1.014	1.015	1.015	1.016	1.016	1.016	1.017	1.017	1.018	1.018	1.019	1.019	1.020
8400	1.010	1.010	1.011	1.011	1.012	1.013	1.014	1.015	1.015	1.015	1.016	1.016	1.016	1.017	1.017	1.018	1.018	1.019	1.019	1.020
8500	1.010	1.011	1.011	1.011	1.012	1.014	1.014	1.015	1.015	1.016	1.016	1.016	1.017	1.017	1.018	1.018	1.019	1.019	1.020	1.021
8600	1.011	1.011	1.011	1.011	1.013	1.014	1.014	1.015	1.015	1.016	1.016	1.017	1.017	1.017	1.018	1.019	1.019	1.020	1.020	1.021
8700	1.011	1.011	1.011	1.012	1.013	1.014	1.014	1.015	1.016	1.016	1.016	1.017	1.017	1.018	1.018	1.019	1.019	1.020	1.021	1.021
8800	1.011	1.011	1.011	1.012	1.013	1.014	1.015	1.015	1.016	1.016	1.017	1.017	1.017	1.018	1.019	1.019	1.020	1.021	1.021	1.022
8900	1.011	1.011	1.011	1.012	1.013	1.014	1.015	1.016	1.016	1.016	1.017	1.017	1.018	1.018	1.019	1.020	1.020	1.021	1.022	1.022
9000	1.011	1.011	1.011	1.012	1.013	1.014	1.015	1.016	1.016	1.017	1.017	1.018	1.018	1.019	1.019	1.020	1.020	1.021	1.022	1.023
9100	1.011	1.011	1.011	1.012	1.013	1.015	1.015	1.016	1.016	1.017	1.017	1.018	1.018	1.019	1.020	1.020	1.021	1.021	1.022	1.023
9200	1.011	1.011	1.012	1.012	1.013	1.015	1.015	1.016	1.017	1.017	1.018	1.018	1.018	1.019	1.020	1.020	1.021	1.022	1.022	1.023
9300	1.011	1.011	1.012	1.012	1.013	1.015	1.015	1.016	1.017	1.017	1.018	1.018	1.019	1.019	1.020	1.021	1.021	1.022	1.023	
9400	1.012	1.012	1.012	1.013	1.014	1.015	1.016	1.017	1.017	1.017	1.018	1.019	1.019	1.020	1.020	1.021	1.022	1.022		
9500	1.012	1.012	1.012	1.013	1.014	1.015	1.016	1.017	1.017	1.018	1.018	1.019	1.019	1.020	1.021	1.021	1.022			
9600	1.012	1.012	1.012	1.013	1.014	1.015	1.016	1.017	1.017	1.018	1.019	1.019	1.019	1.020	1.021	1.021				
9700	1.012	1.012	1.012	1.013	1.014	1.015	1.016	1.017	1.018	1.018	1.019	1.019	1.020	1.021	1.021					
9800	1.012	1.012	1.012	1.013	1.014	1.016	1.016	1.017	1.018	1.019	1.019	1.020	1.020	1.021						
9900	1.012	1.012	1.012	1.013	1.015	1.016	1.017	1.018	1.018	1.019	1.019	1.020	1.020							
10000	1.012	1.012	1.013	1.013	1.015	1.016	1.017	1.018	1.018	1.019	1.020	1.020	1.021							
10100	1.013	1.013	1.013	1.014	1.015	1.016	1.017	1.018	1.018	1.019	1.020	1.020	1.021							
10200	1.013	1.013	1.013	1.014	1.015	1.016	1.017	1.018	1.019	1.019	1.020	1.020	1.021							
10300	1.013	1.013	1.013	1.014	1.015	1.017	1.017	1.018	1.019	1.019	1.020	1.021	1.021							
10400	1.013	1.013	1.013	1.014	1.015	1.017	1.018	1.018	1.019	1.020	1.020	1.021	1.021							
10500	1.013	1.013	1.013	1.014	1.016	1.017	1.018	1.019	1.019	1.020	1.020	1.021	1.021							
10600	1.013	1.013	1.013	1.014	1.016	1.017	1.018	1.019	1.019	1.020	1.020	1.021								
10700	1.013	1.013	1.013	1.015	1.016	1.017	1.018	1.019	1.020	1.020	1.020	1.021								
10800	1.013	1.013	1.014	1.015	1.016	1.018	1.018	1.019	1.020	1.020	1.021	1.021								
10900	1.013	1.013	1.014	1.015	1.016	1.018	1.019	1.019	1.020	1.020	1.021	1.021								
11000	1.014	1.014	1.014	1.015	1.016	1.018	1.019	1.020	1.020	1.021	1.021	1.021								
11100	1.014	1.014	1.014	1.015	1.016	1.018	1.019	1.020	1.020	1.021	1.021									
11200	1.014	1.014	1.014	1.015	1.017	1.018	1.019	1.020	1.020	1.021										
11300	1.014	1.014	1.014	1.015	1.017	1.019	1.019	1.020	1.021	1.021										
11400	1.014	1.014	1.014	1.015	1.017	1.019	1.019	1.020	1.021	1.022										

SIGMA = 890000.0

V-V_c = 1

H_x	3.4	3.2	3.0	2.8	2.6	2.4	2.2	2.0	1.8	1.6	1.4	1.2	1.0	0.8	0.6	0.4	0.2	0.10	0.09	0.08
11500	1.023	1.023	1.023	1.023	1.022	1.022	1.022	1.022	1.021	1.021	1.021	1.020	1.019	1.019	1.018	1.017	1.015	1.014	1.014	1.014
11600	1.023	1.023	1.023	1.023	1.023	1.022	1.022	1.022	1.022	1.021	1.021	1.020	1.020	1.019	1.018	1.017	1.015	1.014	1.014	1.014
11700	1.023	1.023	1.023	1.023	1.023	1.023	1.022	1.022	1.022	1.021	1.021	1.021	1.020	1.019	1.018	1.017	1.015	1.014	1.014	1.014
11800	1.024	1.023	1.023	1.023	1.023	1.023	1.023	1.022	1.022	1.022	1.021	1.021	1.020	1.019	1.018	1.017	1.015	1.014	1.014	1.014
11900	1.024	1.024	1.024	1.024	1.023	1.023	1.023	1.023	1.022	1.022	1.021	1.021	1.020	1.020	1.018	1.017	1.016	1.015	1.014	1.014
12000	1.024	1.024	1.024	1.024	1.024	1.024	1.023	1.023	1.023	1.022	1.022	1.021	1.020	1.020	1.019	1.017	1.016	1.015	1.014	1.014
12100	1.024	1.024	1.024	1.024	1.024	1.024	1.024	1.023	1.023	1.022	1.022	1.021	1.021	1.020	1.019	1.018	1.016	1.015	1.015	1.014
12200	1.024	1.024	1.025	1.025	1.024	1.024	1.024	1.024	1.023	1.023	1.023	1.022	1.021	1.020	1.020	1.018	1.016	1.015	1.015	1.015
12300	1.025	1.025	1.025	1.025	1.025	1.024	1.024	1.024	1.023	1.023	1.023	1.022	1.021	1.021	1.020	1.018	1.016	1.015	1.015	1.015
12400	1.025	1.025	1.025	1.025	1.025	1.025	1.024	1.024	1.024	1.023	1.023	1.023	1.021	1.021	1.021	1.018	1.016	1.015	1.015	1.015
12500	1.025	1.025	1.025	1.025	1.025	1.025	1.025	1.024	1.024	1.023	1.023	1.023	1.022	1.021	1.021	1.018	1.016	1.015	1.015	1.015
12600	1.025	1.025	1.026	1.026	1.026	1.025	1.025	1.025	1.024	1.024	1.024	1.023	1.022	1.021	1.021	1.020	1.017	1.015	1.015	1.015
12700	1.026	1.026	1.026	1.026	1.026	1.026	1.025	1.025	1.024	1.024	1.024	1.023	1.022	1.021	1.021	1.020	1.017	1.016	1.015	1.015
12800	1.026	1.026	1.026	1.026	1.026	1.026	1.025	1.025	1.025	1.024	1.024	1.024	1.022	1.022	1.021	1.020	1.018	1.016	1.016	1.015
12900	1.026	1.026	1.026	1.026	1.027	1.026	1.026	1.026	1.025	1.025	1.024	1.024	1.022	1.022	1.022	1.020	1.018	1.016	1.016	1.015
13000	1.026	1.026	1.027	1.027	1.027	1.026	1.026	1.026	1.025	1.025	1.024	1.024	1.023	1.022	1.022	1.020	1.019	1.016	1.016	1.016
13100	1.027	1.027	1.027	1.027	1.027	1.027	1.026	1.026	1.025	1.025	1.025	1.024	1.023	1.023	1.022	1.021	1.019	1.016	1.016	1.016
13200	1.027	1.027	1.027	1.027	1.027	1.027	1.026	1.026	1.025	1.025	1.025	1.024	1.023	1.023	1.022	1.021	1.019	1.017	1.016	1.016
13300	1.027	1.027	1.027	1.027	1.027	1.027	1.026	1.026	1.026	1.025	1.025	1.025	1.023	1.023	1.022	1.021	1.019	1.017	1.016	1.016
13400	1.027	1.027	1.028	1.028	1.028	1.027	1.027	1.027	1.026	1.026	1.025	1.025	1.023	1.023	1.023	1.021	1.019	1.017	1.016	1.016
13500	1.027	1.028	1.028	1.028	1.028	1.027	1.027	1.027	1.026	1.026	1.026	1.025	1.024	1.023	1.023	1.021	1.020	1.017	1.016	1.016
13600	1.028	1.028	1.028	1.028	1.028	1.028	1.027	1.027	1.026	1.026	1.026	1.025	1.024	1.024	1.023	1.021	1.020	1.017	1.016	1.016
13700	1.028	1.028	1.028	1.028	1.028	1.028	1.027	1.027	1.026	1.026	1.026	1.025	1.024	1.023	1.023	1.022	1.020	1.017	1.016	1.016
13800	1.028	1.028	1.028	1.028	1.028	1.028	1.028	1.027	1.027	1.027	1.026	1.026	1.024	1.023	1.023	1.022	1.020	1.017	1.017	1.016
13900	1.028	1.028	1.029	1.029	1.029	1.028	1.028	1.028	1.027	1.027	1.026	1.026	1.024	1.024	1.024	1.022	1.020	1.017	1.017	1.017
14000	1.029	1.029	1.029	1.029	1.029	1.028	1.028	1.028	1.027	1.027	1.026	1.026	1.025	1.024	1.024	1.022	1.021	1.018	1.017	1.017
14100	1.029	1.029	1.029	1.029	1.029	1.028	1.028	1.028	1.027	1.027	1.027	1.026	1.025	1.024	1.024	1.022	1.021	1.018	1.017	1.017
14200	1.029	1.029	1.029	1.029	1.029	1.029	1.028	1.028	1.027	1.027	1.027	1.026	1.025	1.024	1.024	1.022	1.021	1.018	1.017	1.017
14300	1.029	1.029	1.029	1.029	1.029	1.029	1.028	1.028	1.027	1.027	1.027	1.026	1.025	1.024	1.024	1.023	1.021	1.018	1.017	1.017
14400	1.029	1.029	1.029	1.029	1.029	1.029	1.029	1.029	1.028	1.027	1.027	1.026	1.025	1.024	1.024	1.022	1.021	1.018	1.017	1.017
14500	1.029	1.029	1.030	1.030	1.029	1.029	1.029	1.028	1.028	1.028	1.027	1.027	1.025	1.024	1.024	1.023	1.021	1.018	1.018	1.017
14600	1.030	1.030	1.030	1.030	1.029	1.029	1.029	1.028	1.028	1.028	1.027	1.027	1.025	1.024	1.024	1.023	1.021	1.018	1.018	1.018
14700	1.030	1.030	1.030	1.030	1.030	1.029	1.029	1.028	1.028	1.028	1.027	1.027	1.025	1.024	1.024	1.023	1.021	1.018	1.018	1.018
14800	1.030	1.030	1.030	1.030	1.030	1.029	1.029	1.029	1.028	1.028	1.027	1.027	1.025	1.024	1.024	1.023	1.021	1.018	1.018	1.018
14900	1.030	1.030	1.030	1.029	1.029	1.029	1.029	1.028	1.028	1.028	1.027	1.027	1.025	1.024	1.024	1.023	1.022	1.018	1.018	1.018
15000	1.030	1.030	1.030	1.029	1.029	1.029	1.029	1.028	1.028	1.028	1.027	1.027	1.025	1.024	1.024	1.023	1.022	1.018	1.018	1.018

SIGMA = 222500.0

V-V_c = 2

H	3.4	3.2	3.0	2.8	2.6	2.4	2.2	2.0	1.8	1.6	1.4	1.2	1.0	0.8	0.6	0.4	0.2	0.10	0.09	0.08
0	1.000	1.000	1.000	1.000	1.000	1.000	1.000	1.000	1.000	1.000	1.000	1.000	1.000	1.000	1.000	1.000	1.000	1.000	1.000	1.000
100	1.001	1.001	1.001	1.001	1.001	1.001	1.001	1.001	1.001	1.001	1.001	1.001	1.001	1.001	1.001	1.001	1.001	1.000	1.000	1.000
200	1.002	1.002	1.002	1.002	1.002	1.002	1.002	1.002	1.002	1.002	1.001	1.001	1.001	1.001	1.001	1.001	1.001	1.001	1.001	1.001
300	1.002	1.003	1.003	1.003	1.002	1.002	1.002	1.003	1.003	1.002	1.003	1.002	1.002	1.002	1.002	1.002	1.002	1.002	1.002	1.001
400	1.004	1.004	1.004	1.004	1.003	1.004	1.003	1.004	1.004	1.004	1.004	1.003	1.003	1.003	1.003	1.003	1.003	1.002	1.002	1.002
500	1.004	1.005	1.005	1.005	1.004	1.004	1.004	1.004	1.004	1.004	1.004	1.005	1.004	1.005	1.004	1.004	1.003	1.003	1.003	1.003
600	1.005	1.005	1.006	1.005	1.005	1.005	1.005	1.005	1.005	1.006	1.005	1.005	1.005	1.005	1.005	1.004	1.004	1.004	1.004	1.003
700	1.006	1.006	1.006	1.006	1.006	1.006	1.006	1.006	1.006	1.006	1.006	1.006	1.006	1.006	1.006	1.005	1.005	1.004	1.004	1.004
800	1.006	1.007	1.007	1.007	1.006	1.007	1.007	1.007	1.007	1.007	1.007	1.007	1.007	1.007	1.006	1.006	1.005	1.005	1.005	1.004
900	1.007	1.008	1.008	1.008	1.007	1.008	1.008	1.008	1.008	1.009	1.008	1.008	1.008	1.008	1.007	1.007	1.006	1.005	1.005	1.004
1000	1.008	1.008	1.009	1.009	1.008	1.008	1.008	1.009	1.008	1.009	1.009	1.009	1.009	1.008	1.008	1.007	1.007	1.006	1.006	1.005
1100	1.009	1.009	1.009	1.009	1.009	1.009	1.009	1.009	1.009	1.010	1.010	1.010	1.009	1.009	1.008	1.008	1.007	1.007	1.007	1.005
1200	1.010	1.010	1.010	1.010	1.010	1.010	1.010	1.010	1.010	1.011	1.011	1.010	1.010	1.010	1.009	1.008	1.008	1.007	1.007	1.006
1300	1.011	1.011	1.011	1.011	1.010	1.011	1.011	1.011	1.011	1.012	1.011	1.011	1.011	1.010	1.010	1.009	1.009	1.007	1.007	1.006
1400	1.012	1.012	1.012	1.012	1.011	1.012	1.012	1.012	1.011	1.012	1.012	1.012	1.011	1.011	1.011	1.010	1.009	1.008	1.007	1.007
1500	1.012	1.012	1.013	1.013	1.012	1.012	1.013	1.013	1.013	1.013	1.013	1.013	1.012	1.012	1.011	1.011	1.010	1.008	1.008	1.007
1600	1.013	1.013	1.013	1.013	1.013	1.013	1.013	1.013	1.013	1.014	1.014	1.013	1.013	1.014	1.012	1.011	1.010	1.008	1.008	1.008
1700	1.014	1.014	1.014	1.014	1.014	1.014	1.013	1.014	1.014	1.015	1.015	1.014	1.014	1.014	1.013	1.010	1.011	1.009	1.008	1.008
1800	1.014	1.014	1.015	1.015	1.014	1.015	1.014	1.015	1.015	1.016	1.016	1.014	1.014	1.015	1.014	1.011	1.011	1.009	1.009	1.009
1900	1.015	1.015	1.016	1.016	1.015	1.015	1.015	1.015	1.016	1.016	1.017	1.015	1.015	1.016	1.014	1.012	1.011	1.009	1.009	1.009
2000	1.016	1.016	1.016	1.017	1.016	1.016	1.016	1.016	1.016	1.017	1.018	1.016	1.016	1.017	1.015	1.012	1.011	1.010	1.010	1.010
2100	1.017	1.017	1.017	1.017	1.016	1.017	1.017	1.017	1.017	1.017	1.018	1.017	1.016	1.017	1.016	1.013	1.012	1.010	1.010	1.010
2200	1.018	1.017	1.018	1.018	1.018	1.018	1.017	1.018	1.017	1.018	1.018	1.017	1.017	1.018	1.017	1.014	1.013	1.011	1.011	1.011
2300	1.018	1.018	1.018	1.019	1.018	1.018	1.018	1.018	1.018	1.019	1.019	1.018	1.017	1.018	1.017	1.014	1.013	1.011	1.011	1.011
2400	1.019	1.019	1.019	1.019	1.019	1.019	1.018	1.019	1.019	1.020	1.020	1.019	1.019	1.019	1.018	1.015	1.014	1.012	1.012	1.012
2500	1.020	1.020	1.020	1.020	1.020	1.020	1.020	1.020	1.020	1.020	1.021	1.020	1.020	1.020	1.019	1.015	1.014	1.013	1.013	1.012
2600	1.021	1.021	1.021	1.021	1.020	1.021	1.021	1.020	1.020	1.021	1.021	1.020	1.020	1.020	1.019	1.016	1.015	1.013	1.013	1.013
2700	1.022	1.021	1.021	1.021	1.021	1.022	1.021	1.021	1.021	1.021	1.022	1.021	1.021	1.021	1.020	1.016	1.016	1.013	1.013	1.013
2800	1.022	1.022	1.022	1.022	1.022	1.022	1.022	1.021	1.021	1.022	1.023	1.022	1.022	1.022	1.021	1.017	1.016	1.014	1.014	1.014
2900	1.023	1.023	1.023	1.023	1.023	1.022	1.022	1.022	1.022	1.022	1.023	1.022	1.022	1.022	1.022	1.018	1.017	1.014	1.014	1.014
3000	1.024	1.024	1.024	1.024	1.023	1.023	1.023	1.023	1.022	1.022	1.024	1.023	1.022	1.023	1.023	1.019	1.017	1.015	1.015	1.014

x

SIGMA = 222500.0

$V - V_c = 2$

H / X	3.4	3.2	3.0	2.8	2.6	2.4	2.2	2.0	1.8	1.6	1.4	1.2	1.0	0.8	0.6	0.4	0.2	0.10	0.09	0.08
3100	1.025	1.025	1.025	1.024	1.024	1.024	1.024	1.023	1.023	1.023	1.022	1.022	1.021	1.020	1.019	1.018	1.016	1.015	1.015	1.015
3200	1.026	1.025	1.025	1.025	1.025	1.025	1.024	1.024	1.024	1.023	1.023	1.022	1.022	1.021	1.020	1.019	1.017	1.016	1.016	1.015
3300	1.026	1.026	1.026	1.026	1.026	1.025	1.025	1.025	1.024	1.024	1.024	1.023	1.023	1.022	1.021	1.019	1.018	1.016	1.016	1.016
3400	1.027	1.027	1.027	1.027	1.026	1.026	1.026	1.026	1.025	1.025	1.024	1.024	1.023	1.022	1.021	1.020	1.018	1.017	1.017	1.016
3500	1.028	1.028	1.027	1.027	1.027	1.027	1.027	1.026	1.026	1.026	1.025	1.024	1.024	1.023	1.022	1.021	1.019	1.017	1.017	1.017
3600	1.029	1.029	1.028	1.028	1.028	1.028	1.027	1.027	1.027	1.026	1.026	1.025	1.024	1.024	1.023	1.021	1.019	1.018	1.017	1.017
3700	1.030	1.030	1.029	1.029	1.029	1.028	1.028	1.028	1.027	1.027	1.026	1.026	1.025	1.025	1.024	1.022	1.020	1.019	1.018	1.018
3800	1.031	1.030	1.030	1.030	1.029	1.029	1.029	1.029	1.028	1.028	1.027	1.027	1.026	1.025	1.024	1.022	1.021	1.019	1.018	1.018
3900	1.031	1.031	1.031	1.030	1.030	1.030	1.030	1.029	1.029	1.029	1.028	1.028	1.027	1.026	1.025	1.023	1.022	1.019	1.019	1.019
4000	1.032	1.032	1.032	1.031	1.031	1.031	1.030	1.030	1.030	1.029	1.029	1.028	1.027	1.027	1.025	1.023	1.021	1.020	1.019	1.019
4100	1.033	1.032	1.032	1.032	1.032	1.031	1.031	1.031	1.030	1.030	1.029	1.029	1.028	1.027	1.025	1.023	1.022	1.020	1.019	1.020
4200	1.034	1.034	1.033	1.033	1.032	1.032	1.032	1.031	1.031	1.030	1.030	1.029	1.028	1.027	1.026	1.024	1.022	1.021	1.020	1.020
4300	1.034	1.034	1.034	1.033	1.033	1.033	1.032	1.032	1.031	1.031	1.031	1.030	1.029	1.028	1.026	1.025	1.023	1.021	1.021	1.020
4400	1.035	1.035	1.034	1.034	1.034	1.033	1.033	1.033	1.032	1.032	1.031	1.031	1.030	1.029	1.027	1.025	1.023	1.022	1.021	1.021
4500	1.036	1.036	1.036	1.035	1.035	1.035	1.034	1.034	1.033	1.033	1.032	1.032	1.030	1.030	1.028	1.026	1.023	1.022	1.022	1.021
4600	1.037	1.037	1.036	1.036	1.036	1.035	1.035	1.034	1.034	1.034	1.033	1.032	1.031	1.030	1.029	1.026	1.024	1.023	1.022	1.022
4700	1.038	1.037	1.037	1.037	1.037	1.036	1.036	1.035	1.035	1.034	1.033	1.033	1.032	1.031	1.029	1.027	1.025	1.023	1.023	1.022
4800	1.039	1.038	1.038	1.038	1.037	1.037	1.037	1.036	1.036	1.035	1.034	1.034	1.033	1.031	1.030	1.028	1.025	1.024	1.023	1.023
4900	1.039	1.039	1.039	1.038	1.038	1.038	1.037	1.037	1.036	1.036	1.035	1.034	1.033	1.032	1.031	1.028	1.026	1.024	1.024	1.023
5000	1.040	1.040	1.040	1.039	1.039	1.039	1.038	1.038	1.037	1.037	1.036	1.035	1.034	1.033	1.032	1.029	1.026	1.025	1.025	1.024
5100	1.041	1.041	1.040	1.040	1.040	1.039	1.039	1.039	1.038	1.037	1.037	1.036	1.035	1.034	1.032	1.030	1.027	1.025	1.025	1.024
5200	1.042	1.042	1.041	1.041	1.041	1.040	1.040	1.039	1.039	1.038	1.037	1.037	1.036	1.035	1.033	1.031	1.028	1.026	1.026	1.025
5300	1.043	1.042	1.042	1.042	1.041	1.041	1.041	1.040	1.040	1.039	1.038	1.038	1.037	1.035	1.034	1.031	1.029	1.027	1.026	1.026
5400	1.043	1.043	1.043	1.042	1.042	1.042	1.041	1.041	1.040	1.040	1.039	1.038	1.037	1.036	1.035	1.032	1.029	1.027	1.027	1.026
5500	1.044	1.044	1.044	1.043	1.043	1.043	1.042	1.042	1.041	1.041	1.040	1.039	1.038	1.037	1.035	1.033	1.030	1.028	1.028	1.027
5600	1.045	1.045	1.044	1.044	1.044	1.043	1.043	1.042	1.041	1.041	1.040	1.040	1.038	1.037	1.036	1.034	1.031	1.029	1.028	1.027
5700	1.046	1.046	1.045	1.045	1.044	1.044	1.043	1.043	1.042	1.042	1.041	1.040	1.039	1.038	1.037	1.035	1.032	1.030	1.029	1.028
5800	1.047	1.046	1.046	1.045	1.045	1.045	1.044	1.044	1.043	1.043	1.042	1.041	1.040	1.039	1.037	1.035	1.033	1.031	1.030	1.028
5900	1.048	1.047	1.047	1.046	1.046	1.045	1.045	1.044	1.044	1.043	1.042	1.042	1.041	1.039	1.038	1.036	1.034	1.031	1.030	1.029
6000	1.048	1.048	1.048	1.047	1.047	1.046	1.046	1.045	1.045	1.044	1.043	1.043	1.041	1.040	1.039	1.037	1.034	1.032	1.031	1.029
6100	1.049	1.049	1.048	1.048	1.047	1.047	1.047	1.046	1.046	1.045	1.044	1.043	1.042	1.041	1.040	1.037	1.035	1.032	1.031	1.030
6200	1.050	1.050	1.049	1.049	1.048	1.048	1.047	1.047	1.046	1.045	1.045	1.044	1.043	1.042	1.040	1.038	1.035	1.033	1.032	1.030
6300	1.051	1.050	1.050	1.050	1.049	1.049	1.048	1.048	1.047	1.046	1.046	1.045	1.044	1.042	1.041	1.038	1.036	1.033	1.032	1.030
6400	1.052	1.051	1.051	1.050	1.050	1.049	1.049	1.048	1.048	1.047	1.047	1.046	1.044	1.043	1.042	1.039	1.036	1.034	1.033	1.031
6500	1.053	1.052	1.052	1.051	1.051	1.050	1.050	1.049	1.049	1.048	1.047	1.046	1.045	1.044	1.043	1.039	1.036	1.034	1.033	1.031
6600	1.053	1.053	1.052	1.051	1.051	1.050	1.050	1.049	1.049	1.048	1.047	1.046	1.045	1.043	1.040	1.037	1.033	1.032	1.031	1.031
6700	1.054	1.053	1.053	1.052	1.052	1.051	1.051	1.050	1.049	1.049	1.048	1.047	1.046	1.044	1.041	1.038	1.034	1.032	1.032	1.031
6800	1.055	1.055	1.054	1.053	1.053	1.052	1.052	1.051	1.050	1.050	1.048	1.048	1.046	1.045	1.042	1.039	1.035	1.033	1.032	1.032
6900	1.056	1.056	1.055	1.054	1.054	1.053	1.053	1.052	1.051	1.051	1.050	1.049	1.047	1.045	1.043	1.040	1.036	1.033	1.033	1.033
7000	1.057	1.056	1.056	1.055	1.055	1.055	1.054	1.054	1.053	1.052	1.051	1.050	1.048	1.046	1.044	1.040	1.037	1.034	1.034	1.034
7100	1.057	1.057	1.057	1.056	1.056	1.055	1.055	1.054	1.053	1.052	1.051	1.050	1.048	1.047	1.044	1.041	1.037	1.035	1.035	1.034
7200	1.058	1.058	1.057	1.057	1.056	1.056	1.055	1.055	1.054	1.053	1.052	1.051	1.049	1.047	1.045	1.042	1.038	1.035	1.035	1.035

SIGMA = 222500.0

$V-V_c = 2$

H	0.08	0.09	0.10	0.2	0.4	0.6	0.8	1.0	1.2	1.4	1.6	1.8	2.0	2.2	2.4	2.6	2.8	3.0	3.2	3.4
7300	1.035	1.036	1.036	1.038	1.042	1.046	1.048	1.050	1.051	1.053	1.054	1.055	1.055	1.056	1.057	1.057	1.058	1.058	1.059	1.059
7400	1.036	1.036	1.036	1.039	1.043	1.046	1.049	1.050	1.052	1.053	1.054	1.055	1.056	1.057	1.058	1.058	1.059	1.059	1.059	1.060
7500	1.036	1.037	1.037	1.040	1.044	1.047	1.049	1.051	1.053	1.054	1.055	1.056	1.057	1.058	1.058	1.059	1.059	1.060	1.060	1.061
7600	1.037	1.037	1.037	1.041	1.044	1.047	1.050	1.052	1.054	1.055	1.056	1.057	1.058	1.058	1.059	1.060	1.060	1.061	1.061	1.061
7700	1.037	1.038	1.038	1.041	1.045	1.048	1.051	1.053	1.055	1.056	1.057	1.057	1.058	1.059	1.060	1.060	1.061	1.061	1.062	1.062
7800	1.038	1.038	1.038	1.041	1.045	1.049	1.051	1.053	1.055	1.057	1.057	1.058	1.059	1.060	1.061	1.061	1.062	1.062	1.062	1.063
7900	1.038	1.039	1.039	1.042	1.046	1.050	1.052	1.054	1.056	1.058	1.058	1.059	1.060	1.061	1.061	1.062	1.063	1.063	1.063	1.064
8000	1.039	1.039	1.040	1.043	1.047	1.050	1.053	1.055	1.057	1.058	1.059	1.060	1.061	1.062	1.062	1.063	1.064	1.064	1.064	1.065
8100	1.039	1.040	1.040	1.043	1.047	1.051	1.053	1.055	1.057	1.058	1.060	1.061	1.061	1.062	1.063	1.064	1.064	1.065	1.065	1.065
8200	1.040	1.040	1.041	1.044	1.048	1.051	1.054	1.056	1.058	1.059	1.060	1.061	1.062	1.063	1.064	1.064	1.065	1.066	1.066	1.066
8300	1.040	1.040	1.041	1.044	1.048	1.052	1.055	1.057	1.059	1.060	1.061	1.062	1.063	1.064	1.065	1.065	1.066	1.066	1.067	1.067
8400	1.041	1.041	1.041	1.044	1.049	1.052	1.055	1.057	1.059	1.061	1.062	1.063	1.064	1.065	1.065	1.066	1.067	1.067	1.068	1.068
8500	1.041	1.042	1.042	1.045	1.050	1.053	1.056	1.058	1.060	1.061	1.063	1.064	1.065	1.065	1.066	1.067	1.067	1.068	1.068	1.069
8600	1.042	1.042	1.043	1.045	1.050	1.054	1.056	1.059	1.061	1.062	1.063	1.064	1.065	1.066	1.067	1.068	1.068	1.069	1.069	1.070
8700	1.042	1.043	1.043	1.046	1.051	1.054	1.057	1.060	1.062	1.063	1.064	1.065	1.066	1.067	1.068	1.068	1.069	1.070	1.070	1.070
8800	1.043	1.043	1.044	1.046	1.051	1.055	1.058	1.060	1.062	1.064	1.065	1.066	1.067	1.068	1.069	1.069	1.070	1.070	1.071	1.071
8900	1.043	1.044	1.044	1.047	1.052	1.056	1.058	1.061	1.063	1.064	1.066	1.067	1.068	1.069	1.069	1.070	1.071	1.071	1.072	1.072
9000	1.044	1.044	1.044	1.047	1.052	1.056	1.059	1.061	1.064	1.065	1.066	1.068	1.069	1.069	1.070	1.071	1.072	1.072	1.073	1.073
9100	1.044	1.045	1.045	1.048	1.053	1.057	1.060	1.062	1.065	1.066	1.067	1.068	1.069	1.070	1.071	1.072	1.072	1.073	1.073	1.074
9200	1.045	1.045	1.045	1.049	1.054	1.057	1.060	1.063	1.065	1.067	1.068	1.069	1.070	1.071	1.072	1.072	1.073	1.074	1.074	1.075
9300	1.045	1.046	1.046	1.049	1.054	1.058	1.061	1.064	1.066	1.067	1.069	1.070	1.071	1.072	1.073	1.073	1.074	1.074	1.075	1.075
9400	1.046	1.046	1.046	1.050	1.055	1.059	1.062	1.064	1.066	1.068	1.069	1.071	1.072	1.073	1.073	1.074	1.075	1.075	1.076	1.076
9500	1.046	1.046	1.047	1.050	1.055	1.059	1.062	1.065	1.067	1.069	1.070	1.071	1.072	1.073	1.074	1.075	1.076	1.076	1.077	1.077
9600	1.046	1.047	1.047	1.051	1.056	1.060	1.063	1.066	1.068	1.070	1.071	1.072	1.073	1.074	1.075	1.076	1.076	1.077	1.077	1.078
9700	1.047	1.047	1.048	1.051	1.057	1.061	1.064	1.066	1.069	1.070	1.072	1.073	1.074	1.075	1.076	1.076	1.077	1.078	1.078	1.079
9800	1.047	1.048	1.048	1.052	1.057	1.061	1.064	1.067	1.069	1.071	1.072	1.074	1.075	1.076	1.077	1.077	1.078	1.079	1.079	1.080
9900	1.048	1.048	1.049	1.052	1.058	1.062	1.065	1.068	1.070	1.072	1.073	1.074	1.076	1.076	1.077	1.078	1.079	1.079	1.080	1.080
10000	1.048	1.049	1.049	1.053	1.058	1.063	1.066	1.068	1.071	1.072	1.074	1.075	1.076	1.077	1.078	1.079	1.080	1.080	1.081	1.081
10100	1.049	1.049	1.050	1.053	1.059	1.063	1.066	1.069	1.071	1.073	1.075	1.076	1.077	1.078	1.079	1.080	1.080	1.081	1.082	1.082
10200	1.049	1.050	1.050	1.054	1.060	1.064	1.067	1.070	1.072	1.074	1.075	1.077	1.078	1.079	1.080	1.080	1.081	1.082	1.082	1.083
10300	1.050	1.050	1.051	1.054	1.060	1.064	1.068	1.071	1.073	1.075	1.076	1.077	1.079	1.080	1.080	1.081	1.082	1.083	1.083	1.084
10400	1.050	1.051	1.051	1.055	1.061	1.065	1.068	1.071	1.073	1.075	1.077	1.078	1.079	1.080	1.081	1.082	1.083	1.083	1.084	1.085
10500	1.050	1.051	1.052	1.055	1.061	1.066	1.069	1.072	1.074	1.076	1.078	1.079	1.080	1.081	1.082	1.083	1.084	1.084	1.085	1.085
10600	1.051	1.052	1.052	1.056	1.062	1.066	1.070	1.073	1.075	1.077	1.078	1.080	1.081	1.082	1.083	1.084	1.084	1.085	1.086	1.086
10700	1.052	1.053	1.053	1.056	1.062	1.067	1.070	1.073	1.076	1.077	1.079	1.080	1.082	1.083	1.084	1.084	1.085	1.086	1.087	1.087
10800	1.052	1.053	1.053	1.057	1.063	1.068	1.071	1.074	1.076	1.078	1.080	1.081	1.082	1.084	1.084	1.085	1.086	1.087	1.087	1.088
10900	1.053	1.053	1.054	1.058	1.064	1.068	1.072	1.075	1.077	1.079	1.081	1.082	1.083	1.084	1.085	1.086	1.087	1.088	1.088	1.089
11000	1.053	1.054	1.054	1.058	1.064	1.069	1.072	1.075	1.078	1.080	1.081	1.083	1.084	1.085	1.086	1.087	1.088	1.088	1.089	1.090
11100	1.054	1.054	1.055	1.059	1.065	1.070	1.073	1.076	1.078	1.080	1.082	1.084	1.085	1.086	1.087	1.088	1.088	1.089	1.090	1.090
11200	1.054	1.055	1.055	1.059	1.065	1.070	1.074	1.077	1.079	1.081	1.083	1.084	1.086	1.087	1.088	1.089	1.089	1.090	1.091	1.091
11300	1.055	1.055	1.056	1.060	1.066	1.071	1.075	1.077	1.080	1.082	1.084	1.085	1.086	1.087	1.088	1.089	1.090	1.091	1.091	1.092
11400	1.055	1.056	1.056	1.060	1.067	1.071	1.075	1.078	1.081	1.083	1.084	1.086	1.087	1.088	1.089	1.090	1.091	1.092	1.092	1.093

x

SIGMA = 222500.0

V−V$_c$ = 2

H \ x	0.08	0.09	0.10	0.2	0.4	0.6	0.8	1.0	1.2	1.4	1.6	1.8	2.0	2.2	2.4	2.6	2.8	3.0	3.2	3.4
11500	1.056	1.056	1.057	1.061	1.067	1.072	1.076	1.079	1.081	1.083	1.085	1.087	1.088	1.089	1.090	1.091	1.092	1.092	1.093	1.094
11600	1.056	1.057	1.057	1.061	1.068	1.073	1.077	1.080	1.082	1.084	1.086	1.088	1.089	1.090	1.091	1.092	1.093	1.093	1.094	1.095
11700	1.057	1.057	1.058	1.062	1.068	1.073	1.077	1.080	1.083	1.085	1.087	1.088	1.089	1.091	1.092	1.093	1.093	1.094	1.095	1.096
11800	1.057	1.058	1.058	1.062	1.069	1.074	1.078	1.081	1.083	1.086	1.088	1.089	1.090	1.091	1.093	1.093	1.094	1.095	1.096	1.097
11900	1.058	1.058	1.059	1.063	1.070	1.075	1.079	1.082	1.084	1.086	1.088	1.090	1.091	1.092	1.093	1.094	1.095	1.096	1.097	1.098
12000	1.058	1.059	1.059	1.063	1.070	1.075	1.079	1.082	1.085	1.087	1.089	1.090	1.092	1.093	1.094	1.095	1.096	1.097	1.098	1.099
12100	1.059	1.059	1.060	1.064	1.071	1.076	1.080	1.083	1.086	1.088	1.090	1.091	1.093	1.094	1.095	1.096	1.097	1.097	1.098	1.100
12200	1.059	1.060	1.060	1.064	1.071	1.077	1.080	1.084	1.086	1.089	1.091	1.092	1.094	1.095	1.096	1.096	1.097	1.098	1.099	1.100
12300	1.059	1.060	1.061	1.065	1.072	1.077	1.081	1.084	1.087	1.089	1.091	1.092	1.094	1.096	1.096	1.097	1.098	1.099	1.100	1.101
12400	1.060	1.060	1.061	1.066	1.073	1.078	1.081	1.085	1.088	1.090	1.092	1.093	1.095	1.397	1.097	1.098	1.099	1.100	1.101	1.102
12500	1.060	1.061	1.062	1.066	1.073	1.078	1.082	1.086	1.089	1.091	1.093	1.094	1.096	1.098	1.098	1.099	1.100	1.101	1.101	1.102
12600	1.061	1.061	1.062	1.067	1.074	1.079	1.083	1.087	1.089	1.092	1.093	1.094	1.097	1.099	1.099	1.099	1.100	1.102	1.102	1.103
12700	1.061	1.062	1.063	1.067	1.074	1.080	1.083	1.087	1.090	1.092	1.094	1.095	1.097	1.099	1.100	1.100	1.101	1.102	1.103	1.104
12800	1.061	1.062	1.063	1.068	1.075	1.080	1.084	1.088	1.091	1.093	1.095	1.096	1.098	1.100	1.100	1.101	1.102	1.103	1.104	1.105
12900	1.062	1.062	1.063	1.068	1.076	1.081	1.085	1.089	1.091	1.094	1.096	1.097	1.098	1.100	1.101	1.101	1.102	1.104	1.105	1.106
13000	1.062	1.063	1.064	1.069	1.076	1.082	1.085	1.089	1.092	1.095	1.096	1.097	1.100	1.101	1.102	1.102	1.103	1.105	1.106	1.106
13100	1.063	1.063	1.064	1.069	1.077	1.082	1.086	1.090	1.093	1.095	1.097	1.098	1.100	1.102	1.102	1.103	1.104	1.106	1.106	1.107
13200	1.063	1.064	1.065	1.070	1.077	1.083	1.086	1.091	1.093	1.096	1.098	1.099	1.101	1.103	1.103	1.104	1.105	1.106	1.107	1.108
13300	1.063	1.064	1.065	1.070	1.078	1.084	1.087	1.091	1.094	1.097	1.099	1.100	1.102	1.103	1.104	1.104	1.105	1.107	1.108	1.109
13400	1.064	1.064	1.066	1.071	1.078	1.084	1.088	1.092	1.095	1.097	1.099	1.100	1.102	1.104	1.105	1.105	1.106	1.108	1.109	1.109
13500	1.064	1.065	1.066	1.071	1.079	1.085	1.089	1.093	1.096	1.098	1.100	1.101	1.103	1.105	1.105	1.106	1.107	1.108	1.110	1.110
13600	1.064	1.065	1.067	1.072	1.080	1.085	1.089	1.094	1.097	1.099	1.101	1.102	1.104	1.106	1.106	1.107	1.108	1.109	1.110	1.111
13700	1.065	1.066	1.067	1.072	1.080	1.086	1.090	1.094	1.097	1.100	1.102	1.103	1.105	1.106	1.107	1.108	1.109	1.110	1.111	1.112
13800	1.065	1.066	1.068	1.073	1.081	1.087	1.091	1.095	1.098	1.100	1.103	1.104	1.106	1.107	1.108	1.108	1.109	1.111	1.112	1.113
13900	1.065	1.066	1.068	1.074	1.082	1.087	1.091	1.096	1.099	1.101	1.103	1.104	1.107	1.107	1.108	1.109	1.110	1.111	1.113	1.114
14000	1.066	1.067	1.069	1.074	1.082	1.088	1.092	1.096	1.099	1.102	1.104	1.105	1.108	1.108	1.109	1.110	1.111	1.112	1.114	1.115
14100	1.067	1.067	1.069	1.075	1.083	1.089	1.093	1.097	1.100	1.103	1.105	1.106	1.108	1.109	1.110	1.110	1.111	1.113	1.114	1.115
14200	1.067	1.068	1.070	1.075	1.083	1.089	1.093	1.098	1.101	1.103	1.106	1.107	1.109	1.110	1.111	1.111	1.112	1.114	1.115	1.116
14300	1.068	1.068	1.070	1.076	1.084	1.090	1.094	1.098	1.102	1.104	1.107	1.107	1.110	1.110	1.112	1.112	1.113	1.115	1.116	1.117
14400	1.068	1.069	1.071	1.076	1.085	1.091	1.095	1.099	1.103	1.105	1.108	1.108	1.111	1.111	1.113	1.113	1.114	1.116	1.117	1.118
14500	1.069	1.069	1.072	1.077	1.085	1.091	1.096	1.100	1.104	1.106	1.109	1.109	1.111	1.112	1.113	1.114	1.115	1.116	1.117	1.118
14600	1.069	1.070	1.072	1.077	1.086	1.092	1.096	1.101	1.104	1.106	1.109	1.110	1.112	1.113	1.114	1.115	1.116	1.117	1.118	1.119
14700	1.070	1.070	1.073	1.078	1.087	1.093	1.097	1.102	1.105	1.107	1.110	1.111	1.113	1.114	1.115	1.116	1.117	1.118	1.119	1.120
14800	1.070	1.071	1.073	1.078	1.087	1.093	1.097	1.102	1.106	1.108	1.110	1.112	1.114	1.114	1.116	1.117	1.118	1.119	1.120	1.121
14900	1.072	1.072	1.073	1.079	1.087	1.093	1.098	1.103	1.106	1.109	1.111	1.113	1.115	1.116	1.117	1.118	1.119	1.120	1.121	1.122
15000	1.073	1.073	1.074	1.079	1.088	1.094	1.099	1.103	1.107	1.109	1.112	1.114	1.115	1.117	1.118	1.119	1.120	1.121	1.122	1.123

SIGMA = 98890.0

$v - v_c = 3$

H	x = 3.4	3.2	3.0	2.8	2.6	2.4	2.2	2.0	1.8	1.6	1.4	1.2	1.0	0.8	0.6	0.4	0.2	0.10	0.09	0.08
0	1.000	1.000	1.000	1.000	1.000	1.000	1.000	1.000	1.000	1.000	1.000	1.000	1.000	1.000	1.000	1.000	1.000	1.000	1.000	1.000
20	1.000	1.000	1.000	1.000	1.000	1.000	1.000	1.000	1.000	1.000	1.000	1.000	1.000	1.000	1.000	1.000	1.000	1.000	1.000	1.000
40	1.001	1.001	1.001	1.001	1.001	1.001	1.001	1.001	1.001	1.001	1.001	1.001	1.001	1.001	1.001	1.001	1.000	1.000	1.000	1.000
60	1.001	1.001	1.001	1.001	1.001	1.001	1.001	1.001	1.001	1.001	1.001	1.001	1.001	1.001	1.001	1.001	1.001	1.001	1.001	1.001
80	1.001	1.001	1.001	1.001	1.001	1.001	1.001	1.001	1.001	1.001	1.001	1.001	1.001	1.001	1.001	1.001	1.001	1.001	1.001	1.001
100	1.002	1.002	1.002	1.002	1.002	1.002	1.002	1.002	1.002	1.002	1.002	1.002	1.002	1.002	1.002	1.002	1.002	1.001	1.001	1.001
120	1.002	1.002	1.002	1.002	1.002	1.002	1.002	1.002	1.002	1.002	1.002	1.002	1.002	1.002	1.002	1.002	1.002	1.002	1.002	1.002
140	1.002	1.002	1.002	1.002	1.002	1.002	1.002	1.002	1.002	1.002	1.002	1.002	1.002	1.002	1.002	1.002	1.002	1.002	1.002	1.002
160	1.003	1.003	1.003	1.003	1.003	1.003	1.003	1.003	1.003	1.003	1.003	1.003	1.003	1.003	1.003	1.003	1.002	1.002	1.002	1.002
180	1.003	1.003	1.003	1.003	1.003	1.003	1.003	1.003	1.003	1.003	1.003	1.003	1.003	1.003	1.003	1.003	1.002	1.002	1.002	1.002
200	1.004	1.004	1.004	1.004	1.004	1.004	1.004	1.004	1.004	1.004	1.004	1.004	1.004	1.004	1.004	1.003	1.002	1.002	1.002	1.002
220	1.004	1.004	1.004	1.004	1.004	1.004	1.004	1.004	1.004	1.004	1.004	1.004	1.004	1.004	1.004	1.003	1.002	1.002	1.002	1.002
240	1.004	1.004	1.004	1.004	1.004	1.004	1.004	1.004	1.004	1.004	1.004	1.004	1.004	1.004	1.004	1.004	1.003	1.003	1.003	1.003
260	1.005	1.005	1.005	1.005	1.005	1.005	1.005	1.005	1.005	1.005	1.005	1.005	1.005	1.005	1.005	1.004	1.003	1.003	1.003	1.003
280	1.005	1.005	1.005	1.005	1.005	1.005	1.005	1.005	1.005	1.005	1.005	1.005	1.005	1.005	1.005	1.004	1.003	1.003	1.003	1.003
300	1.005	1.005	1.005	1.005	1.005	1.005	1.005	1.005	1.005	1.005	1.005	1.005	1.005	1.005	1.005	1.004	1.003	1.003	1.003	1.003
320	1.006	1.006	1.006	1.006	1.006	1.006	1.006	1.006	1.006	1.006	1.006	1.006	1.006	1.006	1.005	1.004	1.003	1.003	1.003	1.003
340	1.006	1.006	1.006	1.006	1.006	1.006	1.006	1.006	1.006	1.006	1.006	1.006	1.006	1.006	1.006	1.005	1.004	1.004	1.004	1.004
360	1.006	1.006	1.006	1.006	1.006	1.006	1.006	1.006	1.006	1.006	1.006	1.006	1.006	1.006	1.006	1.005	1.004	1.004	1.004	1.004
380	1.007	1.007	1.007	1.007	1.007	1.007	1.007	1.007	1.007	1.007	1.007	1.007	1.007	1.007	1.006	1.005	1.004	1.004	1.004	1.004
400	1.007	1.007	1.007	1.007	1.007	1.007	1.007	1.007	1.007	1.007	1.007	1.007	1.007	1.007	1.006	1.005	1.004	1.004	1.004	1.004
420	1.007	1.007	1.007	1.007	1.007	1.007	1.007	1.007	1.007	1.007	1.007	1.007	1.007	1.007	1.007	1.006	1.005	1.005	1.005	1.005
440	1.008	1.008	1.008	1.008	1.008	1.008	1.008	1.008	1.008	1.008	1.008	1.008	1.008	1.008	1.007	1.006	1.005	1.005	1.005	1.005
460	1.008	1.008	1.008	1.008	1.008	1.008	1.008	1.008	1.008	1.008	1.008	1.008	1.008	1.008	1.007	1.006	1.005	1.005	1.005	1.005
480	1.009	1.009	1.009	1.009	1.009	1.009	1.009	1.009	1.009	1.009	1.009	1.009	1.008	1.008	1.007	1.006	1.005	1.005	1.005	1.005
500	1.009	1.009	1.009	1.009	1.009	1.009	1.009	1.009	1.009	1.009	1.009	1.009	1.009	1.009	1.008	1.007	1.006	1.006	1.006	1.006
520	1.009	1.009	1.009	1.009	1.009	1.009	1.009	1.009	1.009	1.009	1.009	1.009	1.009	1.009	1.008	1.007	1.006	1.006	1.006	1.006
540	1.010	1.010	1.010	1.010	1.010	1.010	1.010	1.010	1.010	1.010	1.010	1.010	1.009	1.009	1.008	1.007	1.006	1.006	1.006	1.006
560	1.010	1.010	1.010	1.010	1.010	1.010	1.010	1.010	1.010	1.010	1.010	1.010	1.009	1.009	1.008	1.007	1.006	1.006	1.006	1.006
580	1.011	1.011	1.011	1.011	1.011	1.011	1.011	1.011	1.011	1.011	1.011	1.011	1.010	1.010	1.009	1.008	1.007	1.007	1.007	1.007
600	1.011	1.011	1.011	1.011	1.011	1.011	1.011	1.011	1.011	1.011	1.011	1.011	1.010	1.010	1.009	1.008	1.007	1.007	1.007	1.007
620	1.011	1.011	1.011	1.011	1.011	1.011	1.011	1.011	1.011	1.011	1.011	1.011	1.010	1.010	1.009	1.008	1.007	1.007	1.007	1.007
640	1.012	1.012	1.012	1.012	1.012	1.012	1.012	1.012	1.011	1.011	1.011	1.011	1.010	1.010	1.009	1.008	1.007	1.007	1.007	1.007
660	1.012	1.012	1.012	1.012	1.012	1.012	1.012	1.012	1.011	1.011	1.011	1.011	1.010	1.010	1.009	1.008	1.007	1.007	1.007	1.007
680	1.012	1.012	1.012	1.012	1.012	1.012	1.012	1.012	1.012	1.012	1.012	1.012	1.011	1.011	1.010	1.009	1.008	1.008	1.008	1.008
700	1.013	1.013	1.013	1.013	1.013	1.013	1.013	1.013	1.012	1.012	1.012	1.012	1.011	1.011	1.010	1.009	1.008	1.008	1.008	1.008
720	1.013	1.013	1.013	1.013	1.013	1.013	1.013	1.013	1.012	1.012	1.012	1.012	1.011	1.011	1.010	1.009	1.008	1.008	1.008	1.008
740	1.013	1.013	1.013	1.013	1.013	1.013	1.013	1.013	1.012	1.012	1.012	1.012	1.011	1.011	1.010	1.009	1.008	1.008	1.008	1.008
760	1.014	1.014	1.014	1.014	1.014	1.014	1.014	1.014	1.013	1.013	1.013	1.013	1.012	1.012	1.011	1.010	1.009	1.009	1.009	1.008
780	1.014	1.014	1.014	1.014	1.014	1.014	1.014	1.014	1.013	1.013	1.013	1.013	1.012	1.012	1.011	1.010	1.009	1.009	1.009	1.008
800	1.014	1.014	1.014	1.014	1.014	1.014	1.014	1.014	1.013	1.013	1.013	1.013	1.012	1.012	1.011	1.010	1.009	1.009	1.009	1.009

SIGMA =   98890.0

V−V$_c$ = 3

H / x	3.4	3.2	3.0	2.8	2.6	2.4	2.2	2.0	1.8	1.6	1.4	1.2	1.0	0.8	0.6	0.4	0.2	0.10	0.09	0.08
820	1.015	1.015	1.015	1.014	1.014	1.014	1.014	1.014	1.014	1.013	1.013	1.013	1.012	1.012	1.011	1.011	1.010	1.009	1.009	1.009
840	1.015	1.015	1.015	1.015	1.015	1.014	1.014	1.014	1.014	1.014	1.013	1.013	1.013	1.012	1.012	1.011	1.010	1.009	1.009	1.009
860	1.015	1.015	1.015	1.015	1.015	1.015	1.015	1.014	1.014	1.014	1.014	1.013	1.013	1.012	1.012	1.011	1.010	1.009	1.009	1.009
880	1.016	1.016	1.016	1.016	1.016	1.015	1.015	1.015	1.015	1.014	1.014	1.014	1.013	1.013	1.012	1.011	1.010	1.010	1.010	1.010
900	1.016	1.016	1.016	1.016	1.016	1.016	1.015	1.015	1.015	1.015	1.014	1.014	1.014	1.013	1.013	1.012	1.011	1.010	1.010	1.010
920	1.017	1.016	1.016	1.016	1.016	1.016	1.016	1.015	1.015	1.015	1.014	1.014	1.014	1.013	1.013	1.012	1.011	1.010	1.010	1.010
940	1.017	1.017	1.017	1.017	1.016	1.016	1.016	1.016	1.016	1.015	1.015	1.014	1.014	1.014	1.013	1.012	1.011	1.010	1.010	1.010
960	1.017	1.017	1.017	1.017	1.017	1.017	1.016	1.016	1.016	1.015	1.015	1.015	1.014	1.014	1.013	1.012	1.011	1.011	1.011	1.010
980	1.018	1.018	1.017	1.017	1.017	1.017	1.017	1.016	1.016	1.016	1.015	1.015	1.015	1.014	1.013	1.013	1.012	1.011	1.011	1.011
1000	1.018	1.018	1.018	1.018	1.018	1.017	1.017	1.017	1.017	1.016	1.016	1.016	1.015	1.014	1.014	1.013	1.012	1.011	1.011	1.011
1020	1.018	1.018	1.018	1.018	1.018	1.018	1.017	1.017	1.017	1.016	1.016	1.016	1.015	1.015	1.014	1.013	1.012	1.011	1.011	1.011
1040	1.019	1.019	1.018	1.018	1.018	1.018	1.018	1.018	1.017	1.017	1.017	1.016	1.016	1.015	1.014	1.014	1.013	1.012	1.012	1.012
1060	1.019	1.019	1.019	1.019	1.019	1.018	1.018	1.018	1.018	1.017	1.017	1.017	1.016	1.016	1.015	1.014	1.013	1.012	1.012	1.012
1080	1.019	1.019	1.019	1.019	1.019	1.019	1.019	1.018	1.018	1.018	1.017	1.017	1.016	1.016	1.015	1.014	1.013	1.012	1.012	1.012
1100	1.020	1.019	1.019	1.019	1.019	1.019	1.019	1.019	1.018	1.018	1.018	1.017	1.017	1.016	1.015	1.015	1.013	1.013	1.013	1.013
1120	1.020	1.020	1.020	1.020	1.019	1.019	1.019	1.019	1.019	1.018	1.018	1.018	1.016	1.016	1.016	1.015	1.013	1.013	1.012	1.012
1140	1.020	1.020	1.020	1.020	1.020	1.020	1.019	1.019	1.019	1.019	1.018	1.018	1.017	1.017	1.016	1.015	1.014	1.013	1.013	1.013
1160	1.021	1.021	1.021	1.020	1.020	1.020	1.020	1.020	1.020	1.019	1.019	1.018	1.018	1.017	1.016	1.015	1.014	1.013	1.013	1.013
1180	1.021	1.021	1.021	1.021	1.020	1.020	1.020	1.020	1.020	1.019	1.019	1.019	1.018	1.017	1.016	1.015	1.014	1.013	1.013	1.013
1200	1.021	1.021	1.021	1.021	1.021	1.021	1.021	1.020	1.020	1.020	1.019	1.019	1.019	1.018	1.017	1.016	1.014	1.014	1.013	1.013
1220	1.022	1.022	1.021	1.021	1.021	1.021	1.021	1.021	1.020	1.020	1.020	1.019	1.019	1.018	1.017	1.016	1.015	1.014	1.014	1.013
1240	1.022	1.022	1.022	1.022	1.022	1.021	1.021	1.021	1.021	1.020	1.020	1.020	1.019	1.018	1.017	1.016	1.015	1.014	1.014	1.013
1260	1.023	1.023	1.022	1.022	1.022	1.022	1.022	1.021	1.021	1.021	1.020	1.020	1.019	1.018	1.018	1.016	1.015	1.014	1.014	1.014
1280	1.023	1.023	1.023	1.023	1.022	1.022	1.022	1.022	1.021	1.021	1.020	1.020	1.019	1.019	1.017	1.017	1.015	1.014	1.014	1.014
1300	1.023	1.023	1.023	1.023	1.023	1.022	1.022	1.022	1.022	1.021	1.021	1.020	1.020	1.019	1.018	1.017	1.015	1.014	1.014	1.014
1320	1.024	1.024	1.023	1.023	1.023	1.023	1.023	1.022	1.022	1.022	1.021	1.021	1.020	1.019	1.018	1.017	1.016	1.015	1.014	1.014
1340	1.024	1.024	1.024	1.024	1.023	1.023	1.023	1.023	1.022	1.022	1.022	1.021	1.020	1.019	1.018	1.017	1.016	1.015	1.015	1.015
1360	1.025	1.025	1.024	1.024	1.024	1.023	1.023	1.023	1.023	1.022	1.022	1.021	1.021	1.020	1.019	1.018	1.016	1.015	1.015	1.015
1380	1.025	1.025	1.025	1.024	1.024	1.024	1.023	1.023	1.023	1.023	1.022	1.022	1.021	1.020	1.019	1.018	1.016	1.015	1.015	1.015
1400	1.025	1.025	1.025	1.025	1.025	1.024	1.024	1.024	1.024	1.023	1.023	1.022	1.021	1.021	1.020	1.018	1.017	1.015	1.015	1.015
1420	1.026	1.026	1.025	1.025	1.025	1.024	1.024	1.024	1.024	1.023	1.023	1.022	1.022	1.021	1.020	1.019	1.017	1.016	1.016	1.015
1440	1.026	1.026	1.026	1.025	1.025	1.025	1.025	1.024	1.024	1.024	1.023	1.023	1.022	1.021	1.020	1.019	1.017	1.016	1.016	1.016
1460	1.026	1.026	1.026	1.026	1.026	1.025	1.025	1.025	1.025	1.024	1.024	1.023	1.022	1.021	1.020	1.019	1.017	1.016	1.016	1.016
1480	1.027	1.027	1.026	1.026	1.026	1.026	1.025	1.025	1.025	1.024	1.024	1.023	1.022	1.022	1.021	1.019	1.017	1.017	1.016	1.016
1500	1.027	1.027	1.027	1.027	1.026	1.026	1.026	1.026	1.025	1.025	1.024	1.024	1.023	1.022	1.021	1.020	1.018	1.017	1.016	1.016
1520	1.027	1.027	1.027	1.027	1.027	1.026	1.026	1.026	1.025	1.025	1.025	1.024	1.023	1.022	1.021	1.020	1.018	1.017	1.017	1.017
1540	1.028	1.028	1.027	1.027	1.027	1.027	1.027	1.026	1.026	1.026	1.025	1.024	1.024	1.023	1.022	1.020	1.018	1.017	1.017	1.017
1560	1.028	1.028	1.028	1.028	1.027	1.027	1.027	1.026	1.026	1.026	1.025	1.025	1.024	1.023	1.022	1.020	1.019	1.018	1.017	1.017
1580	1.029	1.028	1.028	1.028	1.028	1.027	1.027	1.027	1.026	1.026	1.025	1.025	1.024	1.023	1.022	1.021	1.019	1.018	1.018	1.017
1600	1.029	1.029	1.028	1.028	1.028	1.028	1.027	1.027	1.027	1.026	1.026	1.025	1.024	1.023	1.022	1.021	1.019	1.018	1.018	1.017
1620	1.029	1.029	1.029	1.028	1.028	1.028	1.028	1.027	1.027	1.027	1.026	1.025	1.025	1.024	1.023	1.021	1.019	1.018	1.018	1.018
1640	1.030	1.029	1.029	1.029	1.029	1.028	1.028	1.028	1.027	1.027	1.026	1.026	1.025	1.024	1.023	1.021	1.019	1.018	1.018	1.018

SIGMA = 98890.0

V−V_c = 3

H x	0.08	0.09	0.10	0.2	0.4	0.6	0.8	1.0	1.2	1.4	1.6	1.8	2.0	2.2	2.4	2.6	2.8	3.0	3.2	3.4
1660	1.018	1.018	1.018	1.020	1.022	1.023	1.024	1.025	1.026	1.027	1.027	1.028	1.028	1.029	1.029	1.029	1.029	1.030	1.030	1.030
1680	1.018	1.018	1.019	1.020	1.022	1.023	1.025	1.026	1.026	1.027	1.027	1.028	1.028	1.029	1.029	1.030	1.030	1.030	1.030	1.030
1700	1.018	1.019	1.019	1.020	1.022	1.024	1.025	1.026	1.027	1.027	1.028	1.028	1.029	1.029	1.030	1.030	1.030	1.030	1.031	1.031
1720	1.019	1.019	1.019	1.021	1.022	1.024	1.025	1.026	1.027	1.028	1.028	1.029	1.029	1.030	1.030	1.030	1.030	1.031	1.031	1.031
1740	1.019	1.019	1.019	1.021	1.023	1.024	1.026	1.027	1.027	1.028	1.029	1.029	1.030	1.030	1.031	1.031	1.031	1.031	1.031	1.031
1760	1.019	1.019	1.020	1.021	1.023	1.025	1.026	1.027	1.028	1.029	1.029	1.030	1.030	1.031	1.031	1.031	1.031	1.032	1.032	1.032
1780	1.019	1.020	1.020	1.021	1.023	1.025	1.026	1.027	1.028	1.029	1.030	1.030	1.031	1.031	1.032	1.032	1.032	1.032	1.032	1.032
1800	1.020	1.020	1.020	1.021	1.023	1.025	1.027	1.028	1.029	1.029	1.030	1.031	1.031	1.032	1.032	1.032	1.032	1.032	1.033	1.033
1820	1.020	1.020	1.020	1.022	1.024	1.026	1.027	1.028	1.029	1.030	1.030	1.031	1.032	1.032	1.032	1.033	1.033	1.033	1.033	1.033
1840	1.020	1.020	1.021	1.022	1.024	1.026	1.028	1.028	1.029	1.030	1.031	1.031	1.032	1.032	1.033	1.033	1.033	1.033	1.034	1.034
1860	1.020	1.020	1.021	1.022	1.024	1.026	1.028	1.029	1.030	1.030	1.031	1.032	1.032	1.033	1.033	1.034	1.034	1.034	1.034	1.034
1880	1.020	1.021	1.021	1.022	1.025	1.027	1.028	1.029	1.030	1.031	1.031	1.032	1.032	1.033	1.034	1.034	1.034	1.034	1.035	1.035
1900	1.021	1.021	1.022	1.022	1.025	1.027	1.028	1.030	1.030	1.031	1.032	1.032	1.033	1.033	1.034	1.034	1.035	1.035	1.035	1.035
1920	1.021	1.021	1.022	1.023	1.025	1.027	1.029	1.030	1.031	1.032	1.032	1.033	1.033	1.034	1.034	1.035	1.035	1.035	1.036	1.036
1940	1.021	1.021	1.022	1.023	1.025	1.027	1.029	1.030	1.031	1.032	1.032	1.033	1.034	1.034	1.035	1.035	1.035	1.036	1.036	1.036
1960	1.021	1.021	1.022	1.023	1.026	1.028	1.029	1.030	1.032	1.032	1.033	1.034	1.034	1.035	1.035	1.035	1.036	1.036	1.036	1.036
1980	1.022	1.022	1.022	1.023	1.026	1.028	1.029	1.031	1.032	1.033	1.033	1.034	1.035	1.035	1.036	1.036	1.036	1.036	1.037	1.037
2000	1.022	1.022	1.023	1.024	1.026	1.028	1.030	1.031	1.033	1.033	1.034	1.034	1.035	1.036	1.036	1.036	1.037	1.037	1.037	1.037
2020	1.022	1.022	1.023	1.024	1.026	1.028	1.030	1.032	1.033	1.034	1.034	1.035	1.035	1.036	1.036	1.037	1.037	1.037	1.037	1.037
2040	1.022	1.022	1.023	1.024	1.027	1.028	1.030	1.032	1.033	1.034	1.035	1.035	1.036	1.036	1.037	1.037	1.037	1.037	1.038	1.038
2060	1.022	1.023	1.023	1.024	1.027	1.029	1.031	1.032	1.034	1.034	1.035	1.036	1.036	1.037	1.037	1.038	1.038	1.038	1.038	1.038
2080	1.023	1.023	1.024	1.025	1.027	1.029	1.031	1.033	1.034	1.035	1.035	1.036	1.037	1.037	1.038	1.038	1.038	1.038	1.038	1.038
2100	1.023	1.023	1.024	1.025	1.027	1.030	1.032	1.033	1.035	1.035	1.036	1.036	1.037	1.037	1.038	1.038	1.039	1.039	1.039	1.039
2120	1.023	1.023	1.024	1.025	1.028	1.030	1.032	1.034	1.035	1.036	1.036	1.037	1.037	1.038	1.038	1.039	1.039	1.039	1.039	1.039
2140	1.023	1.024	1.024	1.025	1.028	1.030	1.033	1.034	1.035	1.036	1.037	1.037	1.038	1.038	1.039	1.039	1.040	1.040	1.040	1.040
2160	1.023	1.024	1.024	1.026	1.028	1.030	1.033	1.035	1.036	1.037	1.037	1.038	1.039	1.039	1.040	1.040	1.040	1.040	1.040	1.040
2180	1.024	1.024	1.025	1.026	1.028	1.032	1.033	1.035	1.036	1.037	1.038	1.038	1.039	1.039	1.040	1.040	1.041	1.041	1.041	1.041
2200	1.024	1.025	1.025	1.026	1.029	1.032	1.033	1.035	1.037	1.038	1.038	1.039	1.040	1.040	1.040	1.041	1.041	1.041	1.041	1.041
2220	1.024	1.025	1.025	1.026	1.029	1.032	1.034	1.035	1.037	1.038	1.039	1.039	1.040	1.040	1.041	1.041	1.041	1.041	1.042	1.042
2240	1.024	1.025	1.025	1.027	1.029	1.032	1.035	1.036	1.037	1.038	1.039	1.040	1.040	1.041	1.041	1.042	1.042	1.042	1.042	1.042
2260	1.024	1.025	1.026	1.027	1.030	1.033	1.035	1.036	1.038	1.039	1.040	1.040	1.041	1.041	1.042	1.042	1.042	1.042	1.042	1.042
2280	1.025	1.025	1.026	1.027	1.030	1.034	1.035	1.037	1.038	1.039	1.040	1.040	1.041	1.042	1.042	1.042	1.043	1.043	1.043	1.043
2300	1.025	1.025	1.026	1.027	1.030	1.034	1.036	1.037	1.038	1.039	1.040	1.041	1.042	1.042	1.043	1.043	1.043	1.043	1.043	1.043
2320	1.025	1.026	1.026	1.028	1.031	1.034	1.036	1.038	1.039	1.040	1.041	1.041	1.042	1.043	1.043	1.043	1.043	1.043	1.044	1.044
2340	1.025	1.026	1.026	1.028	1.031	1.034	1.036	1.038	1.039	1.040	1.041	1.042	1.042	1.043	1.043	1.044	1.044	1.044	1.044	1.044
2360	1.026	1.026	1.026	1.028	1.031	1.035	1.037	1.039	1.040	1.041	1.041	1.042	1.043	1.043	1.044	1.044	1.044	1.044	1.044	1.044
2380	1.026	1.026	1.027	1.028	1.032	1.035	1.037	1.039	1.040	1.041	1.042	1.042	1.043	1.044	1.044	1.044	1.045	1.045	1.045	1.045
2400	1.026	1.026	1.027	1.028	1.032	1.036	1.038	1.040	1.041	1.041	1.042	1.043	1.043	1.044	1.044	1.045	1.045	1.045	1.045	1.045
2420	1.026	1.027	1.027	1.029	1.032	1.036	1.038	1.040	1.041	1.042	1.042	1.043	1.044	1.044	1.045	1.045	1.045	1.045	1.045	1.045
2440	1.027	1.027	1.028	1.029	1.032	1.036	1.038	1.041	1.042	1.042	1.043	1.044	1.044	1.045	1.045	1.045	1.044	1.044	1.044	1.045
2460	1.027	1.027	1.029	1.029	1.032	1.036	1.038	1.041	1.042	1.043	1.044	1.044	1.045	1.045	1.045	1.045	1.044	1.044	1.044	1.045
2480	1.027	1.027	1.029	1.029	1.032	1.037	1.038	1.042	1.042	1.043	1.044	1.044	1.045	1.045	1.045	1.045	1.044	1.044	1.045	1.045

SIGMA = 98890.0

$V-V_c = 3$

H \ X	3.4	3.2	3.0	2.8	2.6	2.4	2.2	2.0	1.8	1.6	1.4	1.2	1.0	0.8	0.6	0.4	0.2	0.10	0.09	0.08
2500	1.045	1.045	1.045	1.044	1.044	1.044	1.043	1.043	1.042	1.041	1.040	1.039	1.038	1.037	1.035	1.033	1.030	1.028	1.027	1.027
2520	1.046	1.045	1.045	1.045	1.044	1.044	1.043	1.043	1.042	1.042	1.041	1.040	1.039	1.037	1.035	1.033	1.030	1.026	1.028	1.027
2540	1.046	1.046	1.045	1.045	1.045	1.044	1.044	1.043	1.043	1.042	1.041	1.040	1.039	1.037	1.036	1.033	1.030	1.028	1.028	1.028
2560	1.046	1.046	1.046	1.045	1.045	1.045	1.044	1.044	1.043	1.042	1.041	1.040	1.039	1.038	1.036	1.033	1.030	1.028	1.028	1.028
2580	1.047	1.046	1.046	1.046	1.045	1.045	1.044	1.044	1.043	1.043	1.042	1.041	1.039	1.038	1.036	1.034	1.031	1.029	1.028	1.028
2600	1.047	1.047	1.047	1.046	1.046	1.045	1.045	1.044	1.044	1.043	1.042	1.041	1.039	1.038	1.036	1.034	1.031	1.029	1.029	1.028
2620	1.048	1.047	1.047	1.047	1.046	1.046	1.045	1.045	1.044	1.043	1.042	1.041	1.040	1.039	1.037	1.034	1.031	1.029	1.029	1.029
2640	1.048	1.048	1.047	1.047	1.046	1.046	1.045	1.045	1.044	1.044	1.043	1.042	1.040	1.039	1.037	1.035	1.031	1.029	1.029	1.029
2660	1.048	1.048	1.048	1.048	1.047	1.047	1.046	1.046	1.045	1.044	1.043	1.042	1.041	1.039	1.037	1.035	1.032	1.029	1.029	1.029
2680	1.049	1.049	1.048	1.048	1.047	1.047	1.046	1.046	1.045	1.045	1.043	1.042	1.041	1.039	1.038	1.035	1.032	1.030	1.029	1.029
2700	1.049	1.049	1.049	1.048	1.048	1.048	1.047	1.046	1.046	1.045	1.044	1.043	1.041	1.040	1.038	1.036	1.032	1.030	1.030	1.030
2720	1.050	1.049	1.049	1.049	1.048	1.048	1.047	1.047	1.046	1.045	1.044	1.043	1.042	1.040	1.038	1.036	1.032	1.030	1.030	1.030
2740	1.050	1.050	1.050	1.049	1.049	1.048	1.048	1.047	1.046	1.046	1.044	1.043	1.042	1.040	1.038	1.036	1.033	1.030	1.030	1.030
2760	1.050	1.050	1.050	1.050	1.049	1.049	1.048	1.047	1.047	1.046	1.045	1.044	1.042	1.041	1.039	1.036	1.033	1.031	1.030	1.030
2780	1.051	1.050	1.051	1.050	1.050	1.049	1.048	1.047	1.047	1.046	1.045	1.044	1.042	1.041	1.039	1.037	1.033	1.031	1.030	1.030
2800	1.051	1.051	1.051	1.050	1.050	1.050	1.049	1.048	1.047	1.047	1.046	1.044	1.043	1.041	1.039	1.037	1.033	1.031	1.031	1.031
2820	1.052	1.051	1.051	1.051	1.050	1.050	1.049	1.049	1.048	1.047	1.046	1.045	1.043	1.042	1.040	1.037	1.034	1.031	1.031	1.031
2840	1.052	1.052	1.052	1.051	1.051	1.050	1.050	1.049	1.048	1.047	1.046	1.045	1.043	1.042	1.040	1.037	1.034	1.032	1.031	1.031
2860	1.053	1.052	1.052	1.052	1.051	1.051	1.050	1.049	1.048	1.048	1.047	1.045	1.044	1.042	1.040	1.038	1.034	1.032	1.031	1.031
2880	1.053	1.053	1.052	1.052	1.052	1.051	1.050	1.050	1.049	1.048	1.047	1.046	1.044	1.042	1.041	1.038	1.034	1.032	1.032	1.031
2900	1.053	1.053	1.053	1.052	1.052	1.051	1.051	1.050	1.049	1.048	1.047	1.046	1.044	1.043	1.041	1.038	1.035	1.032	1.032	1.032
2920	1.053	1.053	1.053	1.053	1.052	1.052	1.051	1.050	1.050	1.049	1.048	1.046	1.045	1.043	1.041	1.038	1.035	1.032	1.032	1.032
2940	1.054	1.053	1.053	1.053	1.052	1.052	1.051	1.051	1.050	1.049	1.048	1.047	1.045	1.043	1.041	1.039	1.035	1.033	1.032	1.032
2960	1.054	1.054	1.053	1.053	1.053	1.052	1.052	1.051	1.050	1.049	1.048	1.047	1.046	1.044	1.042	1.039	1.035	1.033	1.032	1.032
2980	1.054	1.054	1.054	1.053	1.053	1.052	1.052	1.051	1.050	1.049	1.048	1.047	1.046	1.044	1.042	1.039	1.035	1.033	1.033	1.032
3000	1.054	1.054	1.054	1.053	1.053	1.052	1.052	1.051	1.051	1.050	1.049	1.047	1.046	1.044	1.042	1.039	1.036	1.033	1.033	1.033

SIGMA = 98890.0

V-Vc = 3

x	3.4	3.2	3.0	2.8	2.6	2.4	2.2	2.0	1.8	1.6	1.4	1.2	1.0	0.8	0.6	0.4	0.2	0.10	0.09	0.08
3100	1.056	1.056	1.056	1.055	1.055	1.054	1.054	1.053	1.052	1.051	1.050	1.049	1.048	1.046	1.043	1.041	1.037	1.034	1.034	1.034
3200	1.058	1.058	1.057	1.057	1.056	1.056	1.055	1.055	1.054	1.053	1.051	1.051	1.049	1.047	1.045	1.042	1.038	1.035	1.035	1.035
3300	1.060	1.060	1.059	1.059	1.058	1.058	1.057	1.056	1.056	1.055	1.053	1.052	1.051	1.049	1.046	1.043	1.039	1.037	1.036	1.036
3400	1.062	1.061	1.061	1.061	1.060	1.059	1.059	1.058	1.057	1.056	1.055	1.054	1.052	1.050	1.048	1.045	1.040	1.038	1.037	1.037
3500	1.064	1.063	1.063	1.062	1.062	1.061	1.061	1.060	1.059	1.058	1.057	1.055	1.054	1.052	1.049	1.047	1.042	1.040	1.039	1.038
3600	1.066	1.065	1.065	1.064	1.063	1.063	1.062	1.062	1.061	1.060	1.058	1.057	1.055	1.053	1.051	1.048	1.043	1.041	1.040	1.039
3700	1.067	1.067	1.067	1.066	1.065	1.065	1.064	1.063	1.062	1.061	1.060	1.059	1.057	1.055	1.052	1.049	1.044	1.042	1.041	1.040
3800	1.069	1.069	1.068	1.068	1.067	1.067	1.066	1.065	1.064	1.063	1.062	1.060	1.058	1.056	1.053	1.050	1.045	1.043	1.042	1.041
3900	1.071	1.071	1.070	1.070	1.069	1.068	1.068	1.067	1.066	1.065	1.063	1.062	1.060	1.058	1.055	1.051	1.046	1.044	1.043	1.042
4000	1.073	1.072	1.072	1.071	1.071	1.070	1.069	1.069	1.068	1.066	1.065	1.063	1.061	1.059	1.056	1.052	1.047	1.045	1.044	1.043
4100	1.075	1.074	1.073	1.073	1.073	1.072	1.071	1.070	1.069	1.068	1.067	1.065	1.063	1.062	1.058	1.054	1.049	1.046	1.045	1.044
4200	1.077	1.076	1.074	1.075	1.074	1.074	1.073	1.072	1.071	1.070	1.068	1.067	1.065	1.063	1.059	1.055	1.050	1.048	1.046	1.045
4300	1.079	1.078	1.076	1.077	1.076	1.076	1.075	1.074	1.073	1.071	1.070	1.068	1.066	1.064	1.060	1.056	1.051	1.049	1.047	1.046
4400	1.080	1.080	1.078	1.079	1.078	1.077	1.077	1.075	1.074	1.073	1.072	1.070	1.068	1.065	1.062	1.058	1.052	1.050	1.048	1.047
4500	1.082	1.082	1.080	1.081	1.080	1.079	1.078	1.077	1.076	1.075	1.073	1.072	1.069	1.067	1.063	1.059	1.053	1.051	1.049	1.048
4600	1.084	1.083	1.081	1.082	1.082	1.081	1.080	1.079	1.078	1.076	1.075	1.073	1.071	1.068	1.065	1.060	1.055	1.052	1.051	1.049
4700	1.086	1.085	1.083	1.084	1.083	1.083	1.082	1.081	1.080	1.078	1.077	1.075	1.072	1.070	1.066	1.063	1.056	1.053	1.052	1.050
4800	1.088	1.087	1.085	1.086	1.084	1.084	1.084	1.082	1.081	1.080	1.078	1.076	1.074	1.071	1.068	1.064	1.057	1.054	1.053	1.051
4900	1.090	1.089	1.087	1.088	1.087	1.086	1.085	1.084	1.083	1.082	1.080	1.078	1.076	1.073	1.069	1.064	1.058	1.055	1.054	1.052
5000	1.092	1.091	1.089	1.090	1.089	1.088	1.087	1.086	1.085	1.083	1.082	1.080	1.077	1.074	1.070	1.066	1.059	1.057	1.055	1.053
5100	1.093	1.093	1.090	1.092	1.091	1.090	1.089	1.088	1.086	1.085	1.083	1.081	1.079	1.076	1.072	1.067	1.061	1.058	1.056	1.054
5200	1.095	1.095	1.092	1.093	1.093	1.091	1.091	1.089	1.088	1.087	1.085	1.083	1.080	1.077	1.073	1.068	1.062	1.059	1.057	1.056
5300	1.097	1.097	1.094	1.095	1.094	1.093	1.092	1.091	1.090	1.088	1.087	1.085	1.082	1.079	1.075	1.071	1.063	1.060	1.058	1.057
5400	1.099	1.098	1.096	1.097	1.096	1.095	1.094	1.093	1.091	1.090	1.088	1.086	1.083	1.081	1.076	1.072	1.064	1.061	1.059	1.058
5500	1.101	1.100	1.098	1.099	1.098	1.097	1.096	1.095	1.093	1.092	1.090	1.088	1.085	1.082	1.078	1.074	1.067	1.063	1.060	1.059
5600	1.103	1.102	1.100	1.101	1.100	1.099	1.098	1.097	1.095	1.093	1.092	1.089	1.087	1.083	1.079	1.075	1.067	1.064	1.062	1.060
5700	1.105	1.104	1.102	1.103	1.102	1.101	1.100	1.098	1.097	1.095	1.093	1.091	1.088	1.085	1.081	1.076	1.068	1.065	1.063	1.061
5800	1.107	1.106	1.103	1.104	1.103	1.102	1.101	1.100	1.098	1.097	1.095	1.093	1.090	1.086	1.082	1.078	1.069	1.066	1.064	1.062
5900	1.109	1.108	1.105	1.106	1.105	1.104	1.103	1.102	1.100	1.099	1.097	1.094	1.091	1.088	1.083	1.079	1.070	1.067	1.065	1.063
6000	1.110	1.110	1.107	1.108	1.107	1.106	1.105	1.104	1.102	1.100	1.098	1.096	1.093	1.089	1.085	1.083	1.071	1.068	1.066	1.064
6100	1.112	1.112	1.109	1.110	1.109	1.108	1.107	1.105	1.104	1.102	1.100	1.097	1.094	1.091	1.086	1.084	1.073	1.069	1.067	1.065
6200	1.114	1.113	1.111	1.112	1.111	1.110	1.109	1.107	1.106	1.104	1.102	1.099	1.096	1.092	1.088	1.082	1.074	1.071	1.068	1.066
6300	1.116	1.115	1.113	1.114	1.113	1.112	1.111	1.109	1.107	1.105	1.103	1.101	1.098	1.094	1.089	1.083	1.075	1.072	1.069	1.067
6400	1.118	1.117	1.115	1.115	1.114	1.113	1.112	1.111	1.109	1.107	1.105	1.103	1.099	1.095	1.091	1.084	1.076	1.073	1.072	1.068
6500	1.120	1.119	1.116	1.117	1.116	1.115	1.114	1.112	1.111	1.109	1.107	1.104	1.101	1.097	1.092	1.086	1.077	1.074	1.073	1.069
6600	1.122	1.121	1.118	1.119	1.118	1.117	1.116	1.114	1.112	1.111	1.108	1.106	1.102	1.098	1.093	1.087	1.079	1.075	1.074	1.070
6700	1.124	1.123	1.120	1.121	1.120	1.119	1.118	1.116	1.114	1.112	1.110	1.107	1.104	1.101	1.096	1.088	1.080	1.077	1.073	1.071
6800	1.126	1.125	1.122	1.123	1.122	1.121	1.119	1.118	1.116	1.114	1.112	1.109	1.106	1.103	1.098	1.090	1.081	1.078	1.075	1.072
6900	1.127	1.127	1.124	1.125	1.124	1.123	1.121	1.120	1.118	1.116	1.113	1.110	1.107	1.104	1.099	1.092	1.082	1.080	1.076	1.073
7000	1.129	1.129	1.126	1.127	1.126	1.124	1.123	1.121	1.120	1.117	1.115	1.112	1.109	1.106	1.101	1.092	1.083	1.081	1.077	1.074
7100	1.131	1.130	1.128	1.128	1.127	1.126	1.125	1.123	1.121	1.119	1.117	1.114	1.110	1.108	1.101	1.094	1.085	1.082	1.078	1.075
7200	1.133	1.132	1.130	1.130	1.129	1.128	1.127	1.125	1.123	1.121	1.118	1.115	1.112	1.108	1.102	1.095	1.086	1.080	1.079	1.079

H

SIGMA = 98890.0

V-V$_c$ = 3

H / x	3.4	3.2	3.0	2.8	2.6	2.4	2.2	2.0	1.8	1.6	1.4	1.2	1.0	0.8	0.6	0.4	0.2	0.10	0.09	0.08
7300	1.135	1.134	1.133	1.132	1.131	1.130	1.128	1.127	1.125	1.123	1.120	1.117	1.113	1.109	1.104	1.096	1.087	1.081	1.080	1.080
7400	1.137	1.136	1.135	1.134	1.133	1.132	1.130	1.128	1.127	1.124	1.122	1.119	1.115	1.111	1.106	1.098	1.089	1.082	1.081	1.081
7500	1.139	1.138	1.137	1.136	1.135	1.133	1.132	1.130	1.128	1.126	1.123	1.120	1.117	1.112	1.106	1.099	1.089	1.083	1.082	1.082
7600	1.141	1.140	1.139	1.138	1.137	1.135	1.134	1.132	1.130	1.128	1.125	1.122	1.118	1.114	1.108	1.100	1.091	1.084	1.084	1.083
7700	1.143	1.142	1.141	1.140	1.138	1.137	1.136	1.134	1.132	1.130	1.127	1.124	1.120	1.115	1.109	1.102	1.092	1.085	1.085	1.083
7800	1.145	1.144	1.143	1.142	1.140	1.139	1.137	1.136	1.134	1.131	1.129	1.125	1.121	1.117	1.111	1.103	1.093	1.087	1.086	1.084
7900	1.147	1.146	1.145	1.143	1.142	1.141	1.139	1.137	1.135	1.133	1.130	1.127	1.123	1.118	1.112	1.105	1.095	1.089	1.087	1.085
8000	1.148	1.148	1.146	1.145	1.144	1.143	1.141	1.139	1.137	1.135	1.132	1.129	1.125	1.120	1.114	1.106	1.095	1.090	1.089	1.086
8100	1.150	1.149	1.148	1.147	1.146	1.144	1.143	1.141	1.139	1.136	1.134	1.130	1.126	1.121	1.115	1.107	1.097	1.090	1.089	1.087
8200	1.152	1.151	1.150	1.149	1.148	1.146	1.145	1.143	1.141	1.138	1.135	1.132	1.128	1.123	1.117	1.109	1.098	1.091	1.090	1.088
8300	1.154	1.153	1.152	1.151	1.150	1.148	1.146	1.145	1.142	1.140	1.137	1.134	1.129	1.124	1.118	1.110	1.099	1.093	1.091	1.090
8400	1.156	1.155	1.154	1.153	1.152	1.150	1.148	1.146	1.144	1.142	1.139	1.135	1.131	1.126	1.120	1.111	1.100	1.093	1.093	1.091
8500	1.158	1.157	1.156	1.155	1.153	1.152	1.150	1.148	1.146	1.143	1.140	1.137	1.133	1.127	1.121	1.113	1.102	1.094	1.094	1.092
8600	1.160	1.159	1.158	1.157	1.155	1.154	1.152	1.150	1.148	1.145	1.142	1.139	1.134	1.129	1.124	1.114	1.103	1.096	1.096	1.093
8700	1.162	1.161	1.160	1.159	1.157	1.156	1.154	1.152	1.150	1.147	1.144	1.140	1.136	1.131	1.124	1.115	1.104	1.097	1.096	1.094
8800	1.164	1.163	1.162	1.160	1.159	1.157	1.156	1.154	1.152	1.149	1.145	1.142	1.137	1.132	1.126	1.117	1.105	1.098	1.097	1.095
8900	1.166	1.165	1.164	1.162	1.161	1.159	1.157	1.155	1.153	1.150	1.147	1.144	1.139	1.135	1.128	1.119	1.108	1.100	1.097	1.096
9000	1.168	1.167	1.165	1.164	1.163	1.161	1.159	1.157	1.155	1.152	1.149	1.145	1.141	1.135	1.129	1.121	1.108	1.100	1.099	1.098
9100	1.170	1.169	1.167	1.166	1.165	1.163	1.161	1.159	1.157	1.154	1.151	1.147	1.142	1.137	1.131	1.122	1.110	1.102	1.101	1.099
9200	1.172	1.171	1.169	1.168	1.166	1.165	1.163	1.161	1.158	1.156	1.152	1.148	1.144	1.138	1.132	1.123	1.111	1.103	1.102	1.100
9300	1.174	1.172	1.171	1.170	1.168	1.167	1.165	1.163	1.160	1.157	1.154	1.150	1.145	1.140	1.133	1.123	1.112	1.103	1.103	1.102
9400	1.175	1.174	1.173	1.172	1.170	1.169	1.167	1.164	1.162	1.159	1.156	1.152	1.147	1.141	1.135	1.125	1.113	1.105	1.104	1.103
9500	1.177	1.176	1.175	1.174	1.172	1.170	1.168	1.166	1.164	1.161	1.157	1.153	1.149	1.143	1.136	1.126	1.114	1.106	1.105	1.103
9600	1.179	1.178	1.177	1.176	1.174	1.172	1.170	1.168	1.166	1.163	1.159	1.155	1.150	1.144	1.137	1.128	1.115	1.107	1.106	1.104
9700	1.181	1.180	1.179	1.177	1.176	1.174	1.172	1.170	1.167	1.164	1.161	1.157	1.152	1.146	1.139	1.129	1.116	1.108	1.107	1.105
9800	1.183	1.182	1.181	1.179	1.178	1.176	1.174	1.172	1.169	1.166	1.163	1.159	1.154	1.148	1.140	1.130	1.117	1.109	1.108	1.106
9900	1.185	1.184	1.183	1.181	1.180	1.178	1.176	1.174	1.171	1.168	1.164	1.160	1.155	1.149	1.141	1.132	1.118	1.110	1.109	1.107
10000	1.187	1.186	1.185	1.183	1.182	1.180	1.178	1.175	1.173	1.170	1.166	1.162	1.157	1.151	1.143	1.133	1.120	1.111	1.110	1.108
10100	1.189	1.188	1.187	1.185	1.183	1.182	1.180	1.177	1.174	1.171	1.168	1.164	1.158	1.152	1.144	1.134	1.121	1.112	1.111	1.110
10200	1.191	1.190	1.188	1.187	1.185	1.183	1.181	1.179	1.176	1.173	1.170	1.165	1.160	1.154	1.146	1.136	1.123	1.113	1.113	1.111
10300	1.193	1.192	1.190	1.189	1.187	1.185	1.183	1.181	1.178	1.175	1.171	1.167	1.162	1.155	1.147	1.137	1.123	1.115	1.114	1.112
10400	1.195	1.194	1.192	1.191	1.189	1.187	1.185	1.183	1.180	1.177	1.173	1.169	1.163	1.157	1.149	1.138	1.125	1.116	1.115	1.113
10500	1.197	1.196	1.194	1.193	1.191	1.189	1.187	1.184	1.182	1.178	1.175	1.170	1.165	1.158	1.150	1.140	1.126	1.117	1.116	1.114
10600	1.199	1.198	1.196	1.195	1.193	1.191	1.189	1.186	1.183	1.180	1.176	1.172	1.167	1.160	1.152	1.141	1.127	1.118	1.117	1.115
10700	1.201	1.200	1.198	1.197	1.195	1.193	1.191	1.188	1.185	1.182	1.178	1.174	1.168	1.161	1.153	1.142	1.128	1.119	1.118	1.116
10800	1.203	1.202	1.200	1.198	1.197	1.195	1.192	1.190	1.187	1.184	1.182	1.175	1.170	1.163	1.155	1.144	1.129	1.120	1.119	1.117
10900	1.205	1.203	1.202	1.200	1.199	1.197	1.194	1.192	1.189	1.186	1.182	1.177	1.173	1.166	1.158	1.147	1.132	1.122	1.121	1.119
11000	1.207	1.205	1.204	1.202	1.201	1.198	1.196	1.194	1.191	1.187	1.183	1.180	1.175	1.168	1.159	1.148	1.133	1.124	1.123	1.120
11100	1.209	1.207	1.206	1.204	1.202	1.200	1.198	1.195	1.193	1.189	1.185	1.182	1.176	1.169	1.161	1.149	1.134	1.125	1.124	1.121
11200	1.211	1.209	1.208	1.206	1.204	1.202	1.200	1.197	1.194	1.191	1.187	1.182	1.178	1.171	1.162	1.151	1.135	1.126	1.125	1.123
11300	1.213	1.211	1.210	1.208	1.206	1.204	1.202	1.199	1.196	1.193	1.189	1.184	1.178	1.171	1.162	1.151	1.135	1.126	1.125	1.124
11400	1.215	1.213	1.212	1.210	1.208	1.206	1.204	1.201	1.198	1.194	1.190	1.185	1.180	1.172	1.163	1.152	1.137	1.127	1.126	1.125

SIGMA = 98890.0

$V - V_c = 3$

H	0.08	0.09	0.10	0.2	0.4	0.6	0.8	1.0	1.2	1.4	1.6	1.8	2.0	2.2	2.4	2.6	2.8	3.0	3.2	3.4
11500	1.126	1.127	1.128	1.138	1.153	1.165	1.174	1.181	1.187	1.192	1.196	1.200	1.203	1.206	1.208	1.210	1.212	1.214	1.215	1.217
11600	1.127	1.128	1.129	1.139	1.155	1.166	1.176	1.183	1.189	1.194	1.198	1.202	1.205	1.207	1.210	1.212	1.214	1.216	1.217	1.219
11700	1.128	1.129	1.130	1.140	1.156	1.168	1.177	1.184	1.190	1.195	1.200	1.203	1.207	1.209	1.212	1.214	1.216	1.218	1.219	1.221
11800	1.129	1.130	1.131	1.142	1.157	1.169	1.179	1.186	1.192	1.197	1.202	1.205	1.208	1.211	1.214	1.216	1.218	1.219	1.221	1.223
11900	1.130	1.131	1.133	1.143	1.159	1.171	1.180	1.188	1.194	1.199	1.203	1.207	1.210	1.213	1.216	1.218	1.220	1.221	1.223	1.225
12000	1.131	1.133	1.134	1.144	1.160	1.172	1.182	1.189	1.196	1.201	1.205	1.209	1.212	1.215	1.217	1.220	1.222	1.223	1.225	1.226
12100	1.131	1.134	1.135	1.145	1.162	1.174	1.183	1.191	1.197	1.202	1.207	1.211	1.214	1.217	1.219	1.221	1.223	1.225	1.227	1.228
12200	1.133	1.135	1.136	1.146	1.163	1.175	1.185	1.193	1.199	1.204	1.209	1.213	1.216	1.219	1.221	1.223	1.225	1.227	1.229	1.230
12300	1.134	1.136	1.137	1.148	1.164	1.177	1.187	1.194	1.201	1.206	1.211	1.215	1.218	1.221	1.223	1.225	1.227	1.229	1.231	1.232
12400	1.135	1.137	1.139	1.149	1.166	1.178	1.188	1.196	1.203	1.208	1.213	1.216	1.219	1.222	1.225	1.227	1.229	1.231	1.233	1.234
12500	1.136	1.138	1.140	1.150	1.167	1.180	1.190	1.198	1.204	1.210	1.215	1.218	1.221	1.224	1.227	1.229	1.231	1.233	1.235	1.236
12600	1.137	1.139	1.141	1.151	1.168	1.181	1.191	1.199	1.206	1.211	1.216	1.220	1.223	1.226	1.229	1.231	1.233	1.235	1.237	1.238
12700	1.138	1.140	1.142	1.153	1.170	1.183	1.193	1.201	1.208	1.213	1.218	1.222	1.225	1.228	1.231	1.233	1.235	1.237	1.239	1.240
12800	1.139	1.142	1.143	1.154	1.171	1.184	1.194	1.203	1.209	1.215	1.220	1.224	1.227	1.230	1.233	1.235	1.237	1.239	1.241	1.242
12900	1.140	1.143	1.144	1.155	1.173	1.186	1.196	1.204	1.211	1.216	1.222	1.225	1.229	1.232	1.235	1.237	1.239	1.241	1.243	1.244
13000	1.141	1.144	1.145	1.156	1.174	1.187	1.198	1.206	1.213	1.218	1.224	1.227	1.231	1.234	1.236	1.239	1.241	1.243	1.245	1.246
13100	1.142	1.145	1.147	1.157	1.175	1.189	1.199	1.207	1.214	1.220	1.225	1.229	1.233	1.236	1.238	1.241	1.243	1.245	1.247	1.248
13200	1.143	1.146	1.148	1.159	1.177	1.190	1.201	1.209	1.216	1.222	1.227	1.231	1.234	1.237	1.240	1.243	1.245	1.247	1.249	1.250
13300	1.144	1.147	1.149	1.160	1.178	1.192	1.202	1.211	1.218	1.223	1.229	1.233	1.236	1.239	1.242	1.245	1.247	1.249	1.251	1.252
13400	1.145	1.148	1.150	1.161	1.179	1.193	1.204	1.212	1.219	1.225	1.231	1.235	1.238	1.241	1.244	1.247	1.249	1.251	1.253	1.254
13500	1.146	1.149	1.151	1.162	1.181	1.195	1.205	1.214	1.221	1.227	1.232	1.236	1.240	1.243	1.246	1.249	1.251	1.253	1.255	1.256
13600	1.147	1.150	1.152	1.164	1.182	1.196	1.207	1.216	1.223	1.229	1.234	1.238	1.242	1.245	1.248	1.250	1.253	1.255	1.257	1.258
13700	1.148	1.151	1.153	1.165	1.184	1.198	1.209	1.217	1.225	1.230	1.236	1.240	1.244	1.247	1.250	1.252	1.255	1.257	1.259	1.260
13800	1.149	1.153	1.154	1.166	1.185	1.199	1.210	1.219	1.226	1.232	1.238	1.242	1.246	1.249	1.252	1.254	1.257	1.259	1.261	1.262
13900	1.150	1.154	1.156	1.167	1.186	1.201	1.212	1.221	1.228	1.234	1.240	1.244	1.247	1.251	1.254	1.256	1.259	1.261	1.263	1.264
14000	1.151	1.155	1.157	1.169	1.188	1.202	1.213	1.222	1.230	1.236	1.241	1.245	1.249	1.253	1.256	1.258	1.261	1.263	1.265	1.266
14100	1.152	1.156	1.159	1.170	1.189	1.204	1.215	1.224	1.231	1.237	1.243	1.247	1.251	1.254	1.258	1.260	1.263	1.265	1.267	1.268
14200	1.154	1.157	1.160	1.171	1.191	1.205	1.217	1.226	1.233	1.239	1.245	1.249	1.253	1.256	1.259	1.262	1.265	1.267	1.269	1.270
14300	1.155	1.158	1.161	1.172	1.192	1.207	1.218	1.227	1.235	1.241	1.247	1.251	1.255	1.258	1.261	1.264	1.267	1.269	1.271	1.273
14400	1.156	1.159	1.162	1.173	1.193	1.208	1.220	1.229	1.237	1.243	1.249	1.253	1.257	1.260	1.263	1.266	1.269	1.271	1.273	1.275
14500	1.157	1.161	1.163	1.175	1.195	1.210	1.221	1.231	1.238	1.245	1.250	1.255	1.258	1.262	1.265	1.268	1.271	1.273	1.275	1.277
14600	1.159	1.162	1.164	1.176	1.196	1.211	1.223	1.232	1.240	1.246	1.252	1.256	1.260	1.264	1.267	1.270	1.272	1.275	1.277	1.279
14700	1.160	1.163	1.165	1.177	1.197	1.213	1.225	1.234	1.242	1.248	1.254	1.258	1.262	1.266	1.269	1.272	1.274	1.277	1.279	1.281
14800	1.162	1.164	1.166	1.178	1.199	1.214	1.226	1.236	1.244	1.250	1.256	1.260	1.264	1.268	1.271	1.274	1.276	1.279	1.281	1.283
14900	1.163	1.165	1.167	1.180	1.200	1.216	1.228	1.237	1.245	1.252	1.257	1.262	1.266	1.270	1.273	1.276	1.278	1.281	1.283	1.285
15000	1.165	1.166	1.168	1.181	1.202	1.217	1.229	1.239	1.247	1.254	1.259	1.264	1.268	1.272	1.275	1.278	1.280	1.283	1.285	1.287

x

SIGMA = 55630.0

$V - V_c = 4$

H x	3.4	3.2	3.0	2.8	2.6	2.4	2.2	2.0	1.8	1.6	1.4	1.2	1.0	0.8	0.6	0.4	0.2	0.10	0.09	0.08
0	1.000	1.000	1.000	1.000	1.000	1.000	1.000	1.000	1.000	1.000	1.000	1.000	1.000	1.000	1.000	1.000	1.000	1.000	1.000	1.000
20	1.001	1.001	1.001	1.001	1.001	1.001	1.001	1.001	1.001	1.001	1.001	1.001	1.001	1.001	1.001	1.000	1.000	1.000	1.000	1.000
40	1.001	1.001	1.001	1.001	1.002	1.001	1.001	1.001	1.001	1.001	1.001	1.001	1.001	1.001	1.001	1.001	1.001	1.001	1.001	1.000
60	1.002	1.002	1.002	1.002	1.002	1.002	1.002	1.002	1.002	1.002	1.002	1.002	1.002	1.002	1.001	1.001	1.001	1.001	1.001	1.001
80	1.003	1.003	1.003	1.003	1.003	1.003	1.003	1.003	1.002	1.002	1.002	1.002	1.002	1.002	1.002	1.002	1.002	1.002	1.002	1.001
100	1.003	1.003	1.004	1.004	1.004	1.003	1.003	1.003	1.003	1.003	1.003	1.003	1.003	1.003	1.002	1.002	1.003	1.002	1.002	1.002
120	1.004	1.004	1.004	1.004	1.004	1.004	1.004	1.004	1.003	1.003	1.003	1.003	1.003	1.003	1.003	1.003	1.003	1.003	1.002	1.002
140	1.004	1.004	1.004	1.005	1.005	1.005	1.004	1.004	1.004	1.004	1.004	1.004	1.004	1.004	1.003	1.003	1.004	1.003	1.003	1.002
160	1.005	1.005	1.005	1.006	1.006	1.005	1.005	1.005	1.005	1.005	1.005	1.005	1.005	1.005	1.004	1.004	1.004	1.004	1.003	1.003
180	1.006	1.006	1.006	1.006	1.006	1.006	1.005	1.005	1.005	1.005	1.005	1.005	1.005	1.005	1.004	1.004	1.004	1.004	1.004	1.003
200	1.006	1.006	1.006	1.007	1.007	1.006	1.006	1.006	1.006	1.006	1.006	1.006	1.005	1.005	1.005	1.005	1.005	1.004	1.004	1.003
220	1.007	1.007	1.008	1.007	1.007	1.007	1.007	1.007	1.007	1.006	1.006	1.006	1.006	1.006	1.006	1.005	1.005	1.005	1.004	1.004
240	1.008	1.008	1.008	1.008	1.008	1.007	1.008	1.007	1.007	1.007	1.007	1.007	1.006	1.006	1.006	1.006	1.005	1.005	1.005	1.004
260	1.008	1.008	1.009	1.008	1.008	1.009	1.008	1.008	1.008	1.008	1.008	1.007	1.007	1.007	1.007	1.006	1.005	1.006	1.005	1.005
280	1.009	1.009	1.009	1.009	1.009	1.009	1.009	1.009	1.008	1.008	1.008	1.008	1.008	1.008	1.007	1.006	1.006	1.006	1.006	1.005
300	1.010	1.010	1.010	1.010	1.010	1.010	1.010	1.010	1.009	1.009	1.009	1.009	1.009	1.009	1.008	1.007	1.006	1.006	1.006	1.005
320	1.010	1.011	1.011	1.011	1.011	1.010	1.010	1.010	1.010	1.009	1.010	1.010	1.009	1.009	1.008	1.008	1.007	1.007	1.006	1.006
340	1.011	1.011	1.011	1.011	1.011	1.012	1.011	1.010	1.010	1.010	1.010	1.011	1.011	1.010	1.009	1.008	1.007	1.007	1.007	1.006
360	1.011	1.011	1.013	1.011	1.011	1.012	1.012	1.011	1.010	1.010	1.010	1.011	1.010	1.009	1.009	1.008	1.008	1.007	1.007	1.007
380	1.012	1.012	1.013	1.012	1.012	1.012	1.012	1.012	1.011	1.011	1.011	1.011	1.010	1.011	1.009	1.009	1.008	1.007	1.007	1.007
400	1.012	1.012	1.013	1.012	1.012	1.013	1.013	1.012	1.012	1.012	1.012	1.011	1.012	1.011	1.009	1.010	1.008	1.008	1.008	1.007
420	1.013	1.013	1.013	1.013	1.013	1.013	1.012	1.013	1.012	1.012	1.012	1.012	1.011	1.011	1.010	1.010	1.009	1.009	1.008	1.008
440	1.013	1.013	1.014	1.013	1.014	1.014	1.014	1.013	1.013	1.013	1.013	1.012	1.014	1.013	1.011	1.010	1.009	1.009	1.009	1.008
460	1.014	1.014	1.015	1.014	1.014	1.014	1.014	1.014	1.013	1.015	1.013	1.013	1.014	1.012	1.011	1.011	1.010	1.010	1.009	1.008
480	1.015	1.015	1.015	1.015	1.016	1.014	1.014	1.015	1.014	1.014	1.014	1.014	1.013	1.012	1.012	1.011	1.011	1.010	1.009	1.009
500	1.016	1.015	1.016	1.016	1.016	1.015	1.015	1.014	1.015	1.015	1.015	1.016	1.015	1.016	1.016	1.012	1.011	1.011	1.010	1.009
520	1.017	1.016	1.017	1.016	1.016	1.017	1.016	1.015	1.015	1.015	1.015	1.017	1.016	1.013	1.013	1.012	1.012	1.011	1.010	1.010
540	1.017	1.017	1.018	1.017	1.017	1.017	1.017	1.016	1.016	1.016	1.016	1.014	1.014	1.014	1.013	1.013	1.011	1.011	1.011	1.010
560	1.018	1.018	1.018	1.018	1.017	1.018	1.017	1.017	1.017	1.017	1.016	1.016	1.015	1.015	1.014	1.013	1.012	1.012	1.011	1.011
580	1.019	1.018	1.019	1.019	1.018	1.018	1.018	1.017	1.018	1.017	1.017	1.017	1.016	1.015	1.015	1.014	1.012	1.013	1.011	1.011
600	1.019	1.019	1.020	1.020	1.019	1.019	1.019	1.020	1.018	1.018	1.018	1.018	1.017	1.016	1.015	1.014	1.014	1.013	1.012	1.011
620	1.020	1.020	1.020	1.019	1.020	1.020	1.019	1.019	1.019	1.019	1.018	1.018	1.016	1.016	1.016	1.015	1.013	1.013	1.012	1.012
640	1.020	1.020	1.021	1.021	1.021	1.021	1.020	1.020	1.020	1.019	1.019	1.019	1.017	1.017	1.016	1.015	1.014	1.014	1.013	1.012
660	1.021	1.021	1.022	1.022	1.022	1.021	1.021	1.021	1.020	1.020	1.020	1.020	1.018	1.017	1.017	1.016	1.015	1.014	1.013	1.012
680	1.022	1.022	1.022	1.023	1.021	1.022	1.022	1.021	1.021	1.021	1.021	1.018	1.019	1.018	1.017	1.015	1.014	1.015	1.013	1.013
700	1.023	1.022	1.023	1.023	1.022	1.022	1.022	1.022	1.022	1.021	1.021	1.019	1.020	1.019	1.018	1.017	1.015	1.015	1.014	1.013
720	1.024	1.023	1.023	1.023	1.023	1.022	1.022	1.023	1.023	1.022	1.022	1.020	1.020	1.020	1.018	1.018	1.016	1.015	1.014	1.013
740	1.024	1.024	1.024	1.024	1.024	1.023	1.023	1.023	1.024	1.022	1.022	1.021	1.021	1.020	1.019	1.017	1.016	1.016	1.014	1.014
760	1.024	1.024	1.025	1.024	1.024	1.024	1.024	1.024	1.023	1.023	1.023	1.021	1.021	1.020	1.019	1.018	1.016	1.016	1.015	1.014
780	1.025	1.025	1.025	1.025	1.025	1.024	1.024	1.024	1.023	1.024	1.023	1.022	1.021	1.020	1.019	1.019	1.016	1.016	1.015	1.015
800	1.026	1.025	1.025	1.025	1.025	1.025	1.024	1.024	1.024	1.023	1.023	1.022	1.022	1.021	1.020	1.019	1.017	1.017	1.016	1.015

SIGMA = 55630.0

$v - V_c = 4$

H \ x	3.4	3.2	3.0	2.8	2.6	2.4	2.2	2.0	1.8	1.6	1.4	1.2	1.0	0.8	0.6	0.4	0.2	0.10	0.09	0.08
820	1.026	1.026	1.026	1.026	1.026	1.025	1.025	1.025	1.024	1.024	1.023	1.023	1.022	1.021	1.020	1.019	1.017	1.016	1.016	1.016
840	1.027	1.027	1.027	1.026	1.026	1.026	1.026	1.025	1.025	1.025	1.024	1.023	1.023	1.022	1.021	1.019	1.018	1.017	1.016	1.016
860	1.028	1.027	1.027	1.027	1.027	1.027	1.026	1.026	1.026	1.025	1.025	1.024	1.023	1.022	1.021	1.020	1.018	1.017	1.017	1.017
880	1.028	1.028	1.028	1.028	1.027	1.027	1.027	1.026	1.026	1.026	1.025	1.025	1.024	1.023	1.022	1.020	1.018	1.017	1.017	1.017
900	1.029	1.029	1.029	1.028	1.028	1.028	1.027	1.027	1.027	1.027	1.026	1.025	1.024	1.024	1.023	1.021	1.019	1.018	1.018	1.018
920	1.029	1.029	1.029	1.029	1.028	1.028	1.028	1.028	1.027	1.027	1.027	1.026	1.025	1.024	1.023	1.021	1.020	1.018	1.018	1.018
940	1.030	1.030	1.030	1.029	1.029	1.029	1.029	1.028	1.028	1.028	1.027	1.027	1.025	1.025	1.024	1.022	1.020	1.019	1.018	1.018
960	1.031	1.031	1.030	1.030	1.030	1.029	1.029	1.029	1.029	1.028	1.028	1.027	1.026	1.025	1.024	1.022	1.021	1.019	1.019	1.019
980	1.031	1.031	1.031	1.031	1.030	1.030	1.030	1.030	1.029	1.029	1.028	1.028	1.027	1.026	1.025	1.023	1.021	1.019	1.019	1.019
1000	1.032	1.032	1.032	1.031	1.031	1.031	1.030	1.030	1.030	1.029	1.029	1.028	1.027	1.027	1.025	1.023	1.021	1.020	1.020	1.019
1020	1.033	1.033	1.032	1.032	1.032	1.031	1.031	1.031	1.030	1.030	1.029	1.029	1.028	1.027	1.026	1.024	1.022	1.020	1.020	1.020
1040	1.033	1.033	1.034	1.032	1.032	1.032	1.031	1.031	1.031	1.030	1.030	1.029	1.028	1.028	1.026	1.024	1.022	1.020	1.020	1.020
1060	1.034	1.034	1.034	1.033	1.033	1.033	1.032	1.032	1.032	1.031	1.030	1.030	1.029	1.028	1.027	1.025	1.023	1.021	1.021	1.020
1080	1.035	1.034	1.035	1.034	1.034	1.033	1.033	1.032	1.032	1.032	1.031	1.030	1.029	1.029	1.027	1.025	1.023	1.021	1.021	1.021
1100	1.035	1.035	1.035	1.034	1.034	1.034	1.033	1.033	1.033	1.032	1.032	1.031	1.030	1.029	1.028	1.026	1.023	1.022	1.021	1.021
1120	1.036	1.036	1.036	1.035	1.035	1.035	1.034	1.034	1.033	1.033	1.032	1.032	1.031	1.030	1.028	1.026	1.024	1.022	1.022	1.021
1140	1.037	1.036	1.037	1.036	1.035	1.035	1.035	1.034	1.034	1.033	1.033	1.032	1.031	1.030	1.029	1.027	1.024	1.022	1.022	1.022
1160	1.038	1.037	1.037	1.036	1.036	1.036	1.035	1.035	1.035	1.034	1.033	1.033	1.032	1.031	1.029	1.027	1.025	1.023	1.023	1.022
1180	1.038	1.038	1.038	1.037	1.037	1.037	1.036	1.036	1.035	1.035	1.034	1.033	1.032	1.032	1.030	1.027	1.025	1.023	1.023	1.023
1200	1.039	1.038	1.038	1.038	1.037	1.037	1.037	1.036	1.036	1.035	1.034	1.034	1.033	1.032	1.030	1.028	1.026	1.024	1.023	1.023
1220	1.039	1.039	1.039	1.038	1.038	1.038	1.037	1.037	1.036	1.036	1.035	1.034	1.033	1.033	1.031	1.028	1.026	1.024	1.024	1.023
1240	1.040	1.040	1.040	1.039	1.039	1.038	1.038	1.037	1.037	1.036	1.035	1.035	1.034	1.033	1.031	1.029	1.027	1.024	1.024	1.024
1260	1.041	1.040	1.040	1.040	1.039	1.039	1.039	1.038	1.038	1.037	1.036	1.036	1.034	1.034	1.032	1.029	1.027	1.025	1.025	1.024
1280	1.041	1.041	1.041	1.040	1.040	1.040	1.039	1.039	1.038	1.037	1.037	1.036	1.035	1.034	1.032	1.030	1.027	1.025	1.025	1.024
1300	1.042	1.042	1.041	1.041	1.040	1.040	1.040	1.039	1.039	1.038	1.037	1.037	1.035	1.035	1.033	1.030	1.028	1.026	1.025	1.025
1320	1.042	1.042	1.042	1.041	1.041	1.041	1.040	1.040	1.039	1.038	1.038	1.037	1.036	1.035	1.033	1.031	1.028	1.026	1.026	1.025
1340	1.043	1.043	1.043	1.042	1.042	1.042	1.041	1.040	1.040	1.039	1.038	1.038	1.036	1.036	1.034	1.031	1.029	1.027	1.026	1.026
1360	1.044	1.043	1.043	1.042	1.042	1.042	1.041	1.041	1.040	1.039	1.039	1.038	1.037	1.036	1.034	1.032	1.029	1.027	1.027	1.026
1380	1.044	1.044	1.044	1.043	1.043	1.043	1.042	1.041	1.041	1.040	1.039	1.039	1.038	1.037	1.035	1.032	1.029	1.027	1.027	1.027
1400	1.045	1.045	1.045	1.044	1.044	1.043	1.043	1.042	1.042	1.040	1.040	1.040	1.038	1.038	1.035	1.033	1.030	1.028	1.027	1.027
1420	1.046	1.045	1.045	1.044	1.044	1.044	1.043	1.043	1.042	1.041	1.041	1.040	1.039	1.038	1.036	1.033	1.030	1.028	1.028	1.027
1440	1.046	1.046	1.046	1.045	1.045	1.045	1.044	1.044	1.043	1.042	1.043	1.041	1.039	1.039	1.036	1.034	1.031	1.028	1.028	1.028
1460	1.047	1.047	1.046	1.046	1.046	1.046	1.045	1.044	1.044	1.043	1.044	1.042	1.040	1.040	1.037	1.035	1.031	1.029	1.028	1.028
1480	1.048	1.047	1.047	1.046	1.046	1.046	1.046	1.045	1.045	1.044	1.045	1.043	1.041	1.040	1.037	1.035	1.032	1.029	1.029	1.028
1500	1.048	1.048	1.048	1.047	1.047	1.047	1.046	1.045	1.045	1.044	1.045	1.043	1.041	1.041	1.038	1.036	1.032	1.030	1.029	1.029
1520	1.049	1.049	1.048	1.048	1.048	1.048	1.047	1.046	1.046	1.045	1.046	1.044	1.042	1.042	1.039	1.036	1.033	1.030	1.030	1.029
1540	1.050	1.049	1.050	1.049	1.048	1.048	1.047	1.047	1.047	1.046	1.046	1.045	1.043	1.042	1.039	1.037	1.033	1.031	1.030	1.030
1560	1.050	1.050	1.050	1.049	1.049	1.049	1.048	1.047	1.047	1.046	1.047	1.045	1.044	1.043	1.040	1.037	1.034	1.031	1.030	1.030
1580	1.051	1.051	1.051	1.050	1.050	1.049	1.049	1.048	1.048	1.047	1.047	1.046	1.044	1.043	1.040	1.038	1.034	1.031	1.031	1.030
1600	1.051	1.051	1.051	1.051	1.050	1.050	1.049	1.048	1.048	1.047	1.048	1.046	1.044	1.044	1.040	1.038	1.034	1.031	1.031	1.031
1620	1.052	1.052	1.052	1.051	1.051	1.050	1.050	1.049	1.049	1.048	1.048	1.046	1.045	1.044	1.041	1.038	1.034	1.032	1.032	1.031
1640	1.053	1.053	1.052	1.052	1.051	1.051	1.050	1.050	1.049	1.048	1.047	1.046	1.045	1.043	1.041	1.038	1.035	1.032	1.032	1.032

SIGMA = 55630.0

$V-V_c = 4$

H (x)	3.4	3.2	3.0	2.8	2.6	2.4	2.2	2.0	1.8	1.6	1.4	1.2	1.0	0.8	0.6	0.4	0.2	0.10	0.09	0.08
1660	1.054	1.053	1.053	1.052	1.052	1.052	1.051	1.050	1.050	1.049	1.048	1.047	1.045	1.043	1.041	1.039	1.035	1.033	1.032	1.032
1680	1.054	1.054	1.054	1.052	1.053	1.053	1.052	1.051	1.050	1.049	1.048	1.047	1.046	1.044	1.042	1.039	1.035	1.033	1.033	1.033
1700	1.055	1.055	1.054	1.054	1.053	1.053	1.052	1.052	1.051	1.050	1.049	1.048	1.046	1.045	1.042	1.040	1.036	1.033	1.033	1.033
1720	1.056	1.055	1.055	1.054	1.054	1.053	1.053	1.052	1.051	1.051	1.050	1.048	1.047	1.045	1.043	1.040	1.036	1.034	1.034	1.033
1740	1.056	1.056	1.055	1.055	1.054	1.054	1.053	1.053	1.052	1.051	1.050	1.049	1.047	1.046	1.043	1.040	1.037	1.034	1.034	1.034
1760	1.057	1.057	1.056	1.056	1.055	1.055	1.054	1.053	1.053	1.052	1.050	1.049	1.047	1.046	1.044	1.041	1.037	1.035	1.034	1.034
1780	1.058	1.057	1.057	1.056	1.056	1.055	1.055	1.054	1.053	1.052	1.051	1.050	1.049	1.047	1.044	1.042	1.038	1.035	1.035	1.034
1800	1.058	1.058	1.057	1.057	1.056	1.056	1.055	1.055	1.054	1.053	1.052	1.051	1.049	1.047	1.045	1.042	1.038	1.035	1.035	1.035
1820	1.059	1.059	1.058	1.058	1.057	1.057	1.056	1.055	1.055	1.054	1.052	1.051	1.049	1.048	1.045	1.043	1.038	1.036	1.035	1.035
1840	1.059	1.059	1.059	1.058	1.058	1.057	1.057	1.056	1.055	1.055	1.053	1.052	1.050	1.048	1.046	1.043	1.039	1.036	1.036	1.035
1860	1.060	1.060	1.060	1.059	1.058	1.058	1.057	1.057	1.056	1.055	1.054	1.052	1.051	1.049	1.046	1.044	1.039	1.037	1.036	1.036
1880	1.061	1.060	1.060	1.060	1.059	1.058	1.058	1.056	1.056	1.055	1.054	1.053	1.051	1.049	1.047	1.044	1.040	1.037	1.037	1.036
1900	1.061	1.061	1.061	1.060	1.060	1.059	1.058	1.058	1.057	1.056	1.055	1.053	1.052	1.050	1.047	1.045	1.040	1.038	1.037	1.037
1920	1.062	1.062	1.061	1.061	1.060	1.060	1.059	1.058	1.057	1.056	1.055	1.054	1.052	1.050	1.048	1.045	1.040	1.038	1.038	1.037
1940	1.063	1.062	1.062	1.061	1.061	1.060	1.060	1.059	1.058	1.057	1.056	1.055	1.053	1.051	1.048	1.046	1.041	1.039	1.038	1.038
1960	1.063	1.063	1.062	1.062	1.061	1.061	1.060	1.059	1.059	1.058	1.057	1.055	1.053	1.052	1.049	1.046	1.042	1.039	1.039	1.038
1980	1.064	1.064	1.063	1.063	1.062	1.062	1.061	1.060	1.059	1.058	1.057	1.056	1.054	1.052	1.050	1.047	1.042	1.039	1.039	1.039
2000	1.065	1.064	1.064	1.063	1.063	1.062	1.061	1.061	1.060	1.059	1.058	1.056	1.055	1.053	1.050	1.047	1.043	1.040	1.039	1.039
2020	1.065	1.065	1.064	1.064	1.063	1.063	1.062	1.061	1.061	1.059	1.058	1.057	1.055	1.053	1.051	1.048	1.043	1.040	1.040	1.039
2040	1.066	1.065	1.065	1.065	1.064	1.064	1.063	1.062	1.061	1.060	1.059	1.057	1.056	1.054	1.051	1.048	1.044	1.041	1.040	1.040
2060	1.067	1.066	1.066	1.065	1.065	1.064	1.063	1.063	1.062	1.061	1.060	1.058	1.056	1.055	1.052	1.049	1.044	1.041	1.041	1.040
2080	1.067	1.067	1.066	1.066	1.065	1.065	1.064	1.063	1.062	1.061	1.060	1.059	1.057	1.055	1.052	1.049	1.045	1.042	1.041	1.041
2100	1.068	1.068	1.067	1.067	1.066	1.065	1.065	1.064	1.063	1.062	1.061	1.059	1.057	1.056	1.053	1.050	1.045	1.043	1.042	1.041
2120	1.069	1.068	1.068	1.067	1.066	1.066	1.065	1.064	1.064	1.062	1.061	1.060	1.058	1.056	1.053	1.050	1.046	1.043	1.042	1.041
2140	1.069	1.069	1.068	1.067	1.067	1.066	1.065	1.065	1.064	1.063	1.062	1.060	1.058	1.057	1.054	1.051	1.046	1.044	1.043	1.042
2160	1.070	1.070	1.069	1.068	1.067	1.067	1.066	1.065	1.065	1.064	1.062	1.061	1.059	1.057	1.054	1.051	1.047	1.044	1.043	1.042
2180	1.071	1.070	1.070	1.069	1.069	1.068	1.067	1.066	1.065	1.064	1.063	1.061	1.060	1.057	1.055	1.052	1.047	1.044	1.044	1.043
2200	1.071	1.071	1.070	1.070	1.069	1.068	1.067	1.066	1.066	1.065	1.063	1.062	1.060	1.058	1.055	1.052	1.048	1.045	1.044	1.043
2220	1.072	1.072	1.071	1.070	1.069	1.069	1.068	1.067	1.067	1.065	1.064	1.062	1.061	1.059	1.056	1.053	1.048	1.046	1.044	1.044
2240	1.073	1.072	1.071	1.071	1.070	1.069	1.069	1.068	1.067	1.066	1.065	1.063	1.061	1.059	1.056	1.053	1.049	1.046	1.045	1.044
2260	1.073	1.073	1.072	1.072	1.071	1.070	1.070	1.068	1.068	1.067	1.065	1.064	1.062	1.060	1.057	1.053	1.049	1.047	1.045	1.044
2280	1.074	1.073	1.073	1.072	1.072	1.071	1.070	1.069	1.068	1.067	1.066	1.064	1.062	1.060	1.057	1.054	1.049	1.047	1.046	1.045
2300	1.075	1.074	1.074	1.073	1.072	1.072	1.071	1.070	1.069	1.068	1.066	1.065	1.063	1.061	1.058	1.054	1.050	1.048	1.046	1.045
2320	1.075	1.075	1.074	1.073	1.073	1.072	1.071	1.071	1.070	1.068	1.067	1.065	1.063	1.062	1.058	1.055	1.050	1.049	1.046	1.045
2340	1.076	1.075	1.075	1.074	1.074	1.073	1.072	1.071	1.071	1.069	1.068	1.066	1.064	1.062	1.059	1.055	1.050	1.049	1.047	1.046
2360	1.077	1.076	1.076	1.075	1.074	1.074	1.073	1.072	1.071	1.070	1.068	1.067	1.065	1.062	1.059	1.056	1.051	1.050	1.047	1.046
2380	1.077	1.077	1.077	1.076	1.075	1.074	1.073	1.073	1.072	1.071	1.069	1.067	1.065	1.063	1.060	1.056	1.051	1.050	1.048	1.046
2400	1.078	1.077	1.077	1.076	1.076	1.075	1.074	1.073	1.072	1.071	1.070	1.068	1.066	1.064	1.061	1.056	1.051	1.051	1.048	1.047
2420	1.079	1.078	1.078	1.077	1.076	1.076	1.074	1.074	1.073	1.072	1.070	1.068	1.066	1.064	1.061	1.057	1.051	1.051	1.048	1.047
2440	1.079	1.079	1.079	1.078	1.077	1.076	1.075	1.074	1.073	1.073	1.071	1.069	1.067	1.065	1.062	1.057	1.052	1.052	1.048	1.047
2460	1.080	1.079	1.079	1.078	1.078	1.077	1.076	1.075	1.074	1.073	1.071	1.069	1.067	1.065	1.062	1.058	1.052	1.052	1.049	1.048
2480	1.081	1.080	1.080	1.079	1.078	1.078	1.077	1.076	1.075	1.073	1.072	1.070	1.068	1.065	1.062	1.058	1.052	1.052	1.049	1.048

SIGMA = 55630.0

V-Vc = 4

H / x	3.4	3.2	3.0	2.8	2.6	2.4	2.2	2.0	1.8	1.6	1.4	1.2	1.0	0.8	0.6	0.4	0.2	0.10	0.09	0.08
2500	1.081	1.081	1.080	1.080	1.079	1.078	1.077	1.076	1.075	1.074	1.072	1.071	1.068	1.066	1.063	1.058	1.053	1.049	1.049	1.048
2520	1.082	1.081	1.081	1.080	1.079	1.079	1.078	1.077	1.076	1.074	1.073	1.071	1.069	1.066	1.063	1.059	1.053	1.050	1.049	1.049
2540	1.083	1.082	1.081	1.081	1.080	1.079	1.078	1.077	1.076	1.075	1.074	1.072	1.070	1.067	1.064	1.059	1.054	1.050	1.050	1.049
2560	1.083	1.083	1.082	1.081	1.081	1.080	1.079	1.078	1.077	1.076	1.074	1.072	1.070	1.067	1.064	1.060	1.054	1.051	1.050	1.050
2580	1.084	1.083	1.082	1.082	1.081	1.081	1.080	1.079	1.078	1.076	1.075	1.073	1.071	1.068	1.065	1.060	1.055	1.051	1.050	1.050
2600	1.085	1.084	1.083	1.083	1.082	1.081	1.080	1.079	1.079	1.077	1.076	1.073	1.071	1.068	1.065	1.061	1.055	1.051	1.051	1.050
2620	1.085	1.085	1.084	1.083	1.083	1.082	1.081	1.080	1.079	1.077	1.076	1.074	1.072	1.069	1.066	1.061	1.055	1.052	1.051	1.051
2640	1.086	1.085	1.085	1.084	1.083	1.083	1.082	1.081	1.080	1.078	1.077	1.075	1.072	1.070	1.066	1.062	1.056	1.052	1.052	1.051
2660	1.087	1.086	1.085	1.084	1.084	1.083	1.082	1.081	1.080	1.079	1.077	1.075	1.073	1.070	1.067	1.062	1.056	1.052	1.052	1.052
2680	1.087	1.087	1.086	1.085	1.085	1.084	1.083	1.082	1.081	1.079	1.078	1.076	1.073	1.071	1.067	1.063	1.057	1.053	1.052	1.052
2700	1.088	1.087	1.087	1.086	1.085	1.084	1.084	1.082	1.081	1.080	1.078	1.076	1.074	1.071	1.068	1.063	1.057	1.053	1.053	1.052
2720	1.089	1.088	1.087	1.086	1.086	1.085	1.084	1.083	1.082	1.080	1.079	1.077	1.075	1.072	1.068	1.064	1.057	1.054	1.053	1.053
2740	1.089	1.089	1.088	1.087	1.087	1.086	1.085	1.084	1.082	1.081	1.079	1.077	1.075	1.072	1.069	1.064	1.058	1.054	1.054	1.053
2760	1.090	1.089	1.089	1.088	1.087	1.086	1.085	1.084	1.083	1.082	1.080	1.078	1.076	1.073	1.069	1.065	1.058	1.054	1.054	1.053
2780	1.091	1.090	1.089	1.088	1.088	1.087	1.086	1.085	1.084	1.082	1.081	1.079	1.076	1.073	1.070	1.065	1.059	1.055	1.054	1.054
2800	1.091	1.091	1.090	1.089	1.089	1.088	1.087	1.086	1.084	1.083	1.081	1.079	1.077	1.074	1.070	1.065	1.059	1.055	1.055	1.054
2820	1.092	1.091	1.091	1.090	1.089	1.088	1.087	1.086	1.085	1.083	1.082	1.080	1.077	1.074	1.071	1.066	1.060	1.056	1.055	1.055
2840	1.093	1.092	1.091	1.090	1.090	1.089	1.088	1.087	1.086	1.084	1.082	1.080	1.078	1.075	1.071	1.066	1.060	1.056	1.055	1.055
2860	1.093	1.093	1.092	1.091	1.090	1.090	1.089	1.087	1.086	1.085	1.083	1.081	1.078	1.075	1.072	1.067	1.060	1.056	1.056	1.055
2880	1.094	1.093	1.092	1.092	1.091	1.090	1.089	1.088	1.087	1.085	1.084	1.081	1.079	1.076	1.072	1.067	1.061	1.057	1.056	1.056
2900	1.094	1.094	1.093	1.092	1.092	1.091	1.090	1.088	1.087	1.085	1.084	1.082	1.080	1.076	1.073	1.068	1.061	1.057	1.057	1.056
2920	1.095	1.094	1.094	1.093	1.092	1.091	1.090	1.089	1.088	1.086	1.085	1.082	1.080	1.077	1.073	1.068	1.062	1.058	1.057	1.057
2940	1.095	1.095	1.094	1.094	1.093	1.092	1.091	1.090	1.089	1.086	1.085	1.083	1.081	1.077	1.074	1.069	1.062	1.058	1.057	1.057
2960	1.096	1.096	1.095	1.094	1.094	1.093	1.092	1.091	1.089	1.088	1.086	1.084	1.081	1.078	1.074	1.069	1.063	1.058	1.058	1.057
2980	1.097	1.096	1.096	1.095	1.094	1.093	1.092	1.091	1.090	1.088	1.086	1.084	1.082	1.079	1.075	1.070	1.063	1.059	1.058	1.058
3000	1.098	1.097	1.097	1.096	1.095	1.094	1.093	1.092	1.090	1.089	1.087	1.085	1.082	1.079	1.075	1.070	1.063	1.059	1.059	1.058

SIGMA = 55630.0

$V - V_c = 4$

H	0.08	0.09	0.10	0.2	0.4	0.6	0.8	1.0	1.2	1.4	1.6	1.8	2.0	2.2	2.4	2.6	2.8	3.0	3.2	3.4
3100	1.060	1.061	1.061	1.066	1.073	1.078	1.082	1.085	1.088	1.090	1.092	1.094	1.095	1.096	1.097	1.098	1.099	1.100	1.101	1.101
3200	1.062	1.063	1.063	1.068	1.075	1.080	1.085	1.088	1.091	1.093	1.095	1.097	1.098	1.099	1.100	1.101	1.102	1.103	1.105	1.105
3300	1.064	1.065	1.065	1.070	1.077	1.083	1.087	1.091	1.094	1.096	1.098	1.100	1.101	1.102	1.104	1.105	1.106	1.106	1.107	1.108
3400	1.066	1.066	1.067	1.072	1.080	1.085	1.090	1.094	1.096	1.099	1.101	1.103	1.104	1.106	1.107	1.108	1.109	1.110	1.111	1.111
3500	1.068	1.068	1.069	1.074	1.082	1.088	1.093	1.096	1.099	1.102	1.104	1.106	1.107	1.109	1.110	1.111	1.112	1.113	1.114	1.115
3600	1.070	1.070	1.071	1.076	1.084	1.091	1.095	1.099	1.102	1.105	1.107	1.109	1.110	1.112	1.114	1.114	1.115	1.116	1.117	1.118
3700	1.072	1.072	1.073	1.078	1.087	1.093	1.098	1.102	1.105	1.108	1.110	1.112	1.114	1.115	1.117	1.118	1.119	1.120	1.121	1.121
3800	1.074	1.074	1.075	1.080	1.089	1.096	1.101	1.105	1.108	1.111	1.113	1.115	1.117	1.118	1.120	1.121	1.122	1.123	1.124	1.125
3900	1.076	1.076	1.077	1.083	1.092	1.098	1.103	1.108	1.111	1.114	1.116	1.118	1.120	1.122	1.123	1.124	1.125	1.126	1.127	1.128
4000	1.078	1.078	1.079	1.085	1.094	1.101	1.106	1.110	1.114	1.117	1.119	1.121	1.123	1.125	1.126	1.128	1.129	1.130	1.131	1.131
4100	1.080	1.080	1.081	1.087	1.096	1.103	1.109	1.113	1.117	1.120	1.122	1.125	1.126	1.128	1.130	1.131	1.132	1.133	1.134	1.135
4200	1.082	1.082	1.083	1.089	1.099	1.106	1.112	1.116	1.120	1.123	1.125	1.128	1.130	1.131	1.133	1.134	1.135	1.136	1.137	1.138
4300	1.083	1.084	1.085	1.091	1.101	1.109	1.114	1.119	1.123	1.126	1.129	1.131	1.133	1.135	1.136	1.137	1.139	1.140	1.141	1.142
4400	1.085	1.086	1.087	1.093	1.103	1.111	1.117	1.122	1.126	1.129	1.132	1.134	1.136	1.138	1.139	1.141	1.142	1.143	1.144	1.145
4500	1.087	1.088	1.089	1.095	1.106	1.114	1.120	1.126	1.129	1.132	1.135	1.137	1.139	1.141	1.143	1.144	1.145	1.146	1.148	1.148
4600	1.089	1.090	1.091	1.098	1.108	1.116	1.122	1.127	1.132	1.135	1.138	1.140	1.142	1.144	1.146	1.147	1.149	1.150	1.151	1.152
4700	1.091	1.092	1.093	1.100	1.111	1.119	1.125	1.130	1.134	1.138	1.141	1.143	1.146	1.147	1.149	1.151	1.152	1.153	1.154	1.155
4800	1.093	1.094	1.095	1.102	1.113	1.121	1.128	1.133	1.137	1.141	1.144	1.147	1.149	1.151	1.152	1.154	1.155	1.157	1.158	1.159
4900	1.095	1.096	1.097	1.104	1.115	1.124	1.131	1.136	1.140	1.144	1.147	1.150	1.152	1.154	1.156	1.157	1.159	1.160	1.161	1.162
5000	1.097	1.098	1.099	1.106	1.118	1.127	1.133	1.139	1.143	1.147	1.150	1.153	1.155	1.157	1.159	1.161	1.162	1.163	1.165	1.166
5100	1.099	1.100	1.101	1.108	1.120	1.129	1.136	1.142	1.146	1.150	1.153	1.156	1.158	1.160	1.162	1.164	1.165	1.167	1.168	1.169
5200	1.101	1.102	1.103	1.111	1.123	1.132	1.139	1.145	1.149	1.153	1.156	1.159	1.162	1.164	1.166	1.167	1.169	1.170	1.171	1.172
5300	1.103	1.104	1.105	1.113	1.125	1.135	1.142	1.147	1.152	1.156	1.159	1.162	1.165	1.167	1.169	1.171	1.172	1.174	1.175	1.176
5400	1.105	1.106	1.107	1.115	1.127	1.137	1.145	1.150	1.155	1.159	1.162	1.165	1.168	1.170	1.172	1.174	1.176	1.177	1.178	1.179
5500	1.107	1.108	1.109	1.117	1.130	1.140	1.147	1.153	1.158	1.162	1.166	1.169	1.171	1.174	1.176	1.177	1.179	1.180	1.182	1.183
5600	1.109	1.110	1.111	1.119	1.132	1.142	1.150	1.156	1.161	1.165	1.169	1.172	1.175	1.177	1.179	1.181	1.182	1.184	1.185	1.186
5700	1.111	1.112	1.113	1.121	1.135	1.145	1.153	1.159	1.164	1.168	1.172	1.175	1.178	1.180	1.182	1.184	1.186	1.187	1.189	1.190
5800	1.113	1.114	1.115	1.123	1.137	1.147	1.155	1.162	1.167	1.171	1.175	1.178	1.181	1.183	1.186	1.187	1.189	1.191	1.192	1.193
5900	1.115	1.116	1.117	1.126	1.140	1.150	1.158	1.165	1.170	1.174	1.178	1.181	1.184	1.187	1.189	1.191	1.192	1.194	1.195	1.197
6000	1.117	1.118	1.119	1.128	1.142	1.153	1.161	1.168	1.173	1.178	1.181	1.185	1.188	1.190	1.192	1.194	1.196	1.197	1.199	1.200
6100	1.119	1.120	1.121	1.130	1.144	1.155	1.164	1.170	1.176	1.181	1.185	1.188	1.191	1.193	1.196	1.198	1.199	1.201	1.202	1.204
6200	1.121	1.122	1.123	1.132	1.147	1.158	1.167	1.173	1.179	1.184	1.188	1.191	1.194	1.197	1.199	1.201	1.203	1.204	1.206	1.207
6300	1.123	1.124	1.125	1.134	1.149	1.161	1.169	1.176	1.182	1.187	1.191	1.194	1.197	1.200	1.202	1.204	1.206	1.208	1.209	1.211
6400	1.125	1.126	1.127	1.136	1.152	1.163	1.172	1.179	1.185	1.190	1.194	1.198	1.201	1.203	1.206	1.208	1.209	1.211	1.213	1.214
6500	1.127	1.128	1.129	1.139	1.154	1.166	1.175	1.182	1.188	1.193	1.197	1.201	1.204	1.207	1.209	1.211	1.213	1.215	1.216	1.218
6600	1.128	1.130	1.131	1.141	1.157	1.168	1.178	1.185	1.191	1.196	1.200	1.204	1.207	1.210	1.212	1.214	1.216	1.218	1.220	1.221
6700	1.130	1.132	1.133	1.143	1.159	1.171	1.180	1.188	1.194	1.199	1.203	1.207	1.210	1.213	1.216	1.218	1.220	1.222	1.223	1.225
6800	1.132	1.134	1.135	1.145	1.161	1.174	1.183	1.191	1.197	1.202	1.207	1.210	1.214	1.217	1.219	1.221	1.223	1.225	1.227	1.228
6900	1.134	1.136	1.137	1.147	1.164	1.176	1.186	1.194	1.200	1.205	1.210	1.214	1.217	1.220	1.222	1.225	1.227	1.229	1.230	1.232
7000	1.136	1.138	1.139	1.149	1.166	1.179	1.189	1.197	1.203	1.208	1.213	1.217	1.220	1.223	1.226	1.228	1.230	1.232	1.234	1.235
7100	1.138	1.140	1.141	1.152	1.169	1.182	1.192	1.200	1.206	1.212	1.216	1.220	1.224	1.227	1.229	1.232	1.234	1.236	1.237	1.239
7200	1.140	1.141	1.143	1.154	1.171	1.184	1.194	1.203	1.209	1.215	1.219	1.223	1.227	1.230	1.233	1.235	1.237	1.239	1.241	1.242

x

SIGMA = 55630.0

V-Vc = 4

H	0.08	0.09	0.10	0.2	0.4	0.6	0.8	1.0	1.2	1.4	1.6	1.8	2.0	2.2	2.4	2.6	2.8	3.0	3.2	3.4
7300	1.142	1.143	1.145	1.156	1.174	1.187	1.197	1.205	1.212	1.218	1.223	1.227	1.230	1.233	1.236	1.239	1.241	1.243	1.244	1.246
7400	1.144	1.145	1.147	1.158	1.176	1.190	1.200	1.208	1.215	1.221	1.226	1.230	1.233	1.237	1.239	1.242	1.244	1.246	1.248	1.249
7500	1.146	1.147	1.149	1.160	1.179	1.192	1.203	1.211	1.218	1.224	1.229	1.233	1.237	1.240	1.243	1.246	1.248	1.250	1.251	1.253
7600	1.148	1.149	1.151	1.162	1.181	1.195	1.206	1.214	1.221	1.227	1.232	1.236	1.240	1.243	1.246	1.249	1.251	1.253	1.255	1.257
7700	1.150	1.151	1.153	1.165	1.183	1.198	1.208	1.217	1.224	1.230	1.235	1.240	1.243	1.247	1.250	1.253	1.255	1.257	1.258	1.260
7800	1.152	1.153	1.155	1.167	1.186	1.200	1.211	1.220	1.227	1.233	1.239	1.243	1.247	1.250	1.253	1.256	1.258	1.260	1.262	1.264
7900	1.154	1.155	1.157	1.169	1.188	1.203	1.214	1.223	1.230	1.237	1.242	1.246	1.250	1.254	1.256	1.259	1.262	1.264	1.266	1.267
8000	1.156	1.157	1.159	1.171	1.191	1.205	1.217	1.226	1.233	1.240	1.245	1.250	1.253	1.257	1.260	1.263	1.265	1.267	1.269	1.271
8100	1.158	1.159	1.161	1.173	1.193	1.208	1.220	1.229	1.237	1.243	1.248	1.253	1.257	1.260	1.263	1.266	1.269	1.271	1.273	1.275
8200	1.160	1.161	1.163	1.176	1.196	1.211	1.223	1.232	1.240	1.246	1.251	1.257	1.260	1.264	1.267	1.270	1.272	1.274	1.276	1.278
8300	1.162	1.163	1.165	1.178	1.198	1.213	1.225	1.235	1.243	1.249	1.255	1.260	1.264	1.267	1.270	1.273	1.276	1.278	1.280	1.282
8400	1.164	1.165	1.167	1.180	1.201	1.216	1.228	1.238	1.246	1.252	1.258	1.263	1.267	1.270	1.274	1.277	1.279	1.281	1.283	1.285
8500	1.166	1.167	1.169	1.182	1.203	1.219	1.231	1.241	1.249	1.256	1.261	1.266	1.270	1.274	1.277	1.280	1.283	1.285	1.287	1.289
8600	1.168	1.169	1.171	1.184	1.206	1.222	1.234	1.244	1.252	1.259	1.264	1.270	1.274	1.277	1.281	1.284	1.286	1.288	1.291	1.293
8700	1.170	1.171	1.173	1.187	1.208	1.224	1.237	1.247	1.255	1.262	1.268	1.273	1.277	1.281	1.284	1.287	1.290	1.292	1.294	1.296
8800	1.172	1.173	1.175	1.189	1.211	1.227	1.240	1.250	1.258	1.265	1.271	1.276	1.280	1.284	1.288	1.291	1.293	1.296	1.298	1.300
8900	1.174	1.175	1.177	1.191	1.213	1.230	1.242	1.253	1.261	1.268	1.274	1.279	1.284	1.288	1.291	1.294	1.297	1.299	1.301	1.303
9000	1.176	1.177	1.179	1.193	1.215	1.232	1.245	1.256	1.264	1.271	1.277	1.283	1.287	1.291	1.294	1.297	1.300	1.303	1.305	1.307
9100	1.178	1.179	1.181	1.195	1.218	1.235	1.248	1.259	1.267	1.274	1.281	1.286	1.290	1.294	1.298	1.301	1.304	1.306	1.309	1.311
9200	1.180	1.181	1.183	1.197	1.220	1.238	1.251	1.262	1.271	1.278	1.284	1.290	1.294	1.298	1.301	1.304	1.307	1.310	1.312	1.314
9300	1.182	1.183	1.185	1.200	1.223	1.240	1.254	1.265	1.274	1.281	1.287	1.293	1.297	1.301	1.305	1.308	1.311	1.314	1.316	1.318
9400	1.184	1.185	1.187	1.202	1.225	1.243	1.257	1.268	1.277	1.284	1.291	1.296	1.301	1.305	1.308	1.312	1.315	1.317	1.319	1.322
9500	1.186	1.187	1.189	1.204	1.228	1.246	1.260	1.271	1.280	1.287	1.294	1.299	1.304	1.308	1.312	1.315	1.318	1.321	1.323	1.325
9600	1.188	1.189	1.191	1.206	1.230	1.248	1.263	1.274	1.283	1.291	1.297	1.303	1.307	1.312	1.315	1.319	1.322	1.324	1.327	1.329
9700	1.190	1.191	1.193	1.208	1.233	1.251	1.265	1.277	1.286	1.294	1.300	1.306	1.311	1.315	1.319	1.322	1.325	1.328	1.330	1.333
9800	1.192	1.193	1.195	1.211	1.235	1.254	1.268	1.280	1.289	1.297	1.304	1.309	1.314	1.319	1.322	1.326	1.329	1.332	1.334	1.336
9900	1.193	1.195	1.197	1.213	1.238	1.257	1.271	1.283	1.292	1.300	1.307	1.313	1.318	1.322	1.326	1.329	1.332	1.335	1.338	1.340
10000	1.195	1.197	1.199	1.215	1.240	1.259	1.274	1.286	1.295	1.303	1.310	1.316	1.321	1.326	1.329	1.333	1.336	1.339	1.341	1.344
10100	1.197	1.199	1.201	1.217	1.243	1.262	1.277	1.289	1.299	1.307	1.314	1.319	1.325	1.329	1.333	1.337	1.340	1.342	1.345	1.347
10200	1.199	1.201	1.203	1.220	1.245	1.265	1.280	1.292	1.302	1.310	1.317	1.323	1.328	1.333	1.337	1.340	1.343	1.346	1.349	1.351
10300	1.201	1.203	1.205	1.222	1.248	1.267	1.283	1.295	1.305	1.313	1.320	1.326	1.331	1.336	1.340	1.344	1.347	1.350	1.352	1.355
10400	1.203	1.205	1.207	1.224	1.250	1.270	1.286	1.298	1.308	1.317	1.324	1.330	1.335	1.340	1.344	1.347	1.350	1.353	1.356	1.358
10500	1.205	1.207	1.209	1.226	1.253	1.273	1.288	1.301	1.311	1.320	1.327	1.333	1.338	1.343	1.347	1.351	1.354	1.357	1.360	1.362
10600	1.207	1.209	1.211	1.228	1.255	1.276	1.291	1.304	1.314	1.323	1.330	1.336	1.342	1.347	1.351	1.355	1.358	1.361	1.363	1.366
10700	1.209	1.211	1.213	1.231	1.258	1.278	1.294	1.307	1.317	1.326	1.333	1.340	1.345	1.350	1.354	1.358	1.361	1.364	1.367	1.370
10800	1.211	1.213	1.215	1.233	1.260	1.281	1.297	1.310	1.321	1.329	1.337	1.343	1.349	1.354	1.358	1.362	1.365	1.368	1.371	1.373
10900	1.213	1.215	1.217	1.235	1.263	1.284	1.300	1.313	1.324	1.333	1.340	1.347	1.352	1.357	1.361	1.365	1.369	1.372	1.374	1.377
11000	1.215	1.217	1.219	1.237	1.265	1.287	1.303	1.316	1.327	1.336	1.343	1.350	1.356	1.361	1.365	1.369	1.372	1.375	1.378	1.381
11100	1.217	1.219	1.221	1.239	1.268	1.289	1.306	1.319	1.330	1.339	1.347	1.353	1.359	1.364	1.368	1.372	1.376	1.379	1.382	1.384
11200	1.219	1.221	1.223	1.242	1.270	1.292	1.309	1.322	1.333	1.342	1.350	1.357	1.363	1.368	1.372	1.376	1.380	1.383	1.386	1.388
11300	1.221	1.223	1.225	1.244	1.273	1.295	1.312	1.325	1.336	1.346	1.354	1.360	1.366	1.371	1.376	1.380	1.383	1.386	1.389	1.392
11400	1.223	1.225	1.227	1.246	1.275	1.298	1.315	1.328	1.340	1.349	1.357	1.364	1.370	1.375	1.379	1.383	1.387	1.390	1.393	1.396

x

SIGMA = 55630.0

V-V_c = 4

X	0.08	0.09	0.10	0.2	0.4	0.6	0.8	1.0	1.2	1.4	1.6	1.8	2.0	2.2	2.4	2.6	2.8	3.0	3.2	3.4
11500	1.225	1.227	1.229	1.248	1.278	1.300	1.318	1.331	1.343	1.352	1.360	1.367	1.373	1.378	1.383	1.387	1.391	1.394	1.397	1.399
11600	1.227	1.229	1.231	1.251	1.281	1.303	1.321	1.335	1.346	1.356	1.364	1.371	1.377	1.382	1.386	1.391	1.394	1.398	1.401	1.403
11700	1.229	1.231	1.233	1.253	1.283	1.306	1.326	1.338	1.349	1.359	1.367	1.374	1.380	1.385	1.390	1.394	1.398	1.401	1.404	1.407
11800	1.231	1.233	1.235	1.255	1.286	1.309	1.329	1.341	1.352	1.362	1.370	1.377	1.384	1.389	1.394	1.398	1.402	1.405	1.408	1.411
11900	1.233	1.235	1.238	1.257	1.288	1.311	1.332	1.344	1.356	1.365	1.374	1.381	1.387	1.392	1.397	1.401	1.405	1.409	1.412	1.415
12000	1.235	1.237	1.240	1.259	1.291	1.314	1.335	1.347	1.359	1.369	1.377	1.384	1.391	1.396	1.401	1.405	1.409	1.412	1.416	1.418
12100	1.237	1.239	1.242	1.262	1.293	1.317	1.338	1.350	1.362	1.372	1.380	1.388	1.394	1.400	1.404	1.409	1.413	1.416	1.419	1.422
12200	1.239	1.241	1.244	1.264	1.296	1.320	1.341	1.353	1.365	1.375	1.384	1.391	1.398	1.403	1.408	1.412	1.416	1.420	1.423	1.426
12300	1.241	1.243	1.246	1.266	1.298	1.322	1.344	1.356	1.368	1.379	1.387	1.395	1.401	1.407	1.412	1.416	1.420	1.424	1.427	1.430
12400	1.243	1.245	1.248	1.268	1.301	1.325	1.347	1.359	1.372	1.382	1.391	1.398	1.405	1.410	1.415	1.420	1.424	1.427	1.431	1.434
12500	1.245	1.247	1.250	1.271	1.303	1.328	1.350	1.362	1.375	1.385	1.394	1.402	1.408	1.414	1.419	1.423	1.427	1.431	1.434	1.437
12600	1.247	1.249	1.252	1.273	1.306	1.331	1.353	1.365	1.378	1.389	1.397	1.405	1.412	1.417	1.423	1.427	1.431	1.435	1.438	1.441
12700	1.249	1.251	1.254	1.275	1.308	1.333	1.356	1.369	1.381	1.392	1.401	1.409	1.415	1.421	1.426	1.431	1.435	1.439	1.442	1.445
12800	1.251	1.253	1.256	1.277	1.311	1.336	1.359	1.372	1.384	1.395	1.404	1.412	1.419	1.425	1.430	1.435	1.439	1.442	1.446	1.449
12900	1.253	1.255	1.258	1.279	1.314	1.339	1.362	1.375	1.388	1.399	1.408	1.416	1.422	1.428	1.434	1.438	1.442	1.446	1.450	1.453
13000	1.255	1.257	1.260	1.282	1.316	1.342	1.365	1.378	1.391	1.402	1.411	1.419	1.426	1.432	1.437	1.442	1.446	1.450	1.453	1.456
13100	1.257	1.260	1.262	1.284	1.319	1.345	1.368	1.381	1.394	1.405	1.415	1.423	1.429	1.436	1.441	1.445	1.450	1.454	1.457	1.460
13200	1.259	1.262	1.264	1.286	1.321	1.347	1.371	1.384	1.397	1.409	1.418	1.426	1.433	1.439	1.445	1.449	1.454	1.457	1.461	1.464
13300	1.261	1.264	1.266	1.288	1.324	1.350	1.374	1.387	1.401	1.412	1.421	1.430	1.437	1.443	1.448	1.453	1.457	1.461	1.465	1.468
13400	1.263	1.266	1.268	1.291	1.326	1.353	1.377	1.390	1.404	1.415	1.425	1.433	1.440	1.446	1.452	1.457	1.461	1.465	1.469	1.472
13500	1.265	1.268	1.270	1.293	1.329	1.356	1.380	1.393	1.407	1.419	1.428	1.437	1.444	1.450	1.456	1.460	1.465	1.469	1.472	1.476
13600	1.267	1.270	1.272	1.295	1.331	1.359	1.383	1.397	1.410	1.422	1.432	1.440	1.447	1.454	1.459	1.464	1.469	1.473	1.476	1.480
13700	1.269	1.272	1.274	1.297	1.334	1.361	1.386	1.400	1.414	1.425	1.435	1.444	1.451	1.457	1.463	1.468	1.472	1.476	1.480	1.483
13800	1.271	1.274	1.276	1.300	1.337	1.364	1.389	1.403	1.417	1.429	1.439	1.447	1.454	1.461	1.467	1.472	1.476	1.480	1.484	1.487
13900	1.273	1.276	1.278	1.302	1.339	1.367	1.392	1.406	1.420	1.432	1.442	1.451	1.458	1.465	1.470	1.475	1.480	1.484	1.488	1.491
14000	1.275	1.278	1.280	1.304	1.342	1.370	1.395	1.409	1.423	1.435	1.446	1.454	1.462	1.468	1.474	1.479	1.483	1.488	1.492	1.495
14100	1.277	1.280	1.282	1.306	1.344	1.373	1.398	1.412	1.427	1.439	1.449	1.458	1.465	1.472	1.478	1.483	1.487	1.492	1.496	1.499
14200	1.279	1.282	1.284	1.309	1.347	1.375	1.401	1.415	1.430	1.442	1.452	1.461	1.469	1.476	1.481	1.487	1.491	1.496	1.499	1.503
14300	1.281	1.284	1.286	1.311	1.349	1.378	1.404	1.419	1.433	1.446	1.456	1.465	1.472	1.479	1.485	1.490	1.495	1.499	1.503	1.507
14400	1.283	1.286	1.288	1.313	1.352	1.381	1.407	1.422	1.437	1.449	1.459	1.468	1.476	1.483	1.489	1.494	1.499	1.503	1.507	1.511
14500	1.285	1.288	1.291	1.315	1.355	1.384	1.410	1.425	1.440	1.452	1.463	1.472	1.480	1.487	1.493	1.498	1.502	1.507	1.511	1.515
14600	1.287	1.290	1.293	1.318	1.357	1.387	1.413	1.428	1.443	1.456	1.466	1.475	1.483	1.490	1.496	1.502	1.506	1.511	1.515	1.518
14700	1.289	1.292	1.295	1.320	1.360	1.390	1.416	1.431	1.446	1.459	1.470	1.479	1.487	1.494	1.500	1.506	1.510	1.515	1.519	1.522
14800	1.291	1.294	1.297	1.322	1.362	1.392	1.419	1.434	1.450	1.463	1.473	1.483	1.491	1.498	1.504	1.509	1.514	1.519	1.523	1.526
14900	1.293	1.296	1.299	1.324	1.365	1.395	1.422	1.438	1.453	1.466	1.477	1.486	1.494	1.501	1.508	1.513	1.518	1.523	1.527	1.530
15000	1.295	1.298	1.301	1.327	1.367	1.398	1.425	1.441	1.456	1.469	1.480	1.490	1.498	1.505	1.511	1.517	1.522	1.526	1.530	1.534

SIGMA = 35600.0

V−V$_c$ = 5

H \ x	3.4	3.2	3.0	2.8	2.6	2.4	2.2	2.0	1.8	1.6	1.4	1.2	1.0	0.8	0.6	0.4	0.2	0.10	0.09	0.08
0	1.000	1.000	1.000	1.000	1.000	1.000	1.000	1.000	1.000	1.000	1.000	1.000	1.000	1.000	1.000	1.000	1.000	1.000	1.000	1.000
20	1.001	1.001	1.001	1.001	1.001	1.001	1.001	1.001	1.001	1.001	1.001	1.001	1.001	1.001	1.001	1.001	1.001	1.001	1.001	1.001
40	1.002	1.002	1.002	1.002	1.002	1.002	1.002	1.002	1.002	1.002	1.002	1.002	1.002	1.002	1.002	1.001	1.001	1.001	1.001	1.001
60	1.003	1.003	1.003	1.003	1.003	1.003	1.003	1.003	1.003	1.003	1.003	1.003	1.003	1.002	1.002	1.002	1.002	1.002	1.002	1.002
80	1.004	1.004	1.004	1.004	1.004	1.004	1.004	1.004	1.004	1.004	1.004	1.004	1.003	1.003	1.003	1.003	1.003	1.003	1.002	1.002
100	1.005	1.005	1.005	1.005	1.005	1.005	1.005	1.005	1.005	1.005	1.005	1.004	1.004	1.004	1.004	1.004	1.003	1.003	1.003	1.003
120	1.006	1.006	1.006	1.006	1.006	1.006	1.006	1.006	1.006	1.006	1.005	1.005	1.005	1.005	1.005	1.004	1.004	1.004	1.004	1.004
140	1.007	1.007	1.007	1.007	1.007	1.007	1.007	1.007	1.006	1.006	1.006	1.006	1.006	1.006	1.005	1.005	1.005	1.004	1.004	1.004
160	1.008	1.008	1.008	1.008	1.008	1.008	1.008	1.008	1.007	1.007	1.007	1.007	1.007	1.007	1.006	1.006	1.005	1.005	1.005	1.005
180	1.009	1.009	1.009	1.009	1.009	1.009	1.009	1.009	1.008	1.008	1.008	1.008	1.008	1.007	1.007	1.007	1.006	1.006	1.005	1.005
200	1.010	1.010	1.010	1.010	1.010	1.010	1.010	1.010	1.009	1.009	1.009	1.009	1.009	1.008	1.008	1.007	1.007	1.006	1.006	1.006
220	1.011	1.011	1.011	1.011	1.011	1.011	1.010	1.010	1.010	1.010	1.010	1.010	1.009	1.009	1.009	1.008	1.007	1.007	1.007	1.007
240	1.012	1.012	1.012	1.012	1.012	1.012	1.011	1.011	1.011	1.011	1.011	1.011	1.010	1.010	1.009	1.009	1.008	1.008	1.007	1.007
260	1.013	1.013	1.013	1.013	1.013	1.013	1.012	1.012	1.012	1.012	1.012	1.011	1.011	1.011	1.010	1.009	1.008	1.008	1.008	1.008
280	1.014	1.014	1.014	1.014	1.014	1.014	1.013	1.013	1.013	1.013	1.013	1.012	1.012	1.012	1.011	1.010	1.009	1.009	1.008	1.008
300	1.015	1.015	1.015	1.015	1.015	1.015	1.014	1.014	1.014	1.014	1.014	1.013	1.013	1.012	1.012	1.011	1.010	1.009	1.009	1.009
320	1.016	1.016	1.016	1.016	1.016	1.016	1.015	1.015	1.015	1.015	1.014	1.014	1.014	1.013	1.012	1.012	1.010	1.010	1.010	1.010
340	1.017	1.017	1.017	1.017	1.017	1.017	1.016	1.016	1.016	1.016	1.015	1.015	1.014	1.014	1.013	1.012	1.011	1.011	1.010	1.010
360	1.018	1.018	1.018	1.018	1.018	1.018	1.017	1.017	1.017	1.017	1.016	1.016	1.015	1.015	1.014	1.013	1.012	1.011	1.011	1.011
380	1.019	1.019	1.019	1.019	1.019	1.019	1.018	1.018	1.018	1.018	1.017	1.017	1.016	1.016	1.015	1.014	1.012	1.012	1.011	1.011
400	1.020	1.020	1.020	1.020	1.020	1.020	1.019	1.019	1.019	1.019	1.018	1.018	1.017	1.017	1.016	1.015	1.013	1.013	1.012	1.012
420	1.021	1.021	1.021	1.020	1.020	1.020	1.020	1.020	1.019	1.019	1.019	1.018	1.018	1.017	1.016	1.015	1.014	1.013	1.013	1.013
440	1.022	1.022	1.022	1.021	1.021	1.021	1.021	1.021	1.020	1.020	1.020	1.019	1.019	1.018	1.017	1.016	1.014	1.014	1.013	1.013
460	1.023	1.023	1.023	1.022	1.022	1.022	1.022	1.022	1.021	1.021	1.021	1.020	1.020	1.019	1.018	1.017	1.015	1.014	1.014	1.014
480	1.024	1.024	1.024	1.023	1.023	1.023	1.023	1.023	1.022	1.022	1.022	1.021	1.020	1.020	1.019	1.017	1.016	1.015	1.014	1.014
500	1.025	1.025	1.025	1.024	1.024	1.024	1.024	1.024	1.023	1.023	1.023	1.022	1.021	1.021	1.019	1.018	1.016	1.016	1.015	1.015
520	1.026	1.026	1.026	1.025	1.025	1.025	1.025	1.025	1.024	1.024	1.023	1.023	1.022	1.021	1.020	1.019	1.017	1.016	1.016	1.016
540	1.027	1.027	1.027	1.026	1.026	1.026	1.026	1.026	1.025	1.025	1.024	1.024	1.023	1.022	1.021	1.020	1.018	1.017	1.016	1.016
560	1.028	1.028	1.028	1.027	1.027	1.027	1.027	1.027	1.026	1.026	1.025	1.025	1.024	1.023	1.022	1.020	1.018	1.018	1.017	1.017
580	1.029	1.029	1.029	1.028	1.028	1.028	1.028	1.028	1.027	1.027	1.026	1.025	1.025	1.024	1.022	1.021	1.019	1.018	1.017	1.017
600	1.030	1.030	1.030	1.029	1.029	1.029	1.029	1.029	1.028	1.028	1.027	1.026	1.026	1.025	1.023	1.022	1.020	1.019	1.018	1.018
620	1.031	1.031	1.031	1.030	1.030	1.030	1.029	1.029	1.029	1.029	1.028	1.027	1.026	1.026	1.024	1.022	1.020	1.019	1.019	1.019
640	1.032	1.032	1.032	1.031	1.031	1.031	1.030	1.030	1.030	1.030	1.029	1.028	1.027	1.026	1.025	1.023	1.021	1.020	1.019	1.019
660	1.033	1.033	1.033	1.032	1.032	1.032	1.031	1.031	1.031	1.031	1.030	1.029	1.028	1.027	1.026	1.024	1.021	1.021	1.020	1.020
680	1.034	1.034	1.034	1.033	1.033	1.033	1.032	1.032	1.031	1.031	1.031	1.030	1.029	1.028	1.026	1.025	1.022	1.021	1.020	1.020
700	1.035	1.035	1.035	1.034	1.034	1.034	1.033	1.033	1.032	1.032	1.032	1.031	1.030	1.029	1.027	1.025	1.023	1.022	1.021	1.021
720	1.036	1.036	1.036	1.035	1.035	1.035	1.034	1.034	1.033	1.033	1.032	1.032	1.031	1.030	1.028	1.026	1.023	1.023	1.022	1.022
740	1.037	1.037	1.037	1.036	1.036	1.036	1.035	1.035	1.034	1.034	1.033	1.032	1.031	1.031	1.029	1.027	1.024	1.023	1.022	1.022
760	1.038	1.038	1.038	1.037	1.037	1.037	1.036	1.036	1.035	1.035	1.034	1.033	1.032	1.031	1.029	1.028	1.025	1.024	1.023	1.023
780	1.039	1.039	1.039	1.038	1.038	1.038	1.037	1.037	1.036	1.036	1.035	1.034	1.033	1.032	1.030	1.028	1.025	1.024	1.023	1.023
800	1.040	1.040	1.040	1.039	1.039	1.039	1.038	1.038	1.037	1.037	1.036	1.035	1.034	1.033	1.031	1.029	1.026	1.025	1.024	1.024

SIGMA = 35600.0

V-V$_c$ = 5

H / x	0.08	0.09	0.10	0.2	0.4	0.6	0.8	1.0	1.2	1.4	1.6	1.8	2.0	2.2	2.4	2.6	2.8	3.0	3.2	3.4
820	1.025	1.025	1.025	1.027	1.030	1.032	1.035	1.035	1.036	1.037	1.038	1.038	1.039	1.039	1.040	1.040	1.040	1.041	1.041	1.041
840	1.025	1.026	1.026	1.028	1.030	1.033	1.036	1.036	1.037	1.038	1.038	1.039	1.040	1.040	1.040	1.041	1.041	1.042	1.042	1.042
860	1.026	1.026	1.026	1.028	1.031	1.033	1.037	1.037	1.038	1.039	1.039	1.040	1.041	1.041	1.042	1.042	1.042	1.043	1.043	1.043
880	1.026	1.027	1.027	1.029	1.032	1.034	1.036	1.037	1.039	1.039	1.040	1.040	1.042	1.042	1.043	1.043	1.043	1.044	1.044	1.044
900	1.027	1.027	1.028	1.030	1.033	1.035	1.038	1.038	1.039	1.040	1.041	1.041	1.042	1.043	1.044	1.044	1.045	1.045	1.045	1.045
920	1.027	1.028	1.028	1.030	1.033	1.036	1.038	1.039	1.040	1.041	1.042	1.042	1.043	1.044	1.045	1.045	1.046	1.046	1.046	1.046
940	1.027	1.029	1.029	1.031	1.034	1.037	1.039	1.040	1.041	1.042	1.043	1.043	1.044	1.045	1.045	1.046	1.047	1.047	1.047	1.047
960	1.028	1.029	1.029	1.032	1.035	1.037	1.038	1.041	1.042	1.043	1.043	1.044	1.045	1.046	1.047	1.047	1.048	1.048	1.048	1.048
980	1.029	1.030	1.030	1.032	1.036	1.038	1.039	1.042	1.043	1.044	1.045	1.045	1.046	1.047	1.047	1.048	1.048	1.049	1.049	1.049
1000	1.030	1.030	1.031	1.033	1.036	1.039	1.040	1.043	1.044	1.045	1.046	1.046	1.047	1.048	1.048	1.049	1.049	1.050	1.050	1.050
1020	1.030	1.030	1.031	1.034	1.037	1.040	1.042	1.043	1.044	1.046	1.046	1.047	1.048	1.049	1.050	1.050	1.050	1.051	1.051	1.051
1040	1.031	1.031	1.032	1.034	1.038	1.040	1.043	1.044	1.045	1.046	1.047	1.048	1.049	1.050	1.050	1.051	1.051	1.052	1.052	1.052
1060	1.031	1.032	1.032	1.035	1.039	1.041	1.043	1.045	1.046	1.047	1.048	1.049	1.050	1.051	1.051	1.052	1.052	1.053	1.053	1.053
1080	1.032	1.032	1.033	1.036	1.039	1.041	1.044	1.045	1.046	1.048	1.049	1.050	1.050	1.051	1.052	1.053	1.053	1.054	1.054	1.054
1100	1.033	1.033	1.034	1.036	1.040	1.042	1.044	1.046	1.047	1.049	1.050	1.050	1.051	1.052	1.053	1.054	1.054	1.055	1.055	1.056
1120	1.033	1.034	1.034	1.037	1.040	1.043	1.045	1.047	1.048	1.049	1.051	1.051	1.052	1.053	1.054	1.055	1.055	1.056	1.056	1.057
1140	1.034	1.034	1.035	1.037	1.041	1.044	1.046	1.048	1.049	1.050	1.052	1.052	1.053	1.054	1.055	1.055	1.056	1.057	1.057	1.058
1160	1.034	1.035	1.035	1.038	1.042	1.044	1.047	1.048	1.050	1.051	1.052	1.053	1.054	1.055	1.056	1.056	1.057	1.058	1.058	1.059
1180	1.035	1.035	1.036	1.039	1.043	1.045	1.048	1.049	1.050	1.052	1.053	1.054	1.055	1.056	1.057	1.057	1.058	1.059	1.059	1.060
1200	1.036	1.036	1.037	1.040	1.043	1.046	1.048	1.050	1.052	1.053	1.054	1.055	1.056	1.057	1.058	1.058	1.059	1.060	1.060	1.061
1220	1.036	1.036	1.037	1.040	1.044	1.047	1.049	1.051	1.053	1.054	1.055	1.056	1.057	1.058	1.059	1.059	1.060	1.061	1.061	1.062
1240	1.037	1.037	1.038	1.041	1.044	1.047	1.050	1.052	1.054	1.055	1.056	1.057	1.058	1.059	1.060	1.060	1.061	1.062	1.062	1.063
1260	1.037	1.037	1.038	1.042	1.045	1.048	1.051	1.053	1.054	1.056	1.057	1.058	1.059	1.060	1.061	1.061	1.062	1.063	1.063	1.064
1280	1.038	1.038	1.039	1.042	1.046	1.049	1.052	1.054	1.055	1.057	1.058	1.059	1.060	1.061	1.062	1.062	1.063	1.064	1.064	1.065
1300	1.039	1.039	1.040	1.043	1.047	1.050	1.053	1.055	1.056	1.058	1.059	1.060	1.061	1.062	1.063	1.064	1.064	1.065	1.065	1.066
1320	1.039	1.040	1.040	1.043	1.047	1.051	1.053	1.055	1.057	1.059	1.060	1.061	1.062	1.063	1.064	1.064	1.065	1.066	1.066	1.067
1340	1.040	1.040	1.041	1.044	1.048	1.051	1.054	1.056	1.058	1.060	1.061	1.062	1.063	1.064	1.065	1.065	1.066	1.067	1.067	1.068
1360	1.041	1.041	1.041	1.044	1.049	1.052	1.055	1.057	1.059	1.061	1.062	1.063	1.064	1.065	1.066	1.067	1.067	1.068	1.068	1.069
1380	1.041	1.042	1.042	1.045	1.050	1.053	1.056	1.059	1.060	1.062	1.063	1.065	1.066	1.066	1.067	1.068	1.068	1.069	1.069	1.070
1400	1.042	1.042	1.043	1.046	1.051	1.054	1.057	1.060	1.061	1.063	1.064	1.066	1.067	1.067	1.068	1.069	1.069	1.070	1.070	1.071
1420	1.042	1.043	1.043	1.047	1.052	1.055	1.058	1.061	1.062	1.064	1.065	1.067	1.068	1.068	1.069	1.070	1.070	1.071	1.071	1.072
1440	1.043	1.043	1.044	1.047	1.052	1.056	1.059	1.061	1.063	1.065	1.066	1.068	1.069	1.069	1.070	1.071	1.071	1.072	1.073	1.073
1460	1.044	1.044	1.045	1.048	1.053	1.057	1.060	1.062	1.064	1.066	1.068	1.069	1.070	1.071	1.072	1.072	1.072	1.073	1.074	1.074
1480	1.044	1.045	1.046	1.049	1.054	1.058	1.061	1.063	1.065	1.067	1.069	1.070	1.071	1.072	1.073	1.073	1.073	1.074	1.075	1.075
1500	1.045	1.046	1.046	1.050	1.055	1.059	1.062	1.064	1.066	1.068	1.070	1.071	1.072	1.073	1.074	1.074	1.074	1.075	1.076	1.076
1520	1.046	1.046	1.047	1.051	1.055	1.060	1.063	1.065	1.067	1.069	1.071	1.072	1.073	1.074	1.075	1.075	1.075	1.076	1.077	1.077
1540	1.047	1.046	1.047	1.052	1.056	1.061	1.064	1.066	1.068	1.070	1.072	1.073	1.074	1.075	1.076	1.076	1.076	1.077	1.078	1.078
1560	1.047	1.047	1.048	1.052	1.057	1.062	1.065	1.067	1.069	1.071	1.073	1.074	1.075	1.076	1.077	1.077	1.077	1.078	1.079	1.079
1580	1.048	1.048	1.049	1.053	1.058	1.063	1.066	1.068	1.070	1.072	1.074	1.075	1.076	1.077	1.078	1.079	1.078	1.079	1.080	1.080
1600	1.048	1.049	1.049	1.054	1.058	1.063	1.067	1.069	1.071	1.073	1.075	1.076	1.077	1.078	1.079	1.080	1.080	1.080	1.081	1.082
1620	1.049	1.049	1.050	1.054	1.059	1.063	1.067	1.069	1.071	1.074	1.076	1.077	1.078	1.079	1.080	1.081	1.081	1.081	1.082	1.082
1640	1.050	1.050	1.050	1.054	1.060	1.064	1.067	1.070	1.072	1.074	1.076	1.077	1.078	1.079	1.080	1.081	1.082	1.082	1.083	1.083

SIGMA = 35600.0

V-Vc = 5

H \ x	0.08	0.09	0.10	0.2	0.4	0.6	0.8	1.0	1.2	1.4	1.6	1.8	2.0	2.2	2.4	2.6	2.8	3.0	3.2	3.4
1660	1.050	1.051	1.051	1.055	1.061	1.065	1.068	1.071	1.073	1.075	1.077	1.078	1.079	1.080	1.081	1.082	1.083	1.083	1.084	1.084
1680	1.051	1.051	1.052	1.055	1.061	1.066	1.069	1.072	1.074	1.076	1.078	1.079	1.080	1.081	1.082	1.083	1.084	1.084	1.085	1.085
1700	1.051	1.052	1.052	1.056	1.062	1.067	1.070	1.073	1.075	1.077	1.079	1.080	1.081	1.082	1.083	1.084	1.085	1.085	1.086	1.086
1720	1.052	1.052	1.053	1.057	1.063	1.067	1.071	1.074	1.076	1.078	1.079	1.081	1.082	1.083	1.084	1.085	1.086	1.086	1.087	1.087
1740	1.053	1.053	1.054	1.057	1.064	1.068	1.072	1.075	1.078	1.079	1.080	1.082	1.083	1.084	1.085	1.086	1.087	1.087	1.088	1.088
1760	1.053	1.054	1.054	1.058	1.064	1.069	1.072	1.076	1.079	1.080	1.081	1.083	1.084	1.085	1.086	1.087	1.088	1.088	1.089	1.090
1780	1.054	1.054	1.055	1.059	1.065	1.070	1.073	1.077	1.080	1.081	1.082	1.084	1.085	1.086	1.087	1.088	1.089	1.090	1.090	1.091
1800	1.054	1.055	1.055	1.059	1.066	1.070	1.074	1.078	1.081	1.082	1.083	1.085	1.086	1.087	1.088	1.089	1.090	1.091	1.091	1.092
1820	1.055	1.055	1.056	1.060	1.066	1.071	1.075	1.079	1.082	1.083	1.084	1.086	1.087	1.088	1.089	1.090	1.091	1.092	1.092	1.093
1840	1.055	1.056	1.057	1.061	1.067	1.072	1.076	1.080	1.083	1.084	1.085	1.087	1.088	1.089	1.090	1.091	1.092	1.093	1.093	1.094
1860	1.056	1.056	1.057	1.061	1.068	1.073	1.077	1.081	1.084	1.085	1.086	1.088	1.089	1.090	1.091	1.092	1.093	1.094	1.094	1.095
1880	1.057	1.057	1.058	1.062	1.069	1.074	1.078	1.082	1.085	1.086	1.087	1.089	1.090	1.091	1.092	1.093	1.094	1.095	1.095	1.096
1900	1.057	1.057	1.058	1.063	1.069	1.074	1.078	1.083	1.086	1.087	1.088	1.090	1.091	1.092	1.093	1.094	1.095	1.096	1.096	1.097
1920	1.058	1.058	1.059	1.063	1.070	1.075	1.079	1.084	1.087	1.088	1.089	1.091	1.092	1.093	1.094	1.095	1.096	1.097	1.097	1.098
1940	1.058	1.058	1.060	1.064	1.071	1.076	1.080	1.085	1.088	1.089	1.090	1.091	1.093	1.094	1.095	1.096	1.097	1.098	1.098	1.099
1960	1.059	1.059	1.060	1.065	1.072	1.077	1.081	1.086	1.089	1.090	1.091	1.092	1.094	1.095	1.096	1.097	1.098	1.099	1.099	1.100
1980	1.059	1.059	1.061	1.065	1.073	1.078	1.082	1.087	1.090	1.091	1.092	1.093	1.095	1.096	1.097	1.098	1.099	1.100	1.100	1.101
2000	1.060	1.060	1.062	1.066	1.073	1.078	1.083	1.088	1.091	1.092	1.093	1.094	1.096	1.097	1.098	1.099	1.100	1.101	1.101	1.102
2020	1.061	1.060	1.062	1.067	1.074	1.079	1.083	1.089	1.092	1.093	1.094	1.095	1.097	1.098	1.099	1.100	1.101	1.102	1.102	1.103
2040	1.061	1.061	1.063	1.067	1.075	1.080	1.084	1.090	1.093	1.094	1.095	1.096	1.098	1.099	1.100	1.101	1.102	1.103	1.103	1.104
2060	1.062	1.062	1.063	1.068	1.075	1.081	1.085	1.091	1.094	1.095	1.096	1.097	1.099	1.100	1.101	1.102	1.103	1.104	1.104	1.105
2080	1.062	1.063	1.064	1.069	1.076	1.081	1.086	1.092	1.095	1.096	1.096	1.098	1.100	1.101	1.102	1.103	1.104	1.105	1.105	1.106
2100	1.063	1.063	1.065	1.069	1.077	1.082	1.087	1.093	1.097	1.097	1.097	1.099	1.101	1.102	1.103	1.104	1.105	1.106	1.106	1.107
2120	1.064	1.064	1.066	1.070	1.078	1.083	1.088	1.094	1.098	1.098	1.098	1.100	1.102	1.103	1.104	1.105	1.106	1.107	1.107	1.108
2140	1.064	1.064	1.067	1.071	1.078	1.084	1.088	1.095	1.099	1.099	1.099	1.101	1.103	1.104	1.105	1.106	1.107	1.108	1.108	1.109
2160	1.065	1.065	1.067	1.071	1.079	1.085	1.089	1.096	1.100	1.100	1.100	1.102	1.104	1.105	1.106	1.107	1.108	1.109	1.109	1.110
2180	1.065	1.065	1.068	1.072	1.080	1.086	1.090	1.097	1.101	1.100	1.101	1.103	1.105	1.106	1.107	1.108	1.109	1.110	1.110	1.111
2200	1.066	1.066	1.069	1.073	1.081	1.087	1.091	1.098	1.102	1.101	1.102	1.104	1.106	1.107	1.108	1.109	1.110	1.111	1.111	1.112
2220	1.067	1.066	1.070	1.074	1.081	1.088	1.092	1.098	1.103	1.102	1.103	1.105	1.107	1.108	1.109	1.110	1.111	1.112	1.112	1.113
2240	1.067	1.067	1.070	1.074	1.082	1.089	1.093	1.099	1.104	1.103	1.104	1.107	1.108	1.109	1.110	1.111	1.112	1.113	1.113	1.114
2260	1.068	1.067	1.071	1.075	1.083	1.089	1.093	1.100	1.105	1.104	1.105	1.108	1.109	1.111	1.111	1.112	1.113	1.114	1.114	1.115
2280	1.068	1.068	1.072	1.076	1.084	1.090	1.094	1.101	1.106	1.105	1.106	1.108	1.110	1.113	1.112	1.113	1.114	1.115	1.115	1.116
2300	1.069	1.069	1.073	1.077	1.085	1.091	1.095	1.102	1.107	1.106	1.108	1.109	1.111	1.114	1.113	1.114	1.115	1.116	1.116	1.117
2320	1.070	1.069	1.074	1.078	1.086	1.092	1.096	1.103	1.108	1.107	1.109	1.110	1.112	1.115	1.114	1.115	1.116	1.117	1.117	1.118
2340	1.071	1.070	1.074	1.079	1.087	1.093	1.097	1.104	1.109	1.108	1.110	1.111	1.113	1.116	1.115	1.116	1.117	1.118	1.118	1.119
2360	1.071	1.070	1.075	1.079	1.087	1.094	1.098	1.105	1.110	1.109	1.111	1.112	1.114	1.117	1.116	1.117	1.118	1.119	1.119	1.120
2380	1.072	1.071	1.075	1.080	1.088	1.094	1.099	1.106	1.111	1.110	1.112	1.113	1.115	1.118	1.117	1.118	1.119	1.120	1.120	1.121
2400	1.072	1.072	1.076	1.081	1.089	1.095	1.100	1.107	1.112	1.111	1.113	1.114	1.116	1.119	1.118	1.119	1.120	1.121	1.121	1.122
2420	1.073	1.072	1.076	1.082	1.090	1.096	1.101	1.108	1.113	1.112	1.114	1.115	1.117	1.120	1.119	1.120	1.121	1.122	1.122	1.124
2440	1.074	1.073	1.076	1.089	1.090	1.097	1.102	1.105	1.109	1.112	1.114	1.116	1.117	1.119	1.120	1.121	1.123	1.124	1.124	1.125
2460	1.075	1.075	1.076	1.081	1.091	1.097	1.102	1.106	1.109	1.112	1.115	1.117	1.118	1.120	1.121	1.122	1.124	1.125	1.125	1.126
2480	1.075	1.076	1.076	1.082	1.091	1.098	1.103	1.107	1.110	1.113	1.116	1.118	1.119	1.121	1.122	1.123	1.125	1.126	1.126	1.127

$V - V_c = 5$ SIGMA = 35600.0

H	x 0.08	0.09	0.10	0.2	0.4	0.6	0.8	1.0	1.2	1.4	1.6	1.8	2.0	2.2	2.4	2.6	2.8	3.0	3.2	3.4
2500	1.076	1.076	1.077	1.083	1.092	1.098	1.104	1.108	1.111	1.114	1.116	1.119	1.120	1.122	1.123	1.125	1.126	1.127	1.127	1.128
2520	1.076	1.077	1.078	1.083	1.092	1.099	1.104	1.109	1.112	1.115	1.117	1.120	1.121	1.123	1.124	1.126	1.127	1.128	1.129	1.129
2540	1.077	1.077	1.078	1.084	1.093	1.100	1.105	1.110	1.113	1.116	1.118	1.120	1.122	1.124	1.125	1.127	1.128	1.129	1.130	1.130
2560	1.077	1.078	1.079	1.085	1.094	1.101	1.106	1.110	1.114	1.117	1.119	1.121	1.123	1.125	1.126	1.128	1.129	1.130	1.131	1.131
2580	1.078	1.079	1.080	1.065	1.095	1.102	1.107	1.111	1.115	1.118	1.120	1.122	1.124	1.126	1.127	1.129	1.130	1.131	1.132	1.133
2600	1.078	1.079	1.080	1.086	1.095	1.102	1.108	1.112	1.116	1.119	1.121	1.123	1.125	1.127	1.128	1.130	1.131	1.132	1.133	1.134
2620	1.078	1.080	1.081	1.087	1.096	1.103	1.109	1.113	1.117	1.120	1.122	1.124	1.126	1.128	1.129	1.131	1.132	1.133	1.134	1.135
2640	1.079	1.080	1.081	1.087	1.097	1.104	1.110	1.114	1.118	1.120	1.123	1.125	1.127	1.129	1.130	1.132	1.133	1.134	1.135	1.136
2660	1.079	1.081	1.081	1.088	1.097	1.104	1.110	1.115	1.118	1.121	1.123	1.126	1.128	1.130	1.131	1.133	1.134	1.135	1.136	1.137
2680	1.080	1.081	1.082	1.089	1.098	1.105	1.111	1.116	1.119	1.122	1.124	1.127	1.129	1.131	1.132	1.134	1.135	1.136	1.137	1.138
2700	1.080	1.082	1.083	1.089	1.098	1.106	1.112	1.117	1.120	1.123	1.125	1.128	1.130	1.132	1.133	1.135	1.136	1.137	1.138	1.139
2720	1.081	1.082	1.083	1.090	1.099	1.106	1.113	1.118	1.121	1.124	1.126	1.129	1.131	1.133	1.134	1.136	1.137	1.138	1.139	1.140
2740	1.081	1.083	1.084	1.090	1.100	1.107	1.114	1.118	1.122	1.125	1.127	1.130	1.132	1.134	1.135	1.137	1.138	1.139	1.140	1.141
2760	1.082	1.083	1.085	1.091	1.100	1.108	1.115	1.119	1.123	1.126	1.128	1.131	1.133	1.135	1.136	1.138	1.139	1.140	1.141	1.142
2780	1.082	1.084	1.085	1.091	1.101	1.109	1.116	1.120	1.124	1.127	1.129	1.132	1.134	1.136	1.138	1.139	1.140	1.141	1.142	1.143
2800	1.083	1.084	1.086	1.092	1.102	1.110	1.116	1.121	1.125	1.128	1.130	1.133	1.135	1.137	1.139	1.140	1.141	1.142	1.143	1.144
2820	1.083	1.085	1.086	1.093	1.102	1.110	1.117	1.122	1.126	1.128	1.131	1.134	1.136	1.138	1.140	1.141	1.142	1.143	1.144	1.145
2840	1.084	1.086	1.087	1.093	1.103	1.111	1.118	1.123	1.126	1.129	1.132	1.135	1.137	1.139	1.141	1.142	1.143	1.144	1.145	1.146
2860	1.085	1.086	1.088	1.094	1.104	1.112	1.119	1.124	1.127	1.130	1.133	1.136	1.138	1.140	1.142	1.143	1.144	1.145	1.146	1.147
2880	1.086	1.087	1.088	1.094	1.104	1.113	1.120	1.125	1.128	1.131	1.134	1.137	1.139	1.141	1.143	1.144	1.145	1.146	1.147	1.148
2900	1.086	1.088	1.089	1.095	1.105	1.114	1.120	1.126	1.129	1.132	1.135	1.138	1.140	1.142	1.144	1.145	1.146	1.147	1.149	1.150
2920	1.087	1.088	1.090	1.096	1.106	1.114	1.121	1.126	1.130	1.133	1.136	1.138	1.141	1.143	1.145	1.146	1.147	1.148	1.150	1.151
2940	1.087	1.089	1.090	1.097	1.107	1.115	1.122	1.127	1.131	1.134	1.137	1.140	1.142	1.144	1.146	1.147	1.148	1.150	1.151	1.152
2960	1.088	1.089	1.091	1.098	1.108	1.116	1.123	1.128	1.132	1.135	1.138	1.141	1.143	1.145	1.147	1.148	1.149	1.151	1.152	1.153
2980	1.090	1.091	1.092	1.099	1.109	1.118	1.124	1.129	1.133	1.137	1.140	1.142	1.144	1.146	1.148	1.149	1.151	1.152	1.153	1.154
3000	1.091	1.092	1.093	1.099	1.110	1.119	1.125	1.130	1.134	1.138	1.140	1.143	1.145	1.147	1.149	1.150	1.152	1.153	1.154	1.155

SIGMA = 35600.0

$V-V_c = 5$

H / x	0.08	0.09	0.10	0.2	0.4	0.6	0.8	1.0	1.2	1.4	1.6	1.8	2.0	2.2	2.4	2.6	2.8	3.0	3.2	3.4
3100	1.094	1.095	1.096	1.103	1.114	1.123	1.129	1.134	1.139	1.142	1.145	1.148	1.150	1.152	1.154	1.155	1.157	1.158	1.159	1.160
3200	1.097	1.098	1.099	1.106	1.118	1.127	1.133	1.139	1.143	1.147	1.150	1.153	1.155	1.157	1.159	1.161	1.162	1.163	1.165	1.166
3300	1.100	1.101	1.102	1.110	1.122	1.131	1.138	1.143	1.148	1.152	1.155	1.158	1.160	1.162	1.164	1.166	1.167	1.169	1.170	1.171
3400	1.103	1.104	1.105	1.113	1.125	1.135	1.142	1.148	1.153	1.157	1.160	1.163	1.165	1.167	1.169	1.171	1.173	1.174	1.175	1.176
3500	1.106	1.107	1.108	1.116	1.129	1.139	1.146	1.152	1.157	1.161	1.165	1.168	1.170	1.173	1.175	1.176	1.178	1.179	1.181	1.182
3600	1.109	1.110	1.111	1.120	1.133	1.143	1.151	1.157	1.162	1.166	1.170	1.173	1.175	1.178	1.180	1.182	1.183	1.185	1.186	1.187
3700	1.112	1.113	1.114	1.123	1.137	1.147	1.155	1.161	1.167	1.171	1.175	1.178	1.180	1.183	1.185	1.187	1.188	1.190	1.191	1.193
3800	1.115	1.116	1.117	1.126	1.140	1.151	1.159	1.166	1.171	1.176	1.179	1.183	1.185	1.188	1.190	1.192	1.194	1.195	1.197	1.198
3900	1.119	1.120	1.121	1.130	1.144	1.155	1.164	1.170	1.176	1.180	1.184	1.188	1.191	1.193	1.195	1.197	1.199	1.201	1.202	1.203
4000	1.122	1.123	1.124	1.133	1.148	1.159	1.168	1.175	1.181	1.185	1.189	1.193	1.196	1.198	1.201	1.203	1.204	1.206	1.208	1.209
4100	1.125	1.126	1.127	1.137	1.152	1.163	1.172	1.179	1.185	1.190	1.194	1.198	1.201	1.203	1.206	1.208	1.210	1.211	1.213	1.214
4200	1.128	1.129	1.130	1.140	1.156	1.167	1.177	1.184	1.190	1.195	1.199	1.203	1.206	1.209	1.211	1.213	1.215	1.217	1.218	1.220
4300	1.131	1.132	1.133	1.143	1.159	1.172	1.181	1.188	1.195	1.200	1.204	1.208	1.211	1.214	1.216	1.219	1.221	1.222	1.224	1.225
4400	1.134	1.135	1.136	1.147	1.163	1.176	1.185	1.193	1.199	1.205	1.209	1.213	1.216	1.219	1.222	1.224	1.226	1.228	1.229	1.231
4500	1.137	1.138	1.139	1.150	1.167	1.180	1.190	1.198	1.204	1.209	1.214	1.218	1.221	1.224	1.227	1.229	1.231	1.233	1.235	1.236
4600	1.140	1.141	1.142	1.154	1.171	1.184	1.194	1.202	1.209	1.214	1.219	1.223	1.226	1.230	1.232	1.235	1.237	1.239	1.240	1.242
4700	1.143	1.144	1.146	1.157	1.175	1.188	1.198	1.207	1.214	1.219	1.224	1.228	1.232	1.235	1.238	1.240	1.242	1.244	1.246	1.248
4800	1.146	1.147	1.149	1.160	1.179	1.192	1.203	1.211	1.218	1.224	1.229	1.233	1.237	1.240	1.243	1.245	1.248	1.250	1.251	1.253
4900	1.149	1.151	1.152	1.164	1.182	1.196	1.207	1.216	1.223	1.229	1.234	1.238	1.242	1.245	1.248	1.251	1.253	1.255	1.257	1.259
5000	1.152	1.154	1.155	1.167	1.186	1.201	1.212	1.221	1.228	1.234	1.239	1.243	1.247	1.251	1.254	1.256	1.258	1.261	1.262	1.264
5100	1.155	1.157	1.158	1.171	1.190	1.205	1.216	1.225	1.233	1.239	1.244	1.249	1.252	1.256	1.259	1.262	1.264	1.266	1.268	1.270
5200	1.158	1.160	1.161	1.174	1.194	1.209	1.220	1.230	1.237	1.244	1.249	1.254	1.258	1.261	1.264	1.267	1.269	1.272	1.274	1.275
5300	1.162	1.163	1.164	1.177	1.198	1.213	1.225	1.234	1.242	1.249	1.254	1.259	1.263	1.266	1.270	1.272	1.275	1.277	1.279	1.281
5400	1.165	1.166	1.168	1.181	1.202	1.217	1.229	1.239	1.247	1.254	1.259	1.264	1.268	1.272	1.275	1.278	1.280	1.283	1.285	1.287
5500	1.168	1.169	1.171	1.184	1.205	1.221	1.234	1.244	1.252	1.259	1.264	1.269	1.273	1.277	1.280	1.283	1.286	1.288	1.290	1.292
5600	1.171	1.172	1.174	1.188	1.209	1.226	1.238	1.248	1.257	1.263	1.269	1.274	1.279	1.282	1.286	1.289	1.291	1.294	1.296	1.298
5700	1.174	1.175	1.177	1.191	1.213	1.230	1.243	1.253	1.261	1.268	1.274	1.280	1.284	1.288	1.291	1.294	1.297	1.299	1.302	1.304
5800	1.177	1.179	1.180	1.194	1.217	1.234	1.247	1.258	1.266	1.273	1.280	1.285	1.289	1.293	1.297	1.300	1.303	1.305	1.307	1.309
5900	1.180	1.182	1.183	1.198	1.221	1.238	1.252	1.262	1.271	1.278	1.285	1.290	1.295	1.299	1.302	1.305	1.308	1.311	1.313	1.315
6000	1.183	1.185	1.186	1.201	1.225	1.242	1.256	1.267	1.276	1.283	1.290	1.295	1.300	1.304	1.308	1.311	1.314	1.316	1.319	1.321
6100	1.186	1.188	1.190	1.205	1.229	1.247	1.261	1.272	1.281	1.288	1.295	1.300	1.305	1.309	1.313	1.316	1.319	1.322	1.324	1.326
6200	1.189	1.191	1.193	1.208	1.233	1.251	1.265	1.276	1.286	1.293	1.300	1.306	1.310	1.315	1.319	1.322	1.325	1.328	1.330	1.332
6300	1.192	1.194	1.196	1.212	1.236	1.255	1.270	1.281	1.291	1.298	1.305	1.311	1.316	1.320	1.324	1.327	1.330	1.333	1.336	1.338
6400	1.195	1.197	1.199	1.215	1.240	1.259	1.274	1.286	1.295	1.303	1.310	1.316	1.321	1.326	1.329	1.333	1.336	1.339	1.341	1.344
6500	1.199	1.200	1.202	1.219	1.244	1.264	1.279	1.291	1.300	1.309	1.315	1.321	1.327	1.331	1.335	1.339	1.342	1.345	1.347	1.349
6600	1.202	1.204	1.205	1.222	1.248	1.268	1.283	1.295	1.305	1.314	1.321	1.327	1.332	1.336	1.341	1.344	1.347	1.350	1.353	1.355
6700	1.205	1.207	1.209	1.225	1.252	1.272	1.288	1.300	1.310	1.319	1.326	1.332	1.337	1.342	1.346	1.350	1.353	1.356	1.359	1.361
6800	1.208	1.210	1.212	1.229	1.256	1.276	1.292	1.305	1.315	1.324	1.331	1.337	1.343	1.347	1.352	1.355	1.359	1.362	1.364	1.367
6900	1.211	1.213	1.215	1.232	1.260	1.281	1.297	1.310	1.320	1.329	1.336	1.343	1.348	1.353	1.357	1.361	1.364	1.367	1.370	1.373
7000	1.214	1.216	1.218	1.236	1.264	1.285	1.301	1.314	1.325	1.334	1.341	1.348	1.353	1.358	1.363	1.367	1.370	1.373	1.376	1.378
7100	1.217	1.219	1.221	1.239	1.268	1.289	1.306	1.319	1.330	1.339	1.347	1.353	1.359	1.364	1.368	1.372	1.376	1.379	1.382	1.384
7200	1.220	1.222	1.224	1.243	1.272	1.293	1.310	1.324	1.335	1.344	1.352	1.359	1.364	1.369	1.374	1.378	1.381	1.385	1.388	1.390

$V - V_c = 5$ SIGMA = 35600.0

H	3.4	3.2	3.0	2.8	2.6	2.4	2.2	2.0	1.8	1.6	1.4	1.2	1.0	0.8	0.6	0.4	0.2	0.10	0.09	0.08
7300	1.396	1.393	1.390	1.387	1.384	1.379	1.375	1.370	1.364	1.357	1.349	1.340	1.329	1.315	1.298	1.276	1.246	1.228	1.225	1.223
7400	1.402	1.399	1.396	1.393	1.389	1.385	1.380	1.375	1.369	1.362	1.354	1.345	1.333	1.319	1.302	1.280	1.250	1.231	1.229	1.226
7500	1.408	1.405	1.402	1.399	1.395	1.391	1.386	1.381	1.375	1.368	1.359	1.350	1.338	1.324	1.306	1.284	1.253	1.234	1.232	1.230
7600	1.414	1.411	1.408	1.404	1.401	1.396	1.392	1.386	1.380	1.373	1.365	1.355	1.343	1.329	1.311	1.287	1.257	1.237	1.235	1.233
7700	1.420	1.417	1.414	1.410	1.406	1.402	1.397	1.392	1.385	1.378	1.370	1.360	1.348	1.333	1.315	1.291	1.260	1.240	1.238	1.236
7800	1.426	1.423	1.419	1.416	1.412	1.408	1.403	1.397	1.391	1.383	1.375	1.365	1.353	1.338	1.319	1.295	1.264	1.243	1.241	1.239
7900	1.431	1.429	1.425	1.422	1.418	1.413	1.408	1.403	1.396	1.389	1.380	1.370	1.357	1.342	1.324	1.299	1.267	1.247	1.244	1.242
8000	1.437	1.434	1.431	1.428	1.423	1.419	1.414	1.408	1.402	1.394	1.385	1.375	1.362	1.347	1.328	1.303	1.271	1.250	1.247	1.245
8100	1.443	1.440	1.437	1.433	1.429	1.425	1.419	1.414	1.407	1.399	1.390	1.380	1.367	1.352	1.332	1.307	1.274	1.253	1.250	1.248
8200	1.449	1.446	1.443	1.439	1.435	1.430	1.425	1.419	1.413	1.405	1.396	1.385	1.372	1.356	1.337	1.311	1.278	1.256	1.254	1.251
8300	1.455	1.452	1.449	1.445	1.441	1.436	1.431	1.425	1.418	1.410	1.401	1.390	1.377	1.361	1.341	1.315	1.281	1.259	1.257	1.254
8400	1.461	1.458	1.455	1.451	1.447	1.442	1.436	1.430	1.423	1.415	1.406	1.395	1.382	1.366	1.345	1.319	1.285	1.262	1.260	1.258
8500	1.467	1.464	1.461	1.457	1.452	1.448	1.442	1.436	1.429	1.421	1.411	1.400	1.387	1.370	1.350	1.323	1.288	1.266	1.263	1.261
8600	1.473	1.470	1.467	1.463	1.458	1.453	1.448	1.442	1.434	1.426	1.417	1.405	1.392	1.375	1.354	1.327	1.292	1.269	1.266	1.264
8800	1.479	1.476	1.472	1.466	1.464	1.459	1.453	1.447	1.440	1.432	1.422	1.410	1.396	1.380	1.358	1.331	1.295	1.272	1.269	1.267
8900	1.485	1.482	1.478	1.474	1.470	1.465	1.459	1.453	1.445	1.437	1.427	1.415	1.401	1.384	1.363	1.335	1.299	1.275	1.273	1.270
9000	1.491	1.488	1.484	1.480	1.476	1.471	1.465	1.458	1.451	1.442	1.432	1.420	1.406	1.389	1.367	1.339	1.302	1.278	1.276	1.273
9100	1.498	1.494	1.490	1.486	1.482	1.476	1.470	1.464	1.456	1.448	1.438	1.426	1.411	1.394	1.372	1.343	1.306	1.282	1.279	1.276
9200	1.504	1.500	1.496	1.492	1.487	1.482	1.476	1.470	1.462	1.453	1.443	1.431	1.416	1.398	1.376	1.347	1.309	1.285	1.282	1.279
9300	1.510	1.506	1.502	1.498	1.493	1.488	1.482	1.475	1.467	1.458	1.448	1.436	1.421	1.403	1.380	1.351	1.313	1.288	1.285	1.283
9400	1.516	1.512	1.508	1.504	1.499	1.494	1.488	1.481	1.473	1.464	1.453	1.441	1.426	1.408	1.385	1.355	1.316	1.291	1.288	1.286
9500	1.522	1.518	1.514	1.510	1.505	1.500	1.494	1.487	1.479	1.470	1.459	1.446	1.431	1.412	1.389	1.359	1.320	1.294	1.292	1.289
9600	1.528	1.524	1.520	1.516	1.511	1.505	1.499	1.492	1.484	1.475	1.464	1.451	1.436	1.417	1.394	1.363	1.323	1.298	1.295	1.292
9700	1.534	1.531	1.526	1.522	1.517	1.511	1.505	1.498	1.490	1.480	1.469	1.456	1.441	1.422	1.398	1.367	1.327	1.301	1.298	1.295
9800	1.540	1.537	1.533	1.528	1.523	1.517	1.511	1.504	1.495	1.486	1.475	1.462	1.446	1.427	1.402	1.372	1.330	1.304	1.301	1.298
9900	1.547	1.543	1.539	1.534	1.529	1.523	1.517	1.509	1.501	1.491	1.480	1.467	1.451	1.431	1.407	1.376	1.334	1.307	1.304	1.301
10000	1.553	1.549	1.545	1.540	1.535	1.529	1.522	1.515	1.507	1.497	1.485	1.472	1.456	1.436	1.411	1.380	1.337	1.310	1.307	1.304
10100	1.559	1.555	1.551	1.546	1.541	1.535	1.528	1.521	1.512	1.502	1.491	1.477	1.461	1.441	1.416	1.384	1.341	1.314	1.311	1.308
10200	1.565	1.561	1.557	1.552	1.547	1.541	1.534	1.527	1.518	1.508	1.496	1.482	1.466	1.446	1.420	1.388	1.344	1.317	1.314	1.311
10300	1.571	1.567	1.563	1.558	1.553	1.547	1.540	1.532	1.523	1.513	1.502	1.488	1.471	1.450	1.425	1.392	1.348	1.320	1.317	1.314
10400	1.578	1.574	1.569	1.564	1.559	1.552	1.546	1.538	1.529	1.519	1.507	1.493	1.476	1.455	1.429	1.396	1.351	1.323	1.320	1.317
10500	1.584	1.580	1.575	1.570	1.565	1.559	1.552	1.544	1.535	1.524	1.512	1.498	1.481	1.460	1.434	1.400	1.355	1.326	1.323	1.320
10600	1.590	1.586	1.581	1.576	1.571	1.565	1.558	1.550	1.540	1.530	1.518	1.503	1.486	1.465	1.438	1.404	1.359	1.330	1.326	1.323
10700	1.596	1.592	1.588	1.583	1.577	1.571	1.563	1.555	1.546	1.536	1.523	1.509	1.491	1.469	1.443	1.408	1.362	1.333	1.330	1.326
10800	1.603	1.598	1.594	1.589	1.583	1.576	1.569	1.561	1.552	1.541	1.529	1.514	1.496	1.474	1.447	1.412	1.366	1.336	1.333	1.330
10900	1.609	1.605	1.600	1.595	1.589	1.582	1.575	1.567	1.558	1.547	1.534	1.519	1.501	1.479	1.452	1.416	1.369	1.339	1.336	1.333
11000	1.615	1.611	1.606	1.601	1.595	1.588	1.582	1.573	1.563	1.552	1.539	1.524	1.506	1.484	1.456	1.420	1.373	1.343	1.339	1.336
11100	1.622	1.617	1.611	1.607	1.601	1.594	1.587	1.579	1.569	1.556	1.545	1.530	1.511	1.489	1.461	1.425	1.376	1.346	1.342	1.339
11200	1.628	1.623	1.619	1.613	1.607	1.601	1.593	1.584	1.575	1.563	1.550	1.535	1.516	1.494	1.465	1.429	1.380	1.349	1.346	1.342
11300	1.634	1.630	1.625	1.619	1.613	1.607	1.599	1.590	1.580	1.569	1.556	1.540	1.521	1.498	1.470	1.433	1.384	1.352	1.349	1.345
11400	1.647	1.642	1.637	1.632	1.626	1.619	1.611	1.602	1.592	1.580	1.567	1.551	1.532	1.508	1.479	1.441	1.391	1.359	1.355	1.352

x

SIGMA = 35600.0

$V - V_c$ = 5

x \ H	0.08	0.09	0.10	0.2	0.4	0.6	0.8	1.0	1.2	1.4	1.6	1.8	2.0	2.2	2.4	2.6	2.8	3.0	3.2	3.4
11500	1.355	1.358	1.362	1.394	1.445	1.483	1.513	1.537	1.556	1.572	1.586	1.598	1.608	1.617	1.625	1.632	1.638	1.644	1.649	1.653
11600	1.358	1.362	1.365	1.398	1.449	1.488	1.518	1.542	1.561	1.578	1.592	1.604	1.614	1.623	1.631	1.638	1.644	1.650	1.655	1.660
11700	1.361	1.365	1.368	1.401	1.453	1.492	1.523	1.547	1.567	1.583	1.597	1.609	1.620	1.629	1.637	1.644	1.650	1.656	1.661	1.666
11800	1.364	1.368	1.372	1.405	1.458	1.497	1.528	1.552	1.572	1.589	1.603	1.615	1.626	1.635	1.643	1.650	1.657	1.662	1.668	1.672
11900	1.367	1.371	1.375	1.409	1.462	1.502	1.532	1.557	1.578	1.594	1.609	1.621	1.632	1.641	1.649	1.656	1.663	1.669	1.674	1.679
12000	1.370	1.374	1.378	1.412	1.466	1.506	1.537	1.562	1.583	1.600	1.614	1.627	1.638	1.647	1.655	1.663	1.669	1.675	1.680	1.685
12100	1.374	1.377	1.381	1.416	1.470	1.511	1.542	1.568	1.588	1.606	1.620	1.633	1.643	1.653	1.661	1.669	1.675	1.681	1.687	1.692
12200	1.377	1.381	1.385	1.419	1.474	1.515	1.547	1.573	1.594	1.611	1.626	1.638	1.649	1.659	1.667	1.675	1.682	1.688	1.693	1.698
12300	1.380	1.384	1.388	1.423	1.478	1.520	1.552	1.578	1.599	1.617	1.632	1.644	1.655	1.665	1.674	1.681	1.688	1.694	1.700	1.705
12400	1.383	1.387	1.391	1.427	1.482	1.524	1.557	1.583	1.604	1.622	1.637	1.650	1.661	1.671	1.680	1.687	1.694	1.701	1.706	1.711
12500	1.386	1.390	1.394	1.430	1.487	1.529	1.562	1.588	1.610	1.628	1.643	1.656	1.667	1.677	1.686	1.694	1.701	1.707	1.713	1.718
12600	1.389	1.393	1.397	1.434	1.491	1.534	1.567	1.594	1.615	1.633	1.649	1.662	1.673	1.683	1.692	1.700	1.707	1.713	1.719	1.724
12700	1.393	1.397	1.401	1.437	1.495	1.538	1.572	1.599	1.621	1.639	1.655	1.668	1.679	1.689	1.698	1.706	1.713	1.720	1.726	1.731
12800	1.396	1.400	1.404	1.441	1.499	1.543	1.577	1.604	1.626	1.645	1.660	1.674	1.685	1.696	1.705	1.713	1.720	1.726	1.732	1.737
12900	1.399	1.403	1.410	1.445	1.503	1.547	1.582	1.609	1.632	1.650	1.666	1.680	1.691	1.702	1.711	1.719	1.726	1.733	1.738	1.744
13000	1.402	1.406	1.410	1.448	1.508	1.552	1.587	1.614	1.637	1.656	1.672	1.686	1.698	1.708	1.717	1.725	1.732	1.739	1.745	1.750
13100	1.405	1.410	1.414	1.452	1.512	1.557	1.592	1.620	1.643	1.662	1.678	1.692	1.704	1.714	1.723	1.732	1.739	1.746	1.752	1.757
13200	1.408	1.413	1.417	1.455	1.516	1.561	1.597	1.625	1.648	1.667	1.684	1.698	1.710	1.720	1.730	1.738	1.745	1.752	1.758	1.764
13300	1.412	1.416	1.420	1.459	1.520	1.566	1.602	1.630	1.654	1.673	1.689	1.703	1.716	1.726	1.736	1.744	1.752	1.758	1.765	1.770
13400	1.415	1.419	1.424	1.463	1.524	1.571	1.607	1.635	1.659	1.679	1.695	1.709	1.722	1.733	1.742	1.751	1.758	1.765	1.771	1.777
13500	1.418	1.422	1.427	1.466	1.529	1.575	1.612	1.641	1.664	1.684	1.701	1.715	1.728	1.739	1.748	1.757	1.765	1.771	1.778	1.783
13600	1.421	1.426	1.430	1.470	1.533	1.580	1.617	1.646	1.670	1.690	1.707	1.721	1.734	1.745	1.755	1.763	1.771	1.778	1.784	1.790
13700	1.424	1.429	1.433	1.474	1.537	1.585	1.622	1.651	1.676	1.696	1.713	1.727	1.740	1.751	1.761	1.770	1.778	1.784	1.791	1.797
13800	1.427	1.432	1.437	1.477	1.541	1.589	1.627	1.657	1.681	1.701	1.719	1.733	1.746	1.757	1.767	1.776	1.784	1.791	1.797	1.803
13900	1.431	1.435	1.440	1.481	1.546	1.594	1.632	1.662	1.687	1.707	1.725	1.739	1.752	1.764	1.774	1.783	1.790	1.798	1.804	1.810
14000	1.434	1.438	1.443	1.485	1.550	1.599	1.637	1.667	1.692	1.713	1.730	1.745	1.759	1.770	1.780	1.789	1.797	1.804	1.811	1.817
14100	1.437	1.442	1.446	1.488	1.554	1.603	1.642	1.673	1.698	1.719	1.736	1.751	1.765	1.776	1.786	1.795	1.803	1.811	1.817	1.823
14200	1.440	1.445	1.450	1.492	1.558	1.608	1.647	1.678	1.703	1.724	1.742	1.758	1.771	1.782	1.793	1.802	1.810	1.817	1.824	1.830
14300	1.443	1.448	1.453	1.496	1.563	1.613	1.652	1.683	1.709	1.730	1.748	1.764	1.777	1.789	1.799	1.808	1.817	1.824	1.831	1.837
14400	1.447	1.451	1.456	1.499	1.567	1.618	1.657	1.689	1.714	1.736	1.754	1.770	1.783	1.795	1.806	1.815	1.823	1.831	1.837	1.843
14500	1.450	1.455	1.459	1.503	1.571	1.622	1.662	1.694	1.720	1.742	1.760	1.776	1.789	1.801	1.812	1.821	1.830	1.837	1.844	1.850
14600	1.453	1.458	1.463	1.506	1.575	1.627	1.667	1.699	1.726	1.747	1.766	1.782	1.796	1.808	1.818	1.828	1.836	1.844	1.851	1.857
14700	1.456	1.461	1.466	1.510	1.580	1.632	1.672	1.705	1.731	1.753	1.772	1.788	1.802	1.814	1.825	1.834	1.843	1.850	1.857	1.864
14800	1.459	1.464	1.469	1.514	1.584	1.636	1.677	1.710	1.737	1.759	1.778	1.794	1.808	1.821	1.831	1.841	1.849	1.857	1.864	1.871
14900	1.463	1.468	1.473	1.517	1.588	1.641	1.682	1.715	1.742	1.765	1.784	1.800	1.814	1.827	1.838	1.847	1.856	1.864	1.871	1.877
15000	1.466	1.471	1.476	1.521	1.592	1.646	1.688	1.721	1.748	1.771	1.790	1.806	1.821	1.833	1.844	1.854	1.863	1.871	1.878	1.884

SIGMA =   24720.0

V-V$_c$ = 6

H x	3.4	3.2	3.0	2.8	2.6	2.4	2.2	2.0	1.8	1.6	1.4	1.2	1.0	0.8	0.6	0.4	0.2	0.10	0.09	0.08
0	1.000	1.000	1.000	1.000	1.000	1.000	1.000	1.000	1.000	1.000	1.000	1.000	1.000	1.000	1.000	1.000	1.000	1.000	1.000	1.000
20	1.001	1.001	1.001	1.001	1.001	1.001	1.001	1.001	1.001	1.001	1.001	1.001	1.001	1.001	1.001	1.001	1.001	1.001	1.001	1.001
40	1.003	1.003	1.003	1.003	1.003	1.003	1.003	1.004	1.003	1.003	1.003	1.003	1.002	1.002	1.002	1.002	1.002	1.002	1.002	1.002
60	1.004	1.004	1.004	1.004	1.004	1.004	1.004	1.005	1.004	1.004	1.004	1.004	1.004	1.004	1.003	1.003	1.003	1.003	1.003	1.003
80	1.006	1.006	1.006	1.006	1.006	1.006	1.005	1.005	1.005	1.005	1.006	1.005	1.006	1.005	1.004	1.004	1.004	1.004	1.004	1.003
100	1.007	1.007	1.007	1.007	1.007	1.007	1.007	1.007	1.007	1.007	1.007	1.006	1.006	1.006	1.005	1.005	1.005	1.004	1.004	1.004
120	1.009	1.009	1.009	1.008	1.008	1.008	1.008	1.008	1.008	1.008	1.008	1.008	1.007	1.007	1.007	1.006	1.006	1.005	1.005	1.004
140	1.010	1.010	1.010	1.010	1.010	1.010	1.010	1.009	1.009	1.009	1.010	1.009	1.009	1.008	1.008	1.007	1.007	1.006	1.006	1.005
160	1.012	1.011	1.011	1.011	1.011	1.011	1.011	1.011	1.011	1.010	1.012	1.010	1.011	1.009	1.009	1.008	1.007	1.007	1.007	1.006
180	1.013	1.013	1.013	1.013	1.013	1.012	1.012	1.012	1.012	1.012	1.013	1.011	1.011	1.011	1.010	1.009	1.008	1.008	1.008	1.007
200	1.014	1.014	1.014	1.014	1.014	1.014	1.014	1.014	1.013	1.013	1.014	1.013	1.013	1.011	1.011	1.009	1.009	1.009	1.009	1.008
220	1.016	1.016	1.016	1.015	1.015	1.015	1.015	1.015	1.015	1.014	1.015	1.014	1.015	1.012	1.012	1.010	1.010	1.010	1.010	1.009
240	1.017	1.017	1.017	1.017	1.017	1.017	1.016	1.016	1.016	1.016	1.017	1.015	1.015	1.013	1.013	1.011	1.011	1.011	1.011	1.010
260	1.019	1.019	1.018	1.018	1.018	1.018	1.018	1.018	1.017	1.017	1.018	1.016	1.016	1.015	1.015	1.013	1.012	1.011	1.011	1.011
280	1.020	1.020	1.020	1.020	1.019	1.019	1.019	1.019	1.019	1.018	1.019	1.018	1.017	1.016	1.016	1.014	1.013	1.012	1.012	1.011
300	1.022	1.021	1.021	1.021	1.021	1.021	1.021	1.020	1.020	1.020	1.021	1.019	1.020	1.017	1.017	1.015	1.013	1.013	1.013	1.012
320	1.023	1.023	1.023	1.023	1.022	1.022	1.022	1.022	1.021	1.021	1.022	1.020	1.022	1.018	1.018	1.016	1.014	1.014	1.013	1.013
340	1.025	1.024	1.024	1.024	1.024	1.024	1.023	1.023	1.023	1.022	1.023	1.021	1.024	1.019	1.019	1.017	1.015	1.014	1.014	1.014
360	1.026	1.026	1.026	1.025	1.025	1.025	1.025	1.024	1.024	1.024	1.024	1.023	1.026	1.020	1.020	1.018	1.017	1.015	1.015	1.015
380	1.027	1.027	1.027	1.027	1.027	1.026	1.026	1.026	1.025	1.025	1.026	1.024	1.027	1.021	1.021	1.019	1.018	1.016	1.016	1.016
400	1.029	1.029	1.029	1.028	1.028	1.028	1.027	1.027	1.027	1.026	1.027	1.025	1.029	1.022	1.022	1.020	1.018	1.017	1.017	1.017
420	1.030	1.030	1.030	1.030	1.029	1.029	1.029	1.029	1.028	1.028	1.028	1.026	1.030	1.023	1.023	1.021	1.019	1.018	1.018	1.017
440	1.032	1.032	1.031	1.031	1.031	1.031	1.030	1.030	1.029	1.029	1.030	1.028	1.031	1.024	1.025	1.022	1.020	1.019	1.018	1.018
460	1.033	1.033	1.033	1.033	1.032	1.032	1.032	1.031	1.031	1.030	1.031	1.029	1.032	1.025	1.026	1.023	1.021	1.019	1.019	1.019
480	1.035	1.035	1.034	1.034	1.034	1.033	1.033	1.033	1.032	1.032	1.032	1.030	1.035	1.027	1.027	1.024	1.023	1.020	1.020	1.020
500	1.036	1.036	1.036	1.035	1.035	1.035	1.034	1.034	1.033	1.033	1.034	1.031	1.031	1.028	1.028	1.025	1.024	1.021	1.021	1.021
520	1.038	1.037	1.037	1.037	1.036	1.036	1.036	1.035	1.035	1.034	1.035	1.033	1.032	1.029	1.029	1.027	1.025	1.023	1.022	1.022
540	1.039	1.039	1.039	1.038	1.038	1.038	1.037	1.037	1.036	1.035	1.036	1.034	1.030	1.031	1.030	1.028	1.026	1.024	1.023	1.023
560	1.041	1.040	1.040	1.040	1.039	1.039	1.039	1.038	1.037	1.036	1.037	1.035	1.034	1.032	1.031	1.029	1.027	1.025	1.024	1.024
580	1.042	1.042	1.041	1.041	1.041	1.040	1.040	1.039	1.038	1.038	1.039	1.038	1.035	1.033	1.032	1.030	1.027	1.026	1.025	1.025
600	1.043	1.043	1.043	1.043	1.042	1.042	1.041	1.041	1.040	1.039	1.040	1.039	1.037	1.035	1.034	1.031	1.029	1.027	1.026	1.026
620	1.045	1.045	1.044	1.044	1.043	1.043	1.043	1.042	1.041	1.040	1.041	1.040	1.038	1.036	1.035	1.032	1.030	1.027	1.026	1.027
640	1.046	1.046	1.046	1.045	1.045	1.045	1.044	1.044	1.042	1.041	1.043	1.042	1.039	1.037	1.036	1.033	1.031	1.028	1.027	1.028
660	1.048	1.048	1.047	1.047	1.046	1.046	1.046	1.045	1.044	1.042	1.044	1.043	1.042	1.039	1.038	1.035	1.032	1.029	1.028	1.029
680	1.049	1.049	1.049	1.048	1.048	1.047	1.047	1.046	1.045	1.044	1.045	1.044	1.043	1.040	1.039	1.036	1.033	1.030	1.029	1.030
700	1.051	1.051	1.050	1.050	1.049	1.049	1.049	1.048	1.047	1.046	1.047	1.045	1.044	1.041	1.041	1.037	1.033	1.031	1.030	1.030
720	1.052	1.052	1.052	1.051	1.051	1.050	1.050	1.049	1.048	1.048	1.048	1.047	1.045	1.042	1.042	1.038	1.034	1.032	1.031	1.031
740	1.054	1.053	1.053	1.053	1.052	1.052	1.051	1.050	1.050	1.049	1.049	1.048	1.047	1.044	1.043	1.039	1.035	1.034	1.032	1.032
760	1.055	1.055	1.055	1.054	1.054	1.053	1.053	1.052	1.051	1.050	1.051	1.049	1.047	1.045	1.044	1.040	1.036	1.035	1.033	1.033
780	1.057	1.056	1.056	1.056	1.055	1.055	1.054	1.053	1.052	1.052	1.052	1.051	1.048	1.046	1.045	1.041	1.037	1.035	1.034	1.034
800	1.058	1.058	1.057	1.057	1.056	1.056	1.055	1.055	1.054	1.053	1.052	1.051	1.049	1.047	1.045	1.042	1.038	1.035	1.035	1.035

SIGMA = 24720.0

$V-V_c = 6$

H x	0.08	0.09	0.10	0.2	0.4	0.6	0.8	1.0	1.2	1.4	1.6	1.8	2.0	2.2	2.4	2.6	2.8	3.0	3.2	3.4
820	1.036	1.036	1.036	1.039	1.043	1.046	1.048	1.050	1.052	1.053	1.054	1.055	1.056	1.057	1.057	1.058	1.058	1.059	1.059	1.060
840	1.037	1.037	1.037	1.040	1.044	1.047	1.050	1.052	1.053	1.054	1.055	1.057	1.057	1.058	1.059	1.059	1.060	1.060	1.061	1.061
860	1.037	1.038	1.038	1.041	1.045	1.048	1.051	1.053	1.054	1.056	1.057	1.058	1.059	1.060	1.060	1.061	1.061	1.062	1.062	1.063
880	1.038	1.039	1.039	1.042	1.046	1.049	1.052	1.055	1.056	1.057	1.058	1.061	1.060	1.061	1.062	1.062	1.063	1.063	1.064	1.064
900	1.039	1.040	1.040	1.043	1.047	1.051	1.053	1.057	1.057	1.058	1.060	1.062	1.061	1.062	1.063	1.064	1.064	1.065	1.066	1.066
920	1.040	1.040	1.041	1.044	1.048	1.052	1.054	1.058	1.058	1.060	1.061	1.063	1.062	1.064	1.064	1.065	1.066	1.066	1.067	1.067
940	1.041	1.041	1.042	1.045	1.049	1.053	1.056	1.059	1.060	1.061	1.062	1.065	1.063	1.065	1.065	1.067	1.067	1.068	1.068	1.069
960	1.041	1.042	1.043	1.046	1.050	1.054	1.057	1.060	1.061	1.062	1.064	1.066	1.064	1.067	1.067	1.068	1.069	1.069	1.070	1.070
980	1.042	1.042	1.043	1.046	1.051	1.056	1.058	1.061	1.062	1.064	1.065	1.068	1.066	1.068	1.069	1.069	1.070	1.071	1.071	1.071
1000	1.043	1.043	1.044	1.047	1.052	1.057	1.060	1.063	1.064	1.065	1.066	1.069	1.067	1.069	1.070	1.071	1.071	1.072	1.073	1.073
1020	1.044	1.044	1.044	1.048	1.054	1.059	1.062	1.064	1.065	1.066	1.068	1.070	1.069	1.071	1.072	1.072	1.073	1.073	1.074	1.074
1040	1.044	1.044	1.045	1.049	1.055	1.060	1.063	1.065	1.066	1.068	1.069	1.072	1.070	1.072	1.073	1.074	1.074	1.075	1.075	1.076
1060	1.045	1.045	1.046	1.050	1.056	1.061	1.064	1.066	1.067	1.069	1.070	1.073	1.071	1.074	1.074	1.075	1.076	1.076	1.077	1.077
1080	1.046	1.046	1.047	1.051	1.057	1.062	1.065	1.068	1.069	1.072	1.072	1.074	1.073	1.075	1.076	1.077	1.077	1.078	1.078	1.079
1100	1.047	1.047	1.048	1.052	1.058	1.063	1.066	1.069	1.070	1.073	1.073	1.076	1.074	1.076	1.077	1.078	1.079	1.079	1.080	1.080
1120	1.048	1.048	1.049	1.053	1.059	1.064	1.068	1.070	1.071	1.074	1.074	1.077	1.075	1.078	1.079	1.080	1.080	1.081	1.081	1.082
1140	1.049	1.049	1.050	1.054	1.060	1.065	1.069	1.071	1.072	1.076	1.076	1.079	1.077	1.079	1.080	1.081	1.082	1.082	1.082	1.083
1160	1.050	1.050	1.051	1.055	1.061	1.066	1.070	1.073	1.074	1.077	1.077	1.080	1.078	1.081	1.082	1.082	1.083	1.084	1.084	1.085
1180	1.050	1.051	1.052	1.055	1.062	1.068	1.071	1.075	1.075	1.078	1.079	1.081	1.080	1.082	1.083	1.084	1.084	1.085	1.085	1.086
1200	1.051	1.051	1.053	1.056	1.063	1.069	1.072	1.077	1.076	1.080	1.080	1.083	1.081	1.083	1.084	1.085	1.086	1.087	1.087	1.088
1220	1.052	1.052	1.054	1.057	1.064	1.070	1.074	1.078	1.078	1.081	1.081	1.084	1.082	1.084	1.086	1.086	1.087	1.088	1.088	1.089
1240	1.053	1.053	1.055	1.058	1.065	1.071	1.075	1.079	1.079	1.082	1.082	1.085	1.084	1.086	1.087	1.088	1.088	1.089	1.089	1.090
1260	1.054	1.054	1.056	1.059	1.066	1.072	1.076	1.080	1.081	1.084	1.084	1.087	1.085	1.088	1.089	1.090	1.090	1.090	1.090	1.091
1280	1.055	1.055	1.057	1.060	1.067	1.073	1.077	1.082	1.083	1.085	1.085	1.088	1.086	1.089	1.090	1.091	1.091	1.091	1.092	1.092
1300	1.056	1.055	1.058	1.061	1.068	1.074	1.079	1.083	1.084	1.086	1.087	1.090	1.087	1.091	1.092	1.093	1.093	1.093	1.093	1.094
1320	1.057	1.056	1.059	1.062	1.069	1.076	1.080	1.084	1.085	1.088	1.088	1.091	1.089	1.092	1.093	1.094	1.094	1.094	1.095	1.095
1340	1.058	1.057	1.060	1.063	1.071	1.077	1.082	1.085	1.086	1.089	1.089	1.092	1.091	1.093	1.095	1.095	1.095	1.096	1.096	1.097
1360	1.059	1.058	1.061	1.064	1.072	1.078	1.083	1.087	1.088	1.092	1.092	1.094	1.092	1.095	1.096	1.097	1.098	1.097	1.098	1.098
1380	1.059	1.059	1.062	1.065	1.073	1.079	1.084	1.088	1.091	1.093	1.093	1.095	1.094	1.096	1.097	1.098	1.099	1.099	1.099	1.100
1400	1.060	1.060	1.063	1.066	1.074	1.080	1.086	1.090	1.092	1.096	1.095	1.098	1.095	1.098	1.100	1.100	1.101	1.100	1.101	1.101
1420	1.061	1.061	1.064	1.067	1.075	1.081	1.087	1.092	1.093	1.097	1.096	1.099	1.098	1.099	1.102	1.101	1.102	1.102	1.102	1.103
1440	1.062	1.062	1.065	1.068	1.076	1.083	1.088	1.093	1.094	1.098	1.098	1.101	1.099	1.101	1.103	1.103	1.104	1.103	1.104	1.104
1460	1.063	1.063	1.066	1.069	1.077	1.084	1.089	1.094	1.097	1.100	1.099	1.102	1.101	1.102	1.104	1.104	1.105	1.104	1.105	1.106
1480	1.064	1.064	1.067	1.070	1.078	1.085	1.090	1.095	1.098	1.101	1.100	1.103	1.102	1.103	1.105	1.107	1.107	1.106	1.107	1.108
1500	1.065	1.065	1.068	1.071	1.079	1.086	1.092	1.097	1.100	1.102	1.102	1.105	1.104	1.105	1.107	1.108	1.108	1.107	1.108	1.109
1520	1.065	1.066	1.068	1.072	1.080	1.087	1.093	1.098	1.102	1.104	1.103	1.106	1.105	1.106	1.108	1.109	1.110	1.109	1.110	1.110
1540	1.066	1.067	1.069	1.073	1.081	1.088	1.094	1.099	1.104	1.105	1.104	1.108	1.106	1.108	1.109	1.110	1.111	1.110	1.111	1.112
1560	1.067	1.068	1.070	1.074	1.082	1.089	1.095	1.100	1.105	1.106	1.106	1.109	1.108	1.109	1.113	1.112	1.114	1.112	1.113	1.113
1580	1.068	1.069	1.071	1.075	1.083	1.091	1.097	1.102	1.105	1.107	1.107	1.110	1.109	1.112	1.114	1.113	1.116	1.113	1.114	1.115
1600	1.070	1.070	1.072	1.076	1.084	1.092	1.097	1.102	1.105	1.108	1.108	1.111	1.111	1.114	1.115	1.116	1.117	1.116	1.117	1.118
1620	1.071	1.071	1.073	1.077	1.085	1.093	1.098	1.102	1.105	1.108	1.109	1.112	1.112	1.115	1.116	1.117	1.118	1.118	1.119	1.119
1640	1.072	1.072	1.073	1.078	1.087	1.093	1.098	1.102	1.105	1.108	1.110	1.112	1.114	1.115	1.116	1.117	1.118	1.119	1.120	1.121

SIGMA = 24720.0

$V-V_c = 6$

H / x	0.08	0.09	0.10	0.2	0.4	0.6	0.8	1.0	1.2	1.4	1.6	1.8	2.0	2.2	2.4	2.6	2.8	3.0	3.2	3.4
1660	1.072	1.073	1.074	1.079	1.088	1.094	1.099	1.103	1.106	1.109	1.111	1.113	1.115	1.116	1.118	1.119	1.120	1.121	1.122	1.123
1680	1.073	1.074	1.075	1.080	1.089	1.095	1.100	1.104	1.107	1.110	1.112	1.114	1.116	1.117	1.119	1.120	1.122	1.123	1.124	1.125
1700	1.074	1.075	1.076	1.081	1.090	1.096	1.101	1.106	1.109	1.112	1.114	1.116	1.118	1.119	1.121	1.122	1.123	1.124	1.125	1.126
1720	1.075	1.076	1.077	1.082	1.091	1.097	1.103	1.107	1.110	1.113	1.115	1.117	1.119	1.121	1.122	1.123	1.124	1.125	1.126	1.127
1740	1.076	1.077	1.077	1.083	1.092	1.099	1.104	1.108	1.111	1.114	1.117	1.119	1.121	1.122	1.124	1.125	1.126	1.128	1.129	1.130
1760	1.077	1.077	1.078	1.084	1.093	1.100	1.105	1.109	1.113	1.116	1.118	1.120	1.122	1.123	1.125	1.126	1.128	1.129	1.130	1.131
1780	1.077	1.078	1.079	1.085	1.094	1.101	1.106	1.111	1.114	1.117	1.119	1.121	1.124	1.125	1.127	1.128	1.129	1.130	1.132	1.132
1800	1.078	1.079	1.080	1.086	1.096	1.102	1.108	1.112	1.115	1.118	1.121	1.123	1.125	1.126	1.128	1.129	1.131	1.132	1.133	1.134
1820	1.079	1.080	1.081	1.087	1.097	1.103	1.109	1.113	1.117	1.120	1.122	1.124	1.127	1.128	1.130	1.131	1.132	1.133	1.134	1.135
1840	1.079	1.081	1.082	1.088	1.098	1.104	1.110	1.114	1.118	1.121	1.124	1.126	1.128	1.129	1.131	1.132	1.134	1.135	1.136	1.136
1860	1.080	1.082	1.083	1.089	1.099	1.106	1.111	1.116	1.119	1.122	1.125	1.127	1.130	1.131	1.132	1.134	1.135	1.136	1.137	1.138
1880	1.081	1.083	1.083	1.090	1.101	1.107	1.112	1.117	1.121	1.124	1.126	1.129	1.131	1.132	1.134	1.135	1.137	1.138	1.139	1.140
1900	1.082	1.084	1.084	1.091	1.101	1.108	1.114	1.118	1.122	1.125	1.128	1.130	1.132	1.134	1.135	1.137	1.138	1.139	1.140	1.141
1920	1.083	1.084	1.085	1.092	1.102	1.109	1.115	1.120	1.123	1.126	1.129	1.131	1.134	1.135	1.137	1.138	1.140	1.141	1.142	1.143
1940	1.084	1.085	1.086	1.093	1.103	1.110	1.116	1.121	1.125	1.128	1.130	1.133	1.135	1.137	1.138	1.140	1.141	1.142	1.144	1.145
1960	1.084	1.086	1.087	1.094	1.104	1.112	1.117	1.122	1.126	1.129	1.132	1.134	1.136	1.138	1.140	1.141	1.143	1.144	1.145	1.146
1980	1.085	1.087	1.088	1.095	1.105	1.113	1.119	1.124	1.127	1.131	1.133	1.136	1.138	1.139	1.141	1.143	1.144	1.145	1.146	1.148
2000	1.086	1.088	1.089	1.096	1.107	1.114	1.120	1.125	1.129	1.132	1.135	1.137	1.139	1.141	1.143	1.144	1.146	1.147	1.148	1.149
2020	1.087	1.089	1.090	1.097	1.107	1.115	1.121	1.126	1.130	1.134	1.136	1.139	1.141	1.142	1.144	1.146	1.147	1.148	1.149	1.151
2040	1.088	1.090	1.091	1.098	1.108	1.116	1.122	1.128	1.131	1.135	1.137	1.140	1.142	1.144	1.146	1.147	1.149	1.150	1.151	1.152
2060	1.089	1.091	1.092	1.099	1.109	1.117	1.123	1.129	1.133	1.136	1.139	1.141	1.144	1.145	1.147	1.149	1.150	1.151	1.152	1.154
2080	1.090	1.091	1.092	1.099	1.111	1.119	1.125	1.130	1.134	1.138	1.140	1.143	1.145	1.147	1.149	1.150	1.152	1.153	1.154	1.155
2100	1.090	1.092	1.093	1.100	1.112	1.120	1.126	1.131	1.136	1.139	1.142	1.144	1.147	1.148	1.150	1.152	1.153	1.154	1.155	1.156
2120	1.091	1.093	1.094	1.101	1.113	1.121	1.127	1.133	1.137	1.140	1.143	1.146	1.148	1.150	1.152	1.153	1.155	1.156	1.157	1.158
2140	1.091	1.094	1.095	1.102	1.114	1.122	1.128	1.134	1.138	1.142	1.144	1.147	1.150	1.151	1.153	1.155	1.156	1.157	1.158	1.160
2160	1.092	1.095	1.096	1.103	1.115	1.123	1.129	1.135	1.140	1.143	1.146	1.148	1.151	1.153	1.155	1.156	1.158	1.159	1.160	1.161
2180	1.093	1.096	1.097	1.104	1.116	1.124	1.131	1.137	1.141	1.145	1.147	1.150	1.152	1.154	1.156	1.158	1.159	1.160	1.161	1.163
2200	1.094	1.097	1.098	1.105	1.117	1.126	1.132	1.138	1.142	1.146	1.149	1.151	1.154	1.156	1.158	1.159	1.161	1.162	1.163	1.164
2220	1.094	1.098	1.099	1.106	1.118	1.127	1.133	1.139	1.144	1.147	1.150	1.153	1.155	1.157	1.159	1.161	1.162	1.163	1.164	1.166
2240	1.095	1.098	1.099	1.107	1.119	1.128	1.134	1.140	1.145	1.149	1.151	1.154	1.157	1.158	1.160	1.162	1.164	1.165	1.166	1.168
2260	1.096	1.099	1.100	1.108	1.120	1.129	1.136	1.142	1.146	1.150	1.153	1.155	1.158	1.160	1.162	1.164	1.165	1.166	1.167	1.169
2280	1.097	1.100	1.101	1.109	1.121	1.130	1.137	1.143	1.148	1.151	1.154	1.157	1.160	1.161	1.163	1.165	1.167	1.168	1.169	1.171
2300	1.098	1.101	1.102	1.110	1.122	1.131	1.138	1.144	1.149	1.153	1.155	1.158	1.161	1.163	1.165	1.167	1.168	1.169	1.170	1.172
2320	1.099	1.102	1.103	1.111	1.123	1.133	1.139	1.146	1.150	1.154	1.157	1.160	1.162	1.164	1.166	1.168	1.170	1.171	1.172	1.174
2340	1.100	1.103	1.104	1.112	1.124	1.134	1.140	1.147	1.152	1.155	1.158	1.161	1.164	1.166	1.168	1.170	1.171	1.172	1.173	1.175
2360	1.101	1.104	1.105	1.113	1.126	1.135	1.142	1.148	1.153	1.157	1.160	1.163	1.165	1.167	1.169	1.171	1.173	1.174	1.175	1.177
2380	1.101	1.105	1.106	1.114	1.127	1.136	1.143	1.150	1.154	1.158	1.161	1.164	1.167	1.169	1.171	1.173	1.174	1.175	1.176	1.178
2400	1.102	1.105	1.106	1.115	1.128	1.137	1.144	1.151	1.156	1.160	1.162	1.165	1.168	1.170	1.172	1.174	1.176	1.177	1.178	1.180
2420	1.103	1.106	1.107	1.116	1.129	1.138	1.145	1.152	1.157	1.161	1.164	1.167	1.170	1.172	1.174	1.176	1.177	1.178	1.179	1.181
2440	1.104	1.107	1.108	1.117	1.130	1.140	1.147	1.153	1.158	1.162	1.165	1.168	1.171	1.173	1.175	1.177	1.179	1.180	1.181	1.183
2460	1.105	1.108	1.109	1.118	1.131	1.141	1.148	1.154	1.159	1.163	1.167	1.170	1.172	1.175	1.177	1.179	1.180	1.182	1.183	1.184
2480	1.106	1.109	1.110	1.119	1.132	1.142	1.149	1.156	1.161	1.165	1.168	1.171	1.174	1.176	1.178	1.180	1.182	1.183	1.184	1.186

SIGMA = 24720.0

$V - V_c = 6$

X \ H	3.4	3.2	3.0	2.8	2.6	2.4	2.2	2.0	1.8	1.6	1.4	1.2	1.0	0.8	0.6	0.4	0.2	0.10	0.09	0.08
2500	1.187	1.186	1.185	1.183	1.182	1.180	1.178	1.175	1.173	1.170	1.166	1.162	1.157	1.151	1.143	1.133	1.120	1.112	1.110	1.109
2520	1.189	1.188	1.186	1.185	1.183	1.181	1.179	1.177	1.174	1.171	1.167	1.163	1.158	1.152	1.144	1.134	1.121	1.112	1.111	1.110
2540	1.190	1.188	1.188	1.186	1.185	1.183	1.181	1.178	1.176	1.172	1.169	1.165	1.159	1.153	1.145	1.135	1.122	1.113	1.112	1.111
2560	1.192	1.191	1.189	1.188	1.186	1.184	1.182	1.180	1.177	1.174	1.170	1.166	1.161	1.154	1.146	1.136	1.123	1.114	1.113	1.112
2580	1.193	1.192	1.191	1.189	1.188	1.186	1.184	1.181	1.178	1.175	1.172	1.167	1.162	1.156	1.148	1.137	1.124	1.116	1.114	1.113
2600	1.195	1.194	1.192	1.191	1.189	1.187	1.185	1.184	1.180	1.177	1.173	1.169	1.163	1.157	1.149	1.138	1.125	1.116	1.115	1.114
2620	1.197	1.195	1.194	1.192	1.191	1.189	1.187	1.185	1.181	1.178	1.174	1.170	1.165	1.158	1.150	1.139	1.126	1.118	1.116	1.115
2640	1.198	1.197	1.195	1.194	1.192	1.190	1.188	1.186	1.183	1.180	1.176	1.171	1.166	1.159	1.151	1.141	1.127	1.118	1.117	1.116
2660	1.200	1.199	1.197	1.195	1.194	1.192	1.190	1.187	1.184	1.181	1.177	1.173	1.167	1.161	1.152	1.142	1.128	1.119	1.118	1.116
2680	1.201	1.200	1.199	1.197	1.195	1.193	1.191	1.189	1.186	1.182	1.179	1.174	1.168	1.162	1.153	1.143	1.129	1.120	1.119	1.117
2700	1.203	1.202	1.200	1.198	1.197	1.195	1.193	1.190	1.187	1.184	1.180	1.176	1.170	1.163	1.155	1.145	1.130	1.121	1.120	1.118
2720	1.204	1.203	1.202	1.200	1.198	1.196	1.194	1.191	1.189	1.185	1.181	1.177	1.171	1.164	1.156	1.146	1.131	1.122	1.121	1.119
2740	1.206	1.205	1.203	1.202	1.200	1.198	1.196	1.193	1.190	1.187	1.183	1.178	1.172	1.166	1.157	1.147	1.132	1.123	1.122	1.120
2760	1.208	1.206	1.205	1.203	1.201	1.199	1.198	1.194	1.191	1.188	1.184	1.180	1.174	1.167	1.158	1.148	1.133	1.124	1.123	1.121
2780	1.209	1.208	1.206	1.205	1.203	1.201	1.199	1.196	1.193	1.189	1.185	1.181	1.175	1.168	1.159	1.149	1.134	1.125	1.124	1.122
2800	1.211	1.209	1.208	1.206	1.204	1.202	1.200	1.197	1.194	1.191	1.187	1.182	1.176	1.169	1.161	1.150	1.135	1.127	1.125	1.123
2820	1.212	1.211	1.209	1.208	1.206	1.204	1.202	1.199	1.196	1.192	1.188	1.184	1.178	1.171	1.162	1.151	1.136	1.127	1.126	1.123
2840	1.214	1.212	1.211	1.209	1.207	1.205	1.203	1.200	1.197	1.194	1.190	1.185	1.179	1.172	1.163	1.153	1.137	1.128	1.127	1.124
2860	1.215	1.214	1.213	1.211	1.209	1.207	1.205	1.202	1.199	1.195	1.191	1.187	1.180	1.173	1.164	1.154	1.138	1.129	1.128	1.125
2880	1.217	1.216	1.214	1.212	1.210	1.208	1.207	1.203	1.200	1.197	1.192	1.188	1.182	1.174	1.165	1.155	1.139	1.130	1.129	1.126
2900	1.219	1.217	1.216	1.214	1.212	1.210	1.209	1.205	1.202	1.198	1.194	1.189	1.183	1.176	1.166	1.156	1.140	1.131	1.130	1.127
2920	1.220	1.219	1.217	1.215	1.214	1.211	1.210	1.206	1.203	1.199	1.195	1.190	1.184	1.177	1.168	1.157	1.141	1.132	1.131	1.128
2940	1.222	1.220	1.219	1.217	1.215	1.213	1.211	1.208	1.205	1.201	1.197	1.192	1.185	1.178	1.169	1.158	1.142	1.133	1.132	1.129
2960	1.223	1.222	1.220	1.219	1.217	1.214	1.212	1.209	1.206	1.202	1.198	1.193	1.187	1.179	1.170	1.159	1.143	1.134	1.133	1.130
2980	1.225	1.223	1.222	1.220	1.218	1.216	1.213	1.211	1.207	1.204	1.199	1.194	1.188	1.181	1.171	1.160	1.144	1.134	1.133	1.131
3000	1.227	1.225	1.223	1.222	1.220	1.217	1.215	1.212	1.209	1.205	1.201	1.196	1.189	1.182	1.172	1.160	1.144	1.134	1.133	1.131

$V - V_c = 6$

SIGMA = 24720.0

H / x	3.4	3.2	3.0	2.8	2.6	2.4	2.2	2.0	1.8	1.6	1.4	1.2	1.0	0.8	0.6	0.4	0.2	0.10	0.09	0.08
3100	1.234	1.233	1.231	1.229	1.227	1.225	1.222	1.219	1.216	1.212	1.208	1.202	1.196	1.188	1.178	1.166	1.149	1.138	1.137	1.136
3200	1.242	1.241	1.239	1.237	1.235	1.233	1.230	1.227	1.223	1.219	1.215	1.209	1.203	1.194	1.184	1.171	1.154	1.143	1.142	1.140
3300	1.250	1.249	1.247	1.245	1.243	1.240	1.238	1.234	1.231	1.227	1.222	1.216	1.209	1.201	1.190	1.177	1.159	1.147	1.146	1.145
3400	1.258	1.257	1.255	1.253	1.251	1.248	1.245	1.242	1.238	1.234	1.229	1.223	1.216	1.207	1.196	1.182	1.164	1.152	1.150	1.149
3500	1.266	1.265	1.263	1.261	1.258	1.256	1.253	1.249	1.245	1.241	1.236	1.230	1.222	1.213	1.202	1.188	1.169	1.156	1.155	1.154
3600	1.275	1.273	1.271	1.269	1.266	1.263	1.260	1.257	1.253	1.248	1.243	1.237	1.229	1.220	1.208	1.193	1.173	1.161	1.159	1.158
3700	1.283	1.281	1.279	1.276	1.274	1.271	1.268	1.264	1.260	1.256	1.250	1.244	1.236	1.226	1.214	1.199	1.178	1.165	1.164	1.162
3800	1.291	1.289	1.287	1.284	1.282	1.279	1.276	1.272	1.268	1.263	1.257	1.250	1.242	1.233	1.220	1.204	1.183	1.170	1.168	1.167
3900	1.299	1.297	1.295	1.292	1.290	1.287	1.283	1.280	1.275	1.270	1.264	1.257	1.249	1.239	1.226	1.210	1.188	1.174	1.173	1.171
4000	1.307	1.305	1.303	1.300	1.298	1.295	1.291	1.287	1.283	1.278	1.271	1.264	1.256	1.245	1.232	1.216	1.193	1.179	1.177	1.176
4100	1.315	1.313	1.311	1.308	1.306	1.302	1.299	1.295	1.290	1.285	1.279	1.271	1.263	1.252	1.238	1.221	1.198	1.183	1.182	1.180
4200	1.323	1.321	1.319	1.316	1.313	1.310	1.307	1.302	1.298	1.292	1.286	1.278	1.269	1.258	1.244	1.227	1.203	1.188	1.186	1.185
4300	1.332	1.330	1.327	1.324	1.321	1.318	1.314	1.310	1.305	1.300	1.293	1.285	1.276	1.265	1.251	1.232	1.208	1.193	1.191	1.189
4400	1.340	1.338	1.335	1.332	1.329	1.326	1.322	1.318	1.313	1.307	1.300	1.292	1.283	1.271	1.257	1.238	1.213	1.197	1.195	1.194
4500	1.348	1.346	1.344	1.341	1.337	1.334	1.330	1.325	1.320	1.314	1.307	1.299	1.290	1.278	1.263	1.244	1.218	1.202	1.200	1.198
4600	1.357	1.354	1.352	1.349	1.345	1.342	1.338	1.333	1.328	1.322	1.315	1.306	1.296	1.284	1.269	1.249	1.223	1.206	1.204	1.202
4700	1.365	1.363	1.360	1.357	1.354	1.350	1.346	1.341	1.336	1.329	1.322	1.314	1.303	1.291	1.275	1.255	1.228	1.211	1.209	1.207
4800	1.373	1.371	1.368	1.365	1.362	1.358	1.354	1.349	1.343	1.337	1.329	1.321	1.310	1.297	1.281	1.260	1.233	1.215	1.213	1.211
4900	1.382	1.379	1.376	1.373	1.370	1.366	1.362	1.357	1.351	1.344	1.337	1.328	1.317	1.304	1.287	1.266	1.238	1.220	1.218	1.216
5000	1.390	1.388	1.385	1.381	1.378	1.374	1.369	1.364	1.359	1.352	1.344	1.335	1.324	1.310	1.293	1.272	1.243	1.225	1.222	1.220
5100	1.399	1.396	1.393	1.390	1.386	1.382	1.377	1.372	1.366	1.359	1.351	1.342	1.331	1.317	1.300	1.277	1.248	1.229	1.227	1.225
5200	1.407	1.404	1.401	1.398	1.394	1.390	1.385	1.380	1.374	1.367	1.359	1.349	1.338	1.324	1.306	1.283	1.253	1.234	1.231	1.229
5300	1.416	1.413	1.410	1.406	1.402	1.398	1.393	1.388	1.382	1.375	1.366	1.356	1.345	1.330	1.312	1.289	1.258	1.238	1.236	1.234
5400	1.424	1.421	1.418	1.415	1.411	1.406	1.401	1.396	1.390	1.382	1.374	1.364	1.352	1.337	1.318	1.294	1.263	1.243	1.240	1.238
5500	1.433	1.430	1.426	1.423	1.419	1.414	1.409	1.404	1.397	1.390	1.381	1.371	1.358	1.343	1.324	1.300	1.268	1.247	1.245	1.243
5600	1.441	1.438	1.435	1.431	1.427	1.423	1.418	1.412	1.405	1.398	1.389	1.378	1.365	1.350	1.331	1.306	1.273	1.252	1.249	1.247
5700	1.450	1.447	1.443	1.440	1.436	1.431	1.426	1.420	1.413	1.405	1.396	1.385	1.372	1.357	1.337	1.312	1.278	1.256	1.254	1.252
5800	1.458	1.455	1.452	1.448	1.444	1.439	1.434	1.428	1.421	1.413	1.404	1.393	1.379	1.363	1.343	1.317	1.283	1.261	1.259	1.256
5900	1.467	1.464	1.460	1.457	1.452	1.447	1.442	1.436	1.429	1.421	1.411	1.400	1.386	1.370	1.350	1.323	1.288	1.266	1.263	1.261
6000	1.476	1.473	1.469	1.465	1.461	1.456	1.450	1.444	1.437	1.429	1.419	1.407	1.394	1.377	1.356	1.329	1.293	1.270	1.268	1.265
6100	1.484	1.481	1.478	1.473	1.469	1.464	1.458	1.452	1.445	1.436	1.426	1.415	1.401	1.383	1.362	1.335	1.298	1.275	1.272	1.270
6200	1.493	1.490	1.486	1.482	1.477	1.472	1.466	1.460	1.452	1.444	1.434	1.422	1.408	1.390	1.368	1.340	1.303	1.279	1.277	1.274
6300	1.502	1.499	1.495	1.491	1.486	1.481	1.475	1.468	1.460	1.452	1.441	1.429	1.415	1.397	1.375	1.346	1.308	1.284	1.281	1.279
6400	1.511	1.507	1.503	1.499	1.494	1.489	1.483	1.476	1.468	1.459	1.449	1.437	1.422	1.404	1.381	1.352	1.313	1.289	1.286	1.283
6500	1.520	1.516	1.512	1.508	1.503	1.497	1.491	1.484	1.476	1.467	1.457	1.444	1.429	1.411	1.387	1.358	1.318	1.293	1.290	1.288
6600	1.528	1.525	1.521	1.516	1.511	1.506	1.500	1.493	1.484	1.475	1.464	1.452	1.436	1.417	1.394	1.364	1.323	1.298	1.295	1.292
6700	1.537	1.534	1.529	1.525	1.520	1.514	1.508	1.501	1.492	1.483	1.472	1.459	1.443	1.424	1.400	1.369	1.328	1.302	1.299	1.297
6800	1.546	1.542	1.538	1.534	1.528	1.523	1.516	1.509	1.501	1.491	1.480	1.466	1.450	1.431	1.407	1.375	1.334	1.307	1.304	1.301
6900	1.555	1.551	1.547	1.542	1.537	1.531	1.525	1.517	1.509	1.499	1.487	1.474	1.458	1.438	1.413	1.381	1.339	1.312	1.309	1.306
7000	1.564	1.560	1.556	1.551	1.545	1.540	1.533	1.525	1.517	1.507	1.495	1.481	1.465	1.445	1.419	1.387	1.344	1.316	1.313	1.310
7100	1.573	1.569	1.564	1.560	1.554	1.548	1.541	1.534	1.525	1.515	1.503	1.489	1.472	1.452	1.426	1.393	1.349	1.321	1.318	1.315
7200	1.582	1.578	1.573	1.568	1.563	1.557	1.550	1.542	1.533	1.523	1.511	1.496	1.479	1.458	1.432	1.399	1.354	1.325	1.322	1.319

SIGMA = 24720.0

V-Vc = 6

H	0.08	0.09	0.10	0.2	0.4	0.6	0.8	1.0	1.2	1.4	1.6	1.8	2.0	2.2	2.4	2.6	2.8	3.0	3.2	3.4
X																				
7300	1.324	1.327	1.330	1.359	1.405	1.439	1.465	1.487	1.504	1.518	1.531	1.541	1.550	1.558	1.565	1.572	1.577	1.582	1.587	1.591
7400	1.328	1.331	1.335	1.364	1.410	1.445	1.472	1.494	1.512	1.526	1.539	1.549	1.559	1.567	1.574	1.580	1.586	1.591	1.596	1.600
7500	1.333	1.336	1.339	1.369	1.416	1.452	1.479	1.501	1.519	1.534	1.547	1.558	1.567	1.575	1.583	1.589	1.595	1.600	1.605	1.609
7600	1.337	1.341	1.344	1.374	1.422	1.458	1.486	1.508	1.527	1.542	1.555	1.566	1.575	1.584	1.591	1.598	1.604	1.609	1.614	1.618
7700	1.342	1.345	1.349	1.380	1.428	1.465	1.493	1.516	1.534	1.550	1.563	1.574	1.584	1.592	1.600	1.607	1.612	1.618	1.623	1.627
7800	1.346	1.350	1.353	1.385	1.434	1.471	1.500	1.523	1.542	1.558	1.571	1.582	1.592	1.601	1.609	1.615	1.621	1.627	1.632	1.636
7900	1.351	1.354	1.358	1.390	1.440	1.478	1.507	1.530	1.550	1.566	1.579	1.591	1.601	1.610	1.617	1.624	1.630	1.636	1.641	1.645
8000	1.355	1.359	1.363	1.395	1.446	1.484	1.514	1.538	1.557	1.573	1.587	1.599	1.609	1.618	1.626	1.633	1.639	1.645	1.650	1.655
8100	1.360	1.364	1.367	1.400	1.452	1.491	1.521	1.545	1.565	1.581	1.595	1.607	1.618	1.627	1.635	1.642	1.648	1.654	1.659	1.664
8200	1.364	1.368	1.372	1.405	1.458	1.497	1.528	1.553	1.573	1.589	1.604	1.616	1.626	1.635	1.644	1.651	1.657	1.663	1.668	1.673
8300	1.369	1.373	1.377	1.411	1.464	1.504	1.535	1.560	1.580	1.597	1.612	1.624	1.635	1.644	1.652	1.660	1.666	1.672	1.677	1.682
8400	1.374	1.377	1.381	1.416	1.470	1.511	1.542	1.567	1.588	1.605	1.620	1.632	1.643	1.653	1.661	1.669	1.675	1.681	1.687	1.692
8500	1.378	1.382	1.386	1.421	1.476	1.517	1.549	1.575	1.596	1.613	1.628	1.641	1.652	1.662	1.670	1.678	1.684	1.690	1.696	1.701
8600	1.383	1.387	1.391	1.426	1.482	1.524	1.556	1.582	1.604	1.621	1.636	1.649	1.661	1.670	1.679	1.687	1.693	1.700	1.705	1.710
8700	1.387	1.391	1.395	1.431	1.488	1.530	1.563	1.590	1.611	1.629	1.645	1.658	1.669	1.679	1.688	1.696	1.703	1.709	1.714	1.720
8800	1.392	1.396	1.400	1.436	1.494	1.537	1.571	1.597	1.619	1.638	1.653	1.666	1.678	1.688	1.697	1.705	1.712	1.718	1.724	1.729
8900	1.396	1.400	1.405	1.442	1.500	1.544	1.578	1.605	1.627	1.646	1.661	1.675	1.686	1.697	1.706	1.714	1.721	1.727	1.733	1.738
9000	1.401	1.405	1.409	1.447	1.506	1.550	1.585	1.612	1.635	1.654	1.670	1.683	1.695	1.706	1.715	1.723	1.730	1.737	1.742	1.748
9100	1.405	1.410	1.414	1.452	1.512	1.557	1.592	1.620	1.643	1.662	1.678	1.692	1.704	1.714	1.724	1.732	1.739	1.746	1.752	1.757
9200	1.410	1.414	1.419	1.457	1.518	1.564	1.599	1.628	1.651	1.670	1.686	1.700	1.713	1.723	1.733	1.741	1.748	1.755	1.761	1.767
9300	1.415	1.419	1.423	1.463	1.524	1.570	1.606	1.635	1.659	1.678	1.695	1.709	1.721	1.732	1.742	1.750	1.758	1.765	1.771	1.776
9400	1.419	1.424	1.428	1.468	1.530	1.577	1.614	1.643	1.667	1.686	1.703	1.718	1.730	1.741	1.751	1.759	1.767	1.774	1.780	1.786
9500	1.424	1.428	1.433	1.473	1.536	1.584	1.621	1.650	1.674	1.695	1.712	1.726	1.739	1.750	1.760	1.769	1.776	1.783	1.790	1.795
9600	1.428	1.433	1.437	1.478	1.542	1.591	1.628	1.658	1.682	1.703	1.720	1.735	1.748	1.759	1.769	1.778	1.786	1.793	1.799	1.805
9700	1.433	1.438	1.442	1.483	1.549	1.597	1.635	1.666	1.690	1.711	1.729	1.744	1.757	1.768	1.778	1.787	1.795	1.802	1.809	1.815
9800	1.437	1.442	1.447	1.489	1.555	1.604	1.642	1.673	1.698	1.719	1.737	1.752	1.766	1.777	1.787	1.796	1.804	1.812	1.818	1.824
9900	1.442	1.447	1.451	1.494	1.561	1.611	1.650	1.681	1.706	1.728	1.746	1.761	1.774	1.787	1.796	1.806	1.814	1.821	1.828	1.834
10000	1.447	1.451	1.456	1.499	1.567	1.618	1.657	1.689	1.714	1.736	1.754	1.770	1.783	1.795	1.806	1.815	1.823	1.831	1.837	1.844
10100	1.451	1.456	1.461	1.504	1.573	1.624	1.664	1.696	1.722	1.744	1.763	1.779	1.792	1.804	1.815	1.824	1.833	1.840	1.847	1.853
10200	1.456	1.461	1.466	1.510	1.579	1.631	1.672	1.704	1.731	1.753	1.771	1.787	1.801	1.813	1.824	1.834	1.842	1.850	1.857	1.863
10300	1.460	1.465	1.470	1.515	1.585	1.638	1.679	1.712	1.739	1.761	1.780	1.796	1.810	1.822	1.834	1.843	1.852	1.859	1.866	1.873
10400	1.465	1.470	1.475	1.520	1.591	1.645	1.686	1.720	1.747	1.769	1.789	1.805	1.819	1.832	1.843	1.852	1.861	1.869	1.876	1.883
10500	1.470	1.475	1.480	1.526	1.598	1.652	1.694	1.727	1.755	1.778	1.797	1.814	1.828	1.841	1.852	1.861	1.871	1.879	1.886	1.892
10600	1.474	1.479	1.485	1.531	1.604	1.659	1.701	1.735	1.763	1.786	1.806	1.823	1.837	1.850	1.861	1.871	1.880	1.888	1.896	1.902
10700	1.479	1.484	1.489	1.536	1.610	1.665	1.708	1.743	1.771	1.795	1.815	1.832	1.846	1.859	1.871	1.880	1.890	1.898	1.905	1.912
10800	1.483	1.489	1.494	1.542	1.616	1.672	1.716	1.751	1.779	1.803	1.823	1.840	1.855	1.868	1.880	1.890	1.899	1.908	1.915	1.922
10900	1.488	1.493	1.499	1.547	1.622	1.679	1.723	1.759	1.788	1.812	1.832	1.849	1.865	1.878	1.890	1.900	1.909	1.917	1.925	1.932
11000	1.493	1.498	1.503	1.552	1.629	1.686	1.731	1.766	1.796	1.820	1.841	1.858	1.874	1.887	1.899	1.909	1.919	1.927	1.935	1.942
11100	1.497	1.503	1.508	1.557	1.635	1.693	1.738	1.774	1.804	1.829	1.849	1.867	1.883	1.896	1.908	1.919	1.928	1.937	1.945	1.952
11200	1.502	1.507	1.513	1.563	1.641	1.700	1.746	1.782	1.812	1.837	1.858	1.876	1.892	1.906	1.918	1.929	1.938	1.947	1.955	1.962
11300	1.506	1.512	1.518	1.568	1.647	1.707	1.753	1.790	1.820	1.846	1.867	1.885	1.901	1.915	1.927	1.938	1.948	1.957	1.965	1.972
11400	1.511	1.517	1.522	1.573	1.654	1.714	1.761	1.798	1.829	1.854	1.876	1.894	1.910	1.924	1.937	1.948	1.958	1.967	1.975	1.982

SIGMA = 24720.0

V − V_c = 6

H	0.08	0.09	0.10	0.2	0.4	0.6	0.8	1.0	1.2	1.4	1.6	1.8	2.0	2.2	2.4	2.6	2.8	3.0	3.2	3.4
11500	1.516	1.521	1.527	1.579	1.660	1.721	1.768	1.806	1.837	1.863	1.885	1.903	1.920	1.934	1.946	1.957	1.967	1.976	1.985	1.992
11600	1.520	1.526	1.532	1.584	1.666	1.728	1.776	1.814	1.845	1.871	1.894	1.912	1.929	1.943	1.956	1.967	1.977	1.986	1.995	2.002
11700	1.525	1.531	1.537	1.590	1.673	1.735	1.783	1.822	1.854	1.880	1.902	1.922	1.938	1.953	1.966	1.977	1.987	1.996	2.005	2.012
11800	1.530	1.536	1.541	1.595	1.679	1.742	1.791	1.830	1.862	1.889	1.911	1.931	1.947	1.962	1.975	1.987	1.997	2.006	2.015	2.022
11900	1.534	1.540	1.546	1.600	1.685	1.749	1.798	1.838	1.870	1.897	1.920	1.940	1.957	1.972	1.985	1.996	2.007	2.016	2.025	2.032
12000	1.539	1.545	1.551	1.606	1.691	1.756	1.806	1.846	1.879	1.906	1.929	1.949	1.966	1.981	1.994	2.006	2.017	2.026	2.035	2.043
12100	1.543	1.550	1.556	1.611	1.698	1.763	1.814	1.854	1.887	1.915	1.938	1.958	1.976	1.991	2.004	2.016	2.027	2.036	2.045	2.053
12200	1.548	1.554	1.561	1.616	1.704	1.770	1.821	1.862	1.896	1.924	1.947	1.967	1.985	2.000	2.014	2.026	2.037	2.046	2.055	2.063
12300	1.553	1.559	1.565	1.622	1.710	1.777	1.829	1.870	1.904	1.932	1.956	1.977	1.994	2.010	2.024	2.036	2.047	2.056	2.065	2.073
12400	1.557	1.564	1.570	1.627	1.717	1.784	1.836	1.878	1.912	1.941	1.965	1.986	2.004	2.019	2.033	2.046	2.057	2.067	2.076	2.084
12500	1.562	1.569	1.575	1.633	1.723	1.791	1.844	1.886	1.921	1.950	1.974	1.995	2.013	2.029	2.043	2.056	2.067	2.077	2.086	2.094
12600	1.567	1.573	1.580	1.638	1.730	1.798	1.852	1.894	1.929	1.959	1.983	2.004	2.023	2.039	2.053	2.065	2.077	2.087	2.096	2.104
12700	1.571	1.578	1.584	1.643	1.736	1.805	1.859	1.903	1.938	1.967	1.992	2.014	2.032	2.048	2.063	2.075	2.087	2.097	2.106	2.115
12800	1.576	1.583	1.589	1.649	1.742	1.813	1.867	1.911	1.946	1.976	2.001	2.023	2.042	2.058	2.073	2.085	2.097	2.107	2.117	2.125
12900	1.581	1.587	1.594	1.654	1.749	1.820	1.875	1.919	1.955	1.985	2.011	2.032	2.051	2.068	2.082	2.095	2.107	2.117	2.127	2.136
13000	1.585	1.592	1.599	1.660	1.755	1.827	1.883	1.927	1.964	1.994	2.020	2.042	2.061	2.078	2.092	2.105	2.117	2.128	2.137	2.146
13100	1.590	1.597	1.604	1.665	1.762	1.834	1.890	1.935	1.972	2.003	2.029	2.051	2.070	2.087	2.102	2.115	2.127	2.138	2.148	2.156
13200	1.595	1.602	1.608	1.670	1.768	1.841	1.898	1.944	1.981	2.012	2.038	2.061	2.080	2.097	2.112	2.126	2.138	2.148	2.158	2.167
13300	1.599	1.606	1.613	1.676	1.774	1.848	1.906	1.952	1.989	2.021	2.047	2.070	2.090	2.107	2.122	2.136	2.148	2.159	2.169	2.177
13400	1.604	1.611	1.618	1.681	1.781	1.856	1.914	1.960	1.998	2.030	2.056	2.079	2.099	2.117	2.132	2.146	2.158	2.169	2.179	2.188
13500	1.609	1.616	1.623	1.687	1.787	1.863	1.921	1.968	2.007	2.039	2.066	2.089	2.109	2.127	2.142	2.156	2.168	2.179	2.189	2.199
13600	1.613	1.621	1.628	1.692	1.794	1.870	1.929	1.977	2.015	2.048	2.075	2.098	2.119	2.136	2.152	2.166	2.179	2.190	2.200	2.209
13700	1.618	1.625	1.633	1.698	1.800	1.877	1.937	1.985	2.024	2.057	2.084	2.108	2.128	2.146	2.162	2.176	2.189	2.200	2.210	2.220
13800	1.623	1.630	1.637	1.703	1.807	1.884	1.945	1.993	2.033	2.066	2.094	2.117	2.138	2.156	2.172	2.187	2.199	2.211	2.221	2.230
13900	1.627	1.635	1.642	1.709	1.813	1.892	1.953	2.002	2.041	2.075	2.103	2.127	2.148	2.166	2.182	2.197	2.210	2.221	2.232	2.241
14000	1.632	1.640	1.647	1.714	1.820	1.899	1.961	2.010	2.050	2.084	2.112	2.137	2.158	2.176	2.193	2.207	2.220	2.232	2.242	2.252
14100	1.637	1.644	1.652	1.720	1.826	1.906	1.968	2.018	2.059	2.093	2.122	2.146	2.168	2.186	2.203	2.217	2.230	2.242	2.253	2.263
14200	1.641	1.649	1.657	1.725	1.833	1.914	1.976	2.027	2.068	2.102	2.131	2.156	2.177	2.196	2.213	2.228	2.241	2.253	2.263	2.273
14300	1.646	1.654	1.661	1.731	1.839	1.921	1.984	2.035	2.077	2.111	2.140	2.165	2.187	2.206	2.223	2.238	2.251	2.263	2.274	2.284
14400	1.651	1.659	1.666	1.736	1.846	1.928	1.992	2.043	2.085	2.120	2.150	2.175	2.197	2.216	2.233	2.248	2.262	2.274	2.285	2.295
14500	1.656	1.663	1.671	1.742	1.852	1.936	2.000	2.052	2.094	2.129	2.159	2.185	2.207	2.226	2.243	2.259	2.272	2.285	2.296	2.306
14600	1.660	1.668	1.676	1.747	1.859	1.943	2.008	2.060	2.103	2.139	2.169	2.195	2.217	2.236	2.254	2.269	2.283	2.295	2.306	2.317
14700	1.665	1.673	1.681	1.753	1.866	1.950	2.016	2.069	2.112	2.148	2.178	2.204	2.227	2.247	2.264	2.280	2.293	2.306	2.317	2.327
14800	1.670	1.678	1.686	1.758	1.872	1.958	2.024	2.077	2.121	2.157	2.188	2.214	2.237	2.257	2.274	2.290	2.304	2.317	2.328	2.338
14900	1.674	1.683	1.691	1.764	1.879	1.965	2.032	2.086	2.130	2.166	2.197	2.224	2.247	2.267	2.285	2.300	2.315	2.327	2.339	2.349
15000	1.679	1.687	1.695	1.769	1.885	1.972	2.040	2.094	2.139	2.176	2.207	2.234	2.257	2.277	2.295	2.311	2.325	2.338	2.350	2.360

SIGMA = 18160.0

$V - V_c = 7$

H/x	0.08	0.09	0.10	0.2	0.4	0.6	0.8	1.0	1.2	1.4	1.6	1.8	2.0	2.2	2.4	2.6	2.8	3.0	3.2	3.4
0	1.000	1.000	1.000	1.000	1.000	1.000	1.000	1.000	1.000	1.000	1.000	1.000	1.000	1.000	1.000	1.000	1.000	1.000	1.000	1.000
20	1.001	1.001	1.001	1.001	1.001	1.002	1.002	1.002	1.002	1.002	1.002	1.002	1.002	1.002	1.002	1.002	1.002	1.002	1.002	1.002
40	1.002	1.002	1.002	1.003	1.003	1.003	1.003	1.003	1.003	1.004	1.004	1.004	1.004	1.004	1.004	1.004	1.004	1.004	1.004	1.004
60	1.004	1.004	1.004	1.004	1.004	1.005	1.005	1.005	1.005	1.005	1.005	1.005	1.006	1.006	1.006	1.006	1.006	1.006	1.006	1.006
80	1.005	1.005	1.005	1.006	1.006	1.006	1.006	1.007	1.007	1.007	1.007	1.007	1.007	1.007	1.008	1.008	1.008	1.008	1.008	1.008
100	1.006	1.006	1.006	1.007	1.007	1.008	1.008	1.008	1.009	1.009	1.009	1.009	1.009	1.009	1.009	1.010	1.010	1.010	1.010	1.010
120	1.007	1.007	1.007	1.008	1.009	1.009	1.010	1.010	1.010	1.010	1.011	1.011	1.011	1.011	1.011	1.011	1.012	1.012	1.012	1.012
140	1.008	1.008	1.008	1.009	1.011	1.011	1.011	1.012	1.012	1.012	1.012	1.013	1.013	1.013	1.013	1.013	1.013	1.014	1.014	1.014
160	1.009	1.010	1.010	1.011	1.012	1.012	1.013	1.013	1.014	1.014	1.014	1.014	1.015	1.015	1.015	1.015	1.015	1.015	1.016	1.016
180	1.011	1.011	1.011	1.012	1.013	1.014	1.014	1.015	1.015	1.016	1.016	1.016	1.017	1.017	1.017	1.017	1.017	1.017	1.018	1.018
200	1.012	1.012	1.012	1.013	1.014	1.015	1.016	1.017	1.017	1.018	1.018	1.018	1.019	1.019	1.019	1.019	1.019	1.020	1.020	1.020
220	1.013	1.013	1.013	1.014	1.016	1.017	1.018	1.018	1.019	1.019	1.020	1.020	1.021	1.021	1.021	1.021	1.021	1.021	1.021	1.022
240	1.014	1.014	1.014	1.016	1.017	1.018	1.019	1.020	1.021	1.021	1.022	1.022	1.022	1.022	1.023	1.023	1.023	1.023	1.023	1.023
260	1.015	1.016	1.016	1.017	1.018	1.020	1.021	1.022	1.022	1.023	1.023	1.024	1.024	1.024	1.025	1.025	1.025	1.025	1.025	1.025
280	1.017	1.017	1.017	1.018	1.020	1.021	1.022	1.023	1.024	1.025	1.025	1.026	1.026	1.026	1.026	1.027	1.027	1.027	1.027	1.028
300	1.018	1.018	1.018	1.020	1.021	1.023	1.024	1.025	1.026	1.026	1.027	1.027	1.028	1.028	1.028	1.029	1.029	1.029	1.029	1.029
320	1.019	1.019	1.019	1.021	1.023	1.024	1.026	1.027	1.027	1.028	1.029	1.029	1.030	1.030	1.030	1.031	1.031	1.031	1.031	1.031
340	1.020	1.020	1.020	1.022	1.024	1.026	1.027	1.028	1.029	1.030	1.030	1.031	1.031	1.032	1.032	1.032	1.033	1.033	1.033	1.033
360	1.022	1.021	1.021	1.023	1.026	1.027	1.029	1.030	1.031	1.032	1.032	1.033	1.033	1.034	1.034	1.034	1.035	1.035	1.035	1.035
380	1.023	1.023	1.023	1.024	1.027	1.029	1.030	1.032	1.033	1.033	1.034	1.035	1.035	1.036	1.036	1.036	1.037	1.037	1.037	1.037
400	1.024	1.024	1.024	1.026	1.028	1.030	1.032	1.033	1.034	1.035	1.036	1.037	1.037	1.038	1.038	1.038	1.039	1.039	1.039	1.039
420	1.025	1.025	1.025	1.027	1.030	1.032	1.034	1.035	1.036	1.037	1.038	1.038	1.039	1.039	1.040	1.040	1.041	1.041	1.041	1.041
440	1.026	1.026	1.026	1.028	1.031	1.034	1.035	1.037	1.037	1.039	1.040	1.040	1.041	1.041	1.042	1.042	1.043	1.043	1.043	1.043
460	1.027	1.027	1.028	1.030	1.033	1.035	1.037	1.038	1.039	1.040	1.041	1.042	1.043	1.043	1.044	1.044	1.044	1.045	1.045	1.045
480	1.028	1.029	1.029	1.031	1.034	1.037	1.038	1.040	1.041	1.042	1.043	1.044	1.045	1.045	1.046	1.046	1.046	1.047	1.047	1.047
500	1.030	1.030	1.030	1.032	1.036	1.038	1.040	1.042	1.042	1.044	1.045	1.046	1.046	1.047	1.048	1.048	1.048	1.049	1.049	1.049
520	1.031	1.031	1.031	1.034	1.037	1.040	1.042	1.043	1.044	1.046	1.047	1.048	1.048	1.049	1.049	1.050	1.050	1.051	1.051	1.051
540	1.032	1.032	1.032	1.035	1.038	1.042	1.043	1.045	1.046	1.048	1.049	1.050	1.050	1.051	1.051	1.052	1.052	1.053	1.053	1.053
560	1.033	1.033	1.034	1.036	1.040	1.043	1.045	1.047	1.048	1.049	1.050	1.051	1.052	1.053	1.053	1.054	1.054	1.055	1.055	1.055
580	1.034	1.035	1.035	1.037	1.041	1.044	1.047	1.048	1.049	1.051	1.052	1.053	1.054	1.055	1.055	1.056	1.056	1.057	1.057	1.057
600	1.035	1.036	1.036	1.039	1.044	1.046	1.048	1.050	1.051	1.053	1.054	1.055	1.056	1.057	1.057	1.058	1.058	1.059	1.059	1.059
620	1.037	1.037	1.037	1.040	1.044	1.047	1.050	1.052	1.053	1.055	1.056	1.057	1.058	1.058	1.059	1.060	1.060	1.061	1.061	1.061
640	1.038	1.038	1.038	1.041	1.046	1.049	1.051	1.053	1.055	1.057	1.058	1.059	1.060	1.060	1.061	1.062	1.062	1.063	1.063	1.063
660	1.039	1.039	1.040	1.043	1.047	1.050	1.053	1.055	1.057	1.058	1.060	1.061	1.061	1.062	1.063	1.064	1.064	1.065	1.065	1.065
680	1.040	1.040	1.041	1.044	1.049	1.052	1.055	1.057	1.058	1.060	1.061	1.062	1.063	1.064	1.065	1.066	1.066	1.067	1.067	1.067
700	1.042	1.042	1.042	1.045	1.050	1.054	1.056	1.059	1.060	1.062	1.063	1.064	1.065	1.066	1.067	1.067	1.068	1.069	1.069	1.069
720	1.043	1.043	1.043	1.047	1.051	1.055	1.058	1.060	1.062	1.064	1.065	1.066	1.067	1.068	1.069	1.069	1.070	1.071	1.071	1.071
740	1.044	1.044	1.045	1.048	1.053	1.057	1.060	1.062	1.064	1.065	1.067	1.068	1.069	1.070	1.071	1.071	1.072	1.073	1.073	1.074
760	1.045	1.045	1.046	1.049	1.054	1.058	1.061	1.064	1.066	1.067	1.069	1.070	1.071	1.072	1.073	1.073	1.074	1.075	1.075	1.076
780	1.047	1.047	1.047	1.050	1.056	1.060	1.063	1.065	1.067	1.069	1.071	1.072	1.073	1.074	1.075	1.075	1.076	1.077	1.077	1.078
800	1.047	1.048	1.048	1.052	1.057	1.061	1.064	1.067	1.069	1.071	1.072	1.074	1.075	1.076	1.077	1.077	1.078	1.079	1.079	1.080

SIGMA = 18160.0

V−V_c = 7

H \ x	0.08	0.09	0.10	0.2	0.4	0.6	0.8	1.0	1.2	1.4	1.6	1.8	2.0	2.2	2.4	2.6	2.8	3.0	3.2	3.4
820	1.049	1.049	1.049	1.053	1.059	1.063	1.066	1.069	1.071	1.073	1.074	1.075	1.077	1.078	1.078	1.079	1.080	1.081	1.081	1.082
840	1.050	1.050	1.051	1.054	1.060	1.064	1.068	1.070	1.073	1.074	1.076	1.077	1.079	1.080	1.080	1.081	1.081	1.083	1.083	1.084
860	1.051	1.051	1.052	1.056	1.062	1.066	1.069	1.072	1.074	1.076	1.078	1.079	1.080	1.081	1.082	1.083	1.084	1.085	1.085	1.086
880	1.052	1.053	1.053	1.057	1.063	1.068	1.071	1.074	1.076	1.078	1.080	1.081	1.082	1.083	1.084	1.085	1.086	1.087	1.087	1.088
900	1.053	1.054	1.054	1.058	1.064	1.069	1.073	1.076	1.078	1.080	1.082	1.083	1.084	1.085	1.086	1.087	1.088	1.089	1.089	1.090
920	1.054	1.055	1.055	1.060	1.066	1.071	1.074	1.077	1.080	1.082	1.083	1.085	1.086	1.087	1.088	1.089	1.090	1.091	1.091	1.092
940	1.055	1.056	1.057	1.061	1.067	1.072	1.076	1.079	1.081	1.084	1.085	1.087	1.088	1.089	1.090	1.091	1.092	1.093	1.093	1.094
960	1.056	1.058	1.058	1.062	1.069	1.074	1.078	1.081	1.083	1.085	1.087	1.089	1.090	1.091	1.092	1.093	1.094	1.095	1.095	1.096
980	1.057	1.059	1.059	1.064	1.070	1.075	1.079	1.082	1.085	1.087	1.089	1.091	1.092	1.093	1.094	1.095	1.096	1.097	1.097	1.098
1000	1.058	1.060	1.060	1.065	1.072	1.077	1.081	1.084	1.087	1.089	1.091	1.092	1.094	1.095	1.096	1.097	1.098	1.099	1.099	1.100
1020	1.059	1.061	1.062	1.066	1.073	1.078	1.083	1.086	1.089	1.091	1.093	1.094	1.096	1.097	1.098	1.099	1.100	1.101	1.101	1.102
1040	1.061	1.062	1.063	1.067	1.075	1.080	1.084	1.088	1.090	1.093	1.095	1.096	1.098	1.099	1.100	1.101	1.102	1.103	1.103	1.104
1060	1.062	1.063	1.064	1.069	1.076	1.082	1.086	1.089	1.092	1.094	1.096	1.098	1.100	1.101	1.102	1.103	1.104	1.105	1.105	1.106
1080	1.063	1.065	1.065	1.070	1.078	1.083	1.087	1.091	1.094	1.096	1.098	1.100	1.101	1.103	1.104	1.105	1.106	1.107	1.107	1.108
1100	1.064	1.066	1.066	1.071	1.079	1.085	1.089	1.093	1.096	1.098	1.100	1.102	1.103	1.105	1.106	1.107	1.108	1.109	1.110	1.110
1120	1.065	1.067	1.067	1.073	1.081	1.086	1.091	1.094	1.097	1.100	1.102	1.104	1.105	1.107	1.108	1.109	1.110	1.111	1.112	1.112
1140	1.066	1.068	1.069	1.074	1.082	1.088	1.092	1.096	1.099	1.102	1.104	1.106	1.107	1.109	1.110	1.111	1.112	1.113	1.114	1.114
1160	1.067	1.069	1.070	1.075	1.084	1.089	1.094	1.098	1.101	1.104	1.106	1.108	1.109	1.111	1.112	1.113	1.114	1.115	1.116	1.116
1180	1.068	1.071	1.071	1.077	1.085	1.091	1.096	1.100	1.103	1.105	1.108	1.109	1.111	1.113	1.114	1.115	1.116	1.117	1.118	1.118
1200	1.069	1.072	1.072	1.078	1.087	1.092	1.097	1.101	1.105	1.107	1.109	1.111	1.113	1.115	1.116	1.117	1.118	1.119	1.120	1.121
1220	1.070	1.073	1.074	1.079	1.088	1.094	1.099	1.103	1.106	1.109	1.111	1.113	1.115	1.116	1.118	1.119	1.120	1.121	1.122	1.123
1240	1.071	1.074	1.075	1.081	1.090	1.096	1.101	1.105	1.108	1.111	1.113	1.115	1.117	1.118	1.120	1.121	1.122	1.123	1.124	1.125
1260	1.072	1.075	1.076	1.082	1.091	1.097	1.102	1.106	1.110	1.113	1.115	1.117	1.119	1.120	1.122	1.123	1.124	1.125	1.126	1.127
1280	1.073	1.077	1.077	1.083	1.093	1.099	1.104	1.108	1.112	1.114	1.117	1.119	1.121	1.122	1.124	1.125	1.126	1.127	1.128	1.129
1300	1.074	1.078	1.079	1.085	1.094	1.100	1.106	1.110	1.113	1.116	1.119	1.121	1.123	1.124	1.126	1.127	1.128	1.129	1.130	1.131
1320	1.075	1.079	1.080	1.086	1.096	1.102	1.107	1.112	1.115	1.118	1.121	1.123	1.125	1.126	1.128	1.129	1.130	1.131	1.132	1.133
1340	1.076	1.080	1.081	1.087	1.097	1.104	1.109	1.113	1.117	1.120	1.123	1.125	1.127	1.128	1.130	1.131	1.132	1.133	1.134	1.135
1360	1.077	1.081	1.082	1.089	1.099	1.105	1.111	1.115	1.119	1.122	1.124	1.127	1.129	1.130	1.132	1.133	1.134	1.135	1.136	1.137
1380	1.078	1.083	1.083	1.090	1.100	1.107	1.112	1.117	1.121	1.124	1.126	1.129	1.131	1.132	1.134	1.135	1.136	1.137	1.138	1.139
1400	1.080	1.084	1.085	1.092	1.102	1.108	1.114	1.119	1.122	1.126	1.128	1.130	1.132	1.134	1.136	1.137	1.138	1.139	1.140	1.141
1420	1.081	1.085	1.086	1.093	1.103	1.110	1.116	1.120	1.124	1.127	1.130	1.132	1.134	1.136	1.138	1.139	1.140	1.141	1.142	1.143
1440	1.082	1.086	1.088	1.094	1.105	1.111	1.117	1.122	1.126	1.129	1.132	1.134	1.136	1.138	1.140	1.141	1.142	1.143	1.145	1.145
1460	1.083	1.088	1.089	1.096	1.106	1.113	1.119	1.124	1.128	1.131	1.134	1.136	1.138	1.140	1.142	1.143	1.144	1.146	1.147	1.148
1480	1.084	1.089	1.090	1.097	1.108	1.115	1.121	1.126	1.130	1.133	1.136	1.138	1.140	1.142	1.144	1.145	1.146	1.148	1.149	1.150
1500	1.086	1.090	1.091	1.098	1.109	1.116	1.122	1.127	1.131	1.135	1.138	1.140	1.142	1.144	1.146	1.147	1.148	1.150	1.151	1.152
1520	1.087	1.091	1.093	1.100	1.111	1.118	1.124	1.129	1.133	1.137	1.140	1.142	1.144	1.146	1.148	1.149	1.151	1.152	1.153	1.154
1540	1.088	1.092	1.094	1.101	1.112	1.119	1.126	1.131	1.135	1.138	1.141	1.144	1.146	1.148	1.150	1.151	1.153	1.154	1.155	1.156
1560	1.089	1.094	1.096	1.103	1.114	1.121	1.127	1.133	1.137	1.140	1.143	1.146	1.148	1.150	1.152	1.153	1.155	1.156	1.157	1.158
1580	1.090	1.095	1.097	1.104	1.115	1.122	1.129	1.134	1.139	1.142	1.145	1.148	1.150	1.152	1.154	1.155	1.157	1.158	1.159	1.160
1600	1.093	1.096	1.098	1.105	1.116	1.124	1.131	1.136	1.140	1.144	1.147	1.150	1.152	1.154	1.156	1.157	1.159	1.160	1.161	1.162
1620	1.096	1.097	1.099	1.106	1.117	1.126	1.132	1.138	1.142	1.146	1.149	1.152	1.154	1.156	1.158	1.159	1.161	1.162	1.163	1.164
1640	1.098	1.098	1.099	1.107	1.118	1.127	1.134	1.140	1.144	1.148	1.151	1.154	1.156	1.158	1.160	1.161	1.163	1.164	1.165	1.166

SIGMA = 18160.0

V-Vc = 7

x \ H	0.08	0.09	0.10	0.2	0.4	0.6	0.8	1.0	1.2	1.4	1.6	1.8	2.0	2.2	2.4	2.6	2.8	3.0	3.2	3.4
1660	1.099	1.100	1.100	1.108	1.120	1.129	1.136	1.141	1.146	1.150	1.153	1.156	1.158	1.160	1.162	1.163	1.165	1.166	1.167	1.169
1680	1.100	1.101	1.102	1.109	1.121	1.130	1.137	1.143	1.148	1.151	1.155	1.157	1.160	1.162	1.164	1.166	1.167	1.168	1.170	1.171
1700	1.101	1.102	1.103	1.111	1.123	1.132	1.139	1.145	1.149	1.153	1.157	1.159	1.162	1.164	1.166	1.168	1.169	1.170	1.172	1.173
1720	1.102	1.104	1.104	1.113	1.124	1.134	1.141	1.147	1.151	1.155	1.159	1.161	1.164	1.166	1.168	1.170	1.171	1.172	1.174	1.175
1740	1.104	1.104	1.105	1.115	1.126	1.135	1.142	1.148	1.153	1.157	1.160	1.163	1.166	1.168	1.170	1.172	1.173	1.174	1.176	1.177
1760	1.105	1.106	1.107	1.117	1.127	1.137	1.144	1.150	1.155	1.159	1.162	1.165	1.168	1.170	1.172	1.174	1.175	1.176	1.178	1.179
1780	1.106	1.107	1.108	1.119	1.129	1.138	1.146	1.152	1.157	1.161	1.164	1.167	1.170	1.172	1.174	1.176	1.177	1.179	1.180	1.181
1800	1.107	1.108	1.109	1.120	1.130	1.140	1.148	1.154	1.159	1.163	1.166	1.169	1.172	1.174	1.176	1.178	1.179	1.181	1.182	1.183
1820	1.108	1.109	1.110	1.121	1.132	1.142	1.149	1.155	1.160	1.165	1.168	1.171	1.174	1.176	1.178	1.180	1.181	1.183	1.184	1.185
1840	1.110	1.110	1.111	1.123	1.133	1.143	1.151	1.157	1.162	1.166	1.170	1.173	1.176	1.178	1.180	1.182	1.184	1.185	1.186	1.188
1860	1.110	1.111	1.113	1.124	1.135	1.145	1.153	1.159	1.164	1.168	1.172	1.175	1.178	1.180	1.182	1.184	1.186	1.187	1.188	1.190
1880	1.111	1.112	1.114	1.126	1.136	1.146	1.154	1.161	1.166	1.170	1.174	1.177	1.180	1.182	1.184	1.186	1.188	1.189	1.191	1.192
1900	1.112	1.113	1.115	1.127	1.138	1.148	1.156	1.162	1.168	1.172	1.176	1.179	1.182	1.184	1.186	1.188	1.190	1.191	1.193	1.194
1920	1.113	1.114	1.116	1.128	1.139	1.150	1.158	1.164	1.169	1.174	1.178	1.181	1.184	1.186	1.188	1.190	1.192	1.193	1.195	1.196
1940	1.114	1.116	1.117	1.130	1.141	1.151	1.159	1.166	1.171	1.176	1.180	1.183	1.186	1.188	1.190	1.192	1.194	1.196	1.197	1.198
1960	1.116	1.117	1.118	1.131	1.142	1.153	1.161	1.168	1.173	1.178	1.182	1.185	1.188	1.190	1.192	1.194	1.196	1.198	1.199	1.200
1980	1.117	1.118	1.119	1.132	1.144	1.154	1.163	1.169	1.175	1.180	1.183	1.187	1.190	1.192	1.194	1.196	1.198	1.200	1.201	1.202
2000	1.118	1.119	1.120	1.133	1.145	1.156	1.164	1.171	1.177	1.181	1.185	1.189	1.192	1.194	1.196	1.198	1.200	1.202	1.203	1.205
2020	1.119	1.120	1.121	1.134	1.147	1.158	1.166	1.173	1.179	1.183	1.187	1.191	1.194	1.196	1.198	1.201	1.202	1.204	1.205	1.207
2040	1.120	1.122	1.122	1.136	1.148	1.159	1.168	1.175	1.180	1.185	1.189	1.193	1.196	1.198	1.201	1.203	1.204	1.206	1.208	1.209
2060	1.122	1.123	1.124	1.137	1.150	1.161	1.170	1.177	1.182	1.187	1.191	1.195	1.198	1.200	1.203	1.205	1.207	1.208	1.210	1.211
2080	1.123	1.124	1.125	1.138	1.151	1.162	1.171	1.178	1.184	1.189	1.193	1.197	1.200	1.202	1.205	1.207	1.209	1.210	1.212	1.213
2100	1.124	1.125	1.126	1.140	1.152	1.164	1.173	1.180	1.186	1.191	1.195	1.199	1.202	1.204	1.207	1.209	1.211	1.212	1.214	1.215
2120	1.125	1.126	1.127	1.142	1.154	1.166	1.175	1.182	1.188	1.193	1.197	1.201	1.204	1.206	1.209	1.211	1.213	1.215	1.216	1.217
2140	1.126	1.128	1.129	1.143	1.155	1.167	1.176	1.184	1.190	1.195	1.199	1.203	1.206	1.208	1.211	1.213	1.215	1.217	1.218	1.220
2160	1.128	1.129	1.130	1.145	1.157	1.169	1.178	1.185	1.192	1.197	1.201	1.205	1.208	1.210	1.213	1.215	1.217	1.219	1.220	1.222
2180	1.129	1.130	1.131	1.146	1.158	1.170	1.180	1.187	1.193	1.198	1.203	1.206	1.210	1.212	1.215	1.217	1.219	1.221	1.222	1.224
2200	1.130	1.131	1.132	1.148	1.160	1.172	1.182	1.189	1.195	1.200	1.205	1.208	1.212	1.214	1.217	1.219	1.221	1.223	1.225	1.226
2220	1.131	1.132	1.133	1.150	1.161	1.174	1.183	1.191	1.197	1.202	1.207	1.210	1.214	1.216	1.219	1.221	1.223	1.225	1.227	1.228
2240	1.132	1.134	1.135	1.152	1.163	1.175	1.185	1.193	1.199	1.204	1.209	1.212	1.216	1.218	1.221	1.223	1.225	1.227	1.229	1.230
2260	1.134	1.135	1.136	1.153	1.164	1.177	1.187	1.194	1.201	1.206	1.211	1.214	1.218	1.221	1.223	1.226	1.228	1.229	1.231	1.233
2280	1.135	1.136	1.137	1.154	1.166	1.179	1.188	1.196	1.203	1.208	1.213	1.216	1.220	1.223	1.225	1.228	1.230	1.232	1.233	1.235
2300	1.136	1.137	1.140	1.156	1.167	1.180	1.190	1.198	1.204	1.210	1.214	1.218	1.222	1.225	1.227	1.230	1.232	1.234	1.235	1.237
2320	1.137	1.138	1.141	1.157	1.169	1.182	1.192	1.200	1.206	1.212	1.216	1.220	1.224	1.227	1.229	1.232	1.234	1.236	1.237	1.239
2340	1.138	1.140	1.142	1.158	1.170	1.183	1.194	1.202	1.208	1.214	1.218	1.222	1.226	1.229	1.232	1.234	1.236	1.238	1.240	1.241
2360	1.140	1.141	1.143	1.160	1.172	1.185	1.195	1.203	1.210	1.216	1.220	1.224	1.228	1.231	1.234	1.236	1.238	1.240	1.242	1.243
2380	1.141	1.142	1.145	1.161	1.173	1.187	1.197	1.205	1.212	1.218	1.222	1.226	1.230	1.233	1.236	1.238	1.240	1.242	1.244	1.246
2400	1.142	1.143	1.146	1.162	1.175	1.188	1.199	1.207	1.214	1.219	1.224	1.228	1.232	1.235	1.238	1.240	1.242	1.244	1.246	1.248
2420	1.143	1.145	1.147	1.163	1.176	1.190	1.200	1.209	1.216	1.221	1.226	1.230	1.234	1.237	1.240	1.242	1.244	1.247	1.248	1.250
2440	1.146	1.147	1.148	1.164	1.178	1.192	1.202	1.211	1.218	1.223	1.228	1.232	1.236	1.239	1.242	1.244	1.247	1.249	1.251	1.253
2460	1.147	1.148	1.149	1.161	1.179	1.193	1.204	1.212	1.219	1.225	1.230	1.234	1.238	1.241	1.244	1.247	1.249	1.251	1.253	1.254
2480	1.148	1.149	1.151	1.162	1.181	1.195	1.206	1.214	1.221	1.227	1.232	1.236	1.240	1.243	1.246	1.249	1.251	1.253	1.255	1.257

SIGMA = 18160.0

V−V$_c$ = 7

H \ x	3.4	3.2	3.0	2.8	2.6	2.4	2.2	2.0	1.8	1.6	1.4	1.2	1.0	0.8	0.6	0.4	0.2	0.10	0.09	0.08
2500	1.259	1.257	1.255	1.253	1.251	1.248	1.245	1.242	1.238	1.234	1.229	1.223	1.216	1.207	1.196	1.182	1.164	1.152	1.151	1.149
2520	1.261	1.259	1.257	1.255	1.253	1.250	1.247	1.244	1.240	1.236	1.231	1.225	1.218	1.209	1.198	1.184	1.165	1.153	1.152	1.150
2540	1.263	1.261	1.259	1.257	1.255	1.252	1.249	1.246	1.242	1.238	1.233	1.227	1.220	1.211	1.200	1.185	1.166	1.154	1.153	1.152
2560	1.265	1.264	1.262	1.259	1.257	1.255	1.252	1.248	1.244	1.240	1.235	1.229	1.221	1.212	1.201	1.187	1.168	1.156	1.154	1.153
2580	1.267	1.266	1.264	1.262	1.259	1.257	1.254	1.250	1.246	1.242	1.237	1.231	1.223	1.214	1.203	1.188	1.169	1.157	1.155	1.154
2600	1.270	1.268	1.266	1.264	1.261	1.259	1.256	1.252	1.248	1.244	1.239	1.232	1.225	1.216	1.205	1.190	1.170	1.158	1.157	1.155
2620	1.272	1.270	1.268	1.266	1.264	1.261	1.258	1.254	1.250	1.246	1.241	1.234	1.227	1.218	1.206	1.191	1.172	1.159	1.158	1.157
2640	1.274	1.272	1.270	1.268	1.266	1.263	1.260	1.256	1.252	1.248	1.243	1.236	1.229	1.219	1.208	1.193	1.173	1.161	1.159	1.158
2660	1.276	1.274	1.272	1.270	1.268	1.265	1.262	1.258	1.254	1.250	1.244	1.238	1.230	1.221	1.209	1.194	1.174	1.162	1.160	1.159
2680	1.278	1.277	1.275	1.272	1.270	1.267	1.264	1.260	1.256	1.252	1.246	1.240	1.232	1.223	1.211	1.196	1.176	1.163	1.162	1.160
2700	1.281	1.279	1.277	1.275	1.272	1.269	1.266	1.263	1.258	1.254	1.248	1.242	1.234	1.225	1.213	1.197	1.177	1.164	1.163	1.161
2720	1.283	1.281	1.279	1.277	1.274	1.271	1.268	1.265	1.260	1.256	1.250	1.244	1.236	1.226	1.214	1.199	1.178	1.165	1.164	1.163
2740	1.285	1.283	1.281	1.279	1.276	1.273	1.270	1.267	1.263	1.258	1.252	1.246	1.238	1.228	1.216	1.200	1.180	1.167	1.165	1.164
2760	1.287	1.285	1.283	1.281	1.278	1.276	1.272	1.269	1.265	1.260	1.254	1.247	1.240	1.230	1.218	1.202	1.181	1.168	1.166	1.165
2780	1.289	1.288	1.285	1.283	1.281	1.278	1.274	1.271	1.267	1.262	1.256	1.249	1.241	1.232	1.219	1.204	1.183	1.169	1.168	1.166
2800	1.292	1.290	1.288	1.285	1.283	1.280	1.277	1.273	1.269	1.264	1.258	1.251	1.243	1.233	1.221	1.205	1.184	1.170	1.169	1.167
2820	1.294	1.292	1.290	1.287	1.285	1.282	1.279	1.275	1.271	1.266	1.260	1.253	1.245	1.235	1.223	1.207	1.185	1.172	1.170	1.169
2840	1.296	1.294	1.292	1.290	1.287	1.284	1.281	1.277	1.273	1.268	1.262	1.255	1.247	1.237	1.224	1.208	1.187	1.173	1.171	1.170
2860	1.298	1.296	1.294	1.292	1.289	1.286	1.283	1.279	1.275	1.270	1.264	1.257	1.249	1.239	1.226	1.210	1.189	1.174	1.173	1.171
2880	1.301	1.299	1.296	1.294	1.291	1.288	1.285	1.281	1.277	1.272	1.266	1.259	1.250	1.240	1.227	1.211	1.191	1.175	1.174	1.172
2900	1.303	1.301	1.299	1.296	1.293	1.290	1.287	1.283	1.279	1.274	1.268	1.261	1.252	1.242	1.229	1.213	1.192	1.177	1.175	1.173
2920	1.305	1.303	1.301	1.298	1.295	1.293	1.289	1.285	1.281	1.276	1.270	1.263	1.254	1.244	1.231	1.214	1.193	1.178	1.177	1.175
2940	1.307	1.305	1.303	1.300	1.297	1.295	1.291	1.287	1.283	1.278	1.272	1.264	1.256	1.245	1.232	1.216	1.195	1.179	1.177	1.176
2960	1.309	1.307	1.305	1.302	1.299	1.297	1.293	1.289	1.285	1.280	1.274	1.266	1.258	1.247	1.234	1.217	1.196	1.180	1.179	1.177
2980	1.312	1.310	1.307	1.305	1.302	1.299	1.295	1.291	1.287	1.282	1.276	1.268	1.260	1.249	1.236	1.219	1.197	1.181	1.180	1.178
3000	1.314	1.312	1.310	1.307	1.304	1.301	1.298	1.294	1.289	1.284	1.277	1.270	1.261	1.251	1.237	1.220	1.197	1.183	1.181	1.179

SIGMA = 18160.0

$V-V_c = 7$

H	0.08	0.09	0.10	0.2	0.4	0.6	0.8	1.0	1.2	1.4	1.6	1.8	2.0	2.2	2.4	2.6	2.8	3.0	3.2	3.4
3100	1.186	1.187	1.189	1.204	1.228	1.246	1.260	1.271	1.280	1.287	1.294	1.299	1.304	1.308	1.312	1.315	1.318	1.321	1.323	1.325
3200	1.192	1.193	1.195	1.211	1.235	1.254	1.268	1.280	1.289	1.297	1.304	1.309	1.314	1.319	1.323	1.326	1.329	1.332	1.334	1.336
3300	1.198	1.199	1.201	1.218	1.243	1.262	1.277	1.289	1.299	1.307	1.314	1.320	1.325	1.329	1.333	1.337	1.340	1.343	1.345	1.348
3400	1.204	1.206	1.207	1.224	1.251	1.271	1.286	1.298	1.308	1.317	1.324	1.330	1.335	1.340	1.344	1.348	1.351	1.354	1.357	1.359
3500	1.210	1.212	1.214	1.231	1.258	1.279	1.295	1.308	1.318	1.327	1.334	1.341	1.346	1.351	1.355	1.359	1.362	1.365	1.368	1.370
3600	1.216	1.218	1.220	1.238	1.266	1.287	1.304	1.317	1.328	1.337	1.344	1.351	1.357	1.362	1.366	1.370	1.373	1.376	1.379	1.382
3700	1.222	1.224	1.226	1.245	1.274	1.296	1.313	1.326	1.338	1.347	1.355	1.361	1.367	1.372	1.377	1.381	1.384	1.388	1.391	1.393
3800	1.228	1.230	1.232	1.251	1.282	1.304	1.322	1.336	1.347	1.357	1.365	1.372	1.378	1.383	1.388	1.392	1.396	1.399	1.402	1.405
3900	1.234	1.236	1.238	1.258	1.289	1.313	1.331	1.345	1.357	1.367	1.375	1.382	1.389	1.394	1.399	1.403	1.407	1.411	1.414	1.416
4000	1.240	1.242	1.245	1.265	1.297	1.321	1.340	1.355	1.367	1.377	1.386	1.393	1.399	1.405	1.410	1.414	1.418	1.422	1.425	1.428
4100	1.246	1.249	1.251	1.272	1.305	1.330	1.349	1.364	1.377	1.387	1.396	1.404	1.410	1.416	1.421	1.426	1.430	1.433	1.437	1.440
4200	1.252	1.255	1.257	1.279	1.313	1.338	1.358	1.374	1.387	1.397	1.407	1.414	1.421	1.427	1.432	1.437	1.441	1.445	1.448	1.451
4300	1.258	1.261	1.263	1.286	1.320	1.347	1.367	1.383	1.397	1.408	1.417	1.425	1.432	1.438	1.444	1.448	1.453	1.456	1.460	1.463
4400	1.265	1.267	1.270	1.292	1.328	1.355	1.376	1.393	1.406	1.418	1.428	1.436	1.443	1.449	1.455	1.460	1.464	1.468	1.472	1.475
4500	1.271	1.273	1.276	1.299	1.336	1.364	1.385	1.402	1.416	1.428	1.438	1.447	1.454	1.460	1.466	1.471	1.476	1.480	1.483	1.487
4600	1.277	1.280	1.282	1.306	1.344	1.372	1.394	1.412	1.426	1.438	1.449	1.457	1.465	1.472	1.477	1.483	1.487	1.491	1.495	1.499
4700	1.283	1.286	1.288	1.313	1.352	1.381	1.404	1.422	1.437	1.449	1.459	1.468	1.476	1.483	1.489	1.494	1.499	1.503	1.507	1.511
4800	1.289	1.292	1.295	1.320	1.360	1.390	1.413	1.431	1.447	1.459	1.470	1.479	1.487	1.494	1.500	1.506	1.511	1.515	1.519	1.523
4900	1.295	1.298	1.301	1.327	1.368	1.398	1.422	1.441	1.457	1.470	1.481	1.490	1.498	1.505	1.512	1.517	1.522	1.527	1.531	1.535
5000	1.301	1.304	1.307	1.334	1.376	1.407	1.431	1.451	1.467	1.480	1.491	1.501	1.509	1.517	1.523	1.529	1.534	1.539	1.543	1.547
5100	1.307	1.311	1.314	1.341	1.384	1.416	1.441	1.461	1.477	1.491	1.502	1.512	1.521	1.528	1.535	1.541	1.546	1.551	1.555	1.559
5200	1.314	1.317	1.320	1.348	1.392	1.424	1.450	1.471	1.487	1.501	1.513	1.523	1.532	1.540	1.546	1.552	1.558	1.563	1.567	1.571
5300	1.320	1.323	1.326	1.355	1.400	1.433	1.459	1.480	1.498	1.512	1.524	1.534	1.543	1.551	1.558	1.564	1.570	1.575	1.579	1.583
5400	1.326	1.329	1.332	1.362	1.408	1.442	1.469	1.490	1.508	1.522	1.535	1.545	1.555	1.563	1.570	1.576	1.582	1.587	1.591	1.595
5500	1.332	1.335	1.339	1.369	1.416	1.451	1.478	1.500	1.518	1.533	1.546	1.557	1.566	1.574	1.581	1.588	1.594	1.599	1.604	1.608
5600	1.338	1.342	1.345	1.376	1.424	1.460	1.488	1.510	1.528	1.544	1.557	1.568	1.577	1.586	1.593	1.600	1.606	1.611	1.616	1.620
5700	1.344	1.348	1.351	1.383	1.432	1.469	1.497	1.520	1.539	1.554	1.568	1.579	1.589	1.597	1.605	1.612	1.618	1.623	1.628	1.633
5800	1.351	1.354	1.358	1.390	1.440	1.477	1.507	1.530	1.549	1.565	1.579	1.590	1.600	1.609	1.617	1.624	1.630	1.635	1.640	1.645
5900	1.357	1.360	1.364	1.397	1.448	1.486	1.516	1.540	1.560	1.576	1.590	1.602	1.612	1.621	1.629	1.636	1.642	1.648	1.653	1.658
6000	1.363	1.367	1.370	1.404	1.456	1.495	1.526	1.550	1.570	1.587	1.601	1.613	1.623	1.633	1.641	1.648	1.654	1.660	1.665	1.670
6100	1.369	1.373	1.377	1.411	1.464	1.504	1.535	1.560	1.581	1.598	1.612	1.623	1.635	1.644	1.653	1.660	1.667	1.672	1.678	1.683
6200	1.375	1.379	1.383	1.418	1.472	1.513	1.545	1.570	1.591	1.609	1.623	1.636	1.647	1.656	1.665	1.672	1.679	1.685	1.690	1.695
6300	1.382	1.385	1.389	1.425	1.480	1.522	1.555	1.581	1.602	1.619	1.634	1.647	1.658	1.668	1.677	1.684	1.691	1.697	1.703	1.708
6400	1.388	1.392	1.396	1.432	1.489	1.531	1.564	1.591	1.612	1.630	1.646	1.658	1.670	1.680	1.689	1.697	1.704	1.710	1.716	1.721
6500	1.394	1.398	1.402	1.439	1.497	1.540	1.574	1.601	1.623	1.641	1.657	1.670	1.682	1.692	1.701	1.709	1.716	1.722	1.728	1.734
6600	1.400	1.404	1.408	1.446	1.505	1.549	1.584	1.611	1.634	1.652	1.668	1.682	1.694	1.704	1.713	1.721	1.729	1.735	1.741	1.746
6700	1.406	1.411	1.415	1.453	1.513	1.558	1.593	1.621	1.644	1.664	1.680	1.694	1.706	1.716	1.725	1.734	1.741	1.748	1.754	1.759
6800	1.413	1.417	1.421	1.460	1.521	1.567	1.603	1.632	1.655	1.675	1.691	1.705	1.718	1.728	1.738	1.746	1.754	1.760	1.767	1.772
6900	1.419	1.423	1.428	1.467	1.530	1.577	1.613	1.642	1.666	1.686	1.703	1.717	1.730	1.740	1.750	1.759	1.766	1.773	1.779	1.785
7000	1.425	1.430	1.434	1.474	1.538	1.586	1.623	1.652	1.677	1.697	1.714	1.729	1.741	1.753	1.762	1.771	1.779	1.786	1.792	1.798
7100	1.431	1.436	1.440	1.482	1.546	1.595	1.633	1.663	1.688	1.708	1.726	1.741	1.754	1.765	1.775	1.784	1.792	1.799	1.805	1.811
7200	1.437	1.442	1.447	1.489	1.555	1.604	1.643	1.673	1.698	1.719	1.737	1.752	1.766	1.777	1.787	1.796	1.804	1.812	1.818	1.824

x

SIGMA = 18160.0

$V-V_c = 7$

H	3.4	3.2	3.0	2.8	2.6	2.4	2.2	2.0	1.8	1.6	1.4	1.2	1.0	0.8	0.6	0.4	0.2	0.10	0.09	0.08
X																				
7300	1.837	1.831	1.825	1.817	1.809	1.800	1.789	1.778	1.764	1.749	1.731	1.709	1.684	1.652	1.613	1.563	1.496	1.453	1.449	1.444
7400	1.851	1.844	1.838	1.830	1.822	1.812	1.802	1.790	1.776	1.760	1.742	1.720	1.694	1.662	1.623	1.571	1.503	1.460	1.455	1.450
7500	1.864	1.858	1.851	1.843	1.834	1.825	1.814	1.802	1.788	1.772	1.753	1.731	1.705	1.672	1.632	1.580	1.510	1.466	1.461	1.456
7600	1.877	1.871	1.864	1.856	1.847	1.838	1.827	1.814	1.800	1.784	1.765	1.742	1.715	1.682	1.641	1.588	1.517	1.472	1.468	1.462
7700	1.891	1.884	1.877	1.869	1.860	1.850	1.839	1.827	1.812	1.796	1.776	1.753	1.726	1.692	1.650	1.597	1.525	1.479	1.474	1.469
7800	1.904	1.897	1.890	1.882	1.873	1.863	1.852	1.839	1.824	1.807	1.788	1.764	1.737	1.702	1.660	1.605	1.532	1.485	1.480	1.475
7900	1.917	1.911	1.903	1.895	1.886	1.876	1.864	1.851	1.836	1.819	1.799	1.776	1.747	1.712	1.669	1.613	1.539	1.492	1.487	1.481
8000	1.931	1.924	1.916	1.908	1.899	1.888	1.877	1.864	1.848	1.831	1.811	1.787	1.758	1.723	1.679	1.622	1.546	1.498	1.493	1.486
8100	1.944	1.937	1.930	1.921	1.912	1.901	1.889	1.876	1.861	1.843	1.822	1.798	1.769	1.733	1.688	1.630	1.554	1.505	1.499	1.493
8200	1.958	1.951	1.943	1.934	1.925	1.914	1.902	1.888	1.873	1.855	1.834	1.809	1.779	1.743	1.697	1.639	1.561	1.511	1.506	1.499
8300	1.972	1.964	1.957	1.948	1.938	1.927	1.915	1.901	1.885	1.867	1.846	1.820	1.790	1.753	1.707	1.647	1.568	1.518	1.512	1.506
8400	1.985	1.978	1.970	1.961	1.951	1.940	1.928	1.914	1.897	1.879	1.857	1.832	1.801	1.763	1.716	1.656	1.575	1.524	1.518	1.513
8500	1.999	1.992	1.983	1.974	1.964	1.953	1.940	1.926	1.910	1.891	1.869	1.843	1.812	1.773	1.726	1.664	1.583	1.531	1.525	1.519
8600	2.013	2.005	1.997	1.988	1.978	1.966	1.953	1.939	1.922	1.903	1.881	1.854	1.822	1.784	1.735	1.673	1.590	1.537	1.531	1.525
8700	2.027	2.019	2.011	2.001	1.991	1.979	1.966	1.951	1.935	1.915	1.892	1.866	1.833	1.794	1.745	1.682	1.597	1.543	1.538	1.532
8800	2.040	2.033	2.024	2.015	2.004	1.992	1.979	1.964	1.947	1.927	1.904	1.877	1.844	1.804	1.754	1.690	1.604	1.550	1.544	1.538
8900	2.054	2.046	2.038	2.028	2.018	2.006	1.992	1.977	1.960	1.939	1.916	1.888	1.855	1.815	1.764	1.699	1.612	1.556	1.550	1.544
9000	2.068	2.060	2.052	2.042	2.031	2.019	2.005	1.990	1.972	1.952	1.928	1.900	1.866	1.825	1.774	1.707	1.619	1.563	1.557	1.550
9100	2.082	2.074	2.065	2.055	2.044	2.032	2.018	2.003	1.985	1.964	1.940	1.911	1.877	1.835	1.783	1.716	1.626	1.569	1.563	1.557
9200	2.096	2.088	2.079	2.069	2.058	2.045	2.031	2.015	1.997	1.976	1.952	1.923	1.888	1.846	1.793	1.725	1.634	1.576	1.570	1.563
9300	2.110	2.102	2.093	2.083	2.071	2.059	2.044	2.028	2.010	1.989	1.964	1.935	1.899	1.856	1.803	1.733	1.641	1.583	1.576	1.569
9400	2.125	2.116	2.107	2.096	2.085	2.072	2.058	2.041	2.023	2.001	1.976	1.946	1.910	1.867	1.812	1.742	1.649	1.589	1.582	1.576
9500	2.139	2.130	2.121	2.110	2.099	2.086	2.071	2.054	2.035	2.013	1.988	1.958	1.922	1.877	1.822	1.751	1.656	1.596	1.589	1.582
9600	2.153	2.144	2.135	2.124	2.112	2.099	2.084	2.067	2.048	2.026	2.000	1.969	1.933	1.888	1.832	1.760	1.663	1.602	1.595	1.588
9700	2.167	2.158	2.149	2.138	2.126	2.113	2.098	2.080	2.061	2.038	2.012	1.981	1.944	1.898	1.841	1.768	1.671	1.609	1.602	1.595
9800	2.182	2.173	2.163	2.152	2.140	2.126	2.111	2.094	2.074	2.051	2.024	1.993	1.955	1.909	1.851	1.777	1.678	1.615	1.608	1.601
9900	2.196	2.187	2.177	2.166	2.154	2.140	2.124	2.107	2.087	2.064	2.037	2.005	1.966	1.920	1.861	1.786	1.686	1.622	1.615	1.608
10000	2.210	2.201	2.191	2.180	2.167	2.153	2.138	2.120	2.100	2.076	2.049	2.016	1.978	1.930	1.871	1.795	1.693	1.628	1.621	1.614
10100	2.225	2.216	2.205	2.194	2.181	2.167	2.151	2.133	2.113	2.089	2.061	2.028	1.989	1.941	1.881	1.803	1.700	1.635	1.628	1.620
10200	2.239	2.230	2.220	2.208	2.195	2.181	2.165	2.146	2.126	2.101	2.073	2.040	2.000	1.952	1.891	1.812	1.708	1.641	1.634	1.627
10300	2.254	2.244	2.234	2.222	2.209	2.195	2.178	2.160	2.139	2.114	2.086	2.052	2.012	1.962	1.901	1.821	1.715	1.648	1.641	1.633
10400	2.269	2.259	2.248	2.236	2.223	2.208	2.192	2.173	2.152	2.127	2.098	2.064	2.023	1.973	1.910	1.830	1.723	1.655	1.647	1.639
10500	2.283	2.273	2.263	2.251	2.237	2.222	2.206	2.187	2.165	2.140	2.110	2.076	2.034	1.984	1.920	1.839	1.730	1.661	1.654	1.646
10600	2.298	2.288	2.277	2.265	2.251	2.236	2.219	2.200	2.178	2.153	2.123	2.088	2.046	1.995	1.930	1.848	1.738	1.668	1.660	1.652
10700	2.313	2.303	2.291	2.279	2.265	2.250	2.233	2.213	2.191	2.165	2.135	2.100	2.057	2.005	1.940	1.857	1.745	1.674	1.667	1.659
10800	2.328	2.317	2.306	2.294	2.280	2.264	2.247	2.227	2.204	2.178	2.148	2.112	2.069	2.016	1.950	1.866	1.753	1.681	1.673	1.665
10900	2.342	2.332	2.321	2.308	2.294	2.278	2.261	2.241	2.218	2.191	2.161	2.124	2.080	2.027	1.960	1.875	1.760	1.688	1.680	1.671
11000	2.357	2.347	2.335	2.322	2.308	2.292	2.274	2.254	2.231	2.204	2.173	2.136	2.092	2.038	1.970	1.884	1.768	1.694	1.686	1.678
11100	2.372	2.362	2.350	2.337	2.323	2.306	2.288	2.268	2.244	2.217	2.186	2.148	2.104	2.049	1.981	1.893	1.775	1.701	1.693	1.684
11200	2.387	2.376	2.365	2.351	2.337	2.321	2.302	2.281	2.258	2.230	2.198	2.161	2.115	2.060	1.991	1.902	1.783	1.707	1.699	1.691
11300	2.402	2.391	2.379	2.366	2.351	2.335	2.316	2.295	2.271	2.243	2.211	2.173	2.127	2.071	2.001	1.911	1.790	1.714	1.706	1.697
11400	2.417	2.406	2.394	2.381	2.366	2.349	2.330	2.309	2.285	2.257	2.224	2.185	2.139	2.082	2.011	1.920	1.798	1.721	1.712	1.703

SIGMA = 18160.0

$V-V_c = 7$

H \ x	0.08	0.09	0.10	0.2	0.4	0.6	0.8	1.0	1.2	1.4	1.6	1.8	2.0	2.2	2.4	2.6	2.8	3.0	3.2	3.4
11500	1.710	1.719	1.727	1.806	1.929	2.021	2.093	2.150	2.197	2.237	2.270	2.298	2.323	2.344	2.363	2.380	2.395	2.409	2.421	2.432
11600	1.716	1.725	1.734	1.813	1.938	2.031	2.104	2.162	2.210	2.249	2.283	2.312	2.337	2.358	2.378	2.395	2.410	2.424	2.436	2.448
11700	1.723	1.732	1.741	1.821	1.947	2.042	2.115	2.174	2.222	2.262	2.296	2.325	2.351	2.373	2.392	2.409	2.425	2.439	2.451	2.463
11800	1.729	1.738	1.747	1.828	1.956	2.052	2.126	2.186	2.235	2.275	2.309	2.339	2.364	2.387	2.406	2.424	2.440	2.454	2.467	2.478
11900	1.736	1.745	1.754	1.836	1.965	2.062	2.137	2.198	2.247	2.288	2.323	2.353	2.378	2.401	2.421	2.439	2.455	2.469	2.482	2.493
12000	1.742	1.751	1.761	1.844	1.974	2.072	2.149	2.210	2.259	2.301	2.336	2.366	2.392	2.415	2.435	2.453	2.469	2.484	2.497	2.509
12100	1.749	1.758	1.767	1.851	1.984	2.083	2.160	2.221	2.272	2.314	2.350	2.380	2.406	2.430	2.450	2.468	2.484	2.499	2.512	2.524
12200	1.755	1.765	1.774	1.859	1.993	2.093	2.171	2.233	2.284	2.327	2.363	2.394	2.421	2.444	2.465	2.483	2.499	2.514	2.528	2.540
12300	1.761	1.771	1.781	1.867	2.002	2.103	2.182	2.245	2.297	2.340	2.376	2.408	2.435	2.458	2.479	2.498	2.514	2.529	2.543	2.555
12400	1.768	1.778	1.787	1.874	2.011	2.114	2.194	2.257	2.310	2.353	2.390	2.422	2.449	2.473	2.494	2.513	2.529	2.545	2.558	2.571
12500	1.774	1.784	1.794	1.882	2.020	2.124	2.205	2.269	2.322	2.366	2.403	2.435	2.463	2.487	2.509	2.528	2.545	2.560	2.574	2.586
12600	1.781	1.791	1.801	1.890	2.030	2.135	2.216	2.281	2.335	2.379	2.417	2.449	2.477	2.502	2.523	2.543	2.560	2.575	2.589	2.602
12700	1.787	1.797	1.807	1.897	2.039	2.145	2.228	2.294	2.348	2.393	2.431	2.463	2.492	2.516	2.538	2.558	2.575	2.591	2.605	2.618
12800	1.794	1.804	1.814	1.905	2.048	2.155	2.239	2.306	2.360	2.406	2.444	2.477	2.506	2.531	2.553	2.573	2.590	2.606	2.620	2.633
12900	1.800	1.811	1.821	1.913	2.057	2.166	2.250	2.318	2.373	2.419	2.458	2.491	2.520	2.546	2.568	2.588	2.606	2.622	2.636	2.649
13000	1.807	1.817	1.828	1.921	2.067	2.176	2.262	2.330	2.386	2.432	2.472	2.505	2.535	2.560	2.583	2.603	2.621	2.637	2.652	2.665
13100	1.813	1.824	1.834	1.928	2.076	2.187	2.273	2.342	2.399	2.446	2.486	2.520	2.549	2.575	2.598	2.618	2.636	2.653	2.667	2.681
13200	1.820	1.831	1.841	1.936	2.086	2.198	2.285	2.354	2.412	2.459	2.499	2.534	2.564	2.590	2.613	2.633	2.652	2.668	2.683	2.697
13300	1.826	1.837	1.848	1.944	2.095	2.208	2.296	2.367	2.424	2.472	2.513	2.548	2.578	2.605	2.628	2.649	2.667	2.684	2.699	2.713
13400	1.833	1.844	1.854	1.952	2.104	2.219	2.308	2.379	2.437	2.486	2.527	2.562	2.593	2.620	2.643	2.664	2.683	2.700	2.715	2.729
13500	1.839	1.850	1.861	1.959	2.114	2.229	2.319	2.391	2.450	2.499	2.541	2.577	2.607	2.634	2.658	2.679	2.698	2.715	2.731	2.745
13600	1.846	1.857	1.868	1.967	2.123	2.240	2.331	2.404	2.463	2.513	2.555	2.591	2.622	2.649	2.673	2.695	2.714	2.731	2.747	2.761
13700	1.852	1.864	1.875	1.975	2.133	2.251	2.343	2.416	2.476	2.526	2.569	2.605	2.637	2.664	2.689	2.710	2.730	2.747	2.763	2.777
13800	1.859	1.870	1.881	1.983	2.142	2.261	2.354	2.429	2.489	2.540	2.583	2.620	2.651	2.679	2.704	2.726	2.746	2.763	2.779	2.793
13900	1.866	1.877	1.888	1.991	2.151	2.272	2.366	2.441	2.502	2.554	2.597	2.634	2.666	2.694	2.719	2.741	2.761	2.779	2.795	2.810
14000	1.872	1.884	1.895	1.998	2.161	2.283	2.378	2.454	2.516	2.567	2.611	2.649	2.681	2.710	2.735	2.757	2.777	2.795	2.811	2.826
14100	1.879	1.890	1.902	2.006	2.171	2.294	2.389	2.466	2.529	2.581	2.625	2.663	2.696	2.725	2.750	2.773	2.793	2.811	2.827	2.842
14200	1.885	1.897	1.909	2.014	2.180	2.304	2.401	2.479	2.542	2.595	2.639	2.678	2.711	2.740	2.765	2.788	2.809	2.827	2.844	2.859
14300	1.892	1.904	1.915	2.022	2.190	2.315	2.413	2.491	2.555	2.608	2.654	2.692	2.726	2.755	2.781	2.804	2.825	2.843	2.860	2.875
14400	1.898	1.910	1.922	2.030	2.199	2.326	2.425	2.504	2.568	2.622	2.668	2.707	2.741	2.770	2.797	2.820	2.841	2.859	2.876	2.892
14500	1.905	1.917	1.929	2.038	2.209	2.337	2.437	2.516	2.582	2.636	2.682	2.722	2.756	2.786	2.812	2.836	2.857	2.875	2.893	2.908
14600	1.911	1.924	1.936	2.046	2.218	2.348	2.449	2.529	2.595	2.650	2.696	2.736	2.771	2.801	2.828	2.851	2.873	2.892	2.909	2.925
14700	1.918	1.930	1.943	2.054	2.228	2.359	2.460	2.542	2.608	2.664	2.711	2.751	2.786	2.816	2.843	2.867	2.889	2.908	2.925	2.941
14800	1.925	1.937	1.949	2.062	2.238	2.370	2.472	2.555	2.622	2.678	2.725	2.766	2.801	2.832	2.859	2.883	2.905	2.924	2.942	2.958
14900	1.931	1.944	1.956	2.069	2.247	2.381	2.484	2.567	2.635	2.692	2.740	2.781	2.816	2.847	2.875	2.899	2.921	2.941	2.959	2.975
15000	1.938	1.951	1.963	2.077	2.257	2.392	2.496	2.580	2.649	2.706	2.754	2.796	2.831	2.863	2.891	2.915	2.937	2.957	2.975	2.991

$$V - V_c = 8 \qquad\qquad \text{SIGMA} = 13910.0$$

H \ x	3.4	3.2	3.0	2.8	2.6	2.4	2.2	2.0	1.8	1.6	1.4	1.2	1.0	0.8	0.6	0.4	0.2	0.10	0.09	0.08
0	1.000	1.000	1.000	1.000	1.000	1.000	1.000	1.000	1.000	1.000	1.000	1.000	1.000	1.000	1.000	1.000	1.000	1.000	1.000	1.000
20	1.003	1.003	1.003	1.002	1.002	1.002	1.002	1.002	1.002	1.002	1.002	1.002	1.002	1.002	1.002	1.002	1.002	1.002	1.002	1.002
40	1.005	1.005	1.005	1.005	1.005	1.005	1.005	1.005	1.005	1.005	1.005	1.004	1.004	1.004	1.004	1.004	1.003	1.003	1.003	1.003
60	1.008	1.008	1.008	1.008	1.007	1.007	1.007	1.007	1.007	1.007	1.007	1.007	1.006	1.006	1.006	1.007	1.005	1.005	1.005	1.005
80	1.010	1.010	1.010	1.010	1.010	1.010	1.010	1.010	1.009	1.009	1.009	1.009	1.009	1.008	1.008	1.009	1.007	1.006	1.006	1.006
100	1.013	1.013	1.013	1.013	1.012	1.012	1.012	1.012	1.012	1.012	1.011	1.011	1.011	1.010	1.010	1.011	1.008	1.008	1.008	1.008
120	1.015	1.015	1.015	1.015	1.015	1.015	1.015	1.014	1.014	1.014	1.014	1.013	1.013	1.012	1.012	1.013	1.010	1.009	1.009	1.009
140	1.018	1.018	1.018	1.018	1.017	1.017	1.017	1.017	1.017	1.016	1.016	1.016	1.015	1.015	1.014	1.015	1.012	1.011	1.011	1.011
160	1.020	1.020	1.020	1.020	1.020	1.020	1.020	1.019	1.019	1.019	1.018	1.018	1.017	1.017	1.016	1.017	1.013	1.013	1.012	1.012
180	1.023	1.023	1.023	1.023	1.022	1.022	1.022	1.021	1.021	1.021	1.021	1.020	1.019	1.019	1.018	1.019	1.015	1.014	1.014	1.014
200	1.026	1.025	1.025	1.025	1.025	1.025	1.024	1.024	1.024	1.023	1.023	1.022	1.022	1.021	1.020	1.020	1.017	1.016	1.016	1.015
220	1.028	1.028	1.028	1.028	1.027	1.027	1.027	1.026	1.026	1.026	1.025	1.025	1.024	1.023	1.022	1.022	1.018	1.017	1.017	1.017
240	1.031	1.031	1.030	1.030	1.030	1.030	1.029	1.029	1.029	1.028	1.027	1.027	1.026	1.025	1.024	1.024	1.020	1.019	1.019	1.019
260	1.033	1.033	1.033	1.033	1.032	1.032	1.032	1.031	1.031	1.030	1.030	1.029	1.028	1.027	1.026	1.026	1.022	1.020	1.020	1.020
280	1.036	1.036	1.036	1.035	1.035	1.035	1.034	1.034	1.033	1.033	1.032	1.031	1.030	1.030	1.028	1.028	1.023	1.022	1.022	1.022
300	1.039	1.038	1.038	1.038	1.037	1.037	1.037	1.036	1.036	1.035	1.034	1.034	1.033	1.032	1.030	1.030	1.025	1.024	1.023	1.023
320	1.041	1.041	1.041	1.040	1.040	1.040	1.039	1.039	1.038	1.037	1.037	1.036	1.035	1.034	1.032	1.032	1.027	1.025	1.025	1.025
340	1.044	1.044	1.043	1.043	1.043	1.042	1.042	1.041	1.041	1.040	1.039	1.038	1.037	1.036	1.034	1.034	1.028	1.027	1.027	1.026
360	1.046	1.046	1.046	1.045	1.045	1.045	1.044	1.044	1.043	1.042	1.041	1.040	1.039	1.038	1.036	1.035	1.030	1.028	1.028	1.028
380	1.049	1.049	1.048	1.048	1.048	1.047	1.047	1.046	1.045	1.045	1.044	1.043	1.041	1.040	1.038	1.037	1.032	1.030	1.030	1.029
400	1.052	1.051	1.051	1.051	1.050	1.050	1.049	1.048	1.048	1.047	1.046	1.045	1.044	1.042	1.040	1.039	1.033	1.031	1.031	1.031
420	1.054	1.054	1.053	1.053	1.053	1.052	1.052	1.051	1.050	1.049	1.048	1.047	1.046	1.044	1.042	1.041	1.035	1.033	1.033	1.032
440	1.057	1.056	1.056	1.056	1.055	1.055	1.054	1.053	1.053	1.052	1.051	1.049	1.048	1.046	1.044	1.043	1.037	1.035	1.034	1.034
460	1.059	1.059	1.059	1.058	1.058	1.057	1.057	1.056	1.055	1.054	1.053	1.052	1.050	1.048	1.046	1.045	1.038	1.036	1.036	1.036
480	1.062	1.062	1.061	1.061	1.060	1.060	1.059	1.058	1.057	1.056	1.055	1.054	1.052	1.050	1.048	1.048	1.040	1.038	1.038	1.037
500	1.065	1.064	1.064	1.063	1.063	1.062	1.062	1.061	1.060	1.059	1.058	1.056	1.055	1.052	1.050	1.050	1.042	1.039	1.039	1.039
520	1.067	1.067	1.066	1.066	1.065	1.065	1.064	1.063	1.062	1.061	1.060	1.059	1.057	1.055	1.052	1.052	1.044	1.041	1.041	1.040
540	1.070	1.070	1.069	1.069	1.068	1.067	1.067	1.066	1.065	1.064	1.062	1.061	1.059	1.057	1.054	1.054	1.046	1.042	1.042	1.042
560	1.073	1.072	1.072	1.071	1.071	1.070	1.069	1.068	1.067	1.066	1.065	1.063	1.061	1.059	1.056	1.056	1.048	1.044	1.044	1.043
580	1.075	1.075	1.074	1.074	1.073	1.073	1.072	1.071	1.070	1.068	1.067	1.065	1.063	1.061	1.058	1.058	1.050	1.046	1.045	1.045
600	1.078	1.077	1.077	1.076	1.076	1.075	1.074	1.073	1.072	1.071	1.069	1.068	1.066	1.063	1.060	1.060	1.052	1.047	1.047	1.046
620	1.081	1.080	1.079	1.079	1.078	1.077	1.077	1.076	1.075	1.073	1.072	1.070	1.068	1.065	1.062	1.062	1.054	1.049	1.048	1.048
640	1.083	1.083	1.082	1.081	1.081	1.080	1.079	1.078	1.077	1.076	1.074	1.072	1.070	1.067	1.064	1.064	1.056	1.050	1.050	1.050
660	1.086	1.085	1.085	1.084	1.083	1.083	1.082	1.081	1.079	1.078	1.076	1.075	1.072	1.070	1.066	1.066	1.058	1.052	1.052	1.051
680	1.089	1.088	1.087	1.087	1.086	1.085	1.084	1.083	1.082	1.080	1.079	1.077	1.075	1.072	1.068	1.068	1.060	1.054	1.053	1.053
700	1.091	1.090	1.090	1.089	1.088	1.088	1.087	1.086	1.084	1.083	1.081	1.079	1.077	1.074	1.070	1.069	1.061	1.055	1.055	1.054
720	1.094	1.093	1.093	1.092	1.091	1.090	1.089	1.088	1.087	1.085	1.084	1.081	1.079	1.076	1.072	1.071	1.063	1.057	1.056	1.056
740	1.096	1.096	1.095	1.094	1.094	1.093	1.092	1.091	1.089	1.088	1.086	1.084	1.081	1.078	1.074	1.073	1.064	1.058	1.058	1.057
760	1.099	1.098	1.098	1.097	1.096	1.095	1.094	1.093	1.092	1.090	1.088	1.086	1.083	1.080	1.076	1.073	1.066	1.060	1.059	1.059
780	1.102	1.101	1.100	1.100	1.099	1.098	1.097	1.096	1.094	1.093	1.091	1.088	1.086	1.082	1.078	1.075	1.067	1.061	1.061	1.060
800	1.105	1.104	1.103	1.102	1.101	1.100	1.099	1.098	1.097	1.095	1.093	1.091	1.088	1.085	1.080	1.075	1.068	1.063	1.062	1.062

SIGMA = 13910.0

$V - V_c = 8$

H	0.08	0.09	0.10	0.2	0.4	0.6	0.8	1.0	1.2	1.4	1.6	1.8	2.0	2.2	2.4	2.6	2.8	3.0	3.2	3.4
820	1.064	1.064	1.065	1.069	1.077	1.082	1.087	1.090	1.093	1.095	1.097	1.099	1.101	1.102	1.103	1.104	1.105	1.106	1.107	1.107
840	1.065	1.066	1.066	1.071	1.079	1.084	1.089	1.092	1.095	1.098	1.100	1.102	1.103	1.104	1.106	1.107	1.108	1.108	1.109	1.110
860	1.067	1.067	1.068	1.073	1.081	1.086	1.091	1.095	1.098	1.100	1.102	1.104	1.106	1.107	1.108	1.109	1.110	1.111	1.112	1.113
880	1.068	1.069	1.069	1.074	1.082	1.088	1.093	1.097	1.100	1.103	1.105	1.107	1.108	1.110	1.111	1.112	1.113	1.114	1.115	1.115
900	1.070	1.070	1.071	1.076	1.084	1.091	1.095	1.099	1.102	1.105	1.107	1.109	1.111	1.112	1.113	1.115	1.115	1.116	1.117	1.118
920	1.071	1.072	1.073	1.078	1.086	1.093	1.097	1.101	1.105	1.107	1.110	1.111	1.113	1.115	1.116	1.117	1.118	1.119	1.120	1.121
940	1.073	1.074	1.074	1.080	1.088	1.095	1.100	1.104	1.107	1.110	1.112	1.114	1.116	1.117	1.119	1.120	1.121	1.122	1.123	1.123
960	1.074	1.075	1.076	1.081	1.090	1.097	1.102	1.106	1.109	1.112	1.114	1.116	1.118	1.120	1.121	1.122	1.123	1.124	1.125	1.126
980	1.076	1.077	1.077	1.083	1.092	1.099	1.104	1.108	1.112	1.114	1.117	1.119	1.121	1.122	1.124	1.125	1.126	1.127	1.128	1.129
1000	1.078	1.078	1.079	1.085	1.094	1.101	1.106	1.110	1.114	1.117	1.119	1.121	1.123	1.125	1.126	1.128	1.129	1.130	1.131	1.131
1020	1.079	1.080	1.080	1.086	1.096	1.103	1.108	1.113	1.116	1.119	1.122	1.124	1.126	1.127	1.129	1.130	1.131	1.132	1.133	1.134
1040	1.081	1.081	1.082	1.088	1.098	1.105	1.110	1.115	1.119	1.122	1.124	1.126	1.128	1.130	1.131	1.133	1.134	1.135	1.136	1.137
1060	1.082	1.083	1.084	1.090	1.100	1.107	1.113	1.117	1.121	1.124	1.127	1.129	1.131	1.133	1.134	1.135	1.137	1.138	1.139	1.140
1080	1.084	1.085	1.085	1.092	1.102	1.109	1.115	1.119	1.123	1.127	1.129	1.131	1.133	1.135	1.137	1.138	1.139	1.140	1.141	1.142
1100	1.086	1.086	1.087	1.093	1.103	1.111	1.117	1.122	1.126	1.129	1.132	1.134	1.135	1.138	1.139	1.141	1.142	1.143	1.144	1.145
1120	1.087	1.088	1.088	1.095	1.105	1.113	1.119	1.124	1.128	1.131	1.134	1.136	1.138	1.140	1.142	1.143	1.145	1.146	1.147	1.148
1140	1.089	1.089	1.090	1.097	1.107	1.115	1.121	1.126	1.130	1.134	1.137	1.139	1.141	1.143	1.145	1.146	1.147	1.148	1.150	1.150
1160	1.090	1.091	1.092	1.098	1.109	1.117	1.124	1.129	1.133	1.136	1.139	1.141	1.144	1.146	1.147	1.149	1.150	1.151	1.152	1.153
1180	1.092	1.092	1.093	1.100	1.111	1.119	1.126	1.131	1.135	1.139	1.141	1.144	1.146	1.148	1.150	1.151	1.153	1.154	1.155	1.156
1200	1.093	1.094	1.095	1.102	1.113	1.121	1.128	1.133	1.137	1.141	1.144	1.147	1.149	1.151	1.152	1.154	1.155	1.157	1.158	1.159
1220	1.095	1.096	1.096	1.104	1.115	1.123	1.130	1.135	1.140	1.143	1.147	1.149	1.151	1.153	1.155	1.157	1.158	1.159	1.160	1.161
1240	1.096	1.097	1.098	1.105	1.117	1.126	1.132	1.138	1.142	1.146	1.149	1.151	1.154	1.156	1.158	1.159	1.161	1.162	1.163	1.164
1260	1.098	1.099	1.100	1.107	1.119	1.128	1.134	1.140	1.144	1.148	1.152	1.154	1.156	1.159	1.160	1.162	1.163	1.165	1.166	1.167
1280	1.099	1.100	1.101	1.109	1.121	1.130	1.137	1.142	1.147	1.151	1.154	1.157	1.159	1.161	1.163	1.165	1.166	1.167	1.169	1.170
1300	1.101	1.102	1.103	1.111	1.123	1.132	1.139	1.145	1.149	1.153	1.156	1.159	1.162	1.164	1.166	1.167	1.169	1.170	1.171	1.172
1320	1.103	1.103	1.104	1.113	1.125	1.134	1.141	1.147	1.152	1.155	1.159	1.162	1.164	1.166	1.168	1.170	1.171	1.173	1.174	1.175
1340	1.104	1.105	1.106	1.114	1.127	1.136	1.143	1.149	1.154	1.158	1.161	1.164	1.167	1.169	1.171	1.173	1.174	1.176	1.177	1.178
1360	1.106	1.107	1.108	1.116	1.128	1.138	1.145	1.151	1.156	1.160	1.164	1.167	1.169	1.172	1.174	1.175	1.177	1.178	1.180	1.181
1380	1.107	1.108	1.109	1.117	1.130	1.140	1.148	1.154	1.159	1.163	1.166	1.169	1.172	1.174	1.176	1.178	1.180	1.181	1.182	1.183
1400	1.109	1.110	1.111	1.119	1.132	1.142	1.150	1.156	1.161	1.165	1.169	1.172	1.174	1.177	1.179	1.181	1.182	1.184	1.185	1.186
1420	1.110	1.111	1.112	1.121	1.134	1.144	1.152	1.158	1.163	1.168	1.171	1.174	1.177	1.179	1.182	1.183	1.185	1.186	1.188	1.189
1440	1.112	1.113	1.114	1.123	1.136	1.146	1.154	1.161	1.166	1.170	1.174	1.177	1.180	1.182	1.184	1.186	1.188	1.189	1.191	1.192
1460	1.114	1.115	1.116	1.124	1.138	1.148	1.157	1.163	1.168	1.173	1.176	1.180	1.182	1.185	1.187	1.189	1.190	1.192	1.193	1.195
1480	1.115	1.116	1.117	1.126	1.140	1.150	1.159	1.165	1.171	1.175	1.179	1.182	1.185	1.187	1.189	1.191	1.193	1.195	1.196	1.197
1500	1.117	1.118	1.119	1.128	1.142	1.153	1.161	1.168	1.173	1.178	1.181	1.185	1.187	1.190	1.192	1.194	1.196	1.197	1.199	1.200
1520	1.118	1.119	1.120	1.129	1.144	1.155	1.163	1.170	1.175	1.180	1.184	1.187	1.190	1.193	1.195	1.197	1.199	1.200	1.202	1.203
1540	1.120	1.121	1.122	1.131	1.146	1.157	1.165	1.172	1.178	1.182	1.186	1.190	1.193	1.195	1.198	1.200	1.201	1.203	1.204	1.206
1560	1.121	1.122	1.123	1.133	1.148	1.159	1.168	1.175	1.180	1.185	1.189	1.192	1.195	1.198	1.200	1.202	1.204	1.206	1.207	1.209
1580	1.123	1.124	1.125	1.135	1.150	1.161	1.170	1.177	1.183	1.187	1.191	1.195	1.198	1.201	1.203	1.205	1.207	1.208	1.210	1.211
1600	1.125	1.126	1.127	1.136	1.152	1.163	1.172	1.179	1.185	1.190	1.194	1.197	1.201	1.203	1.206	1.208	1.210	1.211	1.213	1.214
1620	1.126	1.127	1.128	1.138	1.154	1.165	1.174	1.181	1.187	1.192	1.196	1.200	1.203	1.206	1.208	1.210	1.212	1.214	1.216	1.217
1640	1.128	1.129	1.130	1.140	1.156	1.167	1.176	1.184	1.190	1.195	1.199	1.203	1.206	1.209	1.211	1.213	1.215	1.217	1.218	1.220

x

SIGMA = 13910.0

v–v$_c$ = 8

H	3.4	3.2	3.0	2.8	2.6	2.4	2.2	2.0	1.8	1.6	1.4	1.2	1.0	0.8	0.6	0.4	0.2	0.10	0.09	0.08
X																				
1660	1.223	1.221	1.220	1.218	1.216	1.214	1.211	1.208	1.205	1.202	1.197	1.192	1.186	1.179	1.169	1.158	1.142	1.131	1.130	1.129
1680	1.225	1.224	1.222	1.221	1.219	1.216	1.214	1.211	1.208	1.204	1.200	1.195	1.188	1.181	1.172	1.159	1.143	1.133	1.132	1.131
1700	1.228	1.227	1.225	1.223	1.221	1.219	1.217	1.214	1.210	1.207	1.202	1.197	1.191	1.183	1.174	1.161	1.145	1.135	1.134	1.132
1720	1.231	1.230	1.228	1.226	1.224	1.222	1.219	1.216	1.213	1.209	1.205	1.199	1.193	1.185	1.176	1.163	1.147	1.136	1.135	1.134
1740	1.234	1.232	1.231	1.229	1.227	1.224	1.222	1.219	1.216	1.212	1.207	1.202	1.195	1.188	1.178	1.165	1.149	1.138	1.137	1.135
1760	1.237	1.235	1.233	1.232	1.229	1.227	1.225	1.222	1.218	1.214	1.210	1.204	1.198	1.190	1.180	1.167	1.150	1.139	1.138	1.137
1780	1.240	1.238	1.236	1.234	1.232	1.230	1.227	1.224	1.221	1.217	1.212	1.207	1.200	1.192	1.182	1.169	1.152	1.141	1.140	1.139
1800	1.242	1.241	1.239	1.237	1.235	1.233	1.230	1.227	1.223	1.219	1.215	1.209	1.202	1.194	1.184	1.171	1.154	1.143	1.141	1.140
1820	1.245	1.244	1.242	1.240	1.238	1.235	1.233	1.229	1.226	1.222	1.217	1.212	1.205	1.197	1.186	1.173	1.156	1.144	1.143	1.142
1840	1.248	1.246	1.245	1.243	1.240	1.238	1.235	1.232	1.229	1.224	1.220	1.214	1.207	1.199	1.188	1.175	1.157	1.146	1.145	1.143
1860	1.251	1.249	1.247	1.245	1.243	1.241	1.238	1.235	1.231	1.227	1.222	1.216	1.210	1.201	1.191	1.177	1.159	1.147	1.146	1.145
1880	1.254	1.252	1.250	1.248	1.246	1.243	1.241	1.237	1.234	1.230	1.225	1.219	1.212	1.203	1.193	1.179	1.161	1.149	1.148	1.147
1900	1.257	1.255	1.253	1.251	1.249	1.246	1.243	1.240	1.236	1.232	1.227	1.221	1.214	1.206	1.195	1.181	1.162	1.151	1.149	1.148
1920	1.259	1.258	1.256	1.254	1.251	1.249	1.246	1.243	1.239	1.235	1.230	1.224	1.217	1.208	1.197	1.183	1.164	1.152	1.151	1.150
1940	1.262	1.261	1.259	1.257	1.254	1.252	1.249	1.245	1.242	1.237	1.232	1.226	1.219	1.210	1.199	1.185	1.166	1.154	1.153	1.151
1960	1.265	1.263	1.261	1.259	1.257	1.254	1.251	1.248	1.244	1.240	1.235	1.229	1.221	1.212	1.201	1.187	1.168	1.156	1.154	1.153
1980	1.268	1.266	1.264	1.262	1.260	1.257	1.254	1.251	1.247	1.242	1.237	1.231	1.224	1.215	1.203	1.189	1.169	1.157	1.156	1.154
2000	1.271	1.269	1.267	1.265	1.262	1.260	1.257	1.253	1.250	1.245	1.240	1.233	1.226	1.217	1.205	1.191	1.171	1.159	1.157	1.156
2020	1.274	1.272	1.270	1.268	1.265	1.263	1.260	1.256	1.252	1.248	1.242	1.236	1.228	1.219	1.208	1.193	1.173	1.160	1.159	1.158
2040	1.277	1.275	1.273	1.271	1.268	1.265	1.262	1.259	1.255	1.250	1.245	1.238	1.231	1.221	1.210	1.195	1.175	1.162	1.161	1.159
2060	1.279	1.278	1.276	1.273	1.271	1.268	1.265	1.261	1.257	1.253	1.247	1.241	1.233	1.224	1.212	1.197	1.176	1.164	1.162	1.161
2080	1.282	1.280	1.278	1.276	1.274	1.271	1.268	1.264	1.260	1.255	1.250	1.243	1.235	1.226	1.214	1.199	1.178	1.165	1.164	1.162
2100	1.285	1.283	1.281	1.279	1.276	1.274	1.270	1.267	1.263	1.258	1.252	1.246	1.238	1.228	1.216	1.201	1.180	1.167	1.165	1.164
2120	1.288	1.286	1.284	1.282	1.279	1.276	1.273	1.269	1.265	1.260	1.255	1.248	1.240	1.230	1.218	1.203	1.182	1.168	1.167	1.165
2140	1.291	1.289	1.287	1.285	1.282	1.279	1.276	1.272	1.268	1.263	1.257	1.251	1.243	1.233	1.220	1.205	1.183	1.170	1.169	1.167
2160	1.294	1.292	1.290	1.287	1.285	1.282	1.279	1.275	1.271	1.266	1.260	1.253	1.245	1.235	1.223	1.207	1.185	1.172	1.170	1.169
2180	1.297	1.295	1.293	1.290	1.288	1.285	1.282	1.278	1.273	1.268	1.262	1.255	1.247	1.237	1.225	1.209	1.187	1.173	1.172	1.170
2200	1.300	1.298	1.296	1.293	1.290	1.287	1.284	1.280	1.276	1.271	1.265	1.258	1.250	1.240	1.227	1.210	1.189	1.175	1.173	1.172
2220	1.303	1.301	1.298	1.296	1.293	1.290	1.287	1.283	1.279	1.273	1.268	1.261	1.252	1.242	1.229	1.212	1.190	1.176	1.175	1.173
2240	1.306	1.303	1.301	1.299	1.296	1.293	1.290	1.286	1.281	1.276	1.270	1.263	1.255	1.244	1.231	1.214	1.192	1.178	1.176	1.175
2260	1.308	1.306	1.304	1.302	1.299	1.296	1.292	1.288	1.284	1.279	1.273	1.265	1.257	1.246	1.233	1.216	1.194	1.180	1.178	1.176
2280	1.311	1.309	1.307	1.304	1.302	1.299	1.295	1.291	1.287	1.281	1.275	1.268	1.259	1.249	1.235	1.218	1.196	1.181	1.180	1.178
2300	1.314	1.312	1.310	1.307	1.305	1.301	1.298	1.294	1.289	1.284	1.278	1.270	1.262	1.251	1.238	1.220	1.197	1.183	1.181	1.180
2320	1.317	1.315	1.313	1.310	1.307	1.304	1.301	1.297	1.292	1.287	1.280	1.273	1.264	1.253	1.240	1.222	1.199	1.184	1.183	1.181
2340	1.320	1.318	1.316	1.313	1.310	1.307	1.303	1.299	1.295	1.289	1.283	1.275	1.266	1.256	1.242	1.224	1.201	1.186	1.184	1.183
2360	1.323	1.321	1.318	1.316	1.313	1.310	1.306	1.302	1.297	1.292	1.285	1.278	1.269	1.258	1.244	1.226	1.203	1.188	1.186	1.184
2380	1.326	1.324	1.321	1.319	1.316	1.313	1.309	1.305	1.300	1.294	1.288	1.280	1.271	1.260	1.246	1.228	1.204	1.189	1.188	1.186
2400	1.329	1.327	1.324	1.322	1.319	1.315	1.312	1.307	1.303	1.297	1.291	1.283	1.274	1.262	1.248	1.230	1.206	1.191	1.189	1.188
2420	1.332	1.330	1.327	1.324	1.321	1.318	1.314	1.310	1.305	1.300	1.293	1.285	1.276	1.265	1.251	1.232	1.208	1.193	1.191	1.189
2440	1.335	1.333	1.330	1.327	1.324	1.321	1.317	1.313	1.308	1.302	1.296	1.288	1.279	1.267	1.253	1.234	1.210	1.194	1.192	1.191
2460	1.338	1.335	1.333	1.330	1.327	1.324	1.320	1.316	1.311	1.305	1.298	1.290	1.281	1.269	1.255	1.236	1.212	1.196	1.194	1.192
2480	1.341	1.338	1.336	1.333	1.330	1.327	1.323	1.318	1.313	1.308	1.301	1.293	1.283	1.272	1.257	1.238	1.213	1.197	1.196	1.194

SIGMA = 13910.0

$V - V_c$ = 8

H x	3.4	3.2	3.0	2.8	2.6	2.4	2.2	2.0	1.8	1.6	1.4	1.2	1.0	0.8	0.6	0.4	0.2	0.10	0.09	0.08
2500	1.344	1.341	1.339	1.336	1.333	1.329	1.325	1.321	1.316	1.310	1.303	1.295	1.286	1.274	1.259	1.240	1.215	1.199	1.197	1.195
2520	1.347	1.344	1.342	1.339	1.336	1.332	1.328	1.324	1.319	1.313	1.306	1.298	1.288	1.276	1.261	1.242	1.217	1.201	1.199	1.197
2540	1.349	1.347	1.345	1.342	1.339	1.335	1.331	1.327	1.321	1.315	1.309	1.300	1.291	1.279	1.264	1.244	1.219	1.202	1.200	1.199
2560	1.352	1.350	1.347	1.345	1.341	1.338	1.334	1.329	1.324	1.318	1.311	1.303	1.293	1.281	1.266	1.246	1.220	1.204	1.202	1.200
2580	1.355	1.353	1.350	1.347	1.344	1.341	1.337	1.332	1.327	1.321	1.314	1.305	1.295	1.283	1.268	1.248	1.222	1.205	1.204	1.202
2600	1.358	1.356	1.353	1.350	1.347	1.344	1.339	1.335	1.330	1.323	1.316	1.308	1.298	1.286	1.270	1.250	1.224	1.207	1.205	1.203
2620	1.361	1.359	1.356	1.353	1.350	1.346	1.342	1.338	1.332	1.326	1.319	1.310	1.300	1.288	1.272	1.252	1.226	1.209	1.207	1.205
2640	1.364	1.362	1.359	1.356	1.353	1.349	1.345	1.340	1.335	1.329	1.322	1.313	1.303	1.290	1.274	1.254	1.227	1.210	1.208	1.207
2660	1.367	1.365	1.362	1.359	1.356	1.352	1.348	1.343	1.338	1.331	1.324	1.315	1.305	1.292	1.277	1.256	1.229	1.212	1.210	1.208
2680	1.370	1.368	1.365	1.362	1.359	1.355	1.351	1.346	1.340	1.334	1.327	1.318	1.308	1.295	1.279	1.258	1.231	1.214	1.212	1.210
2700	1.373	1.371	1.368	1.365	1.362	1.358	1.353	1.349	1.343	1.337	1.329	1.321	1.310	1.297	1.281	1.260	1.233	1.215	1.213	1.211
2720	1.376	1.374	1.371	1.368	1.364	1.361	1.356	1.351	1.346	1.339	1.332	1.323	1.312	1.299	1.283	1.262	1.235	1.217	1.215	1.213
2740	1.379	1.377	1.374	1.371	1.367	1.363	1.359	1.354	1.349	1.342	1.335	1.326	1.315	1.302	1.285	1.264	1.236	1.218	1.216	1.214
2760	1.382	1.380	1.377	1.374	1.370	1.366	1.362	1.357	1.351	1.345	1.337	1.328	1.317	1.304	1.288	1.266	1.238	1.220	1.218	1.216
2780	1.385	1.383	1.380	1.377	1.373	1.369	1.365	1.360	1.354	1.347	1.340	1.331	1.320	1.306	1.290	1.268	1.240	1.222	1.220	1.218
2800	1.388	1.386	1.383	1.379	1.376	1.372	1.368	1.363	1.357	1.350	1.342	1.333	1.322	1.309	1.292	1.270	1.243	1.223	1.221	1.219
2820	1.391	1.389	1.386	1.382	1.379	1.375	1.370	1.365	1.359	1.353	1.345	1.336	1.325	1.311	1.294	1.272	1.245	1.225	1.223	1.221
2840	1.394	1.391	1.389	1.385	1.382	1.378	1.373	1.368	1.362	1.355	1.348	1.338	1.327	1.313	1.296	1.274	1.247	1.227	1.224	1.222
2860	1.397	1.394	1.391	1.388	1.385	1.381	1.376	1.371	1.365	1.358	1.350	1.341	1.330	1.316	1.299	1.276	1.249	1.228	1.226	1.224
2880	1.400	1.397	1.394	1.391	1.388	1.383	1.379	1.374	1.368	1.361	1.353	1.343	1.332	1.318	1.301	1.278	1.250	1.230	1.228	1.226
2900	1.403	1.400	1.397	1.394	1.390	1.386	1.382	1.376	1.370	1.364	1.355	1.346	1.334	1.320	1.303	1.280	1.252	1.231	1.229	1.227
2920	1.406	1.403	1.400	1.397	1.393	1.389	1.385	1.379	1.373	1.366	1.358	1.348	1.337	1.323	1.305	1.282	1.254	1.233	1.231	1.229
2940	1.409	1.406	1.403	1.400	1.396	1.392	1.387	1.382	1.376	1.369	1.361	1.351	1.340	1.325	1.307	1.285	1.256	1.235	1.232	1.230
2960	1.412	1.409	1.406	1.403	1.399	1.395	1.390	1.385	1.379	1.372	1.363	1.354	1.342	1.327	1.310	1.287	1.257	1.236	1.234	1.232
2980	1.415	1.412	1.409	1.406	1.402	1.398	1.393	1.388	1.381	1.374	1.366	1.356	1.344	1.330	1.312	1.289	1.258	1.238	1.236	1.234
3000	1.418	1.415	1.412	1.409	1.405	1.401	1.396	1.390	1.384	1.377	1.369	1.359	1.347	1.332	1.314	1.291	1.259	1.240	1.237	1.235

SIGMA = 13910.0

V-V_c = 8

H	3.4	3.2	3.0	2.8	2.6	2.4	2.2	2.0	1.8	1.6	1.4	1.2	1.0	0.8	0.6	0.4	0.2	0.10	0.09	0.08
3100	1.433	1.431	1.427	1.424	1.420	1.415	1.410	1.405	1.398	1.391	1.382	1.372	1.359	1.344	1.325	1.301	1.268	1.248	1.245	1.243
3200	1.449	1.446	1.442	1.439	1.434	1.430	1.425	1.419	1.412	1.404	1.395	1.384	1.372	1.356	1.336	1.311	1.277	1.256	1.253	1.251
3300	1.464	1.461	1.457	1.454	1.449	1.444	1.439	1.433	1.426	1.418	1.408	1.397	1.384	1.368	1.347	1.321	1.286	1.264	1.261	1.259
3400	1.479	1.476	1.473	1.469	1.464	1.459	1.454	1.447	1.440	1.432	1.422	1.410	1.397	1.380	1.358	1.331	1.295	1.272	1.270	1.267
3500	1.495	1.492	1.488	1.484	1.479	1.474	1.468	1.462	1.454	1.445	1.435	1.423	1.409	1.392	1.370	1.342	1.304	1.280	1.278	1.275
3600	1.511	1.507	1.503	1.499	1.494	1.489	1.483	1.476	1.468	1.459	1.449	1.437	1.422	1.404	1.381	1.352	1.313	1.288	1.286	1.283
3700	1.526	1.523	1.519	1.514	1.509	1.504	1.498	1.490	1.482	1.473	1.462	1.450	1.434	1.416	1.392	1.362	1.322	1.297	1.294	1.291
3800	1.542	1.538	1.534	1.529	1.524	1.519	1.512	1.505	1.497	1.487	1.476	1.463	1.447	1.428	1.404	1.373	1.331	1.305	1.302	1.299
3900	1.558	1.554	1.550	1.545	1.540	1.534	1.527	1.520	1.511	1.501	1.490	1.476	1.460	1.440	1.415	1.383	1.340	1.313	1.310	1.307
4000	1.574	1.570	1.565	1.560	1.555	1.549	1.542	1.534	1.526	1.515	1.504	1.490	1.473	1.452	1.426	1.393	1.349	1.321	1.318	1.315
4100	1.590	1.586	1.581	1.576	1.570	1.564	1.557	1.549	1.540	1.530	1.517	1.503	1.486	1.464	1.438	1.404	1.358	1.329	1.326	1.323
4200	1.606	1.601	1.597	1.592	1.586	1.579	1.572	1.564	1.555	1.544	1.531	1.516	1.498	1.477	1.449	1.414	1.367	1.338	1.334	1.331
4300	1.622	1.617	1.613	1.607	1.601	1.595	1.587	1.579	1.569	1.558	1.545	1.530	1.511	1.489	1.461	1.425	1.377	1.346	1.343	1.339
4400	1.638	1.634	1.629	1.623	1.617	1.610	1.603	1.594	1.584	1.573	1.559	1.543	1.525	1.501	1.472	1.435	1.386	1.354	1.351	1.347
4500	1.654	1.650	1.645	1.639	1.633	1.626	1.618	1.609	1.599	1.587	1.573	1.557	1.538	1.514	1.484	1.446	1.395	1.362	1.359	1.355
4600	1.671	1.666	1.661	1.655	1.649	1.641	1.633	1.624	1.614	1.602	1.587	1.571	1.551	1.526	1.496	1.456	1.404	1.371	1.367	1.363
4700	1.687	1.682	1.677	1.671	1.664	1.657	1.649	1.639	1.628	1.616	1.602	1.584	1.564	1.539	1.507	1.467	1.413	1.379	1.375	1.371
4800	1.704	1.699	1.693	1.687	1.680	1.673	1.664	1.655	1.643	1.631	1.616	1.598	1.577	1.551	1.519	1.478	1.422	1.387	1.383	1.379
4900	1.720	1.715	1.710	1.703	1.696	1.689	1.680	1.670	1.658	1.645	1.630	1.612	1.590	1.564	1.531	1.488	1.432	1.396	1.392	1.388
5000	1.737	1.732	1.726	1.720	1.712	1.704	1.695	1.685	1.674	1.660	1.645	1.626	1.604	1.577	1.543	1.499	1.441	1.404	1.400	1.396
5100	1.754	1.748	1.742	1.736	1.729	1.720	1.711	1.701	1.689	1.675	1.659	1.640	1.617	1.589	1.555	1.510	1.450	1.412	1.408	1.404
5200	1.771	1.765	1.759	1.752	1.745	1.736	1.727	1.716	1.704	1.690	1.673	1.654	1.631	1.602	1.566	1.521	1.459	1.421	1.416	1.412
5300	1.788	1.782	1.776	1.769	1.761	1.752	1.743	1.732	1.719	1.705	1.688	1.668	1.644	1.615	1.578	1.531	1.469	1.429	1.424	1.420
5400	1.805	1.799	1.792	1.785	1.777	1.769	1.759	1.747	1.735	1.720	1.703	1.682	1.658	1.628	1.590	1.542	1.478	1.437	1.433	1.428
5500	1.822	1.816	1.809	1.802	1.794	1.785	1.775	1.763	1.750	1.735	1.717	1.696	1.671	1.641	1.602	1.553	1.487	1.446	1.441	1.436
5600	1.839	1.833	1.826	1.819	1.810	1.801	1.791	1.779	1.766	1.750	1.732	1.711	1.685	1.654	1.614	1.564	1.497	1.454	1.449	1.444
5700	1.856	1.850	1.843	1.835	1.827	1.818	1.807	1.795	1.781	1.765	1.747	1.725	1.699	1.667	1.626	1.575	1.506	1.462	1.457	1.453
5800	1.874	1.867	1.860	1.852	1.844	1.834	1.823	1.811	1.797	1.781	1.762	1.739	1.712	1.680	1.639	1.586	1.515	1.471	1.466	1.461
5900	1.891	1.884	1.877	1.869	1.860	1.851	1.839	1.827	1.812	1.796	1.777	1.754	1.726	1.693	1.651	1.597	1.525	1.479	1.474	1.469
6000	1.908	1.902	1.894	1.886	1.877	1.867	1.856	1.843	1.828	1.811	1.791	1.768	1.740	1.706	1.663	1.608	1.534	1.487	1.482	1.477
6100	1.926	1.919	1.912	1.903	1.894	1.884	1.872	1.859	1.844	1.827	1.807	1.783	1.754	1.719	1.675	1.619	1.544	1.496	1.491	1.485
6200	1.944	1.937	1.929	1.921	1.911	1.901	1.889	1.875	1.860	1.842	1.822	1.797	1.768	1.732	1.687	1.630	1.553	1.504	1.499	1.493
6300	1.961	1.954	1.946	1.938	1.928	1.917	1.905	1.892	1.876	1.858	1.837	1.812	1.782	1.745	1.700	1.641	1.563	1.513	1.507	1.502
6400	1.979	1.972	1.964	1.955	1.945	1.934	1.922	1.908	1.892	1.874	1.852	1.827	1.796	1.759	1.712	1.652	1.572	1.521	1.516	1.510
6500	1.997	1.990	1.982	1.973	1.962	1.951	1.939	1.924	1.908	1.889	1.867	1.841	1.810	1.772	1.724	1.663	1.582	1.530	1.524	1.518
6600	2.015	2.008	1.999	1.990	1.980	1.968	1.955	1.941	1.924	1.905	1.883	1.856	1.824	1.785	1.737	1.674	1.591	1.538	1.532	1.526
6700	2.033	2.025	2.017	2.008	1.997	1.985	1.972	1.957	1.940	1.921	1.898	1.871	1.839	1.799	1.749	1.686	1.601	1.547	1.541	1.535
6800	2.051	2.043	2.035	2.025	2.015	2.003	1.989	1.974	1.957	1.937	1.913	1.886	1.853	1.812	1.762	1.697	1.610	1.555	1.549	1.543
6900	2.069	2.061	2.053	2.043	2.032	2.020	2.006	1.991	1.973	1.953	1.929	1.901	1.867	1.826	1.774	1.708	1.620	1.564	1.557	1.551
7000	2.088	2.080	2.071	2.061	2.050	2.037	2.023	2.008	1.990	1.969	1.945	1.916	1.881	1.839	1.787	1.719	1.629	1.572	1.566	1.559
7100	2.106	2.098	2.089	2.078	2.067	2.055	2.040	2.024	2.006	1.985	1.960	1.931	1.896	1.853	1.800	1.731	1.639	1.581	1.574	1.568
7200	2.125	2.116	2.107	2.096	2.085	2.072	2.058	2.041	2.023	2.001	1.976	1.946	1.910	1.867	1.812	1.742	1.649	1.589	1.582	1.576

X

SIGMA = 13910.0

V-Vc = 8

H	3.4	3.2	3.0	2.8	2.6	2.4	2.2	2.0	1.8	1.6	1.4	1.2	1.0	0.8	0.6	0.4	0.2	0.10	0.09	0.08
7300	2.143	2.134	2.125	2.115	2.103	2.090	2.075	2.058	2.039	2.017	1.992	1.961	1.925	1.880	1.825	1.753	1.658	1.598	1.591	1.584
7400	2.162	2.153	2.143	2.133	2.121	2.107	2.092	2.075	2.056	2.034	2.007	1.977	1.939	1.894	1.838	1.765	1.668	1.606	1.599	1.592
7500	2.180	2.171	2.162	2.151	2.139	2.125	2.110	2.092	2.073	2.050	2.023	1.992	1.954	1.908	1.850	1.776	1.677	1.615	1.608	1.601
7600	2.199	2.190	2.180	2.169	2.157	2.143	2.127	2.110	2.090	2.066	2.039	2.007	1.969	1.922	1.863	1.788	1.687	1.623	1.616	1.609
7700	2.218	2.209	2.199	2.187	2.175	2.161	2.145	2.127	2.106	2.083	2.055	2.023	1.984	1.936	1.876	1.799	1.697	1.632	1.625	1.617
7800	2.237	2.228	2.217	2.206	2.193	2.179	2.162	2.144	2.123	2.099	2.071	2.038	1.998	1.950	1.889	1.811	1.707	1.640	1.633	1.626
7900	2.256	2.246	2.236	2.224	2.211	2.197	2.180	2.162	2.140	2.116	2.087	2.054	2.013	1.964	1.902	1.822	1.716	1.649	1.641	1.634
8000	2.275	2.265	2.255	2.243	2.229	2.215	2.198	2.179	2.157	2.133	2.104	2.069	2.028	1.976	1.915	1.834	1.726	1.657	1.650	1.642
8100	2.294	2.284	2.273	2.261	2.248	2.233	2.216	2.197	2.175	2.149	2.120	2.085	2.043	1.992	1.928	1.846	1.736	1.666	1.658	1.651
8200	2.314	2.303	2.292	2.280	2.266	2.251	2.234	2.214	2.192	2.166	2.136	2.101	2.058	2.006	1.941	1.857	1.746	1.675	1.667	1.659
8300	2.333	2.323	2.311	2.299	2.285	2.269	2.252	2.232	2.209	2.183	2.152	2.116	2.073	2.020	1.954	1.869	1.755	1.683	1.675	1.667
8400	2.352	2.342	2.330	2.318	2.303	2.288	2.270	2.250	2.227	2.200	2.169	2.132	2.088	2.034	1.967	1.881	1.765	1.692	1.684	1.676
8500	2.372	2.361	2.349	2.336	2.322	2.306	2.288	2.267	2.244	2.217	2.185	2.148	2.103	2.049	1.980	1.892	1.775	1.701	1.692	1.684
8600	2.391	2.381	2.369	2.355	2.341	2.324	2.306	2.285	2.261	2.234	2.202	2.164	2.119	2.063	1.993	1.904	1.785	1.709	1.701	1.692
8700	2.411	2.400	2.388	2.375	2.360	2.343	2.324	2.303	2.279	2.251	2.218	2.180	2.134	2.077	2.007	1.916	1.795	1.718	1.709	1.701
8800	2.431	2.420	2.407	2.394	2.379	2.362	2.343	2.321	2.297	2.268	2.235	2.196	2.149	2.092	2.020	1.928	1.805	1.727	1.718	1.709
8900	2.451	2.439	2.427	2.413	2.398	2.380	2.361	2.339	2.314	2.285	2.252	2.212	2.164	2.106	2.033	1.940	1.815	1.735	1.726	1.718
9000	2.470	2.459	2.446	2.432	2.417	2.399	2.380	2.357	2.332	2.303	2.269	2.228	2.180	2.121	2.047	1.951	1.825	1.744	1.735	1.726
9100	2.490	2.479	2.466	2.452	2.436	2.418	2.398	2.376	2.350	2.320	2.285	2.244	2.195	2.135	2.060	1.963	1.835	1.753	1.744	1.734
9200	2.510	2.499	2.486	2.471	2.455	2.437	2.417	2.394	2.368	2.338	2.302	2.261	2.211	2.150	2.073	1.975	1.845	1.761	1.752	1.743
9300	2.531	2.519	2.505	2.491	2.474	2.456	2.436	2.412	2.386	2.355	2.319	2.277	2.226	2.164	2.087	1.987	1.855	1.770	1.761	1.751
9400	2.551	2.539	2.525	2.510	2.494	2.475	2.454	2.431	2.404	2.373	2.336	2.293	2.242	2.179	2.100	1.999	1.865	1.779	1.769	1.760
9500	2.571	2.559	2.545	2.530	2.513	2.494	2.473	2.449	2.422	2.390	2.353	2.310	2.258	2.194	2.114	2.011	1.875	1.787	1.778	1.768
9600	2.592	2.579	2.565	2.550	2.533	2.514	2.492	2.468	2.440	2.408	2.371	2.326	2.273	2.209	2.128	2.023	1.885	1.796	1.786	1.777
9700	2.612	2.599	2.585	2.569	2.552	2.533	2.511	2.486	2.458	2.426	2.388	2.343	2.289	2.223	2.141	2.036	1.895	1.805	1.795	1.785
9800	2.632	2.619	2.605	2.589	2.572	2.552	2.530	2.505	2.477	2.444	2.405	2.360	2.305	2.238	2.155	2.048	1.905	1.814	1.804	1.793
9900	2.653	2.640	2.625	2.609	2.592	2.572	2.549	2.524	2.495	2.461	2.422	2.376	2.321	2.253	2.169	2.060	1.915	1.822	1.812	1.802
10000	2.674	2.660	2.646	2.629	2.611	2.591	2.568	2.543	2.513	2.479	2.440	2.393	2.337	2.268	2.182	2.072	1.925	1.831	1.821	1.810
10100	2.695	2.681	2.666	2.650	2.631	2.611	2.588	2.562	2.532	2.497	2.457	2.410	2.353	2.283	2.196	2.084	1.935	1.840	1.830	1.819
10200	2.715	2.702	2.687	2.670	2.651	2.630	2.607	2.581	2.550	2.515	2.475	2.427	2.369	2.298	2.210	2.096	1.945	1.849	1.838	1.827
10300	2.736	2.722	2.707	2.690	2.671	2.650	2.627	2.600	2.569	2.534	2.492	2.443	2.385	2.313	2.224	2.109	1.955	1.858	1.847	1.836
10400	2.757	2.743	2.728	2.710	2.691	2.670	2.646	2.619	2.588	2.552	2.510	2.460	2.401	2.328	2.238	2.121	1.965	1.866	1.856	1.844
10500	2.778	2.764	2.748	2.731	2.712	2.690	2.666	2.638	2.606	2.570	2.528	2.477	2.417	2.344	2.252	2.133	1.976	1.875	1.864	1.853
10600	2.800	2.785	2.769	2.751	2.732	2.710	2.685	2.657	2.625	2.588	2.545	2.494	2.433	2.359	2.266	2.146	1.986	1.884	1.873	1.862
10700	2.821	2.806	2.790	2.772	2.752	2.730	2.705	2.677	2.644	2.607	2.563	2.512	2.450	2.374	2.280	2.158	1.996	1.893	1.882	1.870
10800	2.842	2.827	2.811	2.793	2.773	2.750	2.725	2.696	2.663	2.625	2.581	2.529	2.466	2.389	2.294	2.170	2.006	1.902	1.890	1.879
10900	2.864	2.848	2.832	2.813	2.793	2.770	2.744	2.715	2.682	2.644	2.599	2.546	2.482	2.405	2.308	2.183	2.017	1.911	1.899	1.887
11000	2.885	2.870	2.853	2.834	2.814	2.790	2.764	2.735	2.701	2.662	2.617	2.563	2.499	2.420	2.322	2.195	2.027	1.920	1.908	1.896
11100	2.907	2.891	2.874	2.855	2.834	2.811	2.784	2.755	2.720	2.681	2.635	2.581	2.515	2.436	2.336	2.208	2.037	1.928	1.916	1.904
11200	2.928	2.913	2.895	2.876	2.855	2.831	2.804	2.774	2.740	2.700	2.653	2.598	2.532	2.451	2.350	2.220	2.047	1.937	1.925	1.913
11300	2.950	2.934	2.917	2.897	2.876	2.852	2.825	2.794	2.759	2.718	2.671	2.615	2.549	2.467	2.364	2.233	2.058	1.946	1.934	1.921
11400	2.972	2.956	2.938	2.918	2.897	2.872	2.845	2.814	2.778	2.737	2.689	2.633	2.565	2.482	2.379	2.246	2.068	1.955	1.943	1.930

SIGMA = 13910.0

$V-V_c$ = 8

H	3.4	3.2	3.0	2.8	2.6	2.4	2.2	2.0	1.8	1.6	1.4	1.2	1.0	0.8	0.6	0.4	0.2	0.10	0.09	0.08
x																				
11500	2.994	2.977	2.959	2.940	2.917	2.893	2.865	2.834	2.798	2.756	2.708	2.651	2.582	2.498	2.393	2.258	2.078	1.964	1.951	1.939
11600	3.016	2.999	2.981	2.961	2.938	2.913	2.885	2.854	2.817	2.775	2.726	2.668	2.599	2.514	2.407	2.271	2.089	1.973	1.960	1.947
11700	3.038	3.021	3.003	2.982	2.960	2.934	2.906	2.874	2.837	2.794	2.744	2.686	2.615	2.529	2.422	2.284	2.099	1.982	1.969	1.956
11800	3.060	3.043	3.024	3.004	2.981	2.955	2.926	2.894	2.856	2.813	2.763	2.704	2.632	2.545	2.436	2.296	2.110	1.991	1.978	1.964
11900	3.082	3.065	3.046	3.025	3.002	2.976	2.947	2.914	2.876	2.832	2.781	2.721	2.649	2.561	2.451	2.309	2.120	2.000	1.987	1.973
12000	3.104	3.087	3.068	3.047	3.023	2.997	2.967	2.934	2.896	2.852	2.800	2.739	2.666	2.577	2.465	2.322	2.131	2.009	1.995	1.982
12100	3.127	3.109	3.090	3.068	3.045	3.018	2.988	2.954	2.916	2.871	2.819	2.757	2.683	2.593	2.480	2.335	2.141	2.018	2.004	1.990
12200	3.149	3.131	3.112	3.090	3.066	3.039	3.009	2.975	2.935	2.890	2.837	2.775	2.700	2.609	2.494	2.347	2.151	2.027	2.013	1.999
12300	3.172	3.154	3.134	3.112	3.088	3.060	3.030	2.995	2.955	2.910	2.856	2.793	2.717	2.625	2.509	2.360	2.162	2.036	2.022	2.008
12400	3.194	3.176	3.156	3.134	3.109	3.082	3.051	3.016	2.975	2.929	2.875	2.811	2.735	2.641	2.524	2.373	2.172	2.045	2.031	2.016
12500	3.217	3.199	3.178	3.156	3.131	3.103	3.072	3.036	2.995	2.949	2.894	2.829	2.752	2.657	2.538	2.386	2.183	2.054	2.040	2.025
12600	3.240	3.221	3.201	3.178	3.153	3.124	3.093	3.057	3.016	2.968	2.913	2.847	2.769	2.673	2.553	2.399	2.194	2.063	2.048	2.034
12700	3.263	3.244	3.223	3.200	3.174	3.146	3.114	3.077	3.036	2.988	2.932	2.866	2.786	2.689	2.568	2.412	2.204	2.072	2.057	2.042
12800	3.286	3.266	3.245	3.222	3.196	3.167	3.135	3.098	3.056	3.008	2.951	2.884	2.804	2.706	2.583	2.425	2.215	2.081	2.066	2.051
12900	3.309	3.289	3.268	3.244	3.218	3.189	3.156	3.119	3.076	3.027	2.970	2.902	2.821	2.722	2.598	2.438	2.225	2.090	2.075	2.060
13000	3.332	3.312	3.291	3.267	3.240	3.211	3.178	3.140	3.097	3.047	2.989	2.921	2.839	2.738	2.613	2.451	2.236	2.099	2.084	2.069
13100	3.355	3.335	3.313	3.289	3.262	3.233	3.199	3.161	3.117	3.067	3.009	2.939	2.856	2.755	2.628	2.464	2.247	2.108	2.093	2.077
13200	3.378	3.358	3.336	3.312	3.285	3.254	3.220	3.182	3.138	3.087	3.028	2.958	2.874	2.771	2.643	2.477	2.257	2.117	2.102	2.086
13300	3.401	3.381	3.359	3.334	3.307	3.276	3.242	3.203	3.159	3.107	3.047	2.976	2.891	2.787	2.658	2.491	2.268	2.126	2.111	2.095
13400	3.425	3.404	3.382	3.357	3.329	3.298	3.264	3.224	3.179	3.127	3.067	2.995	2.909	2.804	2.673	2.504	2.279	2.135	2.120	2.103
13500	3.448	3.428	3.405	3.380	3.352	3.320	3.285	3.245	3.200	3.147	3.086	3.014	2.927	2.820	2.688	2.517	2.289	2.144	2.128	2.112
13600	3.472	3.451	3.428	3.402	3.374	3.343	3.307	3.267	3.221	3.168	3.106	3.032	2.945	2.837	2.703	2.530	2.300	2.153	2.137	2.121
13700	3.496	3.474	3.451	3.425	3.397	3.365	3.329	3.288	3.242	3.188	3.125	3.051	2.962	2.854	2.718	2.543	2.311	2.163	2.146	2.130
13800	3.519	3.498	3.474	3.448	3.419	3.387	3.351	3.310	3.263	3.208	3.145	3.070	2.980	2.870	2.733	2.557	2.321	2.172	2.155	2.138
13900	3.543	3.521	3.498	3.471	3.442	3.410	3.373	3.331	3.284	3.229	3.165	3.089	2.998	2.887	2.748	2.570	2.332	2.181	2.164	2.147
14000	3.567	3.545	3.521	3.494	3.465	3.432	3.395	3.353	3.305	3.249	3.184	3.108	3.016	2.904	2.764	2.583	2.343	2.190	2.173	2.156
14100	3.591	3.569	3.545	3.518	3.488	3.454	3.417	3.374	3.326	3.270	3.204	3.127	3.034	2.921	2.779	2.597	2.354	2.199	2.182	2.165
14200	3.615	3.593	3.568	3.541	3.511	3.477	3.439	3.396	3.347	3.290	3.224	3.146	3.052	2.938	2.794	2.610	2.365	2.208	2.191	2.174
14300	3.639	3.617	3.592	3.564	3.534	3.500	3.461	3.418	3.368	3.311	3.244	3.165	3.070	2.955	2.810	2.624	2.376	2.218	2.200	2.182
14400	3.663	3.640	3.615	3.588	3.557	3.522	3.484	3.440	3.390	3.332	3.264	3.184	3.089	2.972	2.825	2.637	2.386	2.227	2.209	2.191
14500	3.688	3.665	3.639	3.611	3.580	3.545	3.506	3.462	3.411	3.353	3.284	3.204	3.107	2.989	2.841	2.651	2.397	2.236	2.218	2.200
14600	3.712	3.689	3.663	3.635	3.603	3.568	3.529	3.484	3.433	3.373	3.305	3.223	3.125	3.006	2.856	2.664	2.408	2.245	2.227	2.209
14700	3.736	3.713	3.687	3.658	3.627	3.591	3.551	3.506	3.454	3.394	3.325	3.242	3.144	3.023	2.872	2.678	2.419	2.254	2.236	2.218
14800	3.761	3.737	3.711	3.682	3.650	3.614	3.574	3.528	3.476	3.415	3.345	3.262	3.162	3.040	2.887	2.691	2.430	2.264	2.245	2.227
14900	3.786	3.762	3.735	3.706	3.673	3.637	3.596	3.550	3.497	3.436	3.365	3.281	3.180	3.057	2.903	2.705	2.441	2.273	2.254	2.236
15000	3.810	3.786	3.759	3.730	3.697	3.660	3.619	3.573	3.519	3.458	3.386	3.301	3.199	3.074	2.919	2.719	2.452	2.282	2.263	2.244

SIGMA = 10990.0

$V - V_c = 9$

x \ H	0.08	0.09	0.10	0.2	0.4	0.6	0.8	1.0	1.2	1.4	1.6	1.8	2.0	2.2	2.4	2.6	2.8	3.0	3.2	3.4
0	1.000	1.000	1.000	1.000	1.000	1.000	1.000	1.000	1.000	1.000	1.000	1.000	1.000	1.000	1.000	1.000	1.000	1.000	1.000	1.000
20	1.002	1.002	1.002	1.002	1.002	1.003	1.003	1.003	1.003	1.003	1.003	1.003	1.003	1.003	1.003	1.003	1.003	1.003	1.003	1.003
40	1.004	1.004	1.004	1.004	1.005	1.005	1.005	1.005	1.006	1.006	1.006	1.006	1.006	1.006	1.006	1.006	1.006	1.006	1.006	1.006
60	1.006	1.006	1.006	1.007	1.007	1.008	1.008	1.008	1.008	1.009	1.009	1.009	1.009	1.009	1.009	1.010	1.010	1.010	1.010	1.010
80	1.008	1.008	1.008	1.009	1.009	1.010	1.011	1.011	1.011	1.012	1.012	1.012	1.012	1.012	1.012	1.013	1.013	1.013	1.013	1.013
100	1.010	1.010	1.010	1.011	1.012	1.013	1.013	1.014	1.014	1.014	1.015	1.015	1.015	1.015	1.016	1.016	1.016	1.016	1.016	1.016
120	1.012	1.012	1.012	1.013	1.014	1.015	1.016	1.016	1.017	1.017	1.017	1.018	1.018	1.019	1.019	1.019	1.019	1.019	1.019	1.019
140	1.014	1.014	1.014	1.015	1.016	1.018	1.018	1.019	1.020	1.020	1.020	1.021	1.021	1.022	1.022	1.022	1.022	1.022	1.023	1.023
160	1.016	1.016	1.016	1.017	1.019	1.020	1.021	1.022	1.023	1.023	1.024	1.024	1.024	1.025	1.025	1.025	1.025	1.026	1.026	1.026
180	1.018	1.018	1.018	1.019	1.021	1.024	1.024	1.025	1.026	1.026	1.027	1.027	1.027	1.028	1.028	1.028	1.029	1.029	1.029	1.029
200	1.020	1.020	1.020	1.021	1.023	1.025	1.026	1.027	1.028	1.029	1.030	1.030	1.031	1.031	1.031	1.032	1.032	1.032	1.032	1.033
220	1.022	1.022	1.022	1.023	1.026	1.028	1.029	1.030	1.031	1.032	1.033	1.033	1.034	1.034	1.034	1.035	1.035	1.035	1.036	1.036
240	1.023	1.024	1.024	1.026	1.028	1.030	1.032	1.033	1.034	1.035	1.036	1.036	1.037	1.037	1.038	1.038	1.038	1.039	1.039	1.039
260	1.025	1.026	1.026	1.028	1.031	1.034	1.034	1.036	1.037	1.038	1.039	1.039	1.040	1.040	1.041	1.041	1.042	1.042	1.042	1.042
280	1.027	1.028	1.028	1.030	1.033	1.035	1.037	1.039	1.040	1.041	1.042	1.042	1.043	1.044	1.044	1.044	1.045	1.045	1.045	1.046
300	1.029	1.030	1.030	1.032	1.035	1.038	1.040	1.041	1.043	1.044	1.045	1.045	1.046	1.047	1.047	1.047	1.048	1.048	1.048	1.049
320	1.031	1.032	1.032	1.034	1.038	1.040	1.042	1.044	1.045	1.047	1.048	1.048	1.049	1.050	1.050	1.050	1.051	1.051	1.052	1.052
340	1.033	1.034	1.034	1.036	1.040	1.043	1.045	1.047	1.050	1.050	1.051	1.051	1.052	1.053	1.053	1.054	1.054	1.055	1.055	1.056
360	1.035	1.036	1.036	1.038	1.042	1.045	1.048	1.050	1.052	1.052	1.054	1.055	1.055	1.056	1.057	1.057	1.058	1.058	1.059	1.059
380	1.037	1.038	1.038	1.041	1.045	1.048	1.050	1.052	1.054	1.055	1.057	1.058	1.058	1.059	1.060	1.060	1.061	1.061	1.062	1.062
400	1.039	1.040	1.040	1.043	1.047	1.051	1.053	1.055	1.057	1.058	1.060	1.061	1.062	1.062	1.063	1.064	1.064	1.065	1.065	1.066
420	1.041	1.042	1.042	1.045	1.050	1.053	1.056	1.058	1.060	1.061	1.063	1.064	1.065	1.065	1.066	1.067	1.067	1.068	1.069	1.069
440	1.043	1.043	1.044	1.047	1.052	1.056	1.059	1.061	1.063	1.064	1.066	1.067	1.068	1.069	1.069	1.070	1.070	1.071	1.072	1.072
460	1.045	1.045	1.046	1.049	1.054	1.058	1.061	1.064	1.066	1.067	1.069	1.070	1.071	1.072	1.073	1.073	1.074	1.075	1.075	1.076
480	1.047	1.047	1.048	1.051	1.057	1.061	1.064	1.066	1.069	1.070	1.072	1.073	1.074	1.075	1.076	1.077	1.077	1.078	1.078	1.079
500	1.049	1.050	1.050	1.053	1.059	1.063	1.066	1.069	1.071	1.073	1.075	1.077	1.077	1.078	1.079	1.080	1.081	1.081	1.082	1.082
520	1.051	1.052	1.052	1.056	1.061	1.066	1.069	1.072	1.074	1.076	1.078	1.079	1.080	1.081	1.082	1.083	1.083	1.084	1.085	1.086
540	1.053	1.053	1.054	1.058	1.064	1.068	1.072	1.075	1.077	1.079	1.081	1.082	1.084	1.085	1.086	1.086	1.087	1.088	1.088	1.089
560	1.055	1.056	1.056	1.060	1.066	1.071	1.075	1.078	1.080	1.082	1.084	1.085	1.087	1.088	1.089	1.090	1.090	1.091	1.092	1.092
580	1.057	1.057	1.058	1.062	1.069	1.074	1.077	1.081	1.083	1.085	1.087	1.088	1.090	1.091	1.092	1.093	1.094	1.094	1.095	1.096
600	1.059	1.059	1.060	1.064	1.071	1.076	1.080	1.083	1.086	1.088	1.090	1.092	1.093	1.094	1.095	1.096	1.097	1.098	1.098	1.099
620	1.061	1.061	1.062	1.066	1.073	1.079	1.083	1.086	1.089	1.091	1.093	1.095	1.096	1.097	1.098	1.099	1.100	1.101	1.102	1.102
640	1.063	1.063	1.064	1.069	1.076	1.081	1.086	1.089	1.092	1.094	1.096	1.098	1.099	1.101	1.102	1.103	1.104	1.104	1.105	1.106
660	1.065	1.065	1.066	1.071	1.078	1.084	1.088	1.092	1.095	1.097	1.099	1.101	1.102	1.104	1.105	1.106	1.107	1.108	1.108	1.109
680	1.067	1.067	1.068	1.073	1.081	1.087	1.091	1.095	1.098	1.100	1.102	1.104	1.106	1.107	1.108	1.109	1.110	1.111	1.111	1.113
700	1.069	1.069	1.070	1.075	1.083	1.089	1.094	1.098	1.101	1.103	1.105	1.107	1.109	1.110	1.112	1.113	1.114	1.115	1.116	1.116
720	1.071	1.071	1.072	1.077	1.085	1.092	1.097	1.100	1.104	1.106	1.108	1.110	1.112	1.114	1.115	1.116	1.117	1.118	1.119	1.119
740	1.073	1.075	1.074	1.079	1.088	1.094	1.099	1.103	1.107	1.109	1.112	1.114	1.115	1.117	1.118	1.119	1.120	1.121	1.122	1.123
760	1.075	1.075	1.076	1.081	1.090	1.097	1.102	1.106	1.109	1.112	1.115	1.117	1.118	1.120	1.121	1.123	1.124	1.125	1.125	1.126
780	1.077	1.077	1.078	1.084	1.093	1.099	1.105	1.109	1.112	1.115	1.118	1.120	1.122	1.123	1.125	1.126	1.127	1.128	1.129	1.130
800	1.079	1.079	1.080	1.086	1.095	1.102	1.108	1.112	1.115	1.118	1.121	1.123	1.125	1.126	1.128	1.129	1.130	1.132	1.131	1.133

SIGMA - 10990.0

$V-V_c = 9$

H_X	0.08	0.09	0.10	0.2	0.4	0.6	0.8	1.0	1.2	1.4	1.6	1.8	2.0	2.2	2.4	2.6	2.8	3.0	3.2	3.4
820	1.081	1.081	1.082	1.088	1.098	1.105	1.110	1.115	1.118	1.121	1.124	1.126	1.128	1.130	1.131	1.133	1.134	1.135	1.136	1.137
840	1.083	1.083	1.084	1.090	1.100	1.107	1.113	1.118	1.121	1.124	1.127	1.129	1.131	1.133	1.135	1.136	1.137	1.138	1.139	1.140
860	1.085	1.085	1.086	1.092	1.102	1.110	1.116	1.120	1.124	1.127	1.130	1.132	1.135	1.136	1.138	1.139	1.140	1.142	1.143	1.143
880	1.086	1.087	1.088	1.094	1.105	1.113	1.119	1.123	1.127	1.131	1.133	1.135	1.138	1.140	1.141	1.143	1.144	1.145	1.146	1.147
900	1.088	1.089	1.090	1.097	1.107	1.115	1.121	1.126	1.130	1.134	1.136	1.139	1.141	1.143	1.144	1.146	1.147	1.148	1.149	1.150
920	1.090	1.091	1.092	1.099	1.110	1.118	1.124	1.129	1.133	1.137	1.140	1.142	1.144	1.146	1.148	1.149	1.151	1.152	1.153	1.154
940	1.092	1.093	1.094	1.101	1.112	1.120	1.127	1.132	1.136	1.140	1.143	1.145	1.147	1.149	1.151	1.153	1.154	1.155	1.156	1.157
960	1.094	1.095	1.096	1.103	1.114	1.123	1.130	1.135	1.139	1.143	1.146	1.148	1.151	1.153	1.154	1.156	1.157	1.159	1.160	1.161
980	1.096	1.097	1.098	1.105	1.117	1.126	1.132	1.138	1.142	1.146	1.149	1.151	1.154	1.156	1.158	1.159	1.161	1.162	1.163	1.164
1000	1.098	1.099	1.100	1.108	1.119	1.128	1.135	1.141	1.145	1.149	1.152	1.155	1.157	1.159	1.161	1.163	1.164	1.166	1.167	1.168
1020	1.100	1.101	1.102	1.110	1.122	1.131	1.138	1.144	1.148	1.152	1.155	1.158	1.160	1.163	1.164	1.166	1.168	1.169	1.170	1.171
1040	1.102	1.103	1.104	1.112	1.124	1.133	1.141	1.146	1.151	1.155	1.158	1.161	1.164	1.166	1.168	1.169	1.171	1.173	1.174	1.175
1060	1.104	1.105	1.106	1.114	1.127	1.136	1.143	1.149	1.154	1.158	1.161	1.164	1.167	1.169	1.171	1.173	1.174	1.176	1.177	1.178
1080	1.106	1.107	1.108	1.116	1.129	1.139	1.146	1.152	1.157	1.161	1.164	1.168	1.170	1.172	1.174	1.176	1.178	1.179	1.181	1.182
1100	1.108	1.109	1.110	1.118	1.132	1.141	1.149	1.155	1.160	1.164	1.167	1.171	1.173	1.176	1.178	1.180	1.181	1.183	1.184	1.185
1120	1.110	1.111	1.112	1.121	1.134	1.144	1.152	1.158	1.163	1.167	1.171	1.174	1.177	1.179	1.181	1.183	1.185	1.186	1.187	1.189
1140	1.112	1.113	1.114	1.123	1.136	1.147	1.155	1.161	1.166	1.171	1.174	1.177	1.180	1.182	1.185	1.186	1.188	1.190	1.191	1.192
1160	1.114	1.115	1.116	1.125	1.139	1.149	1.157	1.164	1.169	1.174	1.177	1.181	1.183	1.186	1.188	1.190	1.192	1.193	1.194	1.196
1180	1.116	1.117	1.118	1.127	1.141	1.152	1.160	1.167	1.172	1.177	1.180	1.184	1.187	1.189	1.191	1.193	1.195	1.197	1.198	1.199
1200	1.118	1.119	1.120	1.129	1.144	1.155	1.163	1.170	1.175	1.180	1.183	1.187	1.190	1.192	1.195	1.197	1.198	1.200	1.201	1.203
1220	1.120	1.121	1.122	1.132	1.146	1.157	1.166	1.173	1.178	1.183	1.187	1.190	1.193	1.196	1.198	1.200	1.202	1.204	1.205	1.206
1240	1.122	1.123	1.124	1.134	1.149	1.160	1.169	1.176	1.181	1.186	1.190	1.194	1.197	1.199	1.201	1.204	1.205	1.207	1.208	1.210
1260	1.124	1.125	1.126	1.136	1.151	1.163	1.171	1.179	1.184	1.189	1.193	1.197	1.200	1.202	1.205	1.207	1.209	1.211	1.212	1.213
1280	1.126	1.127	1.128	1.138	1.154	1.165	1.174	1.181	1.187	1.192	1.196	1.200	1.203	1.206	1.208	1.210	1.212	1.214	1.216	1.217
1300	1.128	1.129	1.130	1.140	1.156	1.168	1.177	1.184	1.190	1.195	1.200	1.203	1.206	1.209	1.212	1.214	1.216	1.218	1.219	1.221
1320	1.130	1.131	1.132	1.143	1.159	1.171	1.180	1.187	1.193	1.199	1.203	1.206	1.210	1.212	1.215	1.217	1.219	1.221	1.223	1.224
1340	1.132	1.133	1.134	1.145	1.161	1.173	1.183	1.190	1.197	1.202	1.206	1.210	1.213	1.216	1.218	1.221	1.223	1.225	1.226	1.228
1360	1.134	1.135	1.136	1.147	1.163	1.176	1.186	1.193	1.200	1.205	1.209	1.213	1.216	1.219	1.222	1.224	1.226	1.228	1.230	1.231
1380	1.136	1.137	1.138	1.149	1.166	1.179	1.188	1.196	1.203	1.208	1.213	1.216	1.220	1.222	1.225	1.228	1.230	1.232	1.233	1.235
1400	1.138	1.139	1.140	1.151	1.168	1.181	1.191	1.199	1.206	1.211	1.216	1.220	1.223	1.226	1.229	1.231	1.233	1.235	1.237	1.238
1420	1.140	1.141	1.142	1.154	1.171	1.184	1.194	1.202	1.209	1.214	1.219	1.223	1.226	1.229	1.232	1.235	1.237	1.239	1.240	1.242
1440	1.142	1.143	1.144	1.156	1.173	1.187	1.197	1.205	1.212	1.217	1.222	1.226	1.230	1.232	1.236	1.238	1.240	1.242	1.244	1.246
1460	1.144	1.145	1.147	1.158	1.176	1.189	1.200	1.208	1.215	1.221	1.225	1.230	1.233	1.236	1.239	1.242	1.244	1.246	1.247	1.249
1480	1.146	1.147	1.149	1.160	1.178	1.192	1.203	1.211	1.218	1.224	1.229	1.233	1.237	1.239	1.243	1.245	1.247	1.249	1.251	1.253
1500	1.148	1.149	1.151	1.162	1.181	1.195	1.205	1.214	1.221	1.227	1.232	1.236	1.240	1.243	1.246	1.249	1.251	1.253	1.255	1.256
1520	1.150	1.151	1.153	1.165	1.183	1.197	1.208	1.217	1.224	1.230	1.235	1.240	1.243	1.246	1.249	1.252	1.254	1.256	1.258	1.260
1540	1.152	1.153	1.155	1.167	1.186	1.200	1.211	1.220	1.227	1.233	1.238	1.243	1.247	1.249	1.253	1.256	1.258	1.260	1.262	1.264
1560	1.154	1.155	1.157	1.169	1.188	1.203	1.214	1.223	1.230	1.237	1.242	1.246	1.250	1.253	1.256	1.259	1.261	1.264	1.265	1.267
1580	1.156	1.157	1.159	1.171	1.191	1.205	1.217	1.226	1.233	1.240	1.245	1.249	1.253	1.256	1.260	1.263	1.265	1.267	1.269	1.271
1600	1.158	1.159	1.161	1.173	1.193	1.208	1.220	1.229	1.237	1.243	1.248	1.253	1.257	1.260	1.263	1.266	1.268	1.271	1.273	1.274
1620	1.160	1.161	1.163	1.176	1.196	1.211	1.223	1.232	1.240	1.246	1.251	1.256	1.260	1.264	1.267	1.270	1.272	1.274	1.276	1.278
1640	1.162	1.163	1.165	1.178	1.198	1.214	1.225	1.235	1.243	1.249	1.255	1.259	1.264	1.267	1.270	1.273	1.276	1.278	1.280	1.282

SIGMA = 10990.0

V-V_c = 9

H	3.4	3.2	3.0	2.8	2.6	2.4	2.2	2.0	1.8	1.6	1.4	1.2	1.0	0.8	0.6	0.4	0.2	0.10	0.09	0.08
1660	1.285	1.284	1.281	1.279	1.277	1.274	1.271	1.267	1.263	1.258	1.252	1.246	1.238	1.228	1.216	1.201	1.180	1.167	1.165	1.164
1680	1.289	1.287	1.285	1.283	1.279	1.277	1.274	1.270	1.266	1.261	1.256	1.249	1.241	1.231	1.219	1.203	1.182	1.169	1.167	1.166
1700	1.293	1.291	1.289	1.286	1.284	1.281	1.277	1.274	1.270	1.265	1.259	1.252	1.244	1.234	1.222	1.206	1.184	1.171	1.169	1.168
1720	1.296	1.294	1.292	1.290	1.287	1.284	1.281	1.277	1.273	1.268	1.262	1.255	1.247	1.237	1.224	1.208	1.187	1.173	1.171	1.170
1740	1.300	1.298	1.296	1.293	1.291	1.288	1.284	1.281	1.276	1.271	1.265	1.258	1.250	1.240	1.227	1.211	1.189	1.175	1.173	1.172
1760	1.304	1.302	1.299	1.297	1.294	1.291	1.288	1.284	1.280	1.274	1.269	1.261	1.253	1.243	1.230	1.213	1.191	1.177	1.175	1.174
1780	1.307	1.305	1.303	1.301	1.298	1.295	1.291	1.287	1.283	1.278	1.272	1.265	1.256	1.246	1.233	1.216	1.193	1.179	1.178	1.176
1800	1.311	1.309	1.307	1.304	1.301	1.298	1.295	1.291	1.286	1.281	1.275	1.268	1.259	1.249	1.235	1.218	1.196	1.181	1.180	1.178
1820	1.315	1.313	1.310	1.308	1.305	1.302	1.298	1.294	1.290	1.284	1.278	1.271	1.262	1.251	1.238	1.221	1.198	1.183	1.182	1.180
1840	1.318	1.316	1.314	1.311	1.309	1.305	1.302	1.298	1.293	1.288	1.281	1.274	1.265	1.254	1.241	1.223	1.200	1.185	1.184	1.180
1860	1.322	1.320	1.318	1.315	1.312	1.309	1.305	1.301	1.296	1.291	1.285	1.277	1.268	1.257	1.243	1.226	1.202	1.187	1.186	1.182
1880	1.326	1.324	1.321	1.319	1.316	1.312	1.309	1.305	1.300	1.294	1.288	1.280	1.271	1.260	1.246	1.228	1.204	1.189	1.188	1.184
1900	1.330	1.327	1.325	1.322	1.319	1.316	1.312	1.308	1.303	1.298	1.291	1.283	1.274	1.263	1.249	1.231	1.207	1.191	1.190	1.186
1920	1.333	1.331	1.329	1.326	1.323	1.320	1.316	1.312	1.307	1.301	1.294	1.287	1.277	1.266	1.252	1.233	1.209	1.193	1.192	1.188
1940	1.337	1.335	1.332	1.330	1.327	1.323	1.319	1.315	1.310	1.304	1.298	1.290	1.280	1.269	1.254	1.236	1.211	1.195	1.194	1.190
1960	1.341	1.338	1.336	1.333	1.330	1.327	1.323	1.318	1.313	1.308	1.301	1.293	1.283	1.272	1.257	1.238	1.213	1.197	1.196	1.192
1980	1.344	1.342	1.340	1.337	1.334	1.330	1.326	1.322	1.317	1.311	1.304	1.296	1.286	1.275	1.260	1.241	1.216	1.199	1.198	1.194
2000	1.348	1.346	1.343	1.340	1.337	1.334	1.330	1.325	1.320	1.314	1.307	1.299	1.290	1.278	1.263	1.243	1.218	1.202	1.200	1.196
2020	1.352	1.350	1.347	1.344	1.341	1.337	1.333	1.329	1.324	1.318	1.311	1.302	1.293	1.281	1.265	1.246	1.220	1.204	1.202	1.198
2040	1.356	1.353	1.351	1.348	1.345	1.341	1.337	1.332	1.327	1.321	1.314	1.306	1.296	1.283	1.268	1.249	1.222	1.206	1.204	1.200
2060	1.359	1.357	1.354	1.351	1.348	1.345	1.341	1.336	1.331	1.324	1.317	1.309	1.299	1.286	1.271	1.251	1.225	1.208	1.206	1.202
2080	1.363	1.361	1.358	1.355	1.352	1.348	1.344	1.339	1.334	1.328	1.321	1.312	1.302	1.289	1.274	1.254	1.227	1.210	1.208	1.204
2100	1.367	1.364	1.362	1.359	1.355	1.352	1.348	1.343	1.337	1.331	1.324	1.315	1.305	1.292	1.276	1.256	1.229	1.212	1.210	1.206
2120	1.371	1.368	1.365	1.362	1.359	1.355	1.351	1.346	1.341	1.335	1.327	1.318	1.308	1.295	1.279	1.259	1.231	1.214	1.212	1.208
2140	1.374	1.372	1.369	1.366	1.363	1.359	1.355	1.350	1.344	1.338	1.330	1.322	1.311	1.298	1.282	1.261	1.233	1.216	1.214	1.210
2160	1.378	1.376	1.373	1.370	1.366	1.363	1.358	1.353	1.348	1.341	1.334	1.325	1.314	1.301	1.285	1.264	1.236	1.218	1.216	1.212
2180	1.382	1.379	1.377	1.374	1.370	1.366	1.362	1.357	1.351	1.345	1.337	1.328	1.317	1.304	1.288	1.266	1.238	1.220	1.218	1.214
2200	1.386	1.383	1.380	1.377	1.374	1.370	1.365	1.360	1.355	1.348	1.340	1.331	1.320	1.307	1.290	1.269	1.240	1.222	1.220	1.216
2220	1.390	1.387	1.384	1.381	1.377	1.373	1.369	1.364	1.358	1.351	1.344	1.334	1.323	1.310	1.293	1.271	1.242	1.224	1.222	1.218
2240	1.393	1.391	1.388	1.385	1.381	1.377	1.373	1.367	1.362	1.355	1.347	1.338	1.327	1.313	1.296	1.274	1.245	1.226	1.224	1.220
2260	1.397	1.395	1.392	1.388	1.385	1.381	1.376	1.371	1.365	1.358	1.350	1.341	1.330	1.316	1.299	1.276	1.247	1.228	1.226	1.222
2280	1.401	1.398	1.395	1.392	1.388	1.384	1.380	1.374	1.368	1.362	1.354	1.344	1.333	1.319	1.301	1.279	1.249	1.230	1.228	1.224
2300	1.405	1.402	1.399	1.396	1.392	1.388	1.383	1.378	1.372	1.365	1.357	1.347	1.336	1.322	1.304	1.282	1.251	1.232	1.230	1.226
2320	1.409	1.406	1.403	1.399	1.396	1.392	1.387	1.381	1.375	1.368	1.360	1.351	1.339	1.325	1.307	1.284	1.254	1.234	1.232	1.228
2340	1.412	1.410	1.407	1.403	1.399	1.395	1.390	1.385	1.379	1.372	1.364	1.354	1.342	1.328	1.310	1.287	1.256	1.236	1.234	1.230
2360	1.416	1.413	1.410	1.407	1.403	1.399	1.394	1.389	1.382	1.375	1.367	1.357	1.345	1.331	1.313	1.289	1.258	1.238	1.236	1.232
2380	1.420	1.417	1.414	1.411	1.407	1.403	1.398	1.392	1.386	1.379	1.370	1.360	1.348	1.334	1.315	1.292	1.260	1.241	1.238	1.234
2400	1.424	1.421	1.418	1.414	1.410	1.406	1.401	1.396	1.389	1.382	1.374	1.364	1.351	1.337	1.318	1.294	1.263	1.243	1.240	1.236
2420	1.428	1.425	1.422	1.418	1.414	1.410	1.405	1.399	1.393	1.386	1.377	1.367	1.355	1.340	1.321	1.297	1.265	1.245	1.242	1.238
2440	1.432	1.429	1.426	1.422	1.418	1.414	1.409	1.403	1.396	1.389	1.380	1.370	1.358	1.343	1.324	1.300	1.267	1.247	1.244	1.240
2460	1.436	1.433	1.429	1.426	1.422	1.417	1.412	1.406	1.400	1.392	1.384	1.373	1.361	1.346	1.327	1.302	1.269	1.249	1.246	1.244
2480	1.439	1.436	1.433	1.429	1.425	1.421	1.416	1.410	1.403	1.396	1.387	1.377	1.364	1.349	1.329	1.305	1.272	1.251	1.248	1.246

x

V-V_c = 9   SIGMA =   10990.0

H	3.4	3.2	3.0	2.8	2.6	2.4	2.2	2.0	1.8	1.6	1.4	1.2	1.0	0.8	0.6	0.4	0.2	0.10	0.09	0.08
2500	1.443	1.440	1.437	1.433	1.429	1.425	1.419	1.414	1.407	1.399	1.390	1.380	1.367	1.352	1.332	1.307	1.274	1.253	1.251	1.248
2520	1.447	1.444	1.441	1.437	1.433	1.428	1.423	1.417	1.411	1.403	1.394	1.383	1.370	1.355	1.335	1.310	1.276	1.255	1.253	1.250
2540	1.451	1.448	1.445	1.441	1.437	1.432	1.427	1.421	1.414	1.406	1.397	1.386	1.373	1.358	1.338	1.312	1.279	1.257	1.255	1.252
2560	1.455	1.452	1.448	1.445	1.440	1.436	1.430	1.424	1.418	1.410	1.400	1.390	1.377	1.361	1.341	1.315	1.281	1.259	1.257	1.254
2580	1.459	1.456	1.452	1.448	1.444	1.439	1.434	1.428	1.421	1.413	1.404	1.393	1.380	1.364	1.343	1.318	1.283	1.261	1.259	1.256
2600	1.463	1.459	1.456	1.452	1.448	1.443	1.438	1.432	1.425	1.417	1.407	1.396	1.383	1.367	1.346	1.320	1.285	1.263	1.261	1.258
2620	1.467	1.463	1.460	1.456	1.452	1.447	1.441	1.435	1.428	1.420	1.411	1.399	1.386	1.370	1.349	1.323	1.288	1.265	1.263	1.260
2640	1.470	1.467	1.464	1.460	1.455	1.451	1.445	1.439	1.432	1.424	1.414	1.403	1.389	1.373	1.352	1.325	1.290	1.267	1.265	1.262
2660	1.474	1.471	1.467	1.464	1.459	1.454	1.449	1.442	1.435	1.427	1.417	1.406	1.392	1.376	1.355	1.328	1.292	1.269	1.267	1.264
2680	1.478	1.475	1.471	1.467	1.463	1.458	1.452	1.446	1.439	1.431	1.421	1.409	1.396	1.379	1.358	1.331	1.294	1.271	1.269	1.266
2700	1.482	1.479	1.475	1.471	1.467	1.462	1.456	1.450	1.442	1.434	1.424	1.413	1.399	1.382	1.360	1.333	1.297	1.273	1.271	1.268
2720	1.486	1.483	1.479	1.475	1.470	1.465	1.460	1.453	1.446	1.437	1.428	1.416	1.402	1.385	1.363	1.336	1.299	1.276	1.273	1.270
2740	1.490	1.487	1.483	1.479	1.474	1.469	1.463	1.457	1.450	1.441	1.431	1.419	1.405	1.388	1.366	1.338	1.301	1.278	1.275	1.272
2760	1.494	1.491	1.487	1.483	1.478	1.473	1.467	1.461	1.453	1.444	1.434	1.423	1.408	1.391	1.369	1.341	1.304	1.280	1.277	1.274
2780	1.498	1.495	1.491	1.487	1.482	1.477	1.471	1.464	1.457	1.448	1.438	1.426	1.411	1.394	1.372	1.344	1.306	1.282	1.279	1.276
2800	1.502	1.498	1.495	1.490	1.486	1.480	1.475	1.468	1.460	1.452	1.441	1.429	1.415	1.397	1.375	1.346	1.308	1.284	1.281	1.278
2820	1.506	1.502	1.498	1.494	1.489	1.484	1.478	1.472	1.464	1.455	1.445	1.432	1.418	1.400	1.378	1.349	1.310	1.286	1.283	1.280
2840	1.510	1.506	1.502	1.498	1.493	1.486	1.482	1.475	1.467	1.459	1.448	1.436	1.421	1.403	1.380	1.351	1.313	1.288	1.285	1.283
2860	1.514	1.510	1.506	1.502	1.497	1.492	1.486	1.479	1.471	1.462	1.452	1.439	1.424	1.406	1.383	1.354	1.315	1.290	1.287	1.285
2880	1.518	1.514	1.510	1.506	1.501	1.496	1.489	1.483	1.475	1.466	1.455	1.442	1.427	1.409	1.386	1.357	1.317	1.292	1.289	1.287
2900	1.522	1.518	1.514	1.510	1.505	1.499	1.493	1.486	1.478	1.469	1.458	1.446	1.431	1.412	1.389	1.359	1.319	1.294	1.291	1.289
2920	1.526	1.522	1.518	1.513	1.509	1.503	1.497	1.490	1.482	1.473	1.462	1.449	1.434	1.415	1.392	1.362	1.322	1.296	1.293	1.291
2940	1.530	1.526	1.522	1.517	1.512	1.507	1.501	1.494	1.485	1.476	1.465	1.452	1.437	1.418	1.395	1.364	1.324	1.298	1.296	1.293
2960	1.534	1.530	1.526	1.521	1.516	1.511	1.504	1.497	1.489	1.480	1.469	1.456	1.440	1.421	1.398	1.367	1.326	1.300	1.298	1.295
2980	1.537	1.534	1.530	1.525	1.520	1.514	1.508	1.501	1.493	1.483	1.472	1.459	1.443	1.424	1.400	1.370	1.329	1.302	1.300	1.297
3000	1.541	1.538	1.534	1.529	1.524	1.518	1.512	1.505	1.496	1.487	1.476	1.463	1.447	1.427	1.403	1.372	1.331	1.305	1.302	1.299

SIGMA = 10990.0

$V - V_c = 9$

H	0.08	0.09	0.10	0.2	0.4	0.6	0.8	1.0	1.2	1.4	1.6	1.8	2.0	2.2	2.4	2.6	2.8	3.0	3.2	3.4
3100	1.309	1.312	1.315	1.342	1.385	1.418	1.443	1.463	1.479	1.493	1.505	1.515	1.523	1.531	1.537	1.543	1.549	1.553	1.558	1.562
3200	1.319	1.322	1.325	1.354	1.399	1.432	1.458	1.479	1.496	1.510	1.523	1.533	1.542	1.550	1.557	1.563	1.568	1.573	1.578	1.582
3300	1.329	1.333	1.336	1.365	1.412	1.447	1.474	1.495	1.513	1.528	1.541	1.551	1.561	1.569	1.576	1.582	1.588	1.593	1.598	1.602
3400	1.339	1.343	1.346	1.377	1.425	1.461	1.489	1.512	1.530	1.546	1.559	1.570	1.579	1.588	1.595	1.602	1.608	1.613	1.618	1.622
3500	1.350	1.353	1.357	1.388	1.438	1.476	1.505	1.528	1.548	1.563	1.577	1.588	1.598	1.607	1.615	1.622	1.628	1.633	1.638	1.643
3600	1.360	1.363	1.367	1.400	1.452	1.491	1.521	1.545	1.565	1.581	1.595	1.607	1.617	1.627	1.635	1.642	1.648	1.654	1.659	1.664
3700	1.370	1.374	1.378	1.412	1.465	1.505	1.537	1.562	1.582	1.599	1.614	1.626	1.637	1.646	1.654	1.662	1.668	1.674	1.680	1.684
3800	1.380	1.384	1.388	1.423	1.479	1.520	1.553	1.578	1.600	1.617	1.632	1.645	1.656	1.666	1.674	1.682	1.689	1.695	1.700	1.705
3900	1.390	1.395	1.399	1.435	1.492	1.535	1.569	1.595	1.617	1.635	1.651	1.664	1.675	1.685	1.694	1.702	1.709	1.715	1.721	1.726
4000	1.401	1.405	1.409	1.447	1.506	1.550	1.585	1.612	1.635	1.654	1.669	1.683	1.695	1.705	1.714	1.723	1.730	1.736	1.742	1.748
4100	1.411	1.415	1.420	1.458	1.519	1.565	1.601	1.629	1.652	1.672	1.688	1.702	1.715	1.725	1.735	1.743	1.751	1.757	1.763	1.769
4200	1.421	1.426	1.430	1.470	1.533	1.580	1.617	1.646	1.670	1.690	1.707	1.722	1.734	1.745	1.755	1.764	1.771	1.778	1.785	1.790
4300	1.432	1.436	1.441	1.482	1.547	1.595	1.633	1.663	1.688	1.709	1.726	1.741	1.754	1.766	1.776	1.784	1.792	1.800	1.806	1.812
4400	1.442	1.447	1.451	1.494	1.561	1.611	1.650	1.681	1.706	1.727	1.745	1.761	1.774	1.786	1.796	1.805	1.813	1.821	1.828	1.834
4500	1.452	1.457	1.462	1.506	1.574	1.626	1.666	1.698	1.724	1.746	1.765	1.780	1.794	1.806	1.817	1.826	1.835	1.842	1.849	1.855
4600	1.463	1.468	1.473	1.518	1.588	1.641	1.682	1.715	1.742	1.765	1.784	1.800	1.814	1.827	1.838	1.847	1.856	1.864	1.871	1.877
4700	1.473	1.478	1.483	1.529	1.602	1.657	1.699	1.733	1.761	1.784	1.803	1.820	1.835	1.847	1.859	1.869	1.878	1.886	1.893	1.899
4800	1.483	1.489	1.494	1.541	1.616	1.672	1.716	1.751	1.779	1.803	1.823	1.840	1.855	1.868	1.880	1.890	1.899	1.907	1.915	1.922
4900	1.494	1.499	1.504	1.553	1.630	1.688	1.732	1.768	1.797	1.822	1.843	1.860	1.876	1.889	1.901	1.911	1.921	1.929	1.937	1.944
5000	1.504	1.510	1.515	1.565	1.644	1.703	1.749	1.786	1.816	1.841	1.862	1.880	1.896	1.910	1.922	1.933	1.943	1.951	1.959	1.966
5100	1.514	1.520	1.526	1.577	1.658	1.719	1.766	1.804	1.835	1.860	1.882	1.901	1.917	1.931	1.944	1.955	1.965	1.974	1.982	1.989
5200	1.525	1.531	1.537	1.589	1.672	1.735	1.783	1.822	1.853	1.880	1.902	1.921	1.938	1.952	1.965	1.977	1.987	1.996	2.004	2.012
5300	1.535	1.541	1.547	1.601	1.686	1.750	1.800	1.840	1.872	1.899	1.922	1.942	1.959	1.974	1.987	1.999	2.009	2.018	2.027	2.035
5400	1.546	1.552	1.558	1.613	1.701	1.766	1.817	1.858	1.891	1.919	1.942	1.962	1.980	1.995	2.009	2.021	2.031	2.041	2.050	2.058
5500	1.556	1.562	1.569	1.626	1.715	1.782	1.834	1.876	1.910	1.938	1.963	1.983	2.001	2.017	2.031	2.043	2.054	2.064	2.073	2.081
5600	1.567	1.573	1.579	1.638	1.729	1.798	1.851	1.894	1.929	1.958	1.983	2.004	2.022	2.038	2.053	2.065	2.076	2.086	2.096	2.104
5700	1.577	1.584	1.590	1.650	1.744	1.814	1.869	1.912	1.948	1.978	2.003	2.025	2.044	2.060	2.075	2.088	2.099	2.109	2.119	2.127
5800	1.587	1.594	1.601	1.662	1.758	1.830	1.886	1.931	1.968	1.998	2.024	2.046	2.065	2.082	2.097	2.110	2.122	2.132	2.142	2.151
5900	1.598	1.605	1.612	1.674	1.773	1.846	1.904	1.949	1.987	2.018	2.045	2.067	2.087	2.104	2.119	2.133	2.145	2.156	2.165	2.174
6000	1.608	1.616	1.623	1.687	1.787	1.862	1.921	1.968	2.006	2.038	2.065	2.089	2.109	2.126	2.142	2.156	2.168	2.179	2.189	2.198
6100	1.619	1.626	1.634	1.699	1.802	1.879	1.939	1.987	2.026	2.059	2.086	2.110	2.130	2.148	2.164	2.178	2.191	2.202	2.213	2.222
6200	1.630	1.637	1.644	1.711	1.816	1.895	1.956	2.005	2.045	2.079	2.107	2.131	2.152	2.171	2.187	2.201	2.214	2.226	2.236	2.246
6300	1.640	1.648	1.655	1.724	1.831	1.911	1.974	2.024	2.065	2.099	2.128	2.153	2.174	2.193	2.210	2.225	2.238	2.250	2.260	2.270
6400	1.651	1.658	1.666	1.736	1.846	1.928	1.992	2.043	2.085	2.120	2.149	2.175	2.197	2.216	2.233	2.248	2.261	2.273	2.284	2.294
6500	1.661	1.669	1.677	1.748	1.860	1.944	2.010	2.062	2.105	2.141	2.171	2.197	2.219	2.239	2.256	2.271	2.285	2.297	2.309	2.319
6600	1.672	1.680	1.688	1.761	1.875	1.961	2.028	2.081	2.125	2.161	2.192	2.218	2.241	2.261	2.279	2.295	2.309	2.321	2.333	2.343
6700	1.682	1.691	1.699	1.773	1.890	1.978	2.046	2.100	2.145	2.182	2.214	2.240	2.264	2.284	2.302	2.318	2.333	2.346	2.357	2.368
6800	1.693	1.701	1.710	1.786	1.905	1.994	2.064	2.120	2.165	2.203	2.235	2.263	2.286	2.307	2.326	2.342	2.357	2.370	2.382	2.393
6900	1.704	1.712	1.721	1.798	1.920	2.011	2.082	2.139	2.185	2.224	2.257	2.285	2.309	2.330	2.349	2.366	2.381	2.394	2.407	2.418
7000	1.714	1.723	1.732	1.811	1.935	2.028	2.100	2.158	2.206	2.245	2.279	2.307	2.332	2.354	2.373	2.390	2.405	2.419	2.431	2.443
7100	1.725	1.734	1.743	1.823	1.950	2.045	2.119	2.178	2.226	2.266	2.300	2.330	2.355	2.377	2.397	2.414	2.430	2.444	2.456	2.468
7200	1.735	1.745	1.754	1.836	1.965	2.062	2.137	2.197	2.247	2.288	2.322	2.352	2.378	2.401	2.421	2.438	2.454	2.468	2.481	2.493

x

SIGMA = 10990.0

V-V_c = 9

H	3.4	3.2	3.0	2.8	2.6	2.4	2.2	2.0	1.8	1.6	1.4	1.2	1.0	0.8	0.6	0.4	0.2	0.10	0.09	0.08
X																				
7300	2.518	2.506	2.493	2.479	2.463	2.445	2.424	2.401	2.375	2.345	2.309	2.267	2.217	2.156	2.079	1.980	1.848	1.765	1.756	1.746
7400	2.544	2.532	2.518	2.504	2.487	2.469	2.448	2.424	2.398	2.367	2.331	2.288	2.237	2.174	2.096	1.995	1.861	1.776	1.766	1.757
7500	2.570	2.557	2.544	2.528	2.512	2.493	2.472	2.448	2.421	2.389	2.352	2.309	2.257	2.193	2.113	2.010	1.874	1.787	1.777	1.767
7600	2.595	2.583	2.569	2.553	2.536	2.517	2.496	2.471	2.444	2.411	2.374	2.330	2.276	2.211	2.130	2.026	1.886	1.798	1.788	1.778
7700	2.621	2.608	2.594	2.579	2.561	2.542	2.520	2.495	2.467	2.434	2.396	2.351	2.296	2.230	2.147	2.041	1.899	1.809	1.799	1.789
7800	2.647	2.634	2.620	2.604	2.586	2.566	2.544	2.519	2.490	2.456	2.418	2.372	2.316	2.249	2.165	2.056	1.912	1.820	1.810	1.800
7900	2.674	2.660	2.646	2.629	2.611	2.591	2.568	2.543	2.513	2.479	2.440	2.393	2.337	2.268	2.182	2.072	1.925	1.831	1.821	1.810
8000	2.700	2.686	2.671	2.655	2.636	2.616	2.593	2.566	2.537	2.502	2.462	2.414	2.357	2.287	2.200	2.087	1.938	1.842	1.832	1.821
8100	2.726	2.712	2.697	2.680	2.662	2.641	2.617	2.591	2.560	2.525	2.484	2.435	2.377	2.306	2.217	2.103	1.950	1.853	1.843	1.832
8200	2.753	2.739	2.723	2.706	2.687	2.666	2.642	2.615	2.584	2.548	2.506	2.457	2.398	2.325	2.235	2.118	1.963	1.865	1.854	1.843
8300	2.780	2.765	2.749	2.732	2.713	2.691	2.667	2.639	2.607	2.571	2.529	2.478	2.418	2.344	2.252	2.134	1.976	1.876	1.865	1.853
8400	2.806	2.792	2.776	2.758	2.738	2.716	2.691	2.663	2.631	2.594	2.551	2.500	2.439	2.364	2.270	2.150	1.989	1.887	1.876	1.864
8500	2.833	2.818	2.802	2.784	2.764	2.742	2.716	2.688	2.655	2.618	2.574	2.522	2.459	2.383	2.288	2.165	2.002	1.898	1.887	1.875
8600	2.860	2.845	2.829	2.810	2.790	2.767	2.742	2.712	2.679	2.641	2.596	2.543	2.480	2.402	2.306	2.181	2.015	1.909	1.898	1.886
8700	2.888	2.872	2.855	2.837	2.816	2.793	2.767	2.737	2.703	2.664	2.619	2.565	2.501	2.422	2.323	2.197	2.028	1.921	1.909	1.897
8800	2.915	2.899	2.882	2.863	2.842	2.819	2.792	2.762	2.728	2.688	2.642	2.587	2.522	2.442	2.341	2.213	2.041	1.932	1.920	1.908
8900	2.942	2.927	2.909	2.890	2.868	2.844	2.817	2.787	2.752	2.712	2.665	2.609	2.543	2.461	2.359	2.229	2.054	1.943	1.931	1.918
9000	2.970	2.954	2.936	2.917	2.895	2.870	2.843	2.812	2.777	2.736	2.688	2.631	2.564	2.481	2.378	2.245	2.067	1.954	1.942	1.929
9100	2.998	2.981	2.963	2.943	2.921	2.896	2.869	2.837	2.801	2.760	2.711	2.654	2.585	2.501	2.396	2.260	2.080	1.966	1.953	1.940
9200	3.026	3.009	2.991	2.970	2.948	2.923	2.894	2.862	2.826	2.784	2.734	2.676	2.606	2.521	2.414	2.277	2.093	1.977	1.964	1.951
9300	3.053	3.037	3.018	2.997	2.975	2.949	2.920	2.888	2.851	2.808	2.758	2.698	2.627	2.541	2.432	2.293	2.107	1.988	1.975	1.962
9400	3.082	3.064	3.045	3.025	3.001	2.975	2.946	2.913	2.875	2.832	2.781	2.721	2.649	2.561	2.450	2.309	2.120	2.000	1.986	1.973
9500	3.110	3.092	3.073	3.052	3.028	3.002	2.972	2.939	2.900	2.856	2.805	2.744	2.670	2.581	2.469	2.325	2.133	2.011	1.997	1.984
9600	3.138	3.120	3.101	3.079	3.055	3.028	2.999	2.965	2.926	2.881	2.828	2.766	2.692	2.601	2.487	2.341	2.146	2.022	2.009	1.995
9700	3.167	3.149	3.129	3.107	3.083	3.056	3.025	2.990	2.951	2.905	2.852	2.789	2.713	2.621	2.506	2.357	2.160	2.034	2.020	2.006
9800	3.195	3.177	3.157	3.135	3.110	3.082	3.051	3.016	2.976	2.930	2.876	2.812	2.735	2.641	2.524	2.374	2.173	2.045	2.031	2.017
9900	3.224	3.205	3.185	3.163	3.137	3.109	3.078	3.042	3.002	2.955	2.900	2.835	2.757	2.662	2.543	2.390	2.186	2.056	2.042	2.028
10000	3.253	3.234	3.213	3.190	3.165	3.137	3.105	3.068	3.027	2.979	2.924	2.858	2.779	2.682	2.562	2.406	2.200	2.068	2.053	2.039
10100	3.282	3.263	3.242	3.219	3.193	3.164	3.131	3.095	3.053	3.004	2.948	2.881	2.801	2.703	2.580	2.423	2.213	2.079	2.065	2.050
10200	3.311	3.292	3.270	3.247	3.221	3.191	3.158	3.121	3.079	3.029	2.972	2.904	2.823	2.723	2.599	2.439	2.226	2.091	2.076	2.061
10300	3.340	3.321	3.299	3.275	3.248	3.219	3.185	3.148	3.104	3.055	2.996	2.928	2.845	2.744	2.618	2.456	2.240	2.102	2.087	2.072
10400	3.370	3.350	3.328	3.303	3.277	3.246	3.213	3.174	3.130	3.080	3.021	2.951	2.867	2.765	2.637	2.473	2.253	2.114	2.098	2.083
10500	3.399	3.379	3.357	3.332	3.305	3.274	3.240	3.201	3.156	3.105	3.045	2.974	2.890	2.786	2.656	2.489	2.267	2.125	2.110	2.094
10600	3.429	3.408	3.386	3.361	3.333	3.302	3.267	3.228	3.183	3.131	3.070	2.998	2.913	2.807	2.675	2.506	2.280	2.137	2.121	2.105
10700	3.458	3.438	3.415	3.389	3.361	3.330	3.295	3.255	3.209	3.156	3.095	3.022	2.934	2.828	2.694	2.523	2.294	2.148	2.132	2.116
10800	3.488	3.467	3.444	3.418	3.390	3.358	3.322	3.282	3.235	3.182	3.119	3.045	2.957	2.849	2.713	2.539	2.307	2.160	2.144	2.127
10900	3.518	3.497	3.473	3.447	3.419	3.386	3.350	3.309	3.262	3.208	3.144	3.069	2.980	2.870	2.733	2.556	2.321	2.171	2.155	2.138
11000	3.548	3.527	3.503	3.477	3.447	3.415	3.378	3.336	3.288	3.233	3.169	3.093	3.002	2.891	2.752	2.573	2.335	2.183	2.166	2.149
11100	3.579	3.557	3.533	3.506	3.476	3.443	3.406	3.363	3.315	3.259	3.194	3.117	3.025	2.912	2.771	2.590	2.348	2.195	2.178	2.160
11200	3.609	3.587	3.562	3.535	3.505	3.472	3.434	3.391	3.342	3.285	3.219	3.141	3.048	2.934	2.791	2.607	2.362	2.206	2.189	2.172
11300	3.640	3.617	3.592	3.565	3.534	3.500	3.462	3.418	3.369	3.312	3.245	3.166	3.071	2.955	2.810	2.624	2.376	2.218	2.200	2.183
11400	3.670	3.647	3.622	3.594	3.564	3.529	3.490	3.446	3.396	3.338	3.270	3.190	3.094	2.977	2.830	2.641	2.390	2.229	2.212	2.194

SIGMA = 10990.0

$V-V_c = 9$

H	0.08	0.09	0.10	0.2	0.4	0.6	0.8	1.0	1.2	1.4	1.6	1.8	2.0	2.2	2.4	2.6	2.8	3.0	3.2	3.4
11500	2.205	2.223	2.241	2.403	2.658	2.849	2.998	3.117	3.214	3.296	3.364	3.423	3.474	3.519	3.558	3.593	3.624	3.652	3.678	3.701
11600	2.216	2.235	2.253	2.417	2.675	2.869	3.020	3.140	3.239	3.321	3.391	3.450	3.502	3.547	3.587	3.622	3.654	3.683	3.709	3.732
11700	2.227	2.246	2.264	2.431	2.693	2.889	3.041	3.164	3.264	3.347	3.417	3.478	3.530	3.576	3.616	3.652	3.684	3.713	3.740	3.763
11800	2.239	2.258	2.276	2.445	2.710	2.909	3.063	3.187	3.288	3.373	3.444	3.505	3.558	3.604	3.645	3.682	3.714	3.744	3.770	3.794
11900	2.250	2.269	2.288	2.459	2.727	2.929	3.085	3.210	3.313	3.398	3.471	3.533	3.586	3.633	3.675	3.712	3.745	3.774	3.801	3.826
12000	2.261	2.281	2.300	2.473	2.745	2.948	3.107	3.234	3.338	3.424	3.498	3.560	3.615	3.662	3.704	3.742	3.775	3.805	3.832	3.857
12100	2.272	2.292	2.311	2.487	2.762	2.968	3.129	3.258	3.363	3.450	3.525	3.588	3.643	3.691	3.734	3.772	3.805	3.836	3.863	3.888
12200	2.284	2.304	2.323	2.501	2.779	2.989	3.151	3.281	3.388	3.477	3.552	3.616	3.672	3.721	3.764	3.802	3.836	3.867	3.895	3.920
12300	2.295	2.315	2.335	2.514	2.797	3.009	3.173	3.305	3.413	3.503	3.579	3.644	3.700	3.750	3.793	3.832	3.867	3.898	3.926	3.952
12400	2.306	2.327	2.347	2.529	2.814	3.029	3.196	3.329	3.438	3.529	3.606	3.672	3.729	3.779	3.823	3.863	3.898	3.929	3.958	3.984
12500	2.317	2.338	2.358	2.543	2.832	3.049	3.218	3.353	3.463	3.556	3.633	3.700	3.758	3.809	3.853	3.893	3.929	3.961	3.990	4.016
12600	2.329	2.350	2.370	2.557	2.850	3.069	3.240	3.377	3.489	3.582	3.661	3.729	3.787	3.838	3.884	3.924	3.960	3.992	4.022	4.048
12700	2.340	2.361	2.382	2.571	2.867	3.090	3.263	3.401	3.514	3.609	3.689	3.757	3.816	3.868	3.914	3.955	3.991	4.024	4.053	4.080
12800	2.351	2.373	2.394	2.585	2.885	3.110	3.285	3.425	3.540	3.635	3.716	3.785	3.846	3.898	3.944	3.986	4.022	4.056	4.086	4.113
12900	2.363	2.384	2.406	2.599	2.903	3.131	3.308	3.450	3.566	3.662	3.744	3.814	3.875	3.928	3.975	4.017	4.054	4.087	4.118	4.145
13000	2.374	2.396	2.418	2.613	2.921	3.151	3.331	3.474	3.591	3.689	3.772	3.843	3.904	3.958	4.006	4.048	4.086	4.119	4.150	4.178
13100	2.386	2.408	2.430	2.627	2.939	3.172	3.353	3.498	3.617	3.716	3.800	3.872	3.934	3.988	4.036	4.079	4.117	4.152	4.183	4.211
13200	2.397	2.419	2.441	2.642	2.956	3.192	3.376	3.523	3.643	3.743	3.828	3.901	3.964	4.019	4.067	4.110	4.149	4.183	4.215	4.244
13300	2.408	2.431	2.453	2.656	2.974	3.213	3.399	3.548	3.669	3.770	3.856	3.930	3.993	4.049	4.098	4.142	4.181	4.215	4.248	4.277
13400	2.420	2.443	2.465	2.670	2.992	3.234	3.422	3.572	3.695	3.798	3.884	3.959	4.023	4.080	4.129	4.174	4.213	4.248	4.281	4.310
13500	2.431	2.454	2.477	2.685	3.010	3.255	3.445	3.597	3.721	3.825	3.913	3.988	4.053	4.110	4.161	4.205	4.245	4.281	4.314	4.344
13600	2.443	2.466	2.489	2.699	3.029	3.276	3.468	3.622	3.748	3.853	3.941	4.017	4.083	4.141	4.192	4.237	4.276	4.314	4.347	4.377
13700	2.454	2.478	2.501	2.713	3.047	3.297	3.491	3.647	3.774	3.880	3.970	4.047	4.114	4.172	4.223	4.269	4.310	4.347	4.380	4.411
13800	2.465	2.490	2.513	2.728	3.065	3.318	3.515	3.672	3.801	3.908	3.999	4.077	4.144	4.203	4.255	4.301	4.343	4.380	4.414	4.444
13900	2.477	2.501	2.525	2.742	3.083	3.339	3.538	3.697	3.827	3.936	4.028	4.106	4.174	4.234	4.287	4.334	4.375	4.413	4.447	4.478
14000	2.488	2.513	2.537	2.757	3.102	3.360	3.561	3.722	3.854	3.964	4.056	4.136	4.205	4.265	4.319	4.366	4.408	4.446	4.481	4.512
14100	2.500	2.525	2.549	2.771	3.120	3.381	3.585	3.748	3.881	3.992	4.085	4.166	4.236	4.297	4.351	4.398	4.441	4.480	4.515	4.546
14200	2.511	2.537	2.561	2.786	3.138	3.403	3.608	3.773	3.907	4.020	4.115	4.196	4.266	4.328	4.383	4.431	4.474	4.513	4.549	4.581
14300	2.523	2.548	2.573	2.800	3.157	3.424	3.632	3.798	3.934	4.048	4.144	4.226	4.297	4.360	4.415	4.464	4.507	4.547	4.583	4.615
14400	2.535	2.560	2.585	2.815	3.175	3.445	3.656	3.824	3.961	4.076	4.173	4.256	4.328	4.391	4.447	4.497	4.541	4.581	4.617	4.649
14500	2.546	2.572	2.598	2.829	3.194	3.467	3.679	3.849	3.989	4.104	4.203	4.287	4.359	4.423	4.479	4.529	4.574	4.614	4.651	4.684
14600	2.558	2.584	2.610	2.844	3.212	3.488	3.703	3.875	4.016	4.133	4.232	4.317	4.391	4.455	4.512	4.563	4.608	4.648	4.685	4.719
14700	2.569	2.596	2.622	2.859	3.231	3.510	3.727	3.901	4.043	4.161	4.262	4.348	4.422	4.487	4.545	4.596	4.641	4.683	4.720	4.754
14800	2.581	2.608	2.634	2.873	3.250	3.532	3.751	3.927	4.070	4.190	4.291	4.378	4.453	4.519	4.577	4.629	4.675	4.717	4.754	4.789
14900	2.592	2.619	2.646	2.888	3.268	3.553	3.775	3.953	4.098	4.219	4.321	4.409	4.485	4.552	4.610	4.662	4.709	4.751	4.789	4.824
15000	2.604	2.631	2.658	2.903	3.287	3.575	3.799	3.979	4.125	4.248	4.351	4.440	4.517	4.584	4.643	4.696	4.743	4.786	4.824	4.859

SIGMA = 8900.0

V−V$_c$ = 10

H X	0.08	0.09	0.10	0.2	0.4	0.6	0.8	1.0	1.2	1.4	1.6	1.8	2.0	2.2	2.4	2.6	2.8	3.0	3.2	3.4
C	1.000	1.000	1.000	1.000	1.000	1.000	1.000	1.000	1.000	1.000	1.000	1.000	1.000	1.000	1.000	1.000	1.000	1.000	1.000	1.000
20	1.002	1.002	1.002	1.003	1.003	1.003	1.003	1.003	1.003	1.004	1.004	1.004	1.004	1.004	1.004	1.004	1.004	1.004	1.004	1.004
40	1.005	1.005	1.005	1.005	1.006	1.006	1.007	1.007	1.007	1.007	1.007	1.007	1.008	1.008	1.008	1.008	1.008	1.008	1.008	1.008
60	1.007	1.007	1.007	1.008	1.009	1.009	1.010	1.010	1.010	1.011	1.011	1.011	1.011	1.011	1.012	1.012	1.012	1.012	1.012	1.012
80	1.010	1.010	1.010	1.011	1.012	1.012	1.013	1.014	1.014	1.014	1.015	1.015	1.015	1.015	1.015	1.016	1.016	1.016	1.016	1.016
100	1.012	1.012	1.012	1.013	1.014	1.016	1.016	1.017	1.017	1.018	1.018	1.019	1.019	1.019	1.019	1.019	1.020	1.020	1.020	1.020
120	1.014	1.015	1.015	1.016	1.017	1.019	1.020	1.020	1.021	1.021	1.022	1.022	1.023	1.023	1.023	1.023	1.024	1.024	1.024	1.024
140	1.017	1.017	1.017	1.018	1.020	1.022	1.023	1.024	1.024	1.025	1.026	1.026	1.026	1.027	1.027	1.027	1.028	1.028	1.028	1.028
160	1.019	1.019	1.020	1.021	1.023	1.025	1.026	1.027	1.028	1.029	1.029	1.030	1.030	1.031	1.031	1.031	1.031	1.032	1.032	1.032
180	1.022	1.022	1.022	1.024	1.026	1.028	1.029	1.031	1.031	1.032	1.033	1.033	1.034	1.034	1.035	1.035	1.035	1.036	1.036	1.036
200	1.024	1.024	1.025	1.026	1.029	1.031	1.033	1.034	1.035	1.036	1.037	1.038	1.038	1.038	1.039	1.039	1.039	1.040	1.040	1.040
220	1.027	1.027	1.027	1.029	1.032	1.034	1.036	1.037	1.039	1.039	1.040	1.042	1.042	1.042	1.043	1.043	1.043	1.044	1.044	1.044
240	1.029	1.029	1.029	1.032	1.035	1.037	1.039	1.041	1.042	1.043	1.044	1.045	1.045	1.046	1.047	1.047	1.047	1.048	1.048	1.048
260	1.031	1.032	1.032	1.034	1.038	1.040	1.043	1.044	1.046	1.047	1.048	1.049	1.049	1.050	1.050	1.051	1.051	1.052	1.052	1.052
280	1.034	1.034	1.034	1.037	1.041	1.043	1.046	1.048	1.049	1.050	1.051	1.052	1.053	1.054	1.054	1.055	1.055	1.056	1.056	1.057
300	1.036	1.037	1.037	1.040	1.044	1.047	1.049	1.051	1.053	1.054	1.055	1.056	1.057	1.058	1.058	1.059	1.059	1.060	1.060	1.061
320	1.039	1.039	1.039	1.042	1.047	1.050	1.053	1.055	1.056	1.058	1.059	1.060	1.061	1.062	1.062	1.063	1.063	1.064	1.064	1.065
340	1.041	1.041	1.042	1.045	1.050	1.053	1.056	1.058	1.060	1.061	1.063	1.064	1.065	1.065	1.066	1.067	1.067	1.068	1.069	1.069
360	1.044	1.044	1.044	1.047	1.052	1.056	1.059	1.061	1.063	1.065	1.066	1.068	1.069	1.069	1.070	1.071	1.071	1.072	1.073	1.073
380	1.046	1.046	1.047	1.050	1.055	1.059	1.062	1.065	1.067	1.069	1.070	1.071	1.072	1.073	1.074	1.075	1.076	1.076	1.077	1.077
400	1.048	1.049	1.049	1.053	1.058	1.063	1.066	1.068	1.071	1.072	1.074	1.076	1.076	1.077	1.078	1.079	1.080	1.080	1.081	1.081
420	1.051	1.051	1.052	1.055	1.061	1.066	1.069	1.072	1.074	1.076	1.078	1.079	1.080	1.081	1.082	1.083	1.084	1.084	1.085	1.085
440	1.053	1.054	1.054	1.058	1.064	1.069	1.072	1.075	1.078	1.080	1.081	1.083	1.084	1.085	1.086	1.087	1.088	1.088	1.089	1.089
460	1.056	1.056	1.057	1.061	1.067	1.072	1.075	1.079	1.081	1.083	1.085	1.087	1.088	1.089	1.090	1.091	1.092	1.092	1.093	1.094
480	1.058	1.059	1.059	1.063	1.070	1.075	1.079	1.082	1.085	1.087	1.089	1.090	1.092	1.093	1.094	1.095	1.096	1.097	1.097	1.098
500	1.061	1.061	1.062	1.066	1.073	1.078	1.083	1.086	1.089	1.091	1.093	1.094	1.096	1.097	1.098	1.099	1.100	1.101	1.101	1.102
520	1.063	1.064	1.064	1.069	1.076	1.082	1.086	1.089	1.092	1.095	1.096	1.098	1.100	1.101	1.102	1.103	1.104	1.105	1.106	1.106
540	1.065	1.066	1.067	1.071	1.079	1.085	1.089	1.093	1.096	1.098	1.100	1.102	1.104	1.105	1.106	1.107	1.108	1.109	1.110	1.110
560	1.068	1.068	1.069	1.074	1.082	1.088	1.093	1.096	1.099	1.102	1.104	1.106	1.108	1.109	1.110	1.111	1.112	1.113	1.114	1.115
580	1.070	1.071	1.071	1.077	1.085	1.091	1.096	1.100	1.103	1.106	1.108	1.110	1.111	1.113	1.114	1.115	1.116	1.117	1.118	1.119
600	1.073	1.073	1.074	1.079	1.088	1.094	1.099	1.103	1.107	1.109	1.112	1.114	1.115	1.117	1.118	1.119	1.120	1.121	1.122	1.123
620	1.075	1.076	1.076	1.082	1.091	1.098	1.103	1.107	1.110	1.113	1.116	1.118	1.119	1.121	1.122	1.123	1.125	1.126	1.126	1.127
640	1.078	1.078	1.079	1.085	1.094	1.101	1.106	1.110	1.114	1.117	1.119	1.121	1.123	1.125	1.126	1.128	1.129	1.130	1.131	1.131
660	1.080	1.081	1.081	1.087	1.097	1.104	1.110	1.114	1.118	1.121	1.123	1.125	1.127	1.129	1.130	1.132	1.133	1.134	1.135	1.136
680	1.082	1.083	1.084	1.090	1.100	1.107	1.113	1.118	1.121	1.124	1.127	1.129	1.131	1.133	1.134	1.136	1.137	1.138	1.139	1.140
700	1.085	1.086	1.086	1.093	1.103	1.110	1.116	1.121	1.125	1.128	1.131	1.133	1.135	1.137	1.139	1.140	1.141	1.142	1.143	1.144
720	1.087	1.088	1.089	1.095	1.106	1.114	1.120	1.125	1.129	1.132	1.135	1.137	1.139	1.141	1.143	1.144	1.145	1.146	1.148	1.148
740	1.090	1.091	1.091	1.098	1.109	1.117	1.123	1.128	1.132	1.136	1.139	1.141	1.143	1.145	1.147	1.148	1.150	1.151	1.152	1.153
760	1.092	1.093	1.094	1.101	1.112	1.120	1.127	1.132	1.136	1.139	1.142	1.145	1.147	1.149	1.151	1.152	1.154	1.155	1.156	1.157
780	1.095	1.096	1.096	1.104	1.115	1.123	1.130	1.135	1.140	1.143	1.146	1.149	1.151	1.153	1.155	1.156	1.158	1.159	1.160	1.161
800	1.097	1.098	1.099	1.106	1.118	1.127	1.133	1.139	1.143	1.147	1.150	1.153	1.155	1.157	1.159	1.161	1.162	1.163	1.165	1.166

SIGMA =   8900.0

V–V$_c$ = 10

H	3.4	3.2	3.0	2.8	2.6	2.4	2.2	2.0	1.8	1.6	1.4	1.2	1.0	0.8	0.6	0.4	0.2	0.10	0.09	0.08
820	1.170	1.169	1.168	1.166	1.165	1.163	1.161	1.159	1.157	1.154	1.151	1.147	1.142	1.137	1.130	1.121	1.109	1.101	1.100	1.100
840	1.174	1.173	1.172	1.170	1.169	1.167	1.165	1.163	1.161	1.158	1.155	1.151	1.146	1.140	1.133	1.124	1.112	1.104	1.103	1.102
860	1.179	1.177	1.176	1.175	1.173	1.171	1.169	1.167	1.165	1.162	1.158	1.154	1.150	1.144	1.136	1.127	1.114	1.106	1.105	1.104
880	1.183	1.182	1.180	1.179	1.177	1.176	1.174	1.171	1.169	1.166	1.162	1.158	1.153	1.147	1.140	1.130	1.117	1.109	1.108	1.107
900	1.187	1.186	1.185	1.183	1.182	1.180	1.178	1.175	1.173	1.170	1.166	1.162	1.157	1.151	1.143	1.133	1.120	1.111	1.110	1.109
920	1.192	1.190	1.189	1.187	1.186	1.184	1.182	1.179	1.177	1.174	1.170	1.166	1.160	1.154	1.146	1.136	1.122	1.114	1.113	1.112
940	1.196	1.195	1.193	1.192	1.190	1.188	1.186	1.183	1.181	1.177	1.174	1.169	1.164	1.158	1.149	1.139	1.125	1.116	1.115	1.114
960	1.200	1.199	1.197	1.196	1.194	1.192	1.190	1.188	1.185	1.181	1.178	1.173	1.168	1.161	1.153	1.142	1.128	1.119	1.118	1.117
980	1.205	1.203	1.202	1.200	1.198	1.196	1.194	1.192	1.189	1.185	1.181	1.177	1.171	1.164	1.156	1.145	1.130	1.121	1.120	1.119
1000	1.209	1.208	1.206	1.204	1.203	1.201	1.198	1.196	1.193	1.189	1.185	1.181	1.175	1.168	1.159	1.148	1.133	1.124	1.123	1.122
1020	1.213	1.212	1.210	1.209	1.207	1.205	1.202	1.200	1.197	1.193	1.189	1.184	1.178	1.171	1.163	1.151	1.136	1.126	1.125	1.124
1040	1.218	1.216	1.215	1.213	1.211	1.209	1.207	1.204	1.201	1.197	1.193	1.188	1.182	1.175	1.166	1.154	1.139	1.129	1.128	1.127
1060	1.222	1.221	1.219	1.217	1.215	1.213	1.211	1.208	1.205	1.201	1.197	1.192	1.186	1.178	1.169	1.157	1.141	1.131	1.130	1.129
1080	1.226	1.225	1.223	1.222	1.220	1.217	1.215	1.212	1.209	1.205	1.201	1.196	1.189	1.182	1.172	1.160	1.144	1.134	1.133	1.131
1100	1.231	1.229	1.228	1.226	1.224	1.222	1.219	1.216	1.213	1.209	1.205	1.199	1.193	1.185	1.176	1.163	1.147	1.136	1.135	1.134
1120	1.235	1.234	1.232	1.230	1.228	1.226	1.223	1.220	1.217	1.213	1.208	1.203	1.197	1.189	1.179	1.166	1.149	1.139	1.138	1.136
1140	1.240	1.238	1.236	1.235	1.232	1.230	1.227	1.224	1.221	1.217	1.212	1.207	1.200	1.192	1.182	1.169	1.152	1.141	1.140	1.139
1160	1.244	1.243	1.241	1.239	1.237	1.234	1.232	1.229	1.225	1.221	1.216	1.211	1.204	1.196	1.186	1.172	1.155	1.144	1.143	1.141
1180	1.249	1.247	1.245	1.243	1.241	1.239	1.236	1.233	1.229	1.225	1.220	1.214	1.208	1.199	1.189	1.175	1.158	1.146	1.145	1.144
1200	1.253	1.251	1.250	1.248	1.245	1.243	1.240	1.237	1.233	1.229	1.224	1.218	1.211	1.203	1.192	1.179	1.160	1.149	1.147	1.146
1220	1.258	1.256	1.254	1.252	1.250	1.247	1.244	1.241	1.237	1.233	1.228	1.222	1.215	1.206	1.196	1.182	1.163	1.151	1.150	1.149
1240	1.262	1.260	1.258	1.256	1.254	1.251	1.248	1.245	1.241	1.237	1.232	1.226	1.219	1.210	1.199	1.185	1.166	1.154	1.152	1.151
1260	1.266	1.265	1.263	1.261	1.258	1.256	1.253	1.249	1.245	1.241	1.236	1.230	1.222	1.213	1.202	1.188	1.169	1.156	1.155	1.154
1280	1.271	1.269	1.267	1.265	1.263	1.260	1.257	1.253	1.250	1.245	1.240	1.234	1.226	1.217	1.206	1.191	1.171	1.159	1.157	1.156
1300	1.275	1.274	1.272	1.269	1.267	1.264	1.261	1.258	1.254	1.249	1.244	1.237	1.230	1.220	1.209	1.194	1.174	1.161	1.160	1.158
1320	1.280	1.278	1.276	1.274	1.271	1.269	1.265	1.262	1.258	1.253	1.248	1.241	1.233	1.224	1.212	1.197	1.177	1.164	1.162	1.161
1340	1.284	1.283	1.280	1.278	1.276	1.273	1.270	1.266	1.262	1.257	1.252	1.245	1.237	1.228	1.216	1.200	1.179	1.166	1.165	1.163
1360	1.289	1.287	1.285	1.283	1.280	1.277	1.274	1.270	1.266	1.261	1.256	1.249	1.241	1.231	1.219	1.203	1.182	1.169	1.167	1.166
1380	1.293	1.292	1.289	1.287	1.284	1.281	1.278	1.274	1.270	1.265	1.260	1.253	1.245	1.235	1.222	1.206	1.185	1.171	1.170	1.168
1400	1.298	1.296	1.294	1.291	1.289	1.286	1.282	1.279	1.274	1.269	1.263	1.257	1.248	1.238	1.226	1.209	1.188	1.174	1.172	1.171
1420	1.303	1.301	1.298	1.296	1.293	1.290	1.287	1.283	1.278	1.273	1.267	1.260	1.252	1.242	1.229	1.212	1.190	1.176	1.175	1.173
1440	1.307	1.305	1.303	1.300	1.298	1.294	1.291	1.287	1.283	1.277	1.271	1.264	1.256	1.245	1.232	1.216	1.193	1.179	1.177	1.176
1460	1.312	1.310	1.307	1.305	1.302	1.299	1.295	1.291	1.287	1.282	1.275	1.268	1.260	1.249	1.236	1.219	1.196	1.181	1.180	1.178
1480	1.316	1.314	1.312	1.309	1.306	1.303	1.300	1.296	1.291	1.286	1.279	1.272	1.263	1.252	1.239	1.222	1.199	1.184	1.182	1.181
1500	1.321	1.319	1.316	1.314	1.311	1.308	1.304	1.300	1.295	1.290	1.283	1.276	1.267	1.256	1.242	1.225	1.201	1.186	1.185	1.183
1520	1.325	1.323	1.321	1.318	1.315	1.312	1.308	1.304	1.299	1.294	1.287	1.280	1.271	1.260	1.246	1.228	1.204	1.189	1.187	1.186
1540	1.330	1.328	1.325	1.323	1.320	1.316	1.313	1.308	1.304	1.298	1.291	1.284	1.275	1.263	1.249	1.231	1.207	1.191	1.190	1.188
1560	1.334	1.332	1.330	1.327	1.324	1.321	1.317	1.313	1.308	1.302	1.295	1.288	1.278	1.267	1.253	1.234	1.210	1.194	1.192	1.191
1580	1.339	1.337	1.334	1.332	1.329	1.325	1.321	1.317	1.312	1.306	1.299	1.292	1.282	1.270	1.256	1.237	1.212	1.196	1.195	1.193
1600	1.344	1.341	1.339	1.336	1.333	1.329	1.326	1.321	1.316	1.310	1.303	1.295	1.286	1.274	1.259	1.240	1.215	1.199	1.197	1.195
1620	1.348	1.346	1.343	1.341	1.337	1.334	1.330	1.325	1.320	1.314	1.308	1.299	1.290	1.278	1.263	1.243	1.218	1.202	1.200	1.198
1640	1.353	1.351	1.348	1.345	1.342	1.338	1.334	1.330	1.325	1.319	1.312	1.303	1.293	1.281	1.266	1.247	1.221	1.204	1.202	1.200

x

$V-V_c = 10$ SIGMA = 8900.0

H (x)	0.08	0.09	0.10	0.2	0.4	0.6	0.8	1.0	1.2	1.4	1.6	1.8	2.0	2.2	2.4	2.6	2.8	3.0	3.2	3.4
1660	1.203	1.205	1.207	1.223	1.250	1.270	1.285	1.297	1.307	1.316	1.323	1.329	1.334	1.339	1.343	1.346	1.350	1.352	1.355	1.358
1680	1.205	1.207	1.209	1.226	1.253	1.273	1.288	1.301	1.311	1.320	1.327	1.333	1.338	1.343	1.347	1.351	1.354	1.357	1.360	1.362
1700	1.208	1.210	1.212	1.229	1.256	1.276	1.292	1.305	1.315	1.324	1.331	1.337	1.343	1.347	1.352	1.355	1.359	1.362	1.364	1.367
1720	1.210	1.212	1.214	1.232	1.259	1.280	1.296	1.309	1.319	1.328	1.335	1.342	1.347	1.352	1.356	1.360	1.363	1.366	1.369	1.371
1740	1.213	1.215	1.217	1.234	1.262	1.283	1.299	1.312	1.323	1.332	1.339	1.346	1.351	1.356	1.360	1.364	1.368	1.371	1.374	1.376
1760	1.215	1.217	1.219	1.237	1.265	1.287	1.303	1.316	1.327	1.336	1.344	1.350	1.356	1.361	1.365	1.369	1.372	1.375	1.378	1.381
1780	1.218	1.220	1.222	1.240	1.269	1.290	1.307	1.320	1.331	1.340	1.348	1.354	1.360	1.365	1.369	1.373	1.377	1.380	1.383	1.385
1800	1.220	1.222	1.224	1.243	1.272	1.293	1.310	1.324	1.335	1.344	1.352	1.359	1.364	1.369	1.373	1.376	1.381	1.385	1.388	1.390
1820	1.223	1.225	1.227	1.246	1.275	1.297	1.314	1.328	1.339	1.348	1.356	1.363	1.369	1.374	1.378	1.382	1.386	1.389	1.392	1.395
1840	1.225	1.227	1.229	1.248	1.278	1.300	1.318	1.331	1.343	1.352	1.360	1.367	1.373	1.378	1.383	1.387	1.391	1.394	1.397	1.400
1860	1.228	1.230	1.232	1.251	1.281	1.304	1.321	1.335	1.347	1.356	1.364	1.371	1.377	1.383	1.387	1.391	1.395	1.398	1.401	1.404
1880	1.230	1.232	1.234	1.254	1.284	1.307	1.325	1.339	1.351	1.360	1.369	1.376	1.382	1.387	1.392	1.396	1.400	1.403	1.406	1.409
1900	1.233	1.235	1.237	1.257	1.287	1.311	1.329	1.343	1.355	1.365	1.373	1.380	1.386	1.392	1.396	1.401	1.404	1.408	1.411	1.414
1920	1.235	1.237	1.240	1.259	1.291	1.314	1.332	1.347	1.359	1.369	1.377	1.384	1.391	1.396	1.401	1.405	1.409	1.412	1.416	1.418
1940	1.238	1.240	1.242	1.262	1.294	1.318	1.336	1.351	1.363	1.373	1.381	1.389	1.395	1.401	1.405	1.410	1.414	1.417	1.420	1.423
1960	1.240	1.242	1.245	1.265	1.297	1.321	1.340	1.355	1.367	1.377	1.386	1.393	1.399	1.405	1.410	1.414	1.418	1.422	1.425	1.428
1980	1.243	1.245	1.247	1.268	1.300	1.324	1.343	1.358	1.371	1.381	1.390	1.397	1.404	1.409	1.414	1.419	1.423	1.426	1.430	1.433
2000	1.245	1.247	1.250	1.271	1.303	1.328	1.347	1.362	1.375	1.385	1.394	1.402	1.408	1.414	1.419	1.423	1.427	1.431	1.434	1.437
2020	1.248	1.250	1.252	1.273	1.307	1.331	1.351	1.366	1.379	1.389	1.398	1.406	1.413	1.418	1.423	1.428	1.432	1.436	1.439	1.442
2040	1.250	1.252	1.255	1.276	1.310	1.335	1.354	1.370	1.383	1.394	1.403	1.410	1.417	1.423	1.428	1.433	1.437	1.441	1.444	1.447
2060	1.253	1.255	1.257	1.279	1.313	1.338	1.358	1.374	1.387	1.398	1.407	1.415	1.421	1.427	1.433	1.437	1.441	1.445	1.449	1.452
2080	1.255	1.258	1.260	1.282	1.316	1.342	1.362	1.378	1.391	1.402	1.411	1.419	1.426	1.432	1.437	1.442	1.446	1.450	1.453	1.457
2100	1.258	1.260	1.262	1.285	1.319	1.345	1.366	1.382	1.395	1.406	1.415	1.423	1.430	1.436	1.442	1.447	1.451	1.455	1.458	1.461
2120	1.260	1.263	1.265	1.287	1.322	1.349	1.369	1.386	1.399	1.410	1.420	1.428	1.435	1.441	1.447	1.451	1.456	1.459	1.463	1.466
2140	1.263	1.265	1.268	1.290	1.326	1.352	1.373	1.390	1.403	1.414	1.424	1.432	1.439	1.446	1.451	1.456	1.460	1.464	1.468	1.471
2160	1.265	1.268	1.270	1.293	1.329	1.356	1.377	1.393	1.407	1.419	1.428	1.437	1.444	1.450	1.456	1.461	1.465	1.469	1.472	1.476
2180	1.268	1.270	1.273	1.296	1.332	1.359	1.380	1.397	1.411	1.423	1.433	1.441	1.448	1.455	1.460	1.465	1.470	1.474	1.477	1.481
2200	1.270	1.273	1.275	1.299	1.335	1.363	1.384	1.401	1.415	1.427	1.437	1.445	1.453	1.459	1.465	1.470	1.474	1.478	1.482	1.485
2220	1.273	1.275	1.278	1.301	1.338	1.366	1.388	1.405	1.419	1.431	1.441	1.450	1.457	1.464	1.469	1.475	1.479	1.483	1.487	1.490
2240	1.275	1.278	1.280	1.304	1.342	1.370	1.392	1.409	1.424	1.435	1.446	1.454	1.462	1.468	1.474	1.479	1.484	1.488	1.492	1.495
2260	1.278	1.280	1.283	1.307	1.345	1.373	1.395	1.413	1.428	1.440	1.450	1.459	1.466	1.473	1.479	1.484	1.489	1.493	1.497	1.500
2280	1.280	1.283	1.285	1.310	1.348	1.377	1.399	1.417	1.432	1.444	1.454	1.463	1.471	1.477	1.483	1.489	1.493	1.498	1.501	1.505
2300	1.283	1.285	1.288	1.313	1.351	1.380	1.403	1.421	1.436	1.448	1.459	1.467	1.475	1.482	1.488	1.493	1.498	1.502	1.506	1.510
2320	1.285	1.288	1.291	1.315	1.355	1.384	1.407	1.425	1.440	1.452	1.463	1.472	1.480	1.487	1.493	1.498	1.503	1.507	1.511	1.515
2340	1.288	1.290	1.293	1.318	1.358	1.387	1.410	1.429	1.444	1.457	1.467	1.476	1.484	1.491	1.497	1.503	1.508	1.512	1.516	1.520
2360	1.290	1.293	1.296	1.321	1.361	1.391	1.414	1.433	1.448	1.461	1.472	1.481	1.489	1.496	1.502	1.507	1.512	1.517	1.521	1.524
2380	1.293	1.295	1.298	1.324	1.364	1.395	1.418	1.437	1.452	1.465	1.476	1.485	1.493	1.500	1.507	1.512	1.517	1.521	1.526	1.529
2400	1.295	1.298	1.301	1.327	1.367	1.398	1.422	1.441	1.456	1.469	1.480	1.490	1.498	1.505	1.511	1.517	1.522	1.526	1.531	1.534
2420	1.298	1.300	1.303	1.330	1.371	1.402	1.426	1.445	1.461	1.474	1.485	1.494	1.502	1.510	1.516	1.522	1.527	1.531	1.535	1.539
2440	1.300	1.303	1.306	1.332	1.374	1.405	1.429	1.449	1.465	1.478	1.489	1.499	1.507	1.514	1.521	1.526	1.532	1.536	1.540	1.544
2460	1.303	1.306	1.308	1.335	1.377	1.409	1.433	1.453	1.469	1.482	1.494	1.503	1.512	1.519	1.525	1.531	1.536	1.541	1.545	1.549
2480	1.305	1.308	1.311	1.338	1.380	1.412	1.437	1.457	1.473	1.486	1.498	1.508	1.516	1.524	1.530	1.536	1.541	1.546	1.550	1.554

SIGMA = 8900.0

$V-V_c = 10$

x	0.08	0.09	0.10	0.2	0.4	0.6	0.8	1.0	1.2	1.4	1.6	1.8	2.0	2.2	2.4	2.6	2.8	3.0	3.2	3.4
2500	1.308	1.311	1.314	1.341	1.384	1.416	1.441	1.461	1.477	1.491	1.502	1.512	1.521	1.528	1.535	1.541	1.546	1.551	1.555	1.559
2520	1.310	1.313	1.316	1.344	1.387	1.419	1.445	1.465	1.481	1.495	1.507	1.517	1.525	1.533	1.540	1.546	1.551	1.556	1.560	1.564
2540	1.313	1.316	1.319	1.347	1.390	1.423	1.448	1.469	1.485	1.499	1.511	1.521	1.530	1.538	1.544	1.550	1.556	1.561	1.565	1.569
2560	1.315	1.318	1.321	1.349	1.393	1.427	1.452	1.473	1.490	1.504	1.516	1.526	1.535	1.542	1.549	1.555	1.561	1.565	1.570	1.574
2580	1.318	1.321	1.324	1.352	1.397	1.430	1.456	1.477	1.494	1.508	1.520	1.530	1.539	1.547	1.554	1.560	1.565	1.570	1.575	1.579
2600	1.320	1.323	1.326	1.355	1.400	1.434	1.460	1.481	1.498	1.512	1.524	1.535	1.544	1.552	1.559	1.565	1.570	1.575	1.580	1.584
2620	1.323	1.326	1.329	1.358	1.403	1.437	1.464	1.485	1.502	1.517	1.529	1.539	1.548	1.556	1.563	1.570	1.575	1.580	1.585	1.589
2640	1.325	1.328	1.332	1.361	1.407	1.441	1.468	1.489	1.506	1.521	1.533	1.544	1.553	1.561	1.568	1.574	1.580	1.585	1.590	1.594
2660	1.328	1.331	1.334	1.364	1.410	1.444	1.471	1.493	1.511	1.525	1.538	1.548	1.558	1.566	1.573	1.579	1.585	1.590	1.595	1.599
2680	1.330	1.333	1.337	1.366	1.413	1.448	1.475	1.497	1.515	1.530	1.542	1.553	1.562	1.570	1.578	1.584	1.590	1.595	1.600	1.604
2700	1.333	1.336	1.339	1.369	1.416	1.452	1.479	1.501	1.519	1.534	1.547	1.558	1.567	1.575	1.582	1.589	1.595	1.600	1.605	1.609
2720	1.335	1.339	1.342	1.372	1.420	1.455	1.483	1.505	1.523	1.538	1.551	1.562	1.572	1.580	1.587	1.594	1.600	1.605	1.610	1.614
2740	1.338	1.341	1.344	1.375	1.423	1.459	1.487	1.509	1.527	1.543	1.556	1.567	1.576	1.585	1.592	1.599	1.605	1.610	1.615	1.619
2760	1.340	1.344	1.347	1.378	1.426	1.462	1.491	1.513	1.532	1.547	1.560	1.571	1.581	1.589	1.597	1.604	1.609	1.615	1.620	1.624
2780	1.343	1.346	1.350	1.381	1.429	1.466	1.495	1.517	1.536	1.551	1.565	1.576	1.586	1.594	1.602	1.608	1.614	1.620	1.625	1.629
2800	1.345	1.349	1.352	1.384	1.433	1.470	1.498	1.521	1.540	1.556	1.569	1.580	1.590	1.599	1.607	1.613	1.619	1.625	1.630	1.634
2820	1.348	1.351	1.355	1.386	1.436	1.473	1.502	1.525	1.544	1.560	1.574	1.585	1.595	1.604	1.611	1.618	1.624	1.630	1.635	1.639
2840	1.350	1.354	1.357	1.389	1.439	1.477	1.506	1.530	1.549	1.565	1.578	1.590	1.600	1.608	1.616	1.623	1.629	1.635	1.640	1.644
2860	1.353	1.356	1.360	1.392	1.443	1.481	1.510	1.534	1.553	1.569	1.583	1.594	1.604	1.613	1.621	1.628	1.634	1.640	1.645	1.649
2880	1.355	1.359	1.363	1.395	1.446	1.484	1.514	1.538	1.557	1.573	1.587	1.599	1.609	1.618	1.626	1.633	1.639	1.645	1.650	1.655
2900	1.358	1.362	1.365	1.398	1.449	1.488	1.518	1.542	1.561	1.578	1.592	1.604	1.614	1.623	1.631	1.638	1.644	1.650	1.655	1.660
2920	1.360	1.364	1.368	1.401	1.453	1.491	1.522	1.546	1.566	1.582	1.596	1.608	1.619	1.628	1.636	1.643	1.649	1.655	1.660	1.665
2940	1.363	1.367	1.370	1.404	1.456	1.495	1.526	1.550	1.570	1.587	1.601	1.613	1.623	1.632	1.641	1.648	1.654	1.660	1.665	1.670
2960	1.365	1.369	1.373	1.406	1.460	1.499	1.530	1.554	1.574	1.591	1.605	1.617	1.628	1.637	1.645	1.653	1.659	1.665	1.670	1.675
2980	1.368	1.372	1.375	1.409	1.463	1.502	1.533	1.558	1.579	1.596	1.610	1.622	1.633	1.642	1.650	1.658	1.664	1.670	1.675	1.680
3000	1.370	1.374	1.378	1.412	1.466	1.506	1.537	1.562	1.583	1.600	1.614	1.627	1.638	1.647	1.655	1.663	1.669	1.675	1.680	1.685

SIGMA =   8900.0

$V-V_c = 10$

H	3.4	3.2	3.0	2.8	2.6	2.4	2.2	2.0	1.8	1.6	1.4	1.2	1.0	0.8	0.6	0.4	0.2	0.10	0.09	0.08
X																				
3100	1.711	1.706	1.701	1.694	1.687	1.680	1.671	1.661	1.650	1.637	1.622	1.604	1.583	1.557	1.524	1.482	1.427	1.391	1.387	1.383
3200	1.737	1.732	1.726	1.720	1.713	1.705	1.696	1.685	1.674	1.660	1.645	1.626	1.604	1.577	1.543	1.499	1.441	1.404	1.400	1.396
3300	1.764	1.758	1.752	1.745	1.738	1.730	1.720	1.710	1.698	1.684	1.667	1.648	1.625	1.597	1.561	1.516	1.455	1.417	1.413	1.408
3400	1.790	1.784	1.778	1.771	1.763	1.755	1.745	1.734	1.721	1.707	1.690	1.670	1.646	1.617	1.580	1.533	1.470	1.430	1.426	1.421
3500	1.817	1.811	1.804	1.797	1.789	1.780	1.770	1.759	1.745	1.730	1.713	1.692	1.667	1.637	1.599	1.550	1.485	1.443	1.438	1.434
3600	1.843	1.837	1.831	1.823	1.815	1.806	1.795	1.783	1.770	1.754	1.736	1.714	1.689	1.657	1.618	1.567	1.499	1.456	1.451	1.447
3700	1.871	1.864	1.857	1.849	1.841	1.831	1.820	1.808	1.794	1.778	1.759	1.737	1.710	1.677	1.636	1.584	1.514	1.469	1.464	1.459
3800	1.898	1.891	1.884	1.876	1.867	1.857	1.846	1.833	1.819	1.802	1.782	1.759	1.732	1.698	1.655	1.601	1.529	1.482	1.477	1.472
3900	1.925	1.918	1.911	1.903	1.893	1.883	1.871	1.858	1.843	1.826	1.806	1.782	1.753	1.718	1.675	1.618	1.543	1.495	1.490	1.485
4000	1.953	1.946	1.938	1.929	1.920	1.909	1.897	1.884	1.868	1.850	1.829	1.805	1.775	1.739	1.694	1.636	1.558	1.509	1.503	1.498
4100	1.981	1.973	1.965	1.956	1.947	1.936	1.923	1.909	1.893	1.875	1.853	1.828	1.797	1.760	1.713	1.653	1.573	1.522	1.516	1.511
4200	2.009	2.001	1.993	1.984	1.974	1.962	1.949	1.935	1.918	1.899	1.877	1.851	1.820	1.781	1.732	1.670	1.588	1.535	1.529	1.523
4300	2.037	2.029	2.021	2.011	2.001	1.989	1.976	1.961	1.944	1.924	1.901	1.874	1.841	1.802	1.752	1.688	1.603	1.548	1.542	1.536
4400	2.065	2.057	2.048	2.039	2.028	2.016	2.002	1.987	1.969	1.949	1.925	1.897	1.864	1.823	1.771	1.705	1.618	1.562	1.555	1.549
4500	2.094	2.086	2.077	2.067	2.055	2.043	2.029	2.013	1.995	1.974	1.950	1.921	1.886	1.844	1.791	1.723	1.632	1.575	1.568	1.562
4600	2.123	2.114	2.105	2.095	2.083	2.070	2.056	2.040	2.021	1.999	1.974	1.944	1.909	1.865	1.811	1.741	1.648	1.588	1.582	1.575
4700	2.152	2.143	2.133	2.123	2.111	2.098	2.083	2.066	2.047	2.025	1.999	1.968	1.932	1.887	1.831	1.759	1.663	1.601	1.595	1.588
4800	2.181	2.172	2.162	2.151	2.139	2.125	2.110	2.093	2.073	2.050	2.024	1.992	1.954	1.908	1.851	1.777	1.678	1.615	1.608	1.601
4900	2.210	2.201	2.191	2.180	2.167	2.153	2.137	2.120	2.099	2.076	2.049	2.016	1.977	1.930	1.871	1.794	1.693	1.628	1.621	1.614
5000	2.240	2.230	2.220	2.208	2.195	2.181	2.165	2.147	2.126	2.102	2.074	2.040	2.001	1.952	1.891	1.812	1.708	1.642	1.634	1.627
5100	2.270	2.260	2.249	2.237	2.224	2.209	2.193	2.174	2.153	2.128	2.099	2.065	2.024	1.974	1.911	1.831	1.723	1.655	1.647	1.640
5200	2.300	2.290	2.279	2.266	2.253	2.238	2.221	2.201	2.179	2.154	2.124	2.089	2.047	1.996	1.931	1.849	1.739	1.668	1.661	1.653
5300	2.330	2.319	2.308	2.296	2.282	2.266	2.249	2.229	2.206	2.180	2.150	2.114	2.071	2.018	1.952	1.867	1.754	1.682	1.674	1.666
5400	2.360	2.350	2.338	2.325	2.311	2.295	2.277	2.257	2.233	2.207	2.175	2.138	2.094	2.040	1.972	1.885	1.769	1.695	1.687	1.679
5500	2.391	2.380	2.368	2.355	2.340	2.324	2.305	2.285	2.261	2.233	2.201	2.163	2.118	2.062	1.993	1.904	1.785	1.709	1.701	1.692
5600	2.421	2.410	2.398	2.385	2.370	2.353	2.334	2.313	2.288	2.260	2.227	2.188	2.142	2.085	2.014	1.922	1.800	1.722	1.714	1.705
5700	2.452	2.441	2.428	2.415	2.399	2.382	2.363	2.341	2.316	2.287	2.253	2.214	2.166	2.107	2.034	1.941	1.816	1.736	1.727	1.718
5800	2.483	2.472	2.459	2.445	2.429	2.411	2.392	2.369	2.344	2.314	2.280	2.239	2.190	2.130	2.055	1.959	1.831	1.750	1.741	1.731
5900	2.515	2.503	2.490	2.475	2.459	2.441	2.421	2.398	2.372	2.341	2.306	2.264	2.214	2.153	2.076	1.978	1.847	1.763	1.754	1.745
6000	2.546	2.534	2.521	2.506	2.489	2.471	2.450	2.427	2.400	2.369	2.333	2.290	2.238	2.176	2.097	1.997	1.862	1.777	1.767	1.758
6100	2.578	2.566	2.552	2.537	2.520	2.501	2.480	2.455	2.428	2.396	2.359	2.315	2.263	2.199	2.119	2.015	1.878	1.790	1.781	1.771
6200	2.610	2.597	2.583	2.568	2.550	2.531	2.509	2.485	2.456	2.424	2.386	2.341	2.288	2.222	2.140	2.034	1.894	1.804	1.794	1.784
6300	2.642	2.629	2.615	2.599	2.581	2.561	2.539	2.514	2.485	2.452	2.413	2.367	2.312	2.245	2.161	2.053	1.909	1.818	1.808	1.797
6400	2.674	2.661	2.646	2.630	2.612	2.592	2.569	2.543	2.514	2.480	2.440	2.393	2.337	2.269	2.183	2.072	1.925	1.831	1.821	1.811
6500	2.707	2.693	2.678	2.662	2.643	2.622	2.599	2.573	2.543	2.508	2.468	2.420	2.362	2.292	2.204	2.091	1.941	1.845	1.835	1.824
6600	2.740	2.726	2.710	2.693	2.674	2.653	2.629	2.603	2.572	2.536	2.495	2.446	2.387	2.316	2.226	2.111	1.957	1.859	1.848	1.837
6700	2.772	2.758	2.742	2.725	2.706	2.684	2.660	2.633	2.601	2.565	2.523	2.473	2.413	2.339	2.248	2.130	1.973	1.873	1.862	1.851
6800	2.806	2.791	2.775	2.757	2.737	2.715	2.691	2.663	2.630	2.593	2.550	2.499	2.438	2.363	2.269	2.149	1.989	1.887	1.875	1.864
6900	2.839	2.824	2.808	2.789	2.769	2.747	2.722	2.693	2.660	2.622	2.578	2.526	2.463	2.387	2.291	2.169	2.005	1.900	1.889	1.877
7000	2.872	2.857	2.840	2.822	2.801	2.778	2.753	2.723	2.690	2.651	2.606	2.553	2.489	2.411	2.313	2.188	2.021	1.914	1.903	1.891
7100	2.906	2.890	2.873	2.855	2.834	2.810	2.784	2.754	2.720	2.680	2.634	2.580	2.515	2.435	2.336	2.208	2.037	1.928	1.916	1.904
7200	2.940	2.924	2.907	2.887	2.866	2.842	2.815	2.785	2.750	2.710	2.663	2.607	2.541	2.459	2.358	2.227	2.053	1.942	1.930	1.917

SIGMA = 8900.0

$V-V_c = 10$

H \ x	0.08	0.09	0.10	0.2	0.4	0.6	0.8	1.0	1.2	1.4	1.6	1.8	2.0	2.2	2.4	2.6	2.8	3.0	3.2	3.4
7300	1.931	1.943	1.956	2.069	2.247	2.380	2.484	2.567	2.635	2.691	2.739	2.780	2.816	2.847	2.874	2.898	2.920	2.940	2.958	2.974
7400	1.944	1.957	1.970	2.085	2.267	2.403	2.508	2.593	2.662	2.720	2.769	2.810	2.847	2.878	2.906	2.931	2.954	2.974	2.992	3.008
7500	1.958	1.971	1.984	2.102	2.286	2.425	2.533	2.619	2.690	2.749	2.798	2.841	2.878	2.910	2.939	2.964	2.987	3.007	3.026	3.043
7600	1.971	1.985	1.998	2.118	2.306	2.448	2.558	2.645	2.717	2.777	2.828	2.872	2.909	2.942	2.971	2.997	3.021	3.041	3.060	3.077
7700	1.985	1.998	2.012	2.134	2.326	2.470	2.582	2.672	2.745	2.806	2.858	2.903	2.941	2.975	3.004	3.031	3.054	3.075	3.095	3.112
7800	1.998	2.012	2.026	2.150	2.346	2.493	2.607	2.699	2.773	2.836	2.888	2.934	2.973	3.007	3.037	3.064	3.088	3.110	3.129	3.147
7900	2.012	2.026	2.040	2.167	2.366	2.516	2.632	2.725	2.802	2.865	2.919	2.965	3.005	3.040	3.070	3.098	3.122	3.144	3.164	3.182
8000	2.025	2.040	2.054	2.183	2.387	2.539	2.657	2.752	2.830	2.895	2.949	2.996	3.037	3.072	3.104	3.132	3.157	3.179	3.199	3.218
8100	2.039	2.054	2.068	2.200	2.407	2.562	2.683	2.779	2.858	2.924	2.980	3.028	3.069	3.105	3.137	3.166	3.191	3.214	3.235	3.253
8200	2.052	2.068	2.082	2.216	2.427	2.585	2.708	2.806	2.887	2.954	3.011	3.059	3.102	3.138	3.171	3.200	3.226	3.249	3.270	3.289
8300	2.066	2.081	2.096	2.233	2.448	2.608	2.734	2.834	2.916	2.984	3.042	3.091	3.134	3.172	3.205	3.234	3.261	3.284	3.306	3.325
8400	2.080	2.095	2.111	2.250	2.468	2.632	2.759	2.861	2.945	3.014	3.073	3.123	3.167	3.205	3.239	3.269	3.296	3.320	3.342	3.361
8500	2.093	2.109	2.125	2.266	2.489	2.655	2.785	2.889	2.974	3.044	3.104	3.155	3.200	3.239	3.273	3.304	3.331	3.355	3.378	3.398
8600	2.107	2.123	2.139	2.283	2.509	2.679	2.811	2.916	3.003	3.075	3.136	3.188	3.233	3.273	3.307	3.339	3.366	3.391	3.414	3.434
8700	2.121	2.137	2.153	2.300	2.530	2.702	2.837	2.944	3.032	3.105	3.167	3.220	3.266	3.307	3.342	3.374	3.402	3.427	3.450	3.471
8800	2.134	2.151	2.168	2.317	2.551	2.726	2.863	2.972	3.061	3.136	3.199	3.253	3.300	3.341	3.377	3.409	3.438	3.464	3.487	3.508
8900	2.148	2.165	2.182	2.333	2.571	2.750	2.889	3.000	3.091	3.167	3.231	3.286	3.333	3.375	3.412	3.444	3.474	3.500	3.524	3.545
9000	2.162	2.179	2.196	2.350	2.592	2.774	2.915	3.028	3.121	3.198	3.263	3.319	3.367	3.409	3.447	3.480	3.510	3.537	3.561	3.583
9100	2.176	2.193	2.210	2.367	2.613	2.798	2.942	3.056	3.150	3.229	3.295	3.352	3.401	3.444	3.482	3.516	3.546	3.573	3.598	3.620
9200	2.189	2.207	2.225	2.384	2.634	2.822	2.968	3.085	3.180	3.260	3.327	3.385	3.435	3.479	3.518	3.552	3.583	3.610	3.635	3.658
9300	2.203	2.221	2.239	2.401	2.655	2.846	2.995	3.113	3.210	3.291	3.360	3.419	3.470	3.514	3.553	3.588	3.619	3.648	3.673	3.696
9400	2.217	2.235	2.254	2.418	2.677	2.871	3.021	3.142	3.241	3.323	3.393	3.452	3.504	3.549	3.589	3.625	3.656	3.685	3.711	3.734
9500	2.231	2.250	2.268	2.435	2.698	2.895	3.048	3.171	3.271	3.355	3.425	3.486	3.539	3.585	3.625	3.661	3.693	3.723	3.749	3.773
9600	2.245	2.264	2.282	2.452	2.719	2.919	3.075	3.200	3.302	3.387	3.458	3.520	3.573	3.620	3.661	3.698	3.731	3.760	3.787	3.811
9700	2.259	2.278	2.297	2.470	2.741	2.944	3.102	3.229	3.332	3.419	3.492	3.554	3.608	3.656	3.698	3.735	3.768	3.798	3.825	3.850
9800	2.273	2.292	2.311	2.487	2.762	2.969	3.129	3.258	3.363	3.451	3.525	3.588	3.644	3.692	3.734	3.772	3.806	3.836	3.864	3.889
9900	2.286	2.306	2.326	2.504	2.784	2.993	3.157	3.287	3.394	3.483	3.558	3.623	3.679	3.728	3.771	3.809	3.844	3.875	3.903	3.928
10000	2.300	2.321	2.341	2.521	2.805	3.018	3.184	3.317	3.425	3.515	3.592	3.657	3.714	3.764	3.808	3.847	3.882	3.913	3.942	3.967
10100	2.314	2.335	2.355	2.539	2.827	3.043	3.212	3.346	3.456	3.548	3.626	3.692	3.750	3.800	3.845	3.885	3.920	3.952	3.981	4.007
10200	2.328	2.349	2.370	2.556	2.849	3.068	3.239	3.376	3.488	3.581	3.660	3.727	3.786	3.837	3.882	3.922	3.958	3.991	4.020	4.047
10300	2.342	2.363	2.384	2.573	2.871	3.094	3.267	3.406	3.519	3.614	3.694	3.762	3.822	3.874	3.920	3.960	3.997	4.030	4.060	4.087
10400	2.356	2.378	2.399	2.591	2.893	3.119	3.295	3.436	3.551	3.647	3.728	3.798	3.858	3.911	3.957	3.999	4.036	4.069	4.099	4.127
10500	2.370	2.392	2.414	2.608	2.915	3.144	3.323	3.466	3.582	3.680	3.762	3.833	3.894	3.948	3.995	4.037	4.075	4.109	4.139	4.167
10600	2.384	2.406	2.428	2.626	2.937	3.170	3.351	3.496	3.614	3.713	3.797	3.869	3.931	3.985	4.033	4.076	4.114	4.148	4.179	4.207
10700	2.398	2.421	2.443	2.644	2.959	3.195	3.379	3.526	3.646	3.747	3.833	3.904	3.967	4.023	4.071	4.114	4.153	4.188	4.220	4.248
10800	2.412	2.435	2.458	2.661	2.981	3.221	3.407	3.556	3.679	3.780	3.866	3.940	4.004	4.060	4.110	4.153	4.193	4.228	4.260	4.289
10900	2.427	2.450	2.472	2.679	3.003	3.246	3.436	3.587	3.711	3.814	3.901	3.976	4.041	4.098	4.148	4.193	4.232	4.268	4.301	4.330
11000	2.441	2.464	2.487	2.697	3.026	3.272	3.464	3.618	3.743	3.848	3.937	4.013	4.078	4.136	4.187	4.232	4.272	4.309	4.342	4.371
11100	2.455	2.479	2.502	2.714	3.048	3.298	3.493	3.649	3.776	3.882	3.972	4.049	4.116	4.174	4.226	4.271	4.312	4.349	4.383	4.413
11200	2.469	2.493	2.517	2.732	3.070	3.324	3.522	3.679	3.809	3.916	4.007	4.085	4.153	4.212	4.265	4.311	4.353	4.390	4.424	4.455
11300	2.483	2.508	2.532	2.750	3.093	3.350	3.550	3.711	3.842	3.951	4.043	4.122	4.191	4.251	4.304	4.351	4.393	4.431	4.465	4.496
11400	2.497	2.522	2.546	2.768	3.116	3.376	3.579	3.742	3.874	3.985	4.079	4.159	4.229	4.289	4.343	4.391	4.434	4.472	4.507	4.538

SIGMA = 8900.0

V-V$_c$ = 10

H x	3.4	3.2	3.0	2.8	2.6	2.4	2.2	2.0	1.8	1.6	1.4	1.2	1.0	0.8	0.6	0.4	0.2	0.10	0.09	0.08
11500	4.581	4.549	4.513	4.474	4.431	4.383	4.328	4.267	4.196	4.115	4.020	3.908	3.773	3.608	3.403	3.138	2.786	2.561	2.537	2.512
11600	4.623	4.591	4.555	4.515	4.472	4.423	4.367	4.305	4.233	4.151	4.055	3.941	3.804	3.638	3.429	3.161	2.804	2.576	2.551	2.526
11700	4.666	4.633	4.597	4.557	4.512	4.462	4.407	4.343	4.271	4.187	4.090	3.974	3.836	3.667	3.456	3.184	2.822	2.591	2.566	2.540
11800	4.709	4.675	4.639	4.598	4.553	4.503	4.446	4.382	4.308	4.224	4.125	4.008	3.868	3.696	3.482	3.207	2.840	2.606	2.580	2.554
11900	4.752	4.718	4.681	4.640	4.594	4.543	4.485	4.420	4.346	4.260	4.160	4.042	3.900	3.726	3.509	3.230	2.858	2.621	2.595	2.569
12000	4.795	4.761	4.723	4.681	4.635	4.583	4.525	4.459	4.384	4.297	4.195	4.075	3.931	3.756	3.536	3.253	2.876	2.636	2.610	2.583
12100	4.838	4.804	4.765	4.723	4.676	4.624	4.565	4.498	4.422	4.334	4.231	4.109	3.964	3.785	3.563	3.276	2.894	2.651	2.624	2.597
12200	4.882	4.847	4.808	4.765	4.718	4.665	4.605	4.537	4.460	4.371	4.267	4.143	3.996	3.815	3.589	3.299	2.912	2.666	2.639	2.612
12300	4.926	4.890	4.851	4.808	4.760	4.706	4.645	4.577	4.498	4.408	4.302	4.178	4.028	3.845	3.617	3.323	2.931	2.681	2.654	2.626
12400	4.970	4.934	4.894	4.850	4.801	4.747	4.686	4.616	4.537	4.445	4.338	4.212	4.060	3.875	3.644	3.346	2.949	2.696	2.669	2.640
12500	5.014	4.978	4.937	4.893	4.844	4.788	4.726	4.656	4.575	4.483	4.374	4.247	4.093	3.905	3.671	3.369	2.967	2.712	2.683	2.655
12600	5.058	5.021	4.981	4.936	4.886	4.830	4.767	4.696	4.614	4.520	4.411	4.281	4.126	3.936	3.698	3.393	2.986	2.727	2.698	2.669
12700	5.103	5.066	5.024	4.979	4.928	4.872	4.808	4.736	4.653	4.558	4.447	4.316	4.159	3.966	3.726	3.416	3.004	2.742	2.713	2.683
12800	5.148	5.110	5.068	5.022	4.971	4.913	4.849	4.776	4.692	4.596	4.484	4.351	4.192	3.997	3.753	3.440	3.023	2.757	2.728	2.698
12900	5.193	5.154	5.112	5.065	5.014	4.956	4.890	4.816	4.732	4.634	4.520	4.386	4.225	4.027	3.781	3.464	3.041	2.772	2.743	2.712
13000	5.238	5.199	5.156	5.109	5.057	4.998	4.932	4.857	4.771	4.673	4.557	4.421	4.258	4.058	3.809	3.488	3.060	2.787	2.757	2.727
13100	5.283	5.244	5.201	5.153	5.100	5.040	4.973	4.898	4.811	4.711	4.594	4.457	4.291	4.089	3.836	3.511	3.078	2.803	2.772	2.741
13200	5.329	5.289	5.245	5.197	5.143	5.083	5.015	4.938	4.851	4.750	4.631	4.492	4.325	4.120	3.864	3.535	3.097	2.818	2.787	2.756
13300	5.375	5.334	5.290	5.241	5.187	5.126	5.057	4.979	4.891	4.788	4.669	4.528	4.358	4.151	3.892	3.559	3.116	2.833	2.802	2.770
13400	5.421	5.380	5.335	5.285	5.230	5.169	5.099	5.021	4.931	4.827	4.706	4.563	4.392	4.182	3.920	3.583	3.134	2.849	2.817	2.785
13500	5.467	5.426	5.380	5.330	5.274	5.212	5.142	5.062	4.971	4.866	4.744	4.599	4.426	4.214	3.948	3.608	3.153	2.864	2.832	2.800
13600	5.513	5.471	5.425	5.375	5.318	5.255	5.184	5.104	5.012	4.905	4.782	4.635	4.460	4.245	3.977	3.632	3.172	2.879	2.847	2.814
13700	5.560	5.517	5.471	5.420	5.362	5.299	5.227	5.145	5.052	4.945	4.819	4.671	4.494	4.277	4.005	3.656	3.191	2.895	2.862	2.829
13800	5.607	5.564	5.517	5.465	5.407	5.342	5.269	5.187	5.093	4.984	4.858	4.708	4.528	4.308	4.034	3.681	3.210	2.910	2.877	2.844
13900	5.653	5.610	5.563	5.510	5.451	5.386	5.312	5.229	5.134	5.024	4.896	4.744	4.562	4.340	4.062	3.705	3.229	2.926	2.892	2.858
14000	5.701	5.657	5.609	5.555	5.496	5.430	5.356	5.271	5.175	5.064	4.934	4.781	4.597	4.372	4.091	3.729	3.248	2.941	2.907	2.873
14100	5.748	5.704	5.655	5.601	5.541	5.474	5.399	5.314	5.216	5.104	4.973	4.817	4.631	4.404	4.120	3.754	3.267	2.956	2.922	2.888
14200	5.795	5.751	5.701	5.647	5.586	5.519	5.443	5.356	5.258	5.144	5.011	4.854	4.666	4.436	4.148	3.779	3.286	2.972	2.937	2.902
14300	5.843	5.798	5.748	5.693	5.632	5.563	5.486	5.399	5.299	5.184	5.050	4.891	4.701	4.468	4.177	3.803	3.305	2.988	2.953	2.917
14400	5.891	5.845	5.795	5.739	5.677	5.608	5.530	5.442	5.341	5.225	5.089	4.928	4.736	4.501	4.206	3.828	3.324	3.003	2.968	2.932
14500	5.939	5.893	5.842	5.786	5.723	5.653	5.574	5.485	5.383	5.265	5.128	4.966	4.771	4.533	4.236	3.853	3.343	3.019	2.983	2.947
14600	5.988	5.941	5.889	5.832	5.769	5.696	5.618	5.528	5.425	5.306	5.167	5.003	4.806	4.566	4.265	3.878	3.362	3.034	2.998	2.961
14700	6.036	5.989	5.936	5.879	5.815	5.743	5.663	5.572	5.467	5.347	5.207	5.041	4.842	4.598	4.294	3.903	3.382	3.050	3.013	2.976
14800	6.085	6.037	5.984	5.926	5.861	5.789	5.707	5.615	5.510	5.388	5.246	5.078	4.877	4.631	4.324	3.928	3.401	3.065	3.029	2.991
14900	6.134	6.085	6.032	5.973	5.908	5.834	5.752	5.659	5.552	5.429	5.286	5.116	4.913	4.664	4.353	3.953	3.420	3.081	3.044	3.006
15000	6.183	6.134	6.080	6.020	5.954	5.880	5.797	5.703	5.595	5.471	5.326	5.154	4.948	4.697	4.383	3.979	3.440	3.097	3.059	3.021

SIGMA = 7355.0

$V - V_c = 11$

H \ x	0.08	0.09	0.10	0.2	0.4	0.6	0.8	1.0	1.2	1.4	1.6	1.8	2.0	2.2	2.4	2.6	2.8	3.0	3.2	3.4
0	1.000	1.000	1.000	1.000	1.000	1.000	1.000	1.000	1.000	1.000	1.000	1.000	1.000	1.000	1.000	1.000	1.000	1.000	1.000	1.000
20	1.003	1.003	1.003	1.003	1.003	1.004	1.004	1.004	1.004	1.004	1.004	1.004	1.005	1.005	1.005	1.005	1.005	1.005	1.005	1.005
40	1.006	1.006	1.006	1.006	1.007	1.007	1.008	1.008	1.008	1.009	1.009	1.009	1.009	1.009	1.009	1.009	1.009	1.010	1.010	1.010
60	1.009	1.009	1.009	1.010	1.011	1.011	1.012	1.012	1.013	1.013	1.013	1.013	1.014	1.014	1.014	1.014	1.014	1.014	1.014	1.015
80	1.012	1.012	1.012	1.013	1.015	1.015	1.016	1.016	1.017	1.017	1.018	1.018	1.018	1.018	1.019	1.019	1.019	1.019	1.019	1.019
100	1.015	1.015	1.015	1.016	1.018	1.019	1.020	1.020	1.021	1.022	1.022	1.022	1.023	1.023	1.023	1.024	1.024	1.024	1.024	1.024
120	1.018	1.018	1.018	1.019	1.022	1.023	1.024	1.025	1.025	1.026	1.027	1.027	1.028	1.028	1.028	1.028	1.029	1.029	1.029	1.029
140	1.020	1.021	1.021	1.022	1.025	1.026	1.028	1.029	1.030	1.030	1.031	1.032	1.032	1.032	1.033	1.033	1.034	1.034	1.034	1.034
160	1.023	1.024	1.024	1.025	1.028	1.030	1.032	1.033	1.034	1.035	1.035	1.036	1.037	1.037	1.037	1.038	1.038	1.038	1.038	1.039
180	1.026	1.027	1.027	1.029	1.032	1.034	1.036	1.037	1.038	1.039	1.040	1.041	1.041	1.042	1.042	1.043	1.043	1.043	1.043	1.044
200	1.029	1.029	1.030	1.032	1.035	1.038	1.040	1.041	1.042	1.043	1.044	1.046	1.046	1.046	1.047	1.047	1.048	1.048	1.048	1.049
220	1.032	1.032	1.033	1.035	1.039	1.041	1.044	1.045	1.047	1.048	1.049	1.050	1.050	1.051	1.052	1.052	1.053	1.053	1.053	1.054
240	1.035	1.035	1.036	1.038	1.042	1.045	1.048	1.049	1.051	1.052	1.053	1.054	1.055	1.056	1.056	1.057	1.057	1.058	1.058	1.059
260	1.038	1.038	1.039	1.041	1.046	1.049	1.052	1.054	1.055	1.057	1.058	1.059	1.060	1.061	1.061	1.061	1.062	1.063	1.063	1.064
280	1.041	1.041	1.042	1.045	1.049	1.053	1.056	1.058	1.060	1.061	1.062	1.063	1.064	1.065	1.066	1.066	1.067	1.067	1.068	1.069
300	1.044	1.044	1.045	1.048	1.053	1.057	1.060	1.062	1.064	1.066	1.067	1.068	1.069	1.070	1.071	1.071	1.072	1.072	1.073	1.074
320	1.047	1.047	1.048	1.051	1.056	1.061	1.064	1.066	1.068	1.070	1.071	1.073	1.074	1.075	1.076	1.076	1.077	1.077	1.078	1.079
340	1.050	1.050	1.051	1.054	1.060	1.064	1.068	1.070	1.073	1.074	1.076	1.077	1.078	1.079	1.080	1.081	1.081	1.082	1.083	1.084
360	1.053	1.053	1.054	1.058	1.064	1.068	1.072	1.075	1.077	1.079	1.081	1.082	1.083	1.084	1.085	1.086	1.086	1.087	1.088	1.089
380	1.056	1.056	1.057	1.061	1.068	1.072	1.076	1.079	1.081	1.083	1.085	1.087	1.088	1.089	1.090	1.091	1.091	1.092	1.093	1.094
400	1.059	1.059	1.060	1.064	1.071	1.076	1.080	1.083	1.086	1.088	1.090	1.091	1.093	1.094	1.095	1.095	1.096	1.097	1.098	1.099
420	1.062	1.062	1.063	1.067	1.074	1.080	1.084	1.087	1.090	1.092	1.094	1.096	1.097	1.099	1.100	1.100	1.101	1.102	1.103	1.104
440	1.065	1.065	1.066	1.070	1.078	1.084	1.088	1.092	1.094	1.097	1.099	1.101	1.102	1.103	1.105	1.105	1.106	1.107	1.108	1.109
460	1.067	1.068	1.069	1.074	1.082	1.087	1.092	1.096	1.099	1.101	1.103	1.105	1.107	1.108	1.109	1.110	1.111	1.112	1.113	1.114
480	1.070	1.071	1.072	1.077	1.085	1.091	1.096	1.100	1.103	1.106	1.108	1.110	1.112	1.113	1.114	1.115	1.116	1.117	1.118	1.119
500	1.073	1.074	1.075	1.080	1.089	1.095	1.100	1.104	1.108	1.110	1.113	1.115	1.116	1.118	1.119	1.120	1.121	1.122	1.123	1.124
520	1.076	1.077	1.078	1.083	1.092	1.099	1.104	1.109	1.112	1.115	1.117	1.119	1.121	1.123	1.124	1.125	1.126	1.127	1.128	1.129
540	1.079	1.080	1.081	1.087	1.096	1.103	1.108	1.113	1.116	1.119	1.122	1.124	1.126	1.128	1.129	1.130	1.131	1.131	1.133	1.134
560	1.082	1.083	1.084	1.090	1.100	1.107	1.113	1.117	1.121	1.124	1.127	1.129	1.131	1.132	1.134	1.135	1.137	1.138	1.138	1.139
580	1.085	1.086	1.087	1.093	1.103	1.111	1.117	1.121	1.125	1.128	1.131	1.134	1.136	1.137	1.139	1.140	1.142	1.143	1.144	1.145
600	1.088	1.089	1.090	1.096	1.107	1.115	1.121	1.126	1.130	1.133	1.136	1.138	1.140	1.142	1.144	1.145	1.147	1.148	1.149	1.150
620	1.091	1.092	1.093	1.100	1.110	1.119	1.125	1.130	1.134	1.138	1.141	1.143	1.145	1.147	1.149	1.150	1.152	1.153	1.154	1.155
640	1.094	1.095	1.096	1.103	1.114	1.122	1.129	1.134	1.139	1.142	1.145	1.148	1.150	1.152	1.154	1.155	1.157	1.158	1.159	1.160
660	1.097	1.098	1.099	1.106	1.118	1.126	1.133	1.139	1.143	1.147	1.150	1.153	1.155	1.157	1.159	1.160	1.162	1.163	1.164	1.165
680	1.100	1.101	1.102	1.109	1.121	1.130	1.137	1.143	1.148	1.151	1.155	1.157	1.160	1.162	1.164	1.165	1.167	1.168	1.169	1.171
700	1.103	1.104	1.105	1.113	1.125	1.134	1.141	1.147	1.152	1.156	1.159	1.162	1.165	1.167	1.169	1.170	1.172	1.173	1.175	1.176
720	1.106	1.107	1.108	1.116	1.129	1.138	1.146	1.152	1.157	1.161	1.164	1.167	1.170	1.172	1.174	1.176	1.177	1.178	1.180	1.181
740	1.109	1.110	1.111	1.119	1.132	1.142	1.150	1.156	1.161	1.165	1.169	1.172	1.174	1.177	1.179	1.181	1.182	1.184	1.185	1.186
760	1.112	1.113	1.114	1.122	1.136	1.146	1.154	1.160	1.166	1.170	1.173	1.177	1.179	1.182	1.184	1.186	1.187	1.189	1.190	1.191
780	1.115	1.116	1.117	1.126	1.140	1.150	1.158	1.165	1.170	1.174	1.178	1.181	1.184	1.187	1.189	1.191	1.192	1.194	1.195	1.197
800	1.118	1.119	1.120	1.129	1.143	1.154	1.162	1.169	1.175	1.179	1.183	1.186	1.189	1.192	1.194	1.196	1.198	1.199	1.201	1.202

SIGMA = 7355.0

$V-V_c = 11$

H	3.4	3.2	3.0	2.8	2.6	2.4	2.2	2.0	1.8	1.6	1.4	1.2	1.0	0.8	0.6	0.4	0.2	0.10	0.09	0.08
820	1.207	1.206	1.204	1.203	1.201	1.199	1.197	1.194	1.191	1.188	1.184	1.179	1.173	1.167	1.158	1.147	1.132	1.123	1.122	1.121
840	1.213	1.211	1.210	1.208	1.206	1.204	1.202	1.199	1.196	1.193	1.188	1.184	1.178	1.171	1.162	1.151	1.135	1.126	1.125	1.124
860	1.218	1.216	1.215	1.213	1.211	1.209	1.207	1.204	1.201	1.197	1.193	1.188	1.182	1.175	1.166	1.154	1.139	1.129	1.128	1.127
880	1.223	1.222	1.220	1.218	1.216	1.214	1.212	1.209	1.206	1.202	1.198	1.193	1.187	1.179	1.170	1.158	1.142	1.132	1.131	1.130
900	1.228	1.227	1.225	1.224	1.222	1.219	1.217	1.214	1.211	1.207	1.202	1.197	1.191	1.183	1.174	1.162	1.145	1.135	1.134	1.133
920	1.234	1.232	1.231	1.229	1.227	1.224	1.222	1.219	1.216	1.212	1.207	1.202	1.195	1.188	1.178	1.165	1.149	1.138	1.137	1.136
940	1.239	1.238	1.236	1.234	1.232	1.230	1.227	1.224	1.220	1.217	1.212	1.206	1.200	1.192	1.182	1.169	1.152	1.141	1.140	1.138
960	1.245	1.243	1.241	1.239	1.237	1.235	1.232	1.229	1.225	1.221	1.217	1.211	1.204	1.196	1.186	1.173	1.155	1.144	1.143	1.141
980	1.250	1.248	1.246	1.245	1.242	1.240	1.237	1.234	1.230	1.226	1.221	1.216	1.209	1.200	1.190	1.176	1.158	1.147	1.146	1.144
1000	1.255	1.254	1.252	1.250	1.248	1.245	1.242	1.239	1.235	1.231	1.226	1.220	1.213	1.205	1.194	1.180	1.162	1.150	1.149	1.147
1020	1.261	1.259	1.257	1.255	1.253	1.250	1.247	1.244	1.240	1.236	1.231	1.225	1.218	1.209	1.198	1.184	1.165	1.153	1.152	1.150
1040	1.266	1.264	1.262	1.260	1.258	1.255	1.252	1.249	1.245	1.241	1.236	1.229	1.222	1.213	1.202	1.188	1.168	1.156	1.155	1.153
1060	1.272	1.270	1.268	1.266	1.263	1.261	1.257	1.254	1.250	1.246	1.240	1.234	1.227	1.217	1.206	1.191	1.172	1.159	1.158	1.156
1080	1.277	1.275	1.273	1.271	1.268	1.266	1.263	1.259	1.255	1.250	1.245	1.239	1.231	1.222	1.210	1.195	1.175	1.162	1.161	1.159
1100	1.282	1.281	1.279	1.276	1.274	1.271	1.268	1.264	1.260	1.255	1.250	1.243	1.236	1.226	1.214	1.199	1.178	1.165	1.164	1.162
1120	1.288	1.286	1.284	1.282	1.279	1.276	1.273	1.269	1.265	1.260	1.255	1.248	1.240	1.230	1.218	1.202	1.182	1.168	1.167	1.165
1140	1.293	1.291	1.289	1.287	1.284	1.281	1.278	1.274	1.270	1.265	1.259	1.253	1.245	1.235	1.222	1.206	1.185	1.171	1.170	1.168
1160	1.299	1.297	1.295	1.292	1.290	1.287	1.283	1.279	1.275	1.270	1.264	1.257	1.249	1.239	1.226	1.210	1.188	1.174	1.173	1.171
1180	1.304	1.302	1.300	1.298	1.295	1.292	1.288	1.285	1.280	1.275	1.269	1.262	1.254	1.243	1.230	1.214	1.191	1.177	1.176	1.174
1200	1.310	1.308	1.305	1.303	1.300	1.297	1.294	1.290	1.285	1.280	1.274	1.267	1.258	1.247	1.234	1.217	1.195	1.180	1.179	1.177
1220	1.315	1.313	1.311	1.308	1.306	1.302	1.299	1.295	1.290	1.285	1.279	1.271	1.263	1.252	1.238	1.221	1.198	1.183	1.182	1.180
1240	1.321	1.319	1.316	1.314	1.311	1.308	1.304	1.300	1.295	1.290	1.284	1.276	1.267	1.256	1.242	1.225	1.201	1.187	1.185	1.183
1260	1.326	1.324	1.322	1.319	1.316	1.313	1.309	1.305	1.300	1.295	1.288	1.281	1.272	1.260	1.247	1.229	1.205	1.190	1.188	1.186
1280	1.332	1.330	1.327	1.325	1.322	1.318	1.314	1.310	1.305	1.300	1.293	1.285	1.276	1.265	1.251	1.232	1.208	1.193	1.191	1.189
1300	1.337	1.335	1.333	1.330	1.327	1.324	1.320	1.315	1.310	1.305	1.298	1.290	1.281	1.269	1.255	1.236	1.211	1.196	1.194	1.192
1320	1.343	1.341	1.338	1.335	1.332	1.329	1.325	1.321	1.316	1.310	1.303	1.295	1.285	1.274	1.259	1.240	1.215	1.199	1.197	1.195
1340	1.349	1.346	1.344	1.341	1.338	1.334	1.330	1.326	1.321	1.315	1.308	1.300	1.290	1.278	1.263	1.244	1.218	1.202	1.200	1.198
1360	1.354	1.352	1.349	1.346	1.343	1.340	1.336	1.331	1.326	1.320	1.313	1.304	1.294	1.282	1.267	1.248	1.221	1.205	1.203	1.201
1380	1.360	1.357	1.355	1.352	1.349	1.345	1.341	1.336	1.331	1.325	1.318	1.309	1.299	1.287	1.271	1.251	1.225	1.208	1.206	1.204
1400	1.365	1.363	1.360	1.357	1.354	1.350	1.346	1.341	1.336	1.330	1.323	1.314	1.304	1.291	1.275	1.255	1.228	1.211	1.209	1.207
1420	1.371	1.369	1.366	1.363	1.359	1.356	1.351	1.347	1.341	1.335	1.327	1.319	1.308	1.295	1.279	1.259	1.231	1.214	1.212	1.210
1440	1.377	1.374	1.371	1.368	1.365	1.361	1.357	1.352	1.347	1.340	1.332	1.323	1.313	1.300	1.284	1.263	1.235	1.217	1.215	1.213
1460	1.382	1.380	1.377	1.374	1.370	1.366	1.362	1.357	1.352	1.345	1.337	1.328	1.317	1.304	1.288	1.266	1.238	1.220	1.218	1.216
1480	1.388	1.385	1.383	1.379	1.376	1.372	1.367	1.362	1.357	1.350	1.342	1.333	1.322	1.309	1.292	1.270	1.242	1.223	1.221	1.219
1500	1.394	1.391	1.388	1.385	1.381	1.377	1.373	1.368	1.362	1.355	1.347	1.338	1.327	1.313	1.296	1.274	1.245	1.226	1.224	1.222
1520	1.399	1.397	1.394	1.390	1.387	1.383	1.378	1.373	1.368	1.360	1.352	1.343	1.331	1.317	1.300	1.278	1.248	1.229	1.227	1.225
1540	1.405	1.402	1.399	1.396	1.392	1.388	1.383	1.378	1.373	1.365	1.357	1.348	1.336	1.322	1.304	1.282	1.252	1.232	1.230	1.228
1560	1.411	1.408	1.405	1.402	1.398	1.394	1.389	1.383	1.378	1.370	1.362	1.352	1.341	1.326	1.309	1.286	1.255	1.235	1.233	1.231
1580	1.416	1.414	1.411	1.407	1.403	1.399	1.394	1.389	1.383	1.375	1.367	1.357	1.345	1.331	1.313	1.289	1.258	1.239	1.236	1.234
1600	1.422	1.419	1.416	1.413	1.409	1.405	1.400	1.394	1.389	1.381	1.372	1.362	1.350	1.335	1.317	1.293	1.262	1.242	1.239	1.237
1620	1.428	1.425	1.422	1.418	1.414	1.410	1.405	1.399	1.393	1.386	1.377	1.367	1.355	1.340	1.321	1.297	1.265	1.245	1.242	1.240
1640	1.434	1.431	1.427	1.424	1.420	1.415	1.410	1.405	1.398	1.391	1.382	1.372	1.359	1.344	1.325	1.301	1.268	1.248	1.245	1.243

x

SIGMA = 7355.0

V-V_c = 11

x	0.08	0.09	0.10	0.2	0.4	0.6	0.8	1.0	1.2	1.4	1.6	1.8	2.0	2.2	2.4	2.6	2.8	3.0	3.2	3.4
1660	1.246	1.249	1.251	1.272	1.305	1.329	1.349	1.364	1.377	1.387	1.396	1.404	1.410	1.416	1.421	1.425	1.430	1.433	1.436	1.439
1680	1.249	1.252	1.254	1.275	1.309	1.334	1.353	1.369	1.381	1.392	1.401	1.409	1.415	1.421	1.426	1.431	1.435	1.439	1.442	1.445
1700	1.252	1.255	1.257	1.279	1.312	1.338	1.358	1.373	1.386	1.397	1.406	1.414	1.421	1.427	1.432	1.437	1.441	1.445	1.448	1.451
1720	1.255	1.258	1.260	1.282	1.316	1.342	1.362	1.378	1.391	1.402	1.411	1.419	1.426	1.432	1.436	1.442	1.446	1.450	1.454	1.457
1740	1.258	1.261	1.263	1.285	1.320	1.346	1.367	1.383	1.396	1.407	1.417	1.425	1.432	1.438	1.443	1.448	1.452	1.456	1.459	1.463
1760	1.261	1.264	1.266	1.289	1.324	1.351	1.371	1.388	1.401	1.412	1.422	1.430	1.437	1.443	1.449	1.453	1.458	1.462	1.465	1.468
1780	1.264	1.267	1.269	1.292	1.328	1.355	1.376	1.392	1.406	1.417	1.427	1.435	1.442	1.449	1.454	1.459	1.463	1.467	1.471	1.474
1800	1.267	1.270	1.272	1.296	1.332	1.359	1.380	1.397	1.411	1.422	1.432	1.441	1.448	1.454	1.460	1.465	1.469	1.473	1.477	1.480
1820	1.270	1.273	1.276	1.299	1.336	1.363	1.385	1.402	1.416	1.428	1.437	1.446	1.453	1.459	1.465	1.470	1.475	1.479	1.483	1.486
1840	1.273	1.276	1.279	1.302	1.340	1.367	1.389	1.407	1.421	1.433	1.443	1.451	1.459	1.465	1.471	1.476	1.481	1.485	1.488	1.492
1860	1.276	1.279	1.282	1.306	1.343	1.372	1.394	1.411	1.426	1.438	1.448	1.457	1.464	1.471	1.477	1.482	1.486	1.491	1.494	1.498
1880	1.279	1.282	1.285	1.309	1.347	1.376	1.398	1.416	1.431	1.443	1.453	1.462	1.470	1.476	1.482	1.487	1.492	1.496	1.500	1.504
1900	1.282	1.285	1.288	1.313	1.351	1.380	1.403	1.421	1.436	1.448	1.458	1.467	1.475	1.482	1.488	1.493	1.498	1.502	1.506	1.510
1920	1.285	1.288	1.291	1.316	1.355	1.384	1.407	1.426	1.441	1.453	1.464	1.473	1.481	1.488	1.493	1.499	1.504	1.508	1.512	1.515
1940	1.288	1.291	1.294	1.319	1.359	1.389	1.412	1.430	1.446	1.458	1.469	1.478	1.486	1.493	1.499	1.505	1.509	1.514	1.518	1.521
1960	1.291	1.294	1.297	1.323	1.363	1.393	1.416	1.435	1.451	1.463	1.474	1.483	1.491	1.499	1.505	1.510	1.515	1.520	1.524	1.527
1980	1.295	1.297	1.300	1.326	1.367	1.397	1.421	1.440	1.456	1.469	1.479	1.489	1.497	1.504	1.510	1.516	1.521	1.525	1.530	1.533
2000	1.298	1.300	1.303	1.330	1.371	1.402	1.426	1.445	1.461	1.474	1.485	1.494	1.503	1.510	1.516	1.522	1.527	1.531	1.535	1.539
2020	1.301	1.304	1.306	1.333	1.375	1.406	1.430	1.450	1.466	1.479	1.490	1.500	1.508	1.515	1.521	1.527	1.533	1.537	1.541	1.545
2040	1.304	1.307	1.310	1.336	1.379	1.410	1.435	1.455	1.471	1.484	1.495	1.505	1.514	1.521	1.527	1.533	1.538	1.543	1.547	1.551
2060	1.307	1.310	1.313	1.340	1.383	1.415	1.439	1.459	1.476	1.489	1.501	1.511	1.519	1.527	1.533	1.539	1.544	1.549	1.553	1.557
2080	1.310	1.313	1.316	1.343	1.386	1.419	1.444	1.464	1.481	1.494	1.506	1.516	1.525	1.532	1.539	1.545	1.550	1.555	1.559	1.563
2100	1.313	1.316	1.319	1.347	1.390	1.423	1.449	1.469	1.486	1.500	1.511	1.521	1.530	1.538	1.545	1.551	1.556	1.561	1.565	1.569
2120	1.316	1.319	1.322	1.350	1.394	1.427	1.453	1.474	1.491	1.505	1.517	1.527	1.536	1.544	1.550	1.556	1.562	1.567	1.571	1.575
2140	1.319	1.322	1.325	1.354	1.398	1.432	1.458	1.479	1.496	1.510	1.522	1.532	1.541	1.549	1.555	1.562	1.568	1.573	1.577	1.581
2160	1.322	1.325	1.328	1.357	1.402	1.436	1.463	1.484	1.501	1.515	1.527	1.538	1.547	1.555	1.562	1.568	1.574	1.579	1.583	1.587
2180	1.325	1.328	1.331	1.360	1.406	1.440	1.467	1.489	1.506	1.521	1.533	1.543	1.553	1.561	1.568	1.574	1.580	1.585	1.589	1.593
2200	1.328	1.331	1.334	1.364	1.410	1.445	1.472	1.493	1.511	1.526	1.538	1.549	1.558	1.566	1.573	1.580	1.585	1.591	1.595	1.599
2220	1.331	1.334	1.338	1.367	1.414	1.449	1.476	1.498	1.516	1.531	1.544	1.554	1.564	1.572	1.579	1.586	1.591	1.597	1.601	1.605
2240	1.334	1.337	1.341	1.371	1.418	1.454	1.481	1.503	1.521	1.536	1.549	1.560	1.569	1.578	1.585	1.591	1.597	1.603	1.607	1.612
2260	1.337	1.340	1.344	1.374	1.422	1.458	1.486	1.508	1.526	1.542	1.554	1.566	1.575	1.583	1.591	1.597	1.603	1.609	1.613	1.618
2280	1.340	1.344	1.347	1.378	1.426	1.462	1.490	1.513	1.531	1.547	1.560	1.571	1.581	1.589	1.597	1.603	1.609	1.615	1.619	1.624
2300	1.343	1.347	1.350	1.381	1.430	1.467	1.495	1.518	1.537	1.552	1.565	1.577	1.586	1.595	1.602	1.609	1.615	1.621	1.625	1.630
2320	1.346	1.350	1.353	1.385	1.434	1.471	1.500	1.523	1.542	1.557	1.571	1.582	1.592	1.601	1.608	1.615	1.621	1.627	1.632	1.636
2340	1.349	1.353	1.356	1.388	1.438	1.475	1.505	1.528	1.547	1.563	1.576	1.588	1.598	1.606	1.614	1.621	1.627	1.633	1.638	1.642
2360	1.352	1.356	1.359	1.392	1.442	1.480	1.509	1.533	1.552	1.568	1.582	1.593	1.603	1.612	1.620	1.627	1.633	1.639	1.644	1.648
2380	1.355	1.359	1.363	1.395	1.446	1.484	1.514	1.538	1.557	1.573	1.587	1.599	1.609	1.618	1.626	1.633	1.639	1.645	1.650	1.655
2400	1.358	1.362	1.366	1.398	1.450	1.489	1.519	1.543	1.562	1.579	1.593	1.605	1.615	1.624	1.632	1.639	1.645	1.651	1.656	1.661
2420	1.361	1.365	1.369	1.402	1.454	1.493	1.523	1.548	1.568	1.584	1.598	1.610	1.621	1.630	1.638	1.645	1.651	1.657	1.662	1.667
2440	1.364	1.368	1.372	1.405	1.458	1.497	1.528	1.553	1.573	1.589	1.604	1.616	1.626	1.635	1.644	1.650	1.657	1.663	1.668	1.673
2460	1.368	1.371	1.375	1.409	1.462	1.502	1.533	1.558	1.578	1.595	1.609	1.621	1.632	1.641	1.650	1.657	1.663	1.669	1.675	1.679
2480	1.371	1.374	1.378	1.412	1.466	1.506	1.538	1.563	1.583	1.600	1.615	1.627	1.638	1.647	1.655	1.663	1.669	1.675	1.681	1.686

SIGMA = 7355.0

$V-V_c = 11$

H	3.4	3.2	3.0	2.8	2.6	2.4	2.2	2.0	1.8	1.6	1.4	1.2	1.0	0.8	0.6	0.4	0.2	0.10	0.09	0.08
2500	1.692	1.687	1.681	1.675	1.669	1.661	1.653	1.644	1.633	1.620	1.606	1.588	1.568	1.542	1.511	1.470	1.416	1.381	1.378	1.374
2520	1.698	1.693	1.688	1.682	1.675	1.667	1.659	1.649	1.638	1.626	1.611	1.594	1.573	1.547	1.515	1.474	1.419	1.384	1.381	1.377
2540	1.704	1.699	1.694	1.688	1.681	1.673	1.665	1.655	1.644	1.631	1.616	1.599	1.578	1.552	1.520	1.478	1.423	1.388	1.384	1.380
2560	1.711	1.706	1.700	1.694	1.687	1.679	1.671	1.661	1.650	1.637	1.622	1.604	1.583	1.557	1.524	1.482	1.426	1.391	1.387	1.383
2580	1.717	1.712	1.706	1.700	1.693	1.685	1.677	1.667	1.655	1.642	1.627	1.609	1.588	1.561	1.528	1.486	1.430	1.394	1.390	1.386
2600	1.723	1.718	1.712	1.706	1.699	1.691	1.682	1.672	1.661	1.648	1.633	1.614	1.593	1.566	1.533	1.490	1.433	1.397	1.393	1.389
2620	1.730	1.724	1.719	1.712	1.705	1.697	1.688	1.678	1.667	1.654	1.638	1.620	1.598	1.571	1.537	1.494	1.437	1.400	1.396	1.392
2640	1.736	1.731	1.725	1.718	1.711	1.703	1.694	1.684	1.673	1.659	1.643	1.625	1.603	1.576	1.542	1.498	1.440	1.403	1.399	1.395
2660	1.742	1.737	1.731	1.725	1.717	1.709	1.700	1.690	1.678	1.665	1.649	1.630	1.608	1.581	1.546	1.502	1.444	1.406	1.402	1.398
2680	1.749	1.743	1.737	1.731	1.723	1.715	1.706	1.696	1.684	1.670	1.654	1.636	1.613	1.585	1.551	1.506	1.447	1.410	1.405	1.401
2700	1.755	1.749	1.743	1.737	1.730	1.721	1.712	1.702	1.690	1.676	1.660	1.641	1.618	1.590	1.555	1.510	1.451	1.413	1.409	1.404
2720	1.761	1.756	1.750	1.743	1.736	1.727	1.718	1.708	1.695	1.682	1.665	1.646	1.623	1.595	1.560	1.515	1.454	1.416	1.412	1.407
2740	1.768	1.762	1.756	1.749	1.742	1.733	1.724	1.713	1.701	1.687	1.671	1.651	1.628	1.600	1.564	1.519	1.458	1.419	1.415	1.410
2760	1.774	1.768	1.762	1.756	1.748	1.740	1.730	1.719	1.707	1.693	1.676	1.657	1.633	1.605	1.569	1.523	1.461	1.422	1.418	1.413
2780	1.780	1.775	1.769	1.762	1.754	1.746	1.736	1.725	1.713	1.698	1.682	1.662	1.638	1.609	1.573	1.527	1.465	1.425	1.421	1.417
2800	1.787	1.781	1.775	1.768	1.760	1.752	1.742	1.731	1.719	1.704	1.687	1.667	1.644	1.614	1.578	1.531	1.468	1.428	1.424	1.420
2820	1.793	1.788	1.781	1.774	1.766	1.758	1.748	1.737	1.724	1.710	1.693	1.673	1.649	1.619	1.582	1.535	1.472	1.432	1.427	1.423
2840	1.800	1.794	1.788	1.781	1.773	1.764	1.754	1.743	1.730	1.716	1.698	1.678	1.654	1.624	1.587	1.539	1.475	1.435	1.430	1.426
2860	1.806	1.800	1.794	1.787	1.779	1.770	1.760	1.749	1.736	1.721	1.704	1.683	1.659	1.629	1.591	1.543	1.479	1.438	1.433	1.429
2880	1.813	1.807	1.800	1.793	1.785	1.776	1.766	1.755	1.742	1.727	1.709	1.689	1.664	1.634	1.596	1.547	1.482	1.441	1.437	1.432
2900	1.819	1.813	1.807	1.799	1.791	1.782	1.772	1.761	1.748	1.733	1.715	1.694	1.669	1.639	1.600	1.551	1.486	1.444	1.440	1.435
2920	1.826	1.820	1.813	1.806	1.798	1.789	1.778	1.767	1.754	1.738	1.721	1.700	1.674	1.644	1.605	1.555	1.489	1.447	1.443	1.438
2940	1.832	1.826	1.819	1.812	1.804	1.795	1.784	1.773	1.759	1.744	1.726	1.705	1.679	1.648	1.610	1.560	1.493	1.451	1.446	1.441
2960	1.839	1.832	1.826	1.818	1.810	1.801	1.790	1.779	1.765	1.750	1.732	1.710	1.685	1.653	1.614	1.564	1.497	1.454	1.449	1.444
2980	1.845	1.839	1.832	1.825	1.816	1.807	1.797	1.785	1.771	1.756	1.737	1.716	1.690	1.658	1.619	1.568	1.500	1.457	1.452	1.447
3000	1.852	1.845	1.839	1.831	1.823	1.813	1.803	1.791	1.777	1.761	1.743	1.721	1.695	1.663	1.623	1.572	1.504	1.460	1.455	1.450

x

$V-V_c = 11$

SIGMA = 7355.5

x (H)	3.4	3.2	3.0	2.8	2.6	2.4	2.2	2.0	1.8	1.6	1.4	1.2	1.0	0.8	0.6	0.4	0.2	0.10	0.09	0.08
3100	1.884	1.878	1.871	1.863	1.854	1.844	1.833	1.821	1.807	1.790	1.771	1.748	1.721	1.688	1.646	1.593	1.521	1.476	1.471	1.466
3200	1.918	1.911	1.903	1.895	1.886	1.876	1.864	1.851	1.836	1.819	1.799	1.776	1.747	1.713	1.669	1.613	1.539	1.492	1.487	1.481
3300	1.951	1.944	1.936	1.928	1.918	1.907	1.896	1.882	1.867	1.849	1.828	1.803	1.774	1.738	1.692	1.634	1.557	1.508	1.502	1.497
3400	1.985	1.977	1.969	1.960	1.950	1.939	1.927	1.913	1.897	1.878	1.857	1.831	1.800	1.763	1.716	1.655	1.575	1.524	1.518	1.512
3500	2.019	2.011	2.003	1.993	1.983	1.972	1.959	1.944	1.927	1.908	1.886	1.859	1.827	1.788	1.739	1.677	1.593	1.540	1.534	1.528
3600	2.053	2.045	2.036	2.027	2.016	2.004	1.991	1.975	1.958	1.938	1.915	1.887	1.854	1.813	1.763	1.698	1.611	1.556	1.550	1.543
3700	2.087	2.079	2.070	2.060	2.049	2.037	2.023	2.007	1.989	1.968	1.944	1.915	1.881	1.839	1.787	1.719	1.629	1.572	1.565	1.559
3800	2.122	2.114	2.104	2.094	2.083	2.070	2.055	2.039	2.020	1.999	1.974	1.944	1.908	1.865	1.811	1.741	1.647	1.588	1.581	1.575
3900	2.157	2.148	2.139	2.128	2.116	2.103	2.088	2.071	2.052	2.030	2.004	1.973	1.936	1.891	1.835	1.762	1.665	1.604	1.597	1.590
4000	2.193	2.184	2.174	2.163	2.150	2.137	2.121	2.104	2.084	2.061	2.034	2.002	1.964	1.917	1.859	1.784	1.684	1.620	1.613	1.606
4100	2.228	2.219	2.209	2.197	2.184	2.170	2.154	2.136	2.116	2.092	2.064	2.031	1.992	1.943	1.883	1.805	1.702	1.636	1.629	1.622
4200	2.264	2.255	2.244	2.232	2.219	2.204	2.188	2.169	2.148	2.123	2.094	2.060	2.020	1.970	1.907	1.827	1.721	1.653	1.645	1.637
4300	2.301	2.290	2.279	2.267	2.254	2.239	2.222	2.202	2.180	2.155	2.125	2.090	2.048	1.996	1.932	1.849	1.739	1.669	1.661	1.653
4400	2.337	2.327	2.315	2.303	2.289	2.273	2.256	2.236	2.213	2.187	2.156	2.120	2.076	2.023	1.957	1.871	1.758	1.685	1.677	1.669
4500	2.374	2.363	2.351	2.338	2.324	2.308	2.290	2.269	2.246	2.219	2.187	2.150	2.105	2.050	1.982	1.894	1.776	1.701	1.693	1.685
4600	2.411	2.400	2.388	2.374	2.360	2.343	2.324	2.303	2.279	2.251	2.218	2.180	2.134	2.077	2.007	1.916	1.795	1.716	1.709	1.701
4700	2.448	2.437	2.425	2.411	2.395	2.378	2.359	2.337	2.312	2.284	2.250	2.210	2.163	2.105	2.032	1.938	1.814	1.734	1.726	1.717
4800	2.486	2.474	2.462	2.447	2.432	2.414	2.394	2.372	2.346	2.316	2.282	2.241	2.192	2.132	2.057	1.961	1.832	1.751	1.742	1.733
4900	2.524	2.512	2.499	2.484	2.468	2.450	2.429	2.406	2.380	2.349	2.314	2.272	2.221	2.160	2.082	1.983	1.851	1.767	1.758	1.748
5000	2.562	2.550	2.536	2.521	2.505	2.486	2.465	2.441	2.414	2.383	2.346	2.303	2.251	2.187	2.106	2.006	1.870	1.784	1.774	1.764
5100	2.601	2.588	2.574	2.559	2.541	2.522	2.501	2.476	2.448	2.416	2.378	2.334	2.281	2.215	2.134	2.029	1.889	1.800	1.790	1.780
5200	2.640	2.627	2.612	2.596	2.579	2.559	2.537	2.512	2.483	2.450	2.411	2.365	2.310	2.243	2.160	2.052	1.908	1.817	1.807	1.796
5300	2.679	2.665	2.650	2.634	2.616	2.596	2.573	2.547	2.518	2.484	2.444	2.397	2.341	2.272	2.186	2.075	1.927	1.833	1.823	1.812
5400	2.718	2.704	2.689	2.672	2.654	2.633	2.610	2.583	2.553	2.518	2.477	2.429	2.371	2.300	2.212	2.098	1.946	1.850	1.839	1.829
5500	2.758	2.744	2.728	2.711	2.692	2.670	2.646	2.619	2.588	2.552	2.510	2.461	2.401	2.329	2.238	2.121	1.966	1.867	1.856	1.845
5600	2.798	2.783	2.767	2.750	2.730	2.708	2.683	2.655	2.624	2.587	2.544	2.493	2.432	2.357	2.264	2.145	1.985	1.883	1.872	1.861
5700	2.838	2.823	2.807	2.789	2.768	2.746	2.721	2.692	2.659	2.621	2.577	2.525	2.463	2.386	2.291	2.168	2.004	1.900	1.889	1.877
5800	2.878	2.863	2.846	2.828	2.807	2.784	2.756	2.729	2.695	2.657	2.611	2.558	2.494	2.415	2.317	2.192	2.024	1.917	1.905	1.893
5900	2.919	2.904	2.886	2.867	2.846	2.823	2.796	2.766	2.732	2.692	2.645	2.591	2.525	2.445	2.344	2.215	2.043	1.934	1.922	1.909
6000	2.960	2.944	2.927	2.907	2.886	2.861	2.834	2.803	2.768	2.727	2.680	2.624	2.556	2.474	2.371	2.239	2.063	1.950	1.938	1.925
6100	3.002	2.985	2.967	2.947	2.925	2.900	2.872	2.841	2.805	2.763	2.714	2.657	2.588	2.504	2.398	2.263	2.082	1.967	1.955	1.942
6200	3.043	3.027	3.008	2.988	2.965	2.940	2.911	2.879	2.842	2.799	2.749	2.690	2.620	2.533	2.426	2.287	2.102	1.984	1.971	1.958
6300	3.085	3.068	3.049	3.028	3.005	2.979	2.950	2.917	2.879	2.835	2.784	2.724	2.652	2.563	2.453	2.311	2.122	2.001	1.988	1.974
6400	3.128	3.110	3.091	3.069	3.045	3.019	2.989	2.955	2.916	2.872	2.819	2.758	2.684	2.593	2.480	2.335	2.141	2.018	2.005	1.991
6500	3.170	3.152	3.132	3.110	3.086	3.059	3.028	2.994	2.954	2.908	2.855	2.792	2.716	2.624	2.508	2.359	2.161	2.035	2.021	2.007
6600	3.213	3.195	3.174	3.152	3.127	3.099	3.068	3.032	2.992	2.945	2.891	2.826	2.749	2.654	2.536	2.384	2.181	2.052	2.038	2.023
6700	3.256	3.237	3.217	3.194	3.168	3.140	3.108	3.071	3.030	2.982	2.926	2.860	2.781	2.685	2.564	2.408	2.201	2.069	2.055	2.040
6800	3.299	3.280	3.259	3.236	3.210	3.181	3.148	3.111	3.068	3.020	2.962	2.895	2.814	2.715	2.592	2.433	2.221	2.086	2.071	2.056
6900	3.343	3.323	3.302	3.278	3.251	3.222	3.188	3.150	3.107	3.057	2.999	2.930	2.847	2.746	2.620	2.458	2.241	2.103	2.088	2.073
7000	3.387	3.367	3.345	3.320	3.293	3.263	3.229	3.190	3.146	3.095	3.035	2.965	2.880	2.777	2.648	2.482	2.261	2.121	2.105	2.089
7100	3.431	3.411	3.388	3.363	3.336	3.305	3.270	3.230	3.185	3.133	3.072	3.000	2.914	2.809	2.677	2.507	2.282	2.138	2.122	2.106
7200	3.476	3.455	3.432	3.406	3.378	3.346	3.311	3.270	3.224	3.171	3.109	3.036	2.948	2.840	2.705	2.532	2.302	2.155	2.139	2.122

V-V_c = 11   SIGMA = 7355.0

H	3.4	3.2	3.0	2.8	2.6	2.4	2.2	2.0	1.8	1.6	1.4	1.2	1.0	0.8	0.6	0.4	0.2	0.10	0.09	0.08
7300	3.521	3.499	3.476	3.450	3.421	3.388	3.352	3.311	3.264	3.210	3.146	3.071	2.981	2.871	2.734	2.558	2.322	2.172	2.156	2.139
7400	3.566	3.544	3.520	3.493	3.464	3.431	3.394	3.352	3.304	3.248	3.184	3.107	3.015	2.903	2.763	2.583	2.343	2.190	2.173	2.156
7500	3.611	3.589	3.564	3.537	3.507	3.474	3.436	3.393	3.344	3.287	3.221	3.143	3.049	2.935	2.792	2.608	2.363	2.207	2.190	2.172
7600	3.657	3.634	3.609	3.581	3.551	3.516	3.478	3.434	3.384	3.326	3.259	3.179	3.084	2.967	2.821	2.634	2.383	2.224	2.207	2.189
7700	3.703	3.680	3.654	3.626	3.595	3.560	3.520	3.476	3.424	3.366	3.297	3.216	3.118	2.999	2.851	2.659	2.404	2.242	2.224	2.206
7800	3.749	3.725	3.699	3.671	3.639	3.603	3.563	3.517	3.465	3.405	3.335	3.252	3.153	3.032	2.880	2.685	2.425	2.259	2.241	2.222
7900	3.796	3.771	3.745	3.716	3.683	3.647	3.606	3.559	3.506	3.445	3.374	3.289	3.188	3.064	2.910	2.711	2.445	2.277	2.258	2.239
8000	3.842	3.818	3.791	3.761	3.728	3.691	3.649	3.602	3.547	3.485	3.412	3.326	3.223	3.097	2.939	2.736	2.466	2.294	2.275	2.256
8100	3.889	3.864	3.837	3.806	3.773	3.735	3.692	3.644	3.589	3.525	3.451	3.364	3.258	3.130	2.969	2.762	2.487	2.312	2.292	2.273
8200	3.937	3.911	3.883	3.852	3.818	3.779	3.736	3.687	3.631	3.566	3.490	3.401	3.294	3.163	2.999	2.789	2.508	2.329	2.310	2.290
8300	3.985	3.958	3.930	3.898	3.863	3.824	3.780	3.730	3.673	3.607	3.530	3.439	3.329	3.196	3.029	2.815	2.529	2.347	2.327	2.306
8400	4.033	4.006	3.977	3.945	3.909	3.869	3.824	3.773	3.715	3.648	3.569	3.476	3.365	3.229	3.059	2.841	2.550	2.364	2.344	2.323
8500	4.081	4.054	4.024	3.991	3.955	3.914	3.868	3.817	3.757	3.689	3.609	3.515	3.401	3.263	3.090	2.867	2.571	2.382	2.361	2.340
8600	4.129	4.102	4.072	4.038	4.001	3.960	3.913	3.860	3.800	3.730	3.649	3.553	3.438	3.297	3.120	2.894	2.592	2.400	2.379	2.357
8700	4.178	4.150	4.119	4.085	4.048	4.005	3.958	3.904	3.843	3.772	3.689	3.591	3.474	3.330	3.151	2.921	2.613	2.418	2.396	2.374
8800	4.227	4.199	4.167	4.133	4.094	4.052	4.003	3.948	3.886	3.814	3.729	3.630	3.510	3.364	3.182	2.947	2.634	2.435	2.413	2.391
8900	4.277	4.248	4.216	4.181	4.142	4.098	4.049	3.993	3.929	3.856	3.770	3.669	3.547	3.399	3.213	2.974	2.656	2.453	2.431	2.408
9000	4.326	4.297	4.264	4.229	4.189	4.144	4.094	4.038	3.973	3.898	3.811	3.708	3.584	3.433	3.244	3.001	2.677	2.471	2.448	2.425
9100	4.376	4.346	4.313	4.277	4.236	4.191	4.140	4.083	4.017	3.941	3.852	3.747	3.621	3.467	3.275	3.028	2.699	2.489	2.466	2.442
9200	4.426	4.396	4.362	4.325	4.284	4.238	4.186	4.128	4.061	3.983	3.893	3.787	3.659	3.502	3.307	3.055	2.720	2.507	2.483	2.459
9300	4.477	4.446	4.412	4.374	4.332	4.286	4.233	4.173	4.105	4.026	3.935	3.826	3.696	3.537	3.338	3.083	2.742	2.525	2.501	2.477
9400	4.528	4.496	4.462	4.423	4.381	4.333	4.280	4.219	4.150	4.070	3.976	3.866	3.734	3.572	3.370	3.110	2.763	2.543	2.518	2.494
9500	4.579	4.547	4.512	4.473	4.429	4.381	4.327	4.265	4.194	4.113	4.018	3.906	3.772	3.607	3.402	3.137	2.785	2.561	2.536	2.511
9600	4.630	4.598	4.562	4.522	4.478	4.429	4.374	4.311	4.240	4.157	4.060	3.946	3.810	3.643	3.434	3.165	2.807	2.579	2.554	2.528
9700	4.682	4.649	4.612	4.572	4.527	4.476	4.421	4.358	4.285	4.201	4.103	3.987	3.848	3.678	3.466	3.193	2.829	2.597	2.571	2.545
9800	4.734	4.700	4.663	4.622	4.577	4.526	4.469	4.404	4.330	4.245	4.145	4.028	3.886	3.714	3.498	3.220	2.850	2.615	2.589	2.563
9900	4.786	4.752	4.714	4.673	4.627	4.575	4.517	4.451	4.376	4.289	4.188	4.068	3.925	3.749	3.530	3.248	2.872	2.633	2.607	2.580
10000	4.839	4.804	4.766	4.724	4.677	4.624	4.565	4.498	4.422	4.334	4.231	4.110	3.964	3.785	3.563	3.276	2.894	2.651	2.625	2.597
10100	4.891	4.856	4.817	4.775	4.727	4.674	4.614	4.546	4.468	4.379	4.274	4.151	4.003	3.822	3.595	3.304	2.916	2.669	2.642	2.615
10200	4.945	4.909	4.869	4.826	4.777	4.723	4.662	4.594	4.515	4.424	4.318	4.192	4.042	3.858	3.628	3.333	2.938	2.688	2.660	2.632
10300	4.998	4.962	4.922	4.877	4.828	4.773	4.711	4.641	4.561	4.469	4.361	4.234	4.081	3.894	3.661	3.361	2.961	2.706	2.678	2.649
10400	5.052	5.015	4.974	4.929	4.879	4.823	4.761	4.690	4.608	4.515	4.405	4.276	4.121	3.931	3.694	3.389	2.983	2.724	2.696	2.667
10500	5.106	5.068	5.027	4.981	4.931	4.874	4.810	4.738	4.656	4.560	4.449	4.318	4.160	3.968	3.727	3.418	3.005	2.743	2.714	2.684
10600	5.160	5.122	5.080	5.034	4.982	4.925	4.860	4.787	4.703	4.606	4.494	4.360	4.200	4.005	3.761	3.446	3.028	2.761	2.732	2.702
10700	5.214	5.176	5.133	5.086	5.034	4.976	4.910	4.836	4.751	4.652	4.538	4.403	4.240	4.042	3.794	3.475	3.050	2.779	2.750	2.719
10800	5.269	5.230	5.187	5.139	5.086	5.027	4.960	4.885	4.798	4.699	4.583	4.445	4.281	4.079	3.828	3.504	3.072	2.798	2.768	2.737
10900	5.324	5.284	5.241	5.192	5.139	5.078	5.011	4.934	4.847	4.746	4.628	4.488	4.321	4.117	3.861	3.533	3.095	2.816	2.786	2.754
11000	5.380	5.339	5.295	5.246	5.191	5.130	5.062	4.984	4.895	4.792	4.673	4.531	4.362	4.154	3.895	3.562	3.118	2.835	2.804	2.772
11100	5.435	5.394	5.349	5.299	5.244	5.182	5.113	5.034	4.944	4.840	4.718	4.575	4.403	4.192	3.929	3.591	3.140	2.853	2.822	2.790
11200	5.491	5.450	5.404	5.353	5.297	5.235	5.164	5.084	4.992	4.887	4.764	4.618	4.444	4.230	3.963	3.620	3.163	2.872	2.840	2.807
11300	5.548	5.505	5.459	5.408	5.351	5.287	5.215	5.134	5.041	4.934	4.810	4.662	4.485	4.268	3.998	3.650	3.186	2.891	2.858	2.825
11400	5.604	5.561	5.514	5.462	5.404	5.340	5.267	5.185	5.091	4.982	4.856	4.706	4.526	4.307	4.032	3.679	3.209	2.909	2.876	2.843

x

SIGMA = 7355.0

$V - V_c = 11$

H	3.4	3.2	3.0	2.8	2.6	2.4	2.2	2.0	1.8	1.6	1.4	1.2	1.0	0.8	0.6	0.4	0.2	0.10	0.09	0.08
11500	5.661	5.617	5.570	5.517	5.458	5.393	5.319	5.236	5.140	5.030	4.902	4.750	4.568	4.345	4.067	3.709	3.232	2.928	2.895	2.860
11600	5.718	5.674	5.626	5.572	5.513	5.446	5.372	5.287	5.190	5.078	4.948	4.794	4.609	4.384	4.101	3.738	3.255	2.947	2.913	2.878
11700	5.775	5.731	5.682	5.627	5.567	5.500	5.424	5.338	5.240	5.127	4.995	4.839	4.651	4.422	4.136	3.768	3.278	2.965	2.931	2.896
11800	5.833	5.788	5.738	5.683	5.622	5.554	5.477	5.390	5.290	5.176	5.042	4.883	4.693	4.461	4.171	3.798	3.301	2.984	2.949	2.914
11900	5.891	5.845	5.795	5.739	5.677	5.608	5.530	5.442	5.341	5.225	5.089	4.928	4.736	4.500	4.206	3.828	3.324	3.003	2.968	2.932
12000	5.949	5.903	5.852	5.795	5.732	5.662	5.583	5.494	5.392	5.274	5.136	4.973	4.778	4.540	4.242	3.858	3.347	3.022	2.986	2.950
12100	6.008	5.961	5.909	5.852	5.788	5.717	5.637	5.546	5.443	5.323	5.184	5.019	4.821	4.579	4.277	3.888	3.370	3.041	3.004	2.967
12200	6.067	6.019	5.966	5.908	5.844	5.772	5.691	5.599	5.494	5.373	5.231	5.064	4.864	4.619	4.313	3.919	3.394	3.060	3.023	2.985
12300	6.126	6.077	6.024	5.965	5.900	5.827	5.745	5.652	5.545	5.423	5.279	5.110	4.907	4.659	4.348	3.949	3.417	3.079	3.041	3.003
12400	6.185	6.136	6.082	6.023	5.956	5.882	5.799	5.705	5.597	5.473	5.327	5.156	4.950	4.698	4.384	3.980	3.441	3.098	3.060	3.021
12500	6.245	6.195	6.140	6.080	6.013	5.938	5.854	5.758	5.649	5.523	5.376	5.202	4.993	4.739	4.420	4.010	3.464	3.117	3.078	3.039
12600	6.305	6.254	6.199	6.138	6.070	5.994	5.909	5.812	5.701	5.573	5.424	5.248	5.037	4.779	4.456	4.041	3.488	3.136	3.097	3.057
12700	6.365	6.314	6.258	6.196	6.127	6.050	5.964	5.866	5.753	5.624	5.473	5.295	5.081	4.819	4.492	4.072	3.511	3.155	3.115	3.075
12800	6.425	6.374	6.317	6.254	6.185	6.107	6.019	5.920	5.806	5.675	5.522	5.342	5.125	4.860	4.529	4.103	3.535	3.174	3.134	3.094
12900	6.486	6.434	6.376	6.313	6.242	6.163	6.075	5.974	5.859	5.726	5.571	5.389	5.169	4.901	4.565	4.134	3.559	3.193	3.153	3.112
13000	6.547	6.494	6.436	6.372	6.300	6.220	6.130	6.029	5.912	5.778	5.621	5.436	5.213	4.942	4.602	4.165	3.583	3.212	3.171	3.130
13100	6.609	6.555	6.496	6.431	6.359	6.277	6.187	6.083	5.965	5.829	5.671	5.483	5.258	4.983	4.639	4.196	3.607	3.231	3.190	3.148
13200	6.670	6.616	6.556	6.490	6.417	6.335	6.243	6.138	6.019	5.881	5.720	5.530	5.303	5.024	4.676	4.228	3.631	3.251	3.209	3.166
13300	6.732	6.677	6.617	6.550	6.476	6.393	6.300	6.194	6.073	5.933	5.771	5.578	5.347	5.065	4.713	4.259	3.655	3.270	3.228	3.184
13400	6.795	6.739	6.678	6.610	6.535	6.451	6.356	6.249	6.127	5.986	5.821	5.626	5.392	5.107	4.750	4.291	3.679	3.289	3.246	3.203
13500	6.857	6.801	6.739	6.670	6.594	6.509	6.414	6.305	6.181	6.038	5.871	5.674	5.438	5.149	4.787	4.322	3.703	3.309	3.265	3.221
13600	6.920	6.863	6.800	6.731	6.654	6.568	6.471	6.361	6.236	6.091	5.922	5.723	5.483	5.190	4.825	4.354	3.727	3.328	3.284	3.239
13700	6.983	6.925	6.862	6.792	6.714	6.627	6.529	6.418	6.291	6.144	5.973	5.771	5.529	5.233	4.862	4.386	3.751	3.347	3.303	3.258
13800	7.046	6.988	6.924	6.853	6.774	6.686	6.586	6.474	6.346	6.197	6.024	5.820	5.575	5.275	4.900	4.418	3.776	3.367	3.322	3.276
13900	7.110	7.051	6.986	6.914	6.834	6.745	6.645	6.531	6.401	6.251	6.076	5.869	5.621	5.317	4.938	4.450	3.800	3.386	3.341	3.294
14000	7.174	7.114	7.048	6.976	6.895	6.805	6.703	6.588	6.456	6.304	6.127	5.918	5.667	5.360	4.976	4.483	3.825	3.406	3.360	3.313
14100	7.238	7.178	7.111	7.038	6.956	6.864	6.762	6.645	6.512	6.358	6.179	5.967	5.713	5.402	5.014	4.515	3.849	3.425	3.379	3.331
14200	7.303	7.242	7.174	7.100	7.017	6.925	6.821	6.703	6.568	6.413	6.231	6.017	5.760	5.445	5.052	4.547	3.874	3.445	3.398	3.350
14300	7.368	7.306	7.238	7.162	7.079	6.985	6.880	6.761	6.624	6.467	6.283	6.067	5.806	5.488	5.091	4.580	3.898	3.465	3.417	3.368
14400	7.433	7.370	7.301	7.225	7.140	7.046	6.939	6.819	6.681	6.522	6.336	6.117	5.853	5.532	5.129	4.612	3.923	3.484	3.436	3.387
14500	7.498	7.435	7.365	7.288	7.202	7.107	6.999	6.877	6.737	6.576	6.389	6.167	5.900	5.575	5.168	4.645	3.948	3.504	3.455	3.405
14600	7.564	7.500	7.429	7.351	7.265	7.168	7.059	6.935	6.794	6.631	6.442	6.217	5.948	5.619	5.207	4.678	3.973	3.524	3.474	3.424
14700	7.630	7.565	7.494	7.415	7.327	7.229	7.119	6.994	6.851	6.687	6.495	6.268	5.995	5.662	5.246	4.711	3.998	3.543	3.493	3.443
14800	7.696	7.630	7.558	7.478	7.390	7.291	7.179	7.053	6.909	6.742	6.548	6.318	6.043	5.706	5.285	4.744	4.022	3.563	3.513	3.461
14900	7.763	7.696	7.623	7.543	7.453	7.353	7.240	7.112	6.966	6.798	6.602	6.369	6.091	5.750	5.325	4.777	4.047	3.583	3.532	3.480
15000	7.829	7.762	7.688	7.607	7.516	7.415	7.301	7.172	7.024	6.854	6.655	6.421	6.139	5.794	5.364	4.810	4.073	3.603	3.551	3.499

$SIGMA = 6181.0$

$V-V_c = 12$

H	3.4	3.2	3.0	2.8	2.6	2.4	2.2	2.0	1.8	1.6	1.4	1.2	1.0	0.8	0.6	0.4	0.2	0.10	0.09	0.08
0	1.000	1.000	1.000	1.000	1.000	1.000	1.000	1.000	1.000	1.000	1.000	1.000	1.000	1.000	1.000	1.000	1.000	1.000	1.000	1.000
20	1.006	1.006	1.006	1.006	1.006	1.006	1.005	1.005	1.005	1.005	1.005	1.005	1.005	1.005	1.004	1.004	1.004	1.004	1.004	1.003
40	1.012	1.011	1.011	1.011	1.011	1.011	1.011	1.011	1.011	1.010	1.010	1.010	1.010	1.009	1.009	1.008	1.008	1.008	1.007	1.007
60	1.017	1.017	1.017	1.017	1.017	1.017	1.016	1.016	1.016	1.016	1.015	1.015	1.015	1.014	1.013	1.013	1.011	1.011	1.011	1.010
80	1.023	1.023	1.023	1.023	1.023	1.022	1.022	1.021	1.021	1.021	1.021	1.020	1.019	1.019	1.018	1.017	1.015	1.015	1.014	1.014
100	1.029	1.029	1.028	1.028	1.028	1.028	1.027	1.027	1.027	1.026	1.026	1.025	1.024	1.023	1.022	1.021	1.019	1.018	1.018	1.017
120	1.035	1.034	1.034	1.034	1.033	1.033	1.033	1.032	1.032	1.031	1.030	1.029	1.029	1.028	1.027	1.025	1.023	1.021	1.021	1.021
140	1.041	1.040	1.040	1.040	1.039	1.039	1.039	1.038	1.037	1.036	1.036	1.035	1.034	1.033	1.031	1.029	1.027	1.025	1.025	1.024
160	1.046	1.046	1.046	1.045	1.045	1.045	1.044	1.044	1.043	1.042	1.041	1.040	1.039	1.038	1.036	1.033	1.030	1.028	1.028	1.028
180	1.052	1.052	1.052	1.051	1.051	1.050	1.050	1.049	1.048	1.047	1.047	1.044	1.044	1.042	1.040	1.038	1.034	1.032	1.032	1.031
200	1.058	1.058	1.057	1.057	1.056	1.056	1.055	1.055	1.054	1.053	1.052	1.051	1.049	1.047	1.045	1.042	1.038	1.035	1.035	1.035
220	1.064	1.064	1.063	1.063	1.062	1.062	1.061	1.061	1.060	1.059	1.058	1.057	1.054	1.054	1.049	1.046	1.042	1.039	1.039	1.038
240	1.070	1.070	1.069	1.069	1.068	1.067	1.067	1.066	1.065	1.064	1.062	1.061	1.059	1.057	1.054	1.050	1.046	1.042	1.042	1.042
260	1.076	1.075	1.075	1.074	1.073	1.073	1.072	1.071	1.070	1.069	1.068	1.066	1.064	1.062	1.059	1.055	1.049	1.046	1.046	1.045
280	1.082	1.081	1.081	1.080	1.079	1.079	1.078	1.077	1.076	1.074	1.073	1.071	1.069	1.066	1.063	1.059	1.053	1.050	1.049	1.049
300	1.088	1.087	1.087	1.086	1.085	1.084	1.084	1.082	1.081	1.080	1.078	1.076	1.074	1.071	1.068	1.063	1.057	1.053	1.053	1.052
320	1.094	1.093	1.093	1.092	1.091	1.090	1.089	1.088	1.087	1.085	1.084	1.081	1.079	1.076	1.072	1.067	1.061	1.057	1.056	1.056
340	1.100	1.099	1.099	1.098	1.097	1.096	1.095	1.094	1.092	1.091	1.089	1.087	1.084	1.081	1.077	1.072	1.065	1.060	1.060	1.059
360	1.106	1.105	1.105	1.104	1.103	1.102	1.101	1.099	1.098	1.096	1.094	1.092	1.089	1.086	1.081	1.076	1.069	1.064	1.063	1.063
380	1.112	1.111	1.111	1.110	1.109	1.108	1.106	1.105	1.103	1.102	1.100	1.097	1.094	1.090	1.086	1.080	1.072	1.067	1.067	1.066
400	1.118	1.117	1.117	1.116	1.115	1.113	1.112	1.111	1.109	1.107	1.105	1.102	1.099	1.095	1.091	1.084	1.076	1.071	1.070	1.070
420	1.124	1.123	1.123	1.122	1.120	1.119	1.118	1.116	1.115	1.113	1.110	1.108	1.104	1.100	1.095	1.089	1.080	1.075	1.074	1.073
440	1.130	1.129	1.128	1.127	1.126	1.125	1.124	1.122	1.120	1.118	1.116	1.113	1.109	1.105	1.100	1.093	1.084	1.078	1.077	1.077
460	1.136	1.135	1.134	1.133	1.132	1.131	1.129	1.128	1.126	1.124	1.121	1.118	1.114	1.110	1.104	1.097	1.088	1.082	1.081	1.080
480	1.142	1.141	1.140	1.139	1.138	1.137	1.135	1.133	1.131	1.129	1.126	1.123	1.120	1.115	1.109	1.102	1.092	1.085	1.085	1.084
500	1.148	1.148	1.146	1.145	1.144	1.143	1.141	1.139	1.137	1.135	1.132	1.129	1.125	1.120	1.114	1.106	1.095	1.089	1.088	1.087
520	1.155	1.154	1.153	1.151	1.150	1.149	1.147	1.145	1.143	1.140	1.137	1.134	1.130	1.125	1.118	1.110	1.099	1.092	1.092	1.091
540	1.161	1.160	1.159	1.157	1.156	1.154	1.153	1.151	1.148	1.146	1.143	1.139	1.135	1.130	1.123	1.115	1.103	1.096	1.095	1.094
560	1.167	1.166	1.165	1.163	1.162	1.160	1.159	1.156	1.154	1.151	1.148	1.144	1.140	1.135	1.128	1.119	1.107	1.100	1.099	1.098
580	1.173	1.172	1.171	1.169	1.168	1.166	1.164	1.162	1.160	1.157	1.154	1.150	1.145	1.139	1.132	1.123	1.111	1.103	1.102	1.101
600	1.179	1.178	1.178	1.176	1.174	1.172	1.170	1.168	1.166	1.163	1.159	1.155	1.150	1.144	1.137	1.127	1.115	1.107	1.106	1.105
620	1.186	1.184	1.183	1.182	1.180	1.178	1.176	1.174	1.171	1.168	1.165	1.161	1.155	1.149	1.142	1.132	1.119	1.110	1.109	1.108
640	1.192	1.191	1.189	1.188	1.186	1.184	1.182	1.180	1.177	1.174	1.170	1.166	1.161	1.154	1.146	1.136	1.123	1.114	1.113	1.112
660	1.198	1.197	1.195	1.194	1.192	1.190	1.188	1.186	1.183	1.180	1.176	1.171	1.166	1.159	1.151	1.141	1.126	1.118	1.117	1.116
680	1.204	1.203	1.202	1.200	1.198	1.196	1.194	1.191	1.189	1.185	1.181	1.177	1.171	1.164	1.156	1.145	1.130	1.121	1.120	1.119
700	1.211	1.209	1.208	1.206	1.204	1.202	1.200	1.197	1.194	1.191	1.187	1.182	1.176	1.170	1.161	1.149	1.134	1.125	1.124	1.123
720	1.217	1.216	1.214	1.212	1.210	1.208	1.206	1.203	1.200	1.197	1.193	1.187	1.182	1.175	1.165	1.154	1.138	1.128	1.127	1.126
740	1.223	1.222	1.220	1.218	1.216	1.214	1.212	1.209	1.206	1.202	1.198	1.193	1.187	1.179	1.170	1.158	1.142	1.132	1.131	1.130
760	1.230	1.228	1.227	1.225	1.223	1.220	1.218	1.215	1.212	1.208	1.204	1.198	1.192	1.184	1.175	1.162	1.146	1.136	1.134	1.133
780	1.236	1.234	1.233	1.231	1.229	1.227	1.224	1.221	1.218	1.214	1.209	1.204	1.197	1.189	1.179	1.167	1.150	1.139	1.138	1.137
800	1.242	1.241	1.239	1.237	1.235	1.233	1.230	1.227	1.223	1.219	1.215	1.209	1.203	1.194	1.184	1.171	1.154	1.143	1.141	1.140

x

SIGMA = 6181.0

V−V$_c$ = 12

H	0.08	0.09	0.10	0.2	0.4	0.6	0.8	1.0	1.2	1.4	1.6	1.8	2.0	2.2	2.4	2.6	2.8	3.0	3.2	3.4
820	1.144	1.145	1.146	1.158	1.176	1.189	1.199	1.208	1.215	1.220	1.225	1.229	1.233	1.236	1.239	1.241	1.243	1.245	1.247	1.249
840	1.147	1.149	1.150	1.162	1.180	1.194	1.205	1.213	1.220	1.226	1.231	1.235	1.239	1.242	1.245	1.247	1.249	1.252	1.254	1.255
860	1.151	1.152	1.154	1.166	1.184	1.199	1.210	1.218	1.226	1.232	1.237	1.241	1.245	1.248	1.251	1.254	1.256	1.258	1.260	1.262
880	1.154	1.156	1.157	1.169	1.189	1.203	1.215	1.224	1.231	1.237	1.242	1.247	1.251	1.254	1.257	1.260	1.262	1.264	1.266	1.268
900	1.158	1.159	1.161	1.173	1.193	1.208	1.220	1.229	1.237	1.243	1.248	1.253	1.257	1.260	1.263	1.266	1.269	1.271	1.273	1.275
920	1.162	1.163	1.164	1.177	1.198	1.213	1.225	1.234	1.242	1.249	1.254	1.259	1.263	1.266	1.270	1.272	1.275	1.277	1.279	1.281
940	1.165	1.167	1.168	1.181	1.202	1.218	1.230	1.240	1.248	1.254	1.260	1.265	1.269	1.273	1.276	1.279	1.281	1.283	1.286	1.287
960	1.169	1.170	1.172	1.185	1.207	1.223	1.235	1.245	1.253	1.260	1.266	1.271	1.275	1.279	1.282	1.285	1.288	1.290	1.292	1.294
980	1.172	1.174	1.175	1.189	1.211	1.227	1.240	1.250	1.259	1.266	1.272	1.277	1.281	1.285	1.288	1.291	1.294	1.296	1.299	1.300
1000	1.176	1.177	1.179	1.193	1.215	1.232	1.245	1.256	1.264	1.271	1.277	1.283	1.287	1.291	1.294	1.298	1.300	1.303	1.305	1.307
1020	1.179	1.181	1.182	1.197	1.220	1.237	1.250	1.261	1.270	1.277	1.283	1.289	1.293	1.297	1.301	1.304	1.307	1.309	1.312	1.314
1040	1.183	1.184	1.186	1.201	1.224	1.242	1.256	1.267	1.275	1.283	1.289	1.295	1.299	1.303	1.307	1.310	1.313	1.316	1.318	1.320
1060	1.186	1.188	1.190	1.205	1.229	1.247	1.261	1.272	1.281	1.289	1.295	1.301	1.305	1.310	1.313	1.317	1.320	1.322	1.325	1.327
1080	1.190	1.192	1.195	1.209	1.233	1.252	1.266	1.277	1.287	1.294	1.301	1.307	1.312	1.316	1.320	1.323	1.326	1.329	1.331	1.333
1100	1.193	1.195	1.197	1.213	1.238	1.257	1.271	1.283	1.292	1.300	1.307	1.313	1.318	1.322	1.326	1.329	1.332	1.335	1.338	1.340
1120	1.197	1.199	1.201	1.217	1.242	1.261	1.276	1.288	1.298	1.306	1.313	1.319	1.324	1.328	1.332	1.336	1.339	1.342	1.344	1.347
1140	1.201	1.202	1.204	1.221	1.247	1.266	1.282	1.294	1.304	1.312	1.319	1.325	1.330	1.335	1.339	1.342	1.345	1.348	1.351	1.353
1160	1.204	1.206	1.208	1.225	1.251	1.271	1.287	1.299	1.309	1.318	1.325	1.331	1.336	1.341	1.345	1.349	1.352	1.355	1.357	1.360
1180	1.208	1.210	1.212	1.229	1.256	1.276	1.292	1.305	1.315	1.324	1.331	1.337	1.342	1.347	1.351	1.355	1.358	1.361	1.364	1.367
1200	1.211	1.213	1.215	1.233	1.260	1.281	1.297	1.310	1.321	1.329	1.337	1.343	1.349	1.354	1.358	1.362	1.365	1.368	1.371	1.373
1220	1.215	1.217	1.219	1.237	1.265	1.286	1.302	1.316	1.326	1.335	1.343	1.349	1.355	1.360	1.364	1.368	1.372	1.375	1.377	1.380
1240	1.218	1.221	1.223	1.241	1.269	1.291	1.308	1.321	1.332	1.341	1.349	1.355	1.361	1.366	1.371	1.375	1.378	1.381	1.384	1.387
1260	1.222	1.224	1.226	1.245	1.274	1.296	1.313	1.327	1.338	1.347	1.355	1.361	1.367	1.373	1.377	1.381	1.385	1.388	1.391	1.393
1280	1.226	1.228	1.230	1.249	1.279	1.301	1.318	1.332	1.343	1.353	1.361	1.368	1.373	1.379	1.384	1.388	1.391	1.395	1.398	1.400
1300	1.229	1.231	1.233	1.253	1.283	1.306	1.323	1.338	1.349	1.359	1.367	1.374	1.380	1.385	1.390	1.394	1.398	1.401	1.404	1.407
1320	1.233	1.235	1.237	1.257	1.288	1.311	1.329	1.343	1.355	1.365	1.373	1.380	1.386	1.392	1.396	1.401	1.405	1.408	1.411	1.414
1340	1.236	1.239	1.241	1.261	1.292	1.316	1.334	1.349	1.361	1.371	1.379	1.386	1.393	1.398	1.403	1.407	1.411	1.415	1.418	1.421
1360	1.240	1.242	1.244	1.265	1.297	1.321	1.339	1.354	1.366	1.377	1.385	1.393	1.399	1.405	1.410	1.414	1.418	1.421	1.425	1.427
1380	1.243	1.246	1.248	1.269	1.301	1.326	1.345	1.360	1.372	1.383	1.391	1.399	1.405	1.411	1.416	1.421	1.425	1.428	1.431	1.434
1400	1.247	1.249	1.252	1.273	1.306	1.331	1.350	1.365	1.378	1.389	1.397	1.405	1.412	1.417	1.423	1.427	1.431	1.435	1.438	1.441
1420	1.251	1.253	1.255	1.277	1.310	1.336	1.355	1.371	1.384	1.395	1.404	1.411	1.418	1.424	1.429	1.434	1.438	1.442	1.445	1.448
1440	1.254	1.257	1.259	1.281	1.315	1.341	1.361	1.377	1.390	1.401	1.410	1.418	1.424	1.430	1.436	1.440	1.445	1.448	1.452	1.455
1460	1.258	1.260	1.263	1.285	1.320	1.346	1.366	1.382	1.395	1.407	1.416	1.424	1.431	1.437	1.442	1.447	1.451	1.455	1.459	1.462
1480	1.262	1.264	1.266	1.289	1.324	1.351	1.371	1.388	1.401	1.413	1.422	1.430	1.437	1.443	1.449	1.454	1.458	1.462	1.466	1.469
1500	1.265	1.268	1.270	1.293	1.329	1.356	1.377	1.393	1.407	1.419	1.428	1.437	1.444	1.450	1.456	1.460	1.465	1.469	1.472	1.476
1520	1.269	1.271	1.274	1.297	1.333	1.361	1.382	1.399	1.413	1.425	1.434	1.443	1.450	1.457	1.462	1.467	1.472	1.476	1.479	1.483
1540	1.272	1.275	1.277	1.301	1.338	1.366	1.387	1.405	1.419	1.431	1.441	1.450	1.457	1.463	1.469	1.474	1.478	1.483	1.486	1.490
1560	1.276	1.278	1.281	1.305	1.343	1.371	1.393	1.410	1.425	1.437	1.447	1.456	1.463	1.470	1.476	1.481	1.485	1.489	1.493	1.497
1580	1.279	1.282	1.285	1.309	1.347	1.376	1.398	1.416	1.431	1.443	1.453	1.463	1.470	1.476	1.482	1.487	1.492	1.496	1.500	1.504
1600	1.283	1.286	1.288	1.313	1.352	1.381	1.404	1.422	1.437	1.449	1.459	1.468	1.476	1.483	1.489	1.494	1.499	1.503	1.507	1.511
1620	1.287	1.289	1.292	1.317	1.357	1.386	1.409	1.427	1.443	1.455	1.466	1.475	1.483	1.490	1.496	1.501	1.506	1.510	1.514	1.518
1640	1.290	1.293	1.296	1.321	1.361	1.391	1.415	1.433	1.448	1.461	1.472	1.481	1.489	1.496	1.502	1.508	1.513	1.517	1.521	1.525

V-V_c = 12

SIGMA = 6181.0

H	3.4	3.2	3.0	2.8	2.6	2.4	2.2	2.0	1.8	1.6	1.4	1.2	1.0	0.8	0.6	0.4	0.2	0.10	0.09	0.08
1660	1.532	1.528	1.524	1.520	1.515	1.509	1.503	1.496	1.488	1.478	1.467	1.454	1.439	1.420	1.396	1.366	1.325	1.300	1.297	1.294
1680	1.539	1.535	1.531	1.527	1.521	1.516	1.509	1.502	1.494	1.485	1.473	1.460	1.445	1.425	1.401	1.371	1.329	1.303	1.300	1.297
1700	1.546	1.542	1.538	1.533	1.528	1.523	1.516	1.509	1.500	1.491	1.480	1.466	1.450	1.431	1.407	1.375	1.333	1.307	1.304	1.301
1720	1.553	1.549	1.545	1.540	1.535	1.529	1.523	1.515	1.507	1.497	1.486	1.472	1.456	1.436	1.412	1.380	1.338	1.311	1.308	1.305
1740	1.560	1.556	1.552	1.547	1.542	1.536	1.530	1.522	1.513	1.504	1.492	1.478	1.462	1.442	1.417	1.385	1.342	1.314	1.311	1.308
1760	1.567	1.563	1.559	1.554	1.549	1.543	1.536	1.529	1.520	1.510	1.498	1.484	1.468	1.447	1.422	1.389	1.346	1.318	1.315	1.312
1780	1.575	1.571	1.566	1.561	1.556	1.550	1.543	1.535	1.526	1.516	1.504	1.490	1.473	1.453	1.427	1.394	1.350	1.322	1.319	1.315
1800	1.582	1.578	1.573	1.568	1.563	1.557	1.550	1.542	1.533	1.523	1.511	1.496	1.479	1.458	1.432	1.399	1.354	1.325	1.322	1.319
1820	1.589	1.585	1.580	1.575	1.570	1.563	1.556	1.549	1.539	1.529	1.517	1.502	1.485	1.464	1.437	1.403	1.358	1.329	1.326	1.323
1840	1.596	1.592	1.587	1.582	1.577	1.570	1.563	1.555	1.546	1.535	1.523	1.508	1.491	1.469	1.443	1.408	1.362	1.333	1.330	1.326
1860	1.603	1.599	1.595	1.589	1.584	1.577	1.570	1.562	1.553	1.542	1.529	1.514	1.497	1.475	1.448	1.413	1.366	1.337	1.333	1.330
1880	1.611	1.606	1.602	1.596	1.591	1.584	1.577	1.569	1.559	1.548	1.536	1.521	1.502	1.480	1.453	1.417	1.370	1.340	1.337	1.334
1900	1.618	1.614	1.609	1.604	1.598	1.591	1.584	1.575	1.566	1.555	1.542	1.527	1.508	1.486	1.458	1.422	1.374	1.344	1.340	1.337
1920	1.625	1.621	1.616	1.611	1.605	1.598	1.591	1.582	1.572	1.561	1.548	1.533	1.514	1.492	1.463	1.427	1.378	1.348	1.344	1.341
1940	1.633	1.628	1.623	1.618	1.612	1.605	1.598	1.589	1.579	1.568	1.554	1.539	1.520	1.497	1.469	1.432	1.383	1.351	1.348	1.344
1960	1.640	1.635	1.630	1.625	1.619	1.612	1.604	1.596	1.586	1.574	1.561	1.545	1.526	1.503	1.474	1.436	1.387	1.355	1.352	1.348
1980	1.647	1.643	1.638	1.632	1.626	1.619	1.611	1.602	1.592	1.581	1.567	1.551	1.532	1.508	1.479	1.441	1.391	1.359	1.355	1.352
2000	1.655	1.650	1.645	1.639	1.633	1.626	1.618	1.609	1.599	1.587	1.573	1.557	1.538	1.514	1.484	1.446	1.395	1.363	1.359	1.355
2020	1.662	1.657	1.652	1.646	1.640	1.633	1.625	1.616	1.606	1.594	1.580	1.563	1.544	1.520	1.489	1.451	1.399	1.366	1.363	1.359
2040	1.669	1.664	1.659	1.653	1.647	1.640	1.632	1.623	1.612	1.600	1.586	1.569	1.550	1.525	1.495	1.455	1.403	1.370	1.366	1.363
2060	1.677	1.672	1.667	1.661	1.654	1.647	1.639	1.630	1.619	1.607	1.592	1.576	1.555	1.531	1.500	1.460	1.407	1.374	1.370	1.366
2080	1.684	1.679	1.674	1.668	1.661	1.654	1.646	1.636	1.626	1.613	1.599	1.582	1.561	1.536	1.505	1.465	1.411	1.377	1.374	1.370
2100	1.691	1.687	1.681	1.675	1.668	1.661	1.653	1.643	1.632	1.620	1.605	1.588	1.567	1.542	1.510	1.470	1.416	1.381	1.377	1.373
2120	1.699	1.694	1.688	1.682	1.676	1.668	1.660	1.650	1.639	1.626	1.612	1.594	1.573	1.548	1.516	1.475	1.420	1.385	1.381	1.377
2140	1.706	1.701	1.696	1.690	1.683	1.675	1.667	1.657	1.646	1.633	1.618	1.600	1.579	1.553	1.521	1.479	1.424	1.389	1.385	1.381
2160	1.714	1.709	1.703	1.697	1.690	1.682	1.674	1.664	1.653	1.640	1.625	1.607	1.585	1.559	1.526	1.484	1.428	1.392	1.388	1.384
2180	1.721	1.716	1.711	1.704	1.697	1.689	1.681	1.671	1.660	1.646	1.631	1.613	1.591	1.565	1.532	1.489	1.432	1.396	1.392	1.388
2200	1.729	1.724	1.718	1.712	1.704	1.697	1.688	1.678	1.666	1.653	1.637	1.619	1.597	1.570	1.537	1.494	1.436	1.400	1.396	1.392
2220	1.736	1.731	1.725	1.719	1.712	1.704	1.695	1.685	1.673	1.660	1.644	1.625	1.603	1.576	1.542	1.499	1.441	1.404	1.399	1.395
2240	1.744	1.739	1.733	1.726	1.719	1.711	1.702	1.692	1.680	1.666	1.650	1.632	1.609	1.582	1.548	1.503	1.445	1.407	1.403	1.399
2260	1.751	1.746	1.740	1.734	1.726	1.718	1.709	1.699	1.687	1.673	1.657	1.638	1.615	1.588	1.553	1.508	1.449	1.411	1.407	1.403
2280	1.759	1.754	1.748	1.741	1.734	1.725	1.716	1.705	1.693	1.680	1.663	1.644	1.621	1.593	1.558	1.513	1.453	1.415	1.411	1.406
2300	1.767	1.761	1.755	1.748	1.741	1.733	1.723	1.712	1.700	1.686	1.670	1.651	1.627	1.599	1.564	1.518	1.457	1.419	1.414	1.410
2320	1.774	1.769	1.763	1.756	1.748	1.740	1.730	1.719	1.707	1.693	1.676	1.657	1.633	1.605	1.569	1.523	1.461	1.422	1.418	1.414
2340	1.782	1.776	1.770	1.763	1.756	1.747	1.737	1.727	1.714	1.700	1.683	1.663	1.640	1.611	1.574	1.528	1.466	1.426	1.422	1.417
2360	1.790	1.784	1.778	1.771	1.763	1.754	1.745	1.734	1.721	1.706	1.690	1.670	1.646	1.616	1.580	1.533	1.470	1.430	1.425	1.421
2380	1.797	1.791	1.785	1.778	1.770	1.762	1.752	1.741	1.728	1.713	1.696	1.676	1.652	1.622	1.585	1.537	1.474	1.434	1.429	1.425
2400	1.805	1.799	1.793	1.785	1.778	1.769	1.759	1.748	1.735	1.720	1.703	1.682	1.658	1.628	1.590	1.542	1.478	1.437	1.433	1.428
2420	1.813	1.807	1.800	1.793	1.785	1.776	1.766	1.755	1.742	1.727	1.709	1.689	1.664	1.634	1.596	1.547	1.482	1.441	1.437	1.432
2440	1.820	1.814	1.808	1.800	1.792	1.783	1.773	1.762	1.749	1.734	1.716	1.695	1.670	1.639	1.601	1.552	1.487	1.445	1.440	1.436
2460	1.828	1.822	1.815	1.808	1.800	1.791	1.781	1.769	1.756	1.740	1.723	1.701	1.676	1.645	1.607	1.557	1.491	1.449	1.444	1.439
2480	1.836	1.830	1.823	1.815	1.807	1.798	1.788	1.776	1.763	1.747	1.729	1.708	1.682	1.651	1.612	1.562	1.495	1.452	1.448	1.443

x

SIGMA = 6181.0

$V - V_c = 12$

H \ x	3.4	3.2	3.0	2.8	2.6	2.4	2.2	2.0	1.8	1.6	1.4	1.2	1.0	0.8	0.6	0.4	0.2	0.10	0.09	0.08
2500	1.843	1.837	1.831	1.823	1.815	1.805	1.795	1.783	1.770	1.754	1.736	1.714	1.688	1.657	1.617	1.567	1.499	1.456	1.451	1.447
2520	1.851	1.845	1.838	1.831	1.822	1.813	1.802	1.790	1.777	1.761	1.742	1.721	1.695	1.663	1.623	1.572	1.503	1.460	1.455	1.450
2540	1.859	1.853	1.846	1.838	1.830	1.820	1.810	1.797	1.784	1.768	1.749	1.727	1.701	1.669	1.628	1.577	1.508	1.464	1.459	1.454
2560	1.867	1.860	1.853	1.846	1.837	1.828	1.817	1.805	1.791	1.775	1.756	1.734	1.707	1.674	1.634	1.582	1.512	1.467	1.463	1.458
2580	1.875	1.868	1.861	1.853	1.845	1.835	1.824	1.812	1.798	1.781	1.763	1.740	1.713	1.680	1.639	1.586	1.516	1.471	1.466	1.461
2600	1.882	1.876	1.869	1.861	1.852	1.842	1.831	1.819	1.805	1.788	1.769	1.747	1.719	1.686	1.645	1.591	1.520	1.475	1.470	1.465
2620	1.890	1.884	1.877	1.869	1.860	1.850	1.839	1.826	1.812	1.795	1.776	1.753	1.726	1.692	1.650	1.596	1.524	1.479	1.474	1.469
2640	1.898	1.892	1.884	1.876	1.867	1.857	1.846	1.833	1.819	1.802	1.783	1.760	1.732	1.698	1.656	1.601	1.529	1.483	1.477	1.472
2660	1.906	1.899	1.892	1.884	1.875	1.865	1.854	1.841	1.826	1.809	1.789	1.766	1.738	1.704	1.661	1.606	1.533	1.486	1.481	1.476
2680	1.914	1.907	1.900	1.892	1.883	1.872	1.861	1.848	1.833	1.816	1.796	1.773	1.744	1.710	1.667	1.611	1.537	1.490	1.485	1.480
2700	1.922	1.915	1.908	1.899	1.890	1.880	1.868	1.855	1.840	1.823	1.803	1.779	1.751	1.716	1.672	1.616	1.541	1.494	1.489	1.483
2720	1.930	1.923	1.915	1.907	1.898	1.887	1.876	1.863	1.847	1.830	1.810	1.786	1.757	1.722	1.678	1.621	1.546	1.498	1.492	1.487
2740	1.938	1.931	1.923	1.915	1.905	1.895	1.883	1.870	1.855	1.837	1.817	1.792	1.763	1.728	1.683	1.626	1.550	1.501	1.496	1.491
2760	1.946	1.939	1.931	1.922	1.913	1.902	1.891	1.877	1.862	1.844	1.823	1.799	1.769	1.734	1.689	1.631	1.554	1.505	1.500	1.494
2780	1.954	1.947	1.939	1.930	1.921	1.910	1.898	1.884	1.869	1.851	1.830	1.805	1.776	1.740	1.694	1.636	1.558	1.509	1.504	1.498
2800	1.962	1.954	1.947	1.938	1.928	1.918	1.906	1.892	1.876	1.858	1.837	1.812	1.782	1.746	1.700	1.641	1.563	1.513	1.507	1.502
2820	1.970	1.962	1.955	1.946	1.936	1.925	1.913	1.899	1.883	1.865	1.844	1.819	1.788	1.752	1.705	1.646	1.567	1.517	1.511	1.505
2840	1.978	1.970	1.962	1.954	1.944	1.933	1.921	1.907	1.891	1.872	1.851	1.825	1.795	1.758	1.711	1.651	1.571	1.520	1.515	1.509
2860	1.986	1.978	1.970	1.961	1.952	1.940	1.928	1.914	1.898	1.879	1.858	1.832	1.801	1.764	1.717	1.656	1.576	1.524	1.519	1.513
2880	1.994	1.986	1.978	1.969	1.959	1.948	1.936	1.921	1.905	1.886	1.864	1.839	1.807	1.770	1.722	1.661	1.580	1.528	1.522	1.517
2900	2.002	1.994	1.986	1.977	1.967	1.956	1.943	1.929	1.912	1.893	1.871	1.845	1.814	1.776	1.728	1.666	1.584	1.532	1.526	1.520
2920	2.010	2.002	1.994	1.985	1.975	1.963	1.951	1.936	1.920	1.900	1.878	1.852	1.820	1.782	1.733	1.671	1.588	1.536	1.530	1.524
2940	2.018	2.010	2.002	1.993	1.983	1.971	1.958	1.944	1.927	1.908	1.885	1.859	1.827	1.788	1.739	1.676	1.593	1.539	1.534	1.528
2960	2.026	2.018	2.010	2.001	1.990	1.979	1.966	1.951	1.934	1.915	1.892	1.865	1.833	1.794	1.744	1.681	1.597	1.543	1.537	1.531
2980	2.034	2.027	2.018	2.009	1.998	1.987	1.973	1.958	1.941	1.922	1.899	1.872	1.839	1.800	1.750	1.686	1.601	1.547	1.541	1.535
3000	2.042	2.035	2.026	2.017	2.006	1.994	1.981	1.966	1.949	1.929	1.906	1.879	1.846	1.806	1.756	1.691	1.606	1.551	1.545	1.539

SIGMA = 6181.0

V−V_c = 12

H x	0.08	0.09	0.10	0.2	0.4	0.6	0.8	1.0	1.2	1.4	1.6	1.8	2.0	2.2	2.4	2.6	2.8	3.0	3.2	3.4
3100	1.557	1.564	1.570	1.627	1.717	1.784	1.836	1.878	1.912	1.941	1.965	1.986	2.004	2.019	2.033	2.045	2.056	2.066	2.075	2.083
3200	1.576	1.583	1.589	1.649	1.742	1.812	1.867	1.911	1.946	1.976	2.001	2.023	2.042	2.058	2.072	2.085	2.097	2.107	2.116	2.125
3300	1.595	1.602	1.608	1.670	1.768	1.841	1.898	1.943	1.981	2.012	2.038	2.060	2.080	2.097	2.112	2.125	2.137	2.148	2.158	2.167
3400	1.613	1.620	1.628	1.692	1.794	1.870	1.929	1.976	2.015	2.047	2.075	2.098	2.119	2.136	2.152	2.166	2.178	2.190	2.200	2.209
3500	1.632	1.639	1.647	1.714	1.820	1.899	1.960	2.010	2.050	2.084	2.112	2.136	2.158	2.176	2.192	2.207	2.220	2.231	2.242	2.252
3600	1.651	1.659	1.666	1.736	1.846	1.928	1.992	2.043	2.085	2.120	2.150	2.175	2.197	2.216	2.233	2.248	2.262	2.274	2.285	2.295
3700	1.669	1.678	1.686	1.758	1.872	1.957	2.024	2.077	2.121	2.157	2.187	2.214	2.237	2.257	2.274	2.290	2.304	2.316	2.328	2.338
3800	1.688	1.697	1.705	1.780	1.898	1.987	2.056	2.111	2.156	2.194	2.226	2.253	2.277	2.297	2.316	2.332	2.346	2.359	2.371	2.382
3900	1.707	1.716	1.725	1.802	1.925	2.017	2.088	2.146	2.192	2.231	2.264	2.293	2.317	2.338	2.357	2.374	2.389	2.403	2.415	2.426
4000	1.726	1.735	1.744	1.825	1.952	2.047	2.121	2.180	2.229	2.269	2.303	2.332	2.358	2.380	2.400	2.417	2.433	2.447	2.459	2.471
4100	1.745	1.754	1.764	1.847	1.979	2.077	2.154	2.215	2.265	2.307	2.342	2.373	2.399	2.422	2.442	2.460	2.476	2.491	2.504	2.516
4200	1.764	1.774	1.783	1.870	2.006	2.107	2.187	2.250	2.302	2.345	2.382	2.413	2.440	2.464	2.485	2.504	2.520	2.535	2.549	2.561
4300	1.783	1.793	1.803	1.892	2.033	2.138	2.220	2.286	2.339	2.384	2.422	2.454	2.482	2.507	2.528	2.548	2.565	2.580	2.594	2.607
4400	1.802	1.812	1.823	1.915	2.060	2.169	2.253	2.321	2.377	2.423	2.462	2.495	2.524	2.550	2.572	2.592	2.610	2.626	2.640	2.654
4500	1.821	1.832	1.842	1.938	2.087	2.200	2.287	2.357	2.414	2.462	2.502	2.537	2.567	2.593	2.616	2.637	2.655	2.672	2.687	2.700
4600	1.840	1.851	1.862	1.961	2.115	2.231	2.321	2.393	2.452	2.501	2.543	2.579	2.610	2.637	2.661	2.682	2.701	2.718	2.733	2.747
4700	1.860	1.871	1.882	1.983	2.143	2.262	2.355	2.430	2.491	2.541	2.584	2.621	2.653	2.681	2.705	2.727	2.747	2.764	2.780	2.795
4800	1.879	1.890	1.902	2.007	2.171	2.294	2.390	2.466	2.529	2.581	2.626	2.663	2.696	2.725	2.750	2.773	2.793	2.811	2.828	2.843
4900	1.898	1.910	1.922	2.030	2.199	2.326	2.424	2.503	2.568	2.622	2.667	2.706	2.740	2.770	2.796	2.819	2.840	2.859	2.876	2.891
5000	1.917	1.930	1.942	2.053	2.227	2.358	2.459	2.541	2.607	2.663	2.709	2.750	2.784	2.815	2.842	2.866	2.887	2.906	2.924	2.940
5100	1.937	1.949	1.962	2.076	2.255	2.390	2.494	2.578	2.647	2.704	2.752	2.793	2.829	2.860	2.888	2.913	2.935	2.955	2.972	2.989
5200	1.956	1.969	1.982	2.099	2.284	2.422	2.530	2.616	2.686	2.745	2.795	2.837	2.874	2.906	2.935	2.960	2.983	3.003	3.022	3.038
5300	1.975	1.989	2.002	2.123	2.313	2.455	2.565	2.654	2.726	2.787	2.838	2.881	2.919	2.952	2.982	3.008	3.031	3.052	3.071	3.088
5400	1.995	2.009	2.022	2.146	2.341	2.487	2.601	2.692	2.766	2.829	2.881	2.926	2.965	2.999	3.029	3.056	3.080	3.101	3.121	3.139
5500	2.014	2.029	2.043	2.170	2.370	2.520	2.637	2.731	2.807	2.871	2.925	2.971	3.011	3.046	3.077	3.104	3.129	3.151	3.171	3.189
5600	2.034	2.049	2.063	2.194	2.399	2.554	2.673	2.769	2.848	2.913	2.969	3.016	3.057	3.093	3.125	3.153	3.178	3.201	3.222	3.240
5700	2.053	2.069	2.083	2.218	2.429	2.587	2.710	2.808	2.889	2.956	3.013	3.062	3.104	3.141	3.173	3.202	3.228	3.252	3.273	3.292
5800	2.073	2.089	2.104	2.242	2.458	2.620	2.747	2.848	2.930	2.999	3.058	3.108	3.151	3.189	3.222	3.252	3.279	3.303	3.324	3.344
5900	2.093	2.109	2.124	2.265	2.488	2.654	2.784	2.887	2.972	3.043	3.103	3.154	3.198	3.237	3.271	3.302	3.329	3.354	3.376	3.396
6000	2.112	2.129	2.145	2.290	2.517	2.688	2.821	2.927	3.014	3.087	3.148	3.201	3.246	3.286	3.321	3.352	3.380	3.405	3.428	3.449
6100	2.132	2.149	2.165	2.314	2.547	2.722	2.858	2.967	3.056	3.131	3.194	3.247	3.294	3.335	3.371	3.403	3.432	3.458	3.481	3.502
6200	2.152	2.169	2.186	2.338	2.577	2.757	2.896	3.008	3.099	3.175	3.240	3.295	3.343	3.384	3.421	3.454	3.484	3.510	3.534	3.556
6300	2.172	2.189	2.206	2.362	2.607	2.791	2.934	3.048	3.142	3.220	3.286	3.342	3.391	3.434	3.472	3.506	3.536	3.563	3.587	3.610
6400	2.192	2.209	2.227	2.387	2.638	2.826	2.972	3.089	3.185	3.265	3.332	3.390	3.440	3.484	3.523	3.558	3.588	3.616	3.641	3.664
6500	2.211	2.230	2.248	2.411	2.668	2.861	3.010	3.130	3.228	3.310	3.379	3.439	3.490	3.535	3.575	3.610	3.641	3.670	3.695	3.719
6600	2.231	2.250	2.269	2.436	2.699	2.896	3.049	3.172	3.272	3.356	3.427	3.487	3.540	3.586	3.626	3.662	3.695	3.724	3.750	3.774
6700	2.251	2.270	2.289	2.460	2.729	2.931	3.088	3.213	3.316	3.402	3.474	3.536	3.590	3.637	3.679	3.715	3.748	3.778	3.805	3.830
6800	2.271	2.291	2.310	2.485	2.760	2.967	3.127	3.255	3.360	3.448	3.522	3.585	3.640	3.689	3.731	3.769	3.803	3.833	3.861	3.886
6900	2.291	2.311	2.331	2.510	2.791	3.002	3.166	3.298	3.405	3.494	3.570	3.635	3.691	3.741	3.784	3.823	3.857	3.888	3.916	3.942
7000	2.311	2.332	2.352	2.535	2.823	3.038	3.206	3.340	3.450	3.541	3.619	3.685	3.743	3.793	3.837	3.877	3.912	3.944	3.973	3.999
7100	2.332	2.352	2.373	2.560	2.854	3.074	3.246	3.383	3.495	3.588	3.668	3.735	3.794	3.846	3.891	3.931	3.967	4.000	4.029	4.056
7200	2.352	2.373	2.394	2.585	2.885	3.111	3.286	3.426	3.540	3.636	3.717	3.786	3.846	3.899	3.945	3.986	4.023	4.056	4.086	4.113

SIGMA = 6181.0

V-V_c = 12

H \ x	0.08	0.09	0.10	0.2	0.4	0.6	0.8	1.0	1.2	1.4	1.6	1.8	2.0	2.2	2.4	2.6	2.8	3.0	3.2	3.4
7300	2.372	2.394	2.415	2.610	2.917	3.147	3.326	3.469	3.586	3.684	3.766	3.837	3.898	3.952	3.999	4.041	4.079	4.113	4.144	4.171
7400	2.392	2.414	2.436	2.636	2.949	3.184	3.366	3.512	3.632	3.732	3.816	3.888	3.951	4.006	4.054	4.097	4.136	4.170	4.201	4.230
7500	2.412	2.435	2.458	2.661	2.981	3.221	3.407	3.556	3.678	3.780	3.866	3.940	4.004	4.060	4.109	4.153	4.192	4.228	4.260	4.289
7600	2.433	2.456	2.479	2.686	3.013	3.258	3.448	3.600	3.725	3.829	3.917	3.992	4.057	4.114	4.165	4.209	4.250	4.286	4.318	4.348
7700	2.453	2.477	2.500	2.712	3.045	3.295	3.489	3.645	3.772	3.878	3.967	4.044	4.111	4.169	4.221	4.266	4.307	4.344	4.377	4.408
7800	2.473	2.498	2.521	2.738	3.077	3.332	3.531	3.689	3.819	3.927	4.018	4.097	4.165	4.224	4.277	4.323	4.365	4.403	4.437	4.468
7900	2.494	2.519	2.543	2.763	3.110	3.370	3.572	3.734	3.866	3.977	4.070	4.150	4.219	4.280	4.333	4.381	4.424	4.462	4.497	4.528
8000	2.514	2.539	2.564	2.789	3.143	3.408	3.614	3.779	3.914	4.026	4.122	4.203	4.274	4.336	4.390	4.439	4.482	4.521	4.557	4.589
8100	2.535	2.560	2.586	2.815	3.176	3.446	3.656	3.824	3.962	4.077	4.174	4.257	4.329	4.392	4.448	4.497	4.541	4.581	4.617	4.650
8200	2.555	2.582	2.607	2.841	3.209	3.484	3.698	3.870	4.010	4.127	4.226	4.311	4.384	4.449	4.505	4.556	4.601	4.642	4.678	4.712
8300	2.576	2.603	2.629	2.867	3.242	3.523	3.741	3.916	4.059	4.178	4.279	4.365	4.440	4.506	4.564	4.615	4.661	4.702	4.740	4.774
8400	2.596	2.624	2.650	2.893	3.275	3.561	3.784	3.962	4.108	4.229	4.332	4.420	4.496	4.563	4.622	4.674	4.721	4.763	4.802	4.836
8500	2.617	2.645	2.672	2.920	3.308	3.600	3.827	4.008	4.157	4.281	4.385	4.475	4.553	4.621	4.681	4.734	4.782	4.825	4.864	4.899
8600	2.638	2.666	2.694	2.946	3.342	3.639	3.870	4.055	4.206	4.332	4.439	4.530	4.610	4.679	4.740	4.794	4.843	4.887	4.926	4.962
8700	2.659	2.687	2.716	2.972	3.376	3.678	3.914	4.102	4.256	4.384	4.493	4.586	4.667	4.737	4.800	4.855	4.904	4.949	4.989	5.026
8800	2.679	2.709	2.737	2.999	3.410	3.718	3.957	4.149	4.306	4.437	4.547	4.642	4.724	4.796	4.859	4.916	4.966	5.012	5.053	5.090
8900	2.700	2.730	2.759	3.025	3.444	3.757	4.001	4.196	4.356	4.489	4.602	4.698	4.782	4.855	4.920	4.977	5.029	5.075	5.115	5.155
9000	2.721	2.751	2.781	3.052	3.478	3.797	4.046	4.244	4.407	4.542	4.657	4.755	4.840	4.915	4.980	5.039	5.091	5.138	5.181	5.220
9100	2.742	2.773	2.803	3.079	3.512	3.837	4.090	4.292	4.458	4.595	4.712	4.812	4.899	4.975	5.042	5.101	5.154	5.202	5.245	5.285
9200	2.763	2.794	2.825	3.106	3.547	3.877	4.135	4.340	4.509	4.649	4.768	4.869	4.958	5.035	5.103	5.163	5.218	5.266	5.310	5.351
9300	2.784	2.816	2.847	3.133	3.581	3.918	4.179	4.389	4.560	4.703	4.824	4.927	5.017	5.095	5.165	5.226	5.281	5.331	5.376	5.417
9400	2.805	2.837	2.869	3.160	3.616	3.958	4.225	4.438	4.612	4.757	4.880	4.985	5.077	5.156	5.227	5.289	5.346	5.396	5.442	5.483
9500	2.826	2.859	2.891	3.187	3.651	3.999	4.270	4.487	4.664	4.812	4.936	5.044	5.136	5.218	5.289	5.353	5.410	5.461	5.508	5.550
9600	2.847	2.881	2.914	3.214	3.686	4.040	4.316	4.536	4.716	4.866	4.993	5.102	5.197	5.279	5.352	5.417	5.475	5.527	5.574	5.617
9700	2.868	2.902	2.936	3.241	3.721	4.081	4.361	4.585	4.769	4.921	5.051	5.161	5.257	5.341	5.416	5.481	5.540	5.593	5.641	5.685
9800	2.889	2.924	2.958	3.269	3.757	4.123	4.407	4.635	4.822	4.977	5.108	5.221	5.318	5.404	5.479	5.546	5.606	5.660	5.709	5.753
9900	2.910	2.946	2.981	3.296	3.792	4.164	4.454	4.685	4.875	5.032	5.166	5.281	5.380	5.467	5.543	5.611	5.672	5.727	5.777	5.822
10000	2.932	2.968	3.003	3.324	3.828	4.206	4.500	4.736	4.928	5.088	5.224	5.341	5.441	5.530	5.608	5.677	5.739	5.794	5.845	5.891
10100	2.953	2.989	3.025	3.351	3.864	4.248	4.547	4.786	4.982	5.145	5.283	5.401	5.503	5.593	5.672	5.743	5.806	5.862	5.913	5.960
10200	2.974	3.011	3.048	3.379	3.900	4.290	4.594	4.837	5.036	5.201	5.342	5.462	5.566	5.657	5.737	5.809	5.873	5.930	5.982	6.030
10300	2.996	3.033	3.070	3.407	3.936	4.333	4.641	4.888	5.090	5.258	5.401	5.523	5.629	5.721	5.803	5.875	5.940	5.999	6.052	6.100
10400	3.017	3.055	3.093	3.435	3.972	4.375	4.689	4.939	5.145	5.316	5.460	5.584	5.692	5.786	5.869	5.942	6.008	6.068	6.122	6.170
10500	3.038	3.077	3.116	3.463	4.009	4.418	4.736	4.991	5.199	5.373	5.520	5.646	5.755	5.851	5.935	6.010	6.077	6.137	6.192	6.241
10600	3.060	3.099	3.138	3.491	4.045	4.461	4.784	5.043	5.255	5.431	5.580	5.708	5.819	5.916	6.001	6.078	6.146	6.207	6.262	6.313
10700	3.081	3.121	3.161	3.519	4.082	4.504	4.832	5.095	5.310	5.489	5.641	5.770	5.883	5.982	6.068	6.146	6.215	6.277	6.333	6.384
10800	3.103	3.144	3.184	3.547	4.119	4.547	4.881	5.147	5.366	5.547	5.701	5.833	5.947	6.048	6.136	6.214	6.284	6.348	6.405	6.457
10900	3.124	3.166	3.206	3.576	4.156	4.591	4.929	5.200	5.422	5.606	5.762	5.896	6.012	6.114	6.203	6.283	6.354	6.418	6.476	6.529
11000	3.146	3.188	3.229	3.604	4.193	4.635	4.978	5.253	5.478	5.665	5.824	5.960	6.078	6.181	6.272	6.352	6.425	6.490	6.549	6.602
11100	3.168	3.210	3.252	3.633	4.230	4.679	5.027	5.306	5.534	5.725	5.886	6.023	6.143	6.248	6.340	6.422	6.495	6.561	6.621	6.676
11200	3.189	3.233	3.275	3.661	4.268	4.723	5.077	5.360	5.591	5.784	5.948	6.088	6.209	6.315	6.409	6.492	6.566	6.634	6.694	6.749
11300	3.211	3.255	3.298	3.690	4.305	4.767	5.126	5.413	5.648	5.844	6.010	6.152	6.275	6.383	6.478	6.562	6.638	6.706	6.768	6.824
11400	3.233	3.278	3.321	3.719	4.343	4.812	5.176	5.467	5.706	5.905	6.073	6.217	6.342	6.451	6.547	6.633	6.710	6.779	6.841	6.898

SIGMA = 6181.0

$V-V_c = 12$

H	3.4	3.2	3.0	2.8	2.6	2.4	2.2	2.0	1.8	1.6	1.4	1.2	1.0	0.8	0.6	0.4	0.2	0.10	0.09	0.08
11500	6.973	6.916	6.852	6.782	6.704	6.617	6.520	6.409	6.282	6.136	5.965	5.764	5.522	5.226	4.856	4.381	3.748	3.344	3.300	3.255
11600	7.049	6.990	6.926	6.855	6.776	6.688	6.588	6.476	6.347	6.199	6.026	5.822	5.576	5.276	4.901	4.419	3.777	3.368	3.322	3.277
11700	7.124	7.065	7.000	6.928	6.848	6.758	6.658	6.544	6.413	6.263	6.087	5.880	5.631	5.327	4.946	4.457	3.806	3.391	3.345	3.299
11800	7.201	7.140	7.074	7.001	6.920	6.829	6.727	6.612	6.479	6.327	6.149	5.938	5.686	5.377	4.992	4.496	3.835	3.414	3.368	3.320
11900	7.277	7.216	7.149	7.075	6.993	6.901	6.797	6.680	6.546	6.391	6.211	5.997	5.741	5.428	5.037	4.534	3.864	3.437	3.390	3.342
12000	7.354	7.292	7.224	7.149	7.066	6.972	6.867	6.748	6.613	6.456	6.273	6.056	5.797	5.479	5.083	4.573	3.893	3.461	3.413	3.364
12100	7.432	7.369	7.300	7.224	7.139	7.045	6.938	6.818	6.680	6.521	6.335	6.116	5.853	5.531	5.129	4.612	3.923	3.484	3.436	3.386
12200	7.509	7.446	7.376	7.299	7.213	7.117	7.009	6.887	6.747	6.586	6.398	6.175	5.909	5.583	5.175	4.651	3.952	3.507	3.458	3.409
12300	7.588	7.523	7.452	7.374	7.287	7.190	7.081	6.957	6.815	6.651	6.461	6.235	5.965	5.634	5.221	4.690	3.982	3.531	3.481	3.431
12400	7.666	7.601	7.529	7.450	7.362	7.263	7.152	7.027	6.883	6.717	6.524	6.296	6.022	5.686	5.268	4.729	4.011	3.554	3.504	3.453
12500	7.745	7.679	7.606	7.526	7.437	7.337	7.224	7.097	6.952	6.784	6.588	6.356	6.078	5.739	5.314	4.769	4.041	3.578	3.527	3.475
12600	7.825	7.758	7.684	7.603	7.512	7.411	7.297	7.168	7.020	6.850	6.652	6.417	6.136	5.791	5.361	4.808	4.071	3.602	3.550	3.497
12700	7.905	7.837	7.762	7.679	7.588	7.485	7.370	7.239	7.090	6.917	6.716	6.478	6.193	5.844	5.408	4.848	4.101	3.625	3.573	3.520
12800	7.985	7.916	7.840	7.757	7.664	7.560	7.443	7.310	7.159	6.984	6.780	6.540	6.251	5.897	5.456	4.888	4.131	3.649	3.596	3.542
12900	8.066	7.996	7.918	7.834	7.740	7.635	7.516	7.382	7.229	7.052	6.845	6.601	6.308	5.950	5.503	4.928	4.161	3.673	3.619	3.564
13000	8.147	8.076	7.998	7.912	7.817	7.710	7.590	7.454	7.299	7.120	6.910	6.663	6.367	6.004	5.551	4.968	4.191	3.697	3.642	3.587
13100	8.228	8.157	8.078	7.991	7.894	7.786	7.665	7.527	7.369	7.188	6.976	6.726	6.425	6.058	5.599	5.008	4.221	3.720	3.665	3.609
13200	8.310	8.237	8.158	8.070	7.972	7.862	7.739	7.600	7.440	7.256	7.042	6.788	6.484	6.112	5.647	5.049	4.252	3.744	3.688	3.632
13300	8.392	8.319	8.238	8.149	8.050	7.939	7.814	7.673	7.511	7.325	7.108	6.851	6.543	6.166	5.695	5.089	4.282	3.768	3.712	3.654
13400	8.475	8.401	8.319	8.228	8.128	8.016	7.890	7.747	7.583	7.394	7.174	6.914	6.602	6.220	5.743	5.130	4.313	3.792	3.735	3.677
13500	8.558	8.483	8.400	8.308	8.207	8.093	7.965	7.820	7.655	7.464	7.241	6.977	6.661	6.275	5.792	5.171	4.343	3.816	3.758	3.699
13600	8.642	8.565	8.481	8.389	8.286	8.171	8.041	7.895	7.727	7.534	7.308	7.041	6.721	6.330	5.841	5.212	4.374	3.840	3.782	3.722
13700	8.725	8.648	8.563	8.470	8.365	8.249	8.118	7.969	7.800	7.604	7.375	7.105	6.781	6.385	5.890	5.253	4.405	3.865	3.805	3.745
13800	8.810	8.732	8.646	8.551	8.445	8.327	8.195	8.044	7.872	7.674	7.443	7.169	6.841	6.440	5.939	5.295	4.436	3.889	3.829	3.767
13900	8.894	8.815	8.728	8.632	8.525	8.406	8.272	8.120	7.946	7.745	7.511	7.234	6.902	6.496	5.989	5.336	4.467	3.913	3.852	3.790
14000	8.980	8.899	8.811	8.714	8.606	8.485	8.349	8.195	8.019	7.816	7.579	7.299	6.963	6.552	6.038	5.378	4.498	3.937	3.876	3.813
14100	9.065	8.984	8.895	8.796	8.687	8.565	8.427	8.271	8.093	7.887	7.647	7.364	7.024	6.608	6.088	5.420	4.529	3.962	3.899	3.836
14200	9.151	9.069	8.979	8.879	8.768	8.645	8.505	8.348	8.167	7.959	7.716	7.429	7.085	6.664	6.138	5.462	4.560	3.986	3.923	3.858
14300	9.237	9.154	9.063	8.962	8.850	8.725	8.584	8.424	8.242	8.031	7.785	7.495	7.147	6.721	6.188	5.504	4.591	4.010	3.947	3.881
14400	9.324	9.240	9.148	9.046	8.932	8.805	8.663	8.501	8.317	8.103	7.855	7.561	7.208	6.777	6.239	5.546	4.623	4.035	3.970	3.904
14500	9.411	9.326	9.233	9.129	9.015	8.886	8.742	8.579	8.392	8.176	7.925	7.627	7.270	6.834	6.289	5.589	4.654	4.059	3.994	3.927
14600	9.499	9.413	9.318	9.214	9.096	8.968	8.822	8.656	8.467	8.249	7.995	7.694	7.333	6.892	6.340	5.631	4.686	4.084	4.018	3.950
14700	9.587	9.500	9.404	9.298	9.181	9.050	8.902	8.734	8.543	8.322	8.065	7.761	7.395	6.949	6.391	5.674	4.717	4.109	4.042	3.973
14800	9.675	9.587	9.490	9.383	9.264	9.132	8.982	8.813	8.619	8.396	8.136	7.828	7.458	7.007	6.442	5.717	4.749	4.133	4.066	3.996
14900	9.764	9.675	9.577	9.469	9.348	9.214	9.063	8.892	8.696	8.470	8.207	7.895	7.521	7.065	6.494	5.760	4.781	4.158	4.089	4.020
15000	9.853	9.763	9.664	9.554	9.433	9.297	9.144	8.971	8.773	8.544	8.278	7.963	7.585	7.123	6.545	5.803	4.813	4.183	4.113	4.043

x

$$V - V_c = 13 \qquad\qquad \text{SIGMA} = 5266.0$$

H / x	3.4	3.2	3.0	2.8	2.6	2.4	2.2	2.0	1.8	1.6	1.4	1.2	1.0	0.8	0.6	0.4	0.2	0.10	0.09	0.08
0	1.000	1.000	1.000	1.000	1.000	1.000	1.000	1.000	1.000	1.000	1.000	1.000	1.000	1.000	1.000	1.000	1.000	1.000	1.000	1.000
20	1.007	1.007	1.007	1.007	1.006	1.006	1.006	1.006	1.006	1.006	1.006	1.006	1.006	1.005	1.005	1.005	1.004	1.004	1.004	1.004
40	1.014	1.013	1.013	1.013	1.013	1.013	1.013	1.013	1.013	1.012	1.012	1.012	1.011	1.011	1.010	1.010	1.009	1.008	1.008	1.008
60	1.020	1.020	1.020	1.020	1.019	1.019	1.019	1.019	1.019	1.018	1.018	1.018	1.017	1.017	1.016	1.015	1.013	1.012	1.012	1.012
80	1.027	1.027	1.027	1.026	1.025	1.026	1.025	1.025	1.025	1.025	1.024	1.024	1.023	1.022	1.021	1.020	1.018	1.017	1.016	1.016
100	1.034	1.034	1.034	1.033	1.032	1.033	1.032	1.032	1.031	1.031	1.030	1.030	1.029	1.028	1.026	1.025	1.022	1.021	1.021	1.020
120	1.041	1.041	1.040	1.040	1.039	1.039	1.039	1.038	1.038	1.037	1.036	1.036	1.034	1.033	1.032	1.029	1.027	1.025	1.025	1.025
140	1.048	1.047	1.047	1.047	1.045	1.046	1.045	1.045	1.044	1.043	1.043	1.042	1.040	1.039	1.037	1.034	1.031	1.029	1.029	1.029
160	1.055	1.054	1.054	1.053	1.052	1.052	1.052	1.051	1.051	1.050	1.049	1.047	1.046	1.044	1.042	1.039	1.036	1.033	1.033	1.033
180	1.061	1.061	1.061	1.060	1.059	1.059	1.058	1.058	1.057	1.056	1.055	1.053	1.052	1.050	1.047	1.044	1.040	1.037	1.037	1.037
200	1.068	1.068	1.068	1.066	1.065	1.066	1.065	1.064	1.063	1.062	1.061	1.059	1.058	1.056	1.053	1.049	1.045	1.042	1.041	1.041
220	1.075	1.075	1.074	1.073	1.072	1.072	1.072	1.071	1.070	1.069	1.067	1.066	1.064	1.061	1.058	1.054	1.049	1.046	1.045	1.045
240	1.082	1.082	1.081	1.080	1.079	1.079	1.078	1.077	1.076	1.075	1.073	1.072	1.069	1.067	1.063	1.059	1.054	1.050	1.050	1.049
260	1.089	1.089	1.088	1.087	1.086	1.086	1.085	1.084	1.083	1.081	1.080	1.078	1.075	1.072	1.069	1.064	1.058	1.054	1.054	1.053
280	1.096	1.096	1.095	1.094	1.093	1.093	1.092	1.091	1.089	1.088	1.086	1.084	1.081	1.078	1.074	1.069	1.063	1.058	1.058	1.057
300	1.103	1.103	1.102	1.100	1.100	1.100	1.098	1.097	1.096	1.094	1.092	1.090	1.087	1.084	1.080	1.074	1.067	1.062	1.062	1.061
320	1.111	1.110	1.109	1.107	1.106	1.106	1.105	1.104	1.102	1.100	1.098	1.096	1.093	1.089	1.085	1.079	1.072	1.067	1.066	1.066
340	1.118	1.117	1.116	1.114	1.113	1.113	1.112	1.110	1.109	1.107	1.105	1.102	1.099	1.095	1.090	1.084	1.076	1.071	1.070	1.070
360	1.125	1.124	1.123	1.121	1.120	1.120	1.119	1.117	1.115	1.113	1.111	1.108	1.105	1.101	1.096	1.089	1.081	1.075	1.074	1.074
380	1.132	1.131	1.130	1.128	1.127	1.127	1.125	1.124	1.122	1.120	1.117	1.114	1.111	1.107	1.101	1.094	1.085	1.079	1.079	1.078
400	1.139	1.138	1.137	1.135	1.134	1.134	1.132	1.130	1.128	1.126	1.124	1.121	1.117	1.112	1.107	1.099	1.090	1.083	1.083	1.082
420	1.146	1.145	1.144	1.142	1.141	1.141	1.139	1.137	1.135	1.133	1.130	1.127	1.123	1.118	1.112	1.104	1.094	1.088	1.087	1.086
440	1.154	1.153	1.151	1.149	1.148	1.147	1.146	1.144	1.142	1.139	1.136	1.133	1.129	1.124	1.118	1.109	1.099	1.092	1.091	1.090
460	1.161	1.160	1.159	1.157	1.155	1.154	1.153	1.151	1.148	1.146	1.143	1.139	1.135	1.130	1.123	1.114	1.103	1.096	1.095	1.094
480	1.168	1.167	1.166	1.163	1.161	1.161	1.160	1.157	1.155	1.152	1.149	1.145	1.141	1.135	1.128	1.120	1.108	1.100	1.099	1.099
500	1.175	1.174	1.173	1.170	1.168	1.168	1.166	1.164	1.162	1.159	1.156	1.152	1.147	1.141	1.134	1.125	1.112	1.104	1.104	1.103
520	1.183	1.181	1.180	1.177	1.175	1.175	1.173	1.171	1.168	1.166	1.162	1.158	1.153	1.147	1.139	1.130	1.117	1.109	1.108	1.107
540	1.190	1.187	1.187	1.184	1.182	1.182	1.180	1.178	1.175	1.172	1.168	1.164	1.159	1.153	1.145	1.135	1.121	1.113	1.112	1.111
560	1.197	1.196	1.195	1.191	1.190	1.189	1.189	1.185	1.182	1.179	1.175	1.171	1.165	1.159	1.150	1.140	1.126	1.117	1.116	1.115
580	1.205	1.203	1.202	1.198	1.196	1.196	1.194	1.192	1.189	1.185	1.181	1.177	1.171	1.164	1.156	1.145	1.131	1.121	1.120	1.119
600	1.212	1.211	1.209	1.206	1.204	1.204	1.201	1.199	1.196	1.192	1.188	1.183	1.177	1.170	1.162	1.150	1.135	1.125	1.124	1.123
620	1.219	1.218	1.216	1.213	1.211	1.211	1.208	1.205	1.202	1.199	1.195	1.190	1.184	1.176	1.167	1.155	1.140	1.130	1.129	1.127
640	1.227	1.224	1.224	1.220	1.218	1.218	1.215	1.212	1.209	1.205	1.201	1.196	1.190	1.182	1.173	1.160	1.144	1.134	1.133	1.132
660	1.234	1.233	1.231	1.227	1.225	1.225	1.222	1.219	1.216	1.212	1.208	1.202	1.196	1.188	1.178	1.166	1.149	1.138	1.137	1.136
680	1.242	1.240	1.238	1.234	1.232	1.232	1.229	1.226	1.223	1.219	1.214	1.209	1.202	1.194	1.184	1.171	1.153	1.142	1.141	1.140
700	1.249	1.248	1.246	1.242	1.239	1.239	1.236	1.233	1.230	1.226	1.221	1.215	1.208	1.200	1.189	1.176	1.158	1.147	1.145	1.144
720	1.257	1.255	1.253	1.249	1.246	1.246	1.244	1.240	1.237	1.232	1.227	1.221	1.214	1.206	1.195	1.181	1.163	1.151	1.150	1.148
740	1.264	1.263	1.261	1.256	1.254	1.251	1.251	1.247	1.244	1.239	1.234	1.228	1.221	1.212	1.201	1.186	1.167	1.155	1.154	1.152
760	1.272	1.270	1.268	1.264	1.261	1.258	1.258	1.254	1.250	1.246	1.241	1.234	1.227	1.218	1.206	1.192	1.172	1.159	1.158	1.157
780	1.280	1.278	1.276	1.271	1.271	1.264	1.265	1.261	1.257	1.253	1.247	1.241	1.233	1.224	1.212	1.197	1.176	1.164	1.162	1.161
800	1.287	1.285	1.283	1.281	1.278	1.275	1.272	1.269	1.264	1.260	1.254	1.247	1.239	1.230	1.218	1.202	1.181	1.168	1.166	1.165

SIGMA = 5266.0

$V - V_c = 13$

H x	0.08	0.09	0.10	0.2	0.4	0.6	0.8	1.0	1.2	1.4	1.6	1.8	2.0	2.2	2.4	2.6	2.8	3.0	3.2	3.4
820	1.169	1.171	1.172	1.186	1.207	1.223	1.236	1.246	1.254	1.261	1.266	1.271	1.276	1.279	1.283	1.286	1.288	1.291	1.293	1.295
840	1.173	1.175	1.176	1.190	1.212	1.229	1.242	1.252	1.260	1.267	1.273	1.278	1.283	1.287	1.290	1.293	1.296	1.298	1.300	1.302
860	1.177	1.179	1.181	1.195	1.218	1.235	1.248	1.258	1.267	1.274	1.280	1.285	1.290	1.294	1.297	1.301	1.303	1.306	1.308	1.310
880	1.182	1.183	1.185	1.200	1.223	1.240	1.254	1.265	1.273	1.281	1.287	1.292	1.297	1.301	1.305	1.308	1.311	1.313	1.316	1.318
900	1.186	1.187	1.189	1.204	1.228	1.246	1.260	1.271	1.280	1.288	1.294	1.300	1.304	1.308	1.312	1.315	1.318	1.321	1.323	1.326
920	1.190	1.192	1.193	1.209	1.233	1.252	1.266	1.277	1.287	1.294	1.301	1.307	1.312	1.316	1.320	1.323	1.326	1.329	1.331	1.333
940	1.194	1.196	1.198	1.214	1.239	1.257	1.272	1.284	1.293	1.301	1.308	1.314	1.319	1.323	1.327	1.330	1.334	1.336	1.339	1.341
960	1.198	1.200	1.202	1.218	1.244	1.263	1.278	1.290	1.300	1.308	1.315	1.321	1.326	1.330	1.334	1.338	1.341	1.344	1.347	1.349
980	1.202	1.204	1.206	1.223	1.249	1.269	1.284	1.296	1.306	1.315	1.322	1.328	1.333	1.338	1.342	1.346	1.349	1.352	1.354	1.357
1000	1.207	1.209	1.210	1.228	1.254	1.275	1.290	1.303	1.313	1.322	1.329	1.335	1.341	1.345	1.349	1.353	1.356	1.359	1.362	1.365
1020	1.211	1.213	1.215	1.232	1.260	1.280	1.296	1.309	1.320	1.329	1.336	1.342	1.348	1.353	1.357	1.361	1.364	1.367	1.370	1.372
1040	1.215	1.217	1.219	1.237	1.265	1.286	1.303	1.316	1.326	1.335	1.343	1.350	1.355	1.360	1.364	1.368	1.372	1.375	1.378	1.380
1060	1.219	1.221	1.223	1.242	1.270	1.292	1.309	1.322	1.333	1.342	1.350	1.357	1.362	1.368	1.372	1.376	1.379	1.383	1.386	1.388
1080	1.223	1.225	1.228	1.246	1.276	1.298	1.315	1.329	1.340	1.349	1.357	1.364	1.370	1.375	1.380	1.384	1.387	1.390	1.393	1.396
1100	1.228	1.230	1.232	1.251	1.281	1.304	1.321	1.335	1.347	1.356	1.364	1.371	1.377	1.382	1.387	1.391	1.395	1.398	1.401	1.404
1120	1.232	1.234	1.236	1.256	1.286	1.309	1.327	1.342	1.353	1.363	1.371	1.378	1.385	1.390	1.395	1.399	1.403	1.406	1.409	1.412
1140	1.236	1.238	1.240	1.260	1.292	1.315	1.334	1.348	1.360	1.370	1.379	1.386	1.392	1.398	1.402	1.407	1.411	1.414	1.417	1.420
1160	1.240	1.242	1.245	1.265	1.297	1.321	1.340	1.355	1.367	1.377	1.386	1.393	1.399	1.405	1.410	1.414	1.418	1.422	1.425	1.428
1180	1.244	1.247	1.249	1.270	1.302	1.327	1.346	1.361	1.374	1.384	1.393	1.400	1.407	1.413	1.418	1.422	1.426	1.430	1.433	1.436
1200	1.249	1.251	1.253	1.275	1.308	1.333	1.352	1.368	1.380	1.391	1.400	1.408	1.414	1.420	1.425	1.430	1.434	1.438	1.441	1.444
1220	1.253	1.255	1.258	1.279	1.313	1.339	1.358	1.374	1.387	1.398	1.407	1.415	1.422	1.428	1.433	1.438	1.442	1.446	1.449	1.452
1240	1.257	1.259	1.262	1.284	1.319	1.345	1.365	1.381	1.394	1.405	1.415	1.422	1.429	1.435	1.441	1.446	1.450	1.454	1.457	1.460
1260	1.261	1.264	1.266	1.289	1.324	1.350	1.371	1.388	1.401	1.412	1.422	1.430	1.437	1.443	1.449	1.453	1.458	1.462	1.465	1.468
1280	1.265	1.268	1.271	1.293	1.329	1.356	1.377	1.394	1.408	1.419	1.429	1.437	1.445	1.451	1.456	1.461	1.466	1.470	1.473	1.477
1300	1.270	1.272	1.275	1.298	1.335	1.362	1.384	1.401	1.415	1.426	1.436	1.445	1.452	1.458	1.464	1.469	1.474	1.478	1.481	1.485
1320	1.274	1.277	1.279	1.303	1.340	1.368	1.390	1.407	1.422	1.434	1.444	1.452	1.460	1.466	1.472	1.477	1.482	1.486	1.490	1.493
1340	1.278	1.281	1.283	1.308	1.346	1.374	1.396	1.414	1.429	1.441	1.451	1.460	1.467	1.474	1.480	1.485	1.490	1.494	1.498	1.501
1360	1.282	1.285	1.288	1.312	1.351	1.380	1.403	1.421	1.436	1.448	1.458	1.467	1.475	1.482	1.488	1.493	1.498	1.502	1.506	1.509
1380	1.287	1.289	1.292	1.317	1.357	1.386	1.409	1.427	1.442	1.455	1.466	1.475	1.483	1.489	1.496	1.501	1.506	1.510	1.514	1.518
1400	1.291	1.294	1.296	1.322	1.362	1.392	1.415	1.434	1.449	1.462	1.473	1.482	1.490	1.497	1.503	1.509	1.514	1.518	1.522	1.526
1420	1.295	1.298	1.301	1.327	1.367	1.398	1.422	1.441	1.456	1.469	1.480	1.490	1.498	1.505	1.511	1.517	1.522	1.526	1.531	1.534
1440	1.299	1.302	1.305	1.331	1.373	1.404	1.428	1.448	1.463	1.477	1.488	1.497	1.506	1.513	1.519	1.525	1.530	1.535	1.539	1.543
1460	1.303	1.306	1.309	1.336	1.378	1.410	1.435	1.454	1.470	1.484	1.495	1.505	1.513	1.521	1.527	1.533	1.538	1.543	1.547	1.551
1480	1.308	1.311	1.314	1.341	1.384	1.416	1.441	1.461	1.477	1.491	1.503	1.513	1.521	1.529	1.535	1.541	1.546	1.551	1.555	1.559
1500	1.312	1.315	1.318	1.346	1.389	1.422	1.448	1.468	1.484	1.498	1.510	1.520	1.529	1.536	1.543	1.549	1.555	1.559	1.564	1.568
1520	1.316	1.319	1.322	1.351	1.395	1.428	1.454	1.475	1.492	1.506	1.518	1.528	1.537	1.544	1.551	1.557	1.563	1.568	1.572	1.576
1540	1.320	1.324	1.327	1.355	1.400	1.434	1.460	1.481	1.499	1.513	1.525	1.535	1.544	1.552	1.559	1.565	1.571	1.576	1.580	1.585
1560	1.325	1.328	1.331	1.360	1.406	1.440	1.467	1.488	1.506	1.520	1.533	1.543	1.552	1.560	1.567	1.574	1.579	1.584	1.589	1.593
1580	1.329	1.332	1.335	1.365	1.411	1.446	1.473	1.495	1.513	1.528	1.540	1.551	1.560	1.568	1.575	1.582	1.587	1.593	1.597	1.601
1600	1.333	1.337	1.340	1.370	1.417	1.452	1.480	1.502	1.520	1.535	1.548	1.559	1.568	1.576	1.583	1.590	1.596	1.601	1.606	1.610
1620	1.337	1.341	1.344	1.375	1.423	1.458	1.486	1.509	1.527	1.542	1.555	1.566	1.576	1.584	1.592	1.598	1.604	1.609	1.614	1.618
1640	1.342	1.345	1.349	1.380	1.428	1.465	1.493	1.516	1.534	1.550	1.563	1.574	1.584	1.592	1.600	1.606	1.612	1.618	1.623	1.627

SIGMA = 5266.0

V-V_c = 13

x \ H	3.4	3.2	3.0	2.8	2.6	2.4	2.2	2.0	1.8	1.6	1.4	1.2	1.0	0.8	0.6	0.4	0.2	0.10	0.09	0.08
1660	1.636	1.631	1.626	1.621	1.615	1.608	1.600	1.592	1.582	1.570	1.557	1.541	1.523	1.499	1.471	1.434	1.384	1.353	1.349	1.346
1680	1.644	1.640	1.635	1.629	1.623	1.616	1.608	1.600	1.590	1.578	1.564	1.549	1.529	1.506	1.477	1.439	1.389	1.357	1.354	1.350
1700	1.653	1.648	1.643	1.637	1.631	1.624	1.616	1.608	1.597	1.586	1.572	1.556	1.536	1.513	1.483	1.445	1.394	1.362	1.358	1.354
1720	1.661	1.657	1.652	1.646	1.640	1.632	1.625	1.615	1.605	1.593	1.579	1.563	1.543	1.519	1.489	1.450	1.399	1.366	1.362	1.359
1740	1.670	1.665	1.660	1.654	1.648	1.641	1.633	1.623	1.613	1.601	1.587	1.570	1.550	1.526	1.495	1.456	1.404	1.370	1.367	1.363
1760	1.679	1.674	1.669	1.663	1.656	1.649	1.641	1.632	1.621	1.609	1.594	1.577	1.557	1.532	1.501	1.462	1.409	1.375	1.371	1.367
1780	1.688	1.683	1.677	1.671	1.665	1.657	1.649	1.640	1.629	1.616	1.602	1.585	1.564	1.539	1.508	1.467	1.413	1.379	1.375	1.372
1800	1.696	1.691	1.686	1.680	1.673	1.666	1.657	1.648	1.637	1.624	1.609	1.592	1.571	1.546	1.514	1.473	1.418	1.384	1.379	1.376
1820	1.705	1.700	1.694	1.688	1.681	1.674	1.665	1.656	1.645	1.632	1.617	1.599	1.578	1.552	1.520	1.478	1.423	1.388	1.384	1.380
1840	1.714	1.709	1.703	1.697	1.690	1.682	1.674	1.664	1.653	1.640	1.624	1.607	1.585	1.559	1.526	1.484	1.428	1.392	1.388	1.384
1860	1.723	1.717	1.712	1.705	1.698	1.691	1.682	1.672	1.660	1.647	1.632	1.614	1.592	1.566	1.532	1.490	1.433	1.397	1.393	1.389
1880	1.731	1.726	1.720	1.714	1.707	1.699	1.690	1.680	1.668	1.655	1.640	1.621	1.599	1.572	1.539	1.495	1.438	1.401	1.397	1.393
1900	1.740	1.735	1.729	1.723	1.715	1.707	1.698	1.688	1.676	1.663	1.647	1.629	1.606	1.579	1.545	1.501	1.443	1.405	1.401	1.397
1920	1.749	1.744	1.738	1.731	1.724	1.716	1.707	1.696	1.684	1.671	1.655	1.636	1.613	1.586	1.551	1.507	1.448	1.410	1.406	1.401
1940	1.758	1.752	1.746	1.740	1.732	1.724	1.715	1.704	1.692	1.679	1.662	1.643	1.620	1.592	1.557	1.512	1.452	1.414	1.410	1.406
1960	1.767	1.761	1.755	1.749	1.741	1.733	1.723	1.713	1.701	1.686	1.670	1.651	1.628	1.599	1.564	1.518	1.457	1.419	1.414	1.410
1980	1.776	1.770	1.764	1.757	1.750	1.741	1.732	1.721	1.709	1.694	1.678	1.658	1.635	1.606	1.570	1.524	1.462	1.423	1.419	1.414
2000	1.785	1.779	1.773	1.766	1.758	1.750	1.740	1.729	1.717	1.702	1.685	1.666	1.642	1.613	1.576	1.530	1.467	1.427	1.423	1.419
2020	1.794	1.788	1.782	1.775	1.767	1.758	1.748	1.737	1.725	1.710	1.693	1.673	1.649	1.619	1.583	1.535	1.472	1.432	1.427	1.423
2040	1.803	1.797	1.790	1.783	1.776	1.767	1.757	1.746	1.733	1.718	1.701	1.681	1.656	1.626	1.589	1.541	1.477	1.436	1.432	1.427
2060	1.812	1.806	1.799	1.792	1.784	1.775	1.765	1.754	1.741	1.726	1.709	1.688	1.663	1.633	1.595	1.547	1.482	1.441	1.436	1.432
2080	1.821	1.815	1.808	1.801	1.793	1.784	1.774	1.762	1.749	1.734	1.716	1.696	1.670	1.640	1.602	1.552	1.487	1.445	1.440	1.436
2100	1.830	1.824	1.817	1.810	1.802	1.793	1.782	1.771	1.757	1.742	1.724	1.703	1.678	1.647	1.608	1.558	1.492	1.449	1.445	1.440
2120	1.839	1.833	1.826	1.819	1.810	1.801	1.791	1.779	1.766	1.750	1.732	1.711	1.685	1.654	1.614	1.564	1.497	1.454	1.449	1.444
2140	1.848	1.842	1.835	1.828	1.819	1.810	1.799	1.787	1.774	1.758	1.740	1.718	1.692	1.660	1.621	1.570	1.502	1.458	1.454	1.449
2160	1.857	1.851	1.844	1.836	1.828	1.818	1.808	1.796	1.782	1.766	1.748	1.726	1.699	1.667	1.627	1.575	1.507	1.463	1.458	1.453
2180	1.866	1.860	1.853	1.845	1.837	1.827	1.816	1.804	1.790	1.774	1.755	1.733	1.707	1.674	1.633	1.581	1.512	1.467	1.462	1.457
2200	1.875	1.869	1.862	1.854	1.846	1.836	1.825	1.813	1.799	1.782	1.763	1.741	1.714	1.681	1.640	1.587	1.516	1.472	1.467	1.462
2220	1.885	1.878	1.871	1.863	1.854	1.845	1.834	1.821	1.807	1.790	1.771	1.748	1.721	1.688	1.646	1.593	1.521	1.476	1.471	1.466
2240	1.894	1.887	1.880	1.872	1.863	1.853	1.842	1.830	1.815	1.798	1.779	1.756	1.729	1.695	1.653	1.599	1.526	1.480	1.475	1.470
2260	1.903	1.896	1.889	1.881	1.872	1.862	1.851	1.838	1.823	1.807	1.787	1.764	1.736	1.702	1.659	1.604	1.531	1.485	1.480	1.475
2280	1.912	1.906	1.898	1.890	1.881	1.871	1.860	1.847	1.832	1.815	1.795	1.771	1.743	1.709	1.666	1.610	1.536	1.489	1.484	1.479
2300	1.922	1.915	1.907	1.899	1.890	1.880	1.868	1.855	1.840	1.823	1.803	1.779	1.751	1.716	1.672	1.616	1.541	1.494	1.489	1.483
2320	1.931	1.924	1.917	1.908	1.899	1.889	1.877	1.864	1.849	1.831	1.811	1.787	1.758	1.723	1.679	1.622	1.546	1.498	1.493	1.488
2340	1.940	1.933	1.926	1.917	1.908	1.897	1.886	1.872	1.857	1.839	1.819	1.794	1.765	1.730	1.685	1.628	1.551	1.503	1.497	1.492
2360	1.950	1.943	1.935	1.926	1.917	1.906	1.894	1.881	1.865	1.848	1.827	1.802	1.773	1.737	1.692	1.634	1.556	1.507	1.502	1.496
2380	1.959	1.952	1.944	1.935	1.926	1.915	1.903	1.889	1.874	1.856	1.835	1.810	1.780	1.744	1.698	1.639	1.561	1.512	1.506	1.501
2400	1.968	1.961	1.953	1.945	1.935	1.924	1.912	1.898	1.882	1.864	1.843	1.818	1.787	1.751	1.705	1.645	1.566	1.516	1.511	1.505
2420	1.978	1.971	1.963	1.953	1.944	1.933	1.921	1.907	1.891	1.872	1.851	1.825	1.795	1.758	1.711	1.651	1.571	1.521	1.515	1.509
2440	1.987	1.980	1.972	1.963	1.953	1.942	1.930	1.915	1.899	1.881	1.859	1.833	1.802	1.765	1.718	1.657	1.576	1.525	1.519	1.514
2460	1.997	1.989	1.981	1.972	1.962	1.951	1.938	1.924	1.908	1.889	1.867	1.841	1.810	1.772	1.724	1.663	1.581	1.529	1.524	1.518
2480	2.006	1.999	1.990	1.981	1.971	1.960	1.947	1.933	1.916	1.897	1.875	1.849	1.817	1.779	1.731	1.669	1.586	1.534	1.528	1.522

SIGMA = 5266.0

V-V$_c$ = 13

H	3.4	3.2	3.0	2.8	2.6	2.4	2.2	2.0	1.8	1.6	1.4	1.2	1.0	0.8	0.6	0.4	0.2	0.10	0.09	0.08
x																				
2500	2.016	2.008	2.000	1.991	1.980	1.969	1.956	1.941	1.925	1.906	1.883	1.857	1.825	1.786	1.737	1.675	1.591	1.538	1.533	1.527
2520	2.025	2.018	2.009	2.000	1.990	1.978	1.965	1.950	1.933	1.914	1.891	1.864	1.832	1.793	1.744	1.681	1.596	1.543	1.537	1.531
2540	2.035	2.027	2.019	2.009	1.999	1.987	1.974	1.959	1.942	1.922	1.899	1.872	1.840	1.800	1.750	1.687	1.602	1.547	1.541	1.535
2560	2.044	2.037	2.028	2.018	2.008	1.996	1.983	1.968	1.951	1.931	1.908	1.880	1.847	1.807	1.757	1.693	1.607	1.552	1.546	1.540
2580	2.054	2.046	2.037	2.028	2.017	2.005	1.992	1.977	1.959	1.939	1.916	1.888	1.855	1.814	1.764	1.698	1.612	1.556	1.550	1.544
2600	2.064	2.056	2.047	2.037	2.026	2.014	2.001	1.985	1.968	1.948	1.924	1.896	1.862	1.822	1.770	1.704	1.617	1.561	1.555	1.548
2620	2.073	2.065	2.056	2.047	2.036	2.023	2.010	1.994	1.977	1.956	1.932	1.904	1.870	1.829	1.777	1.710	1.622	1.565	1.559	1.553
2640	2.083	2.075	2.066	2.056	2.045	2.033	2.019	2.003	1.985	1.965	1.940	1.912	1.878	1.836	1.784	1.716	1.627	1.570	1.563	1.557
2660	2.093	2.084	2.075	2.065	2.054	2.042	2.028	2.012	1.994	1.973	1.949	1.920	1.885	1.843	1.790	1.722	1.632	1.574	1.568	1.561
2680	2.102	2.094	2.085	2.075	2.064	2.051	2.037	2.021	2.003	1.981	1.957	1.928	1.893	1.850	1.797	1.728	1.637	1.579	1.572	1.566
2700	2.112	2.104	2.094	2.084	2.073	2.060	2.046	2.030	2.011	1.990	1.965	1.936	1.901	1.857	1.804	1.734	1.642	1.583	1.577	1.570
2720	2.122	2.113	2.104	2.094	2.082	2.069	2.055	2.039	2.020	1.999	1.973	1.944	1.908	1.865	1.810	1.740	1.647	1.588	1.581	1.575
2740	2.132	2.123	2.114	2.103	2.092	2.079	2.064	2.048	2.029	2.007	1.982	1.952	1.916	1.872	1.817	1.746	1.652	1.592	1.586	1.579
2760	2.141	2.133	2.123	2.113	2.101	2.088	2.073	2.057	2.038	2.016	1.990	1.960	1.924	1.879	1.824	1.752	1.657	1.597	1.590	1.583
2780	2.151	2.142	2.133	2.122	2.110	2.097	2.082	2.066	2.046	2.024	1.998	1.968	1.931	1.886	1.830	1.758	1.662	1.601	1.595	1.588
2800	2.161	2.152	2.143	2.132	2.120	2.107	2.092	2.075	2.055	2.033	2.007	1.976	1.939	1.894	1.837	1.764	1.667	1.606	1.599	1.592
2820	2.171	2.162	2.152	2.141	2.129	2.116	2.101	2.084	2.064	2.042	2.015	1.984	1.947	1.901	1.844	1.770	1.673	1.610	1.603	1.596
2840	2.181	2.172	2.162	2.151	2.139	2.125	2.110	2.093	2.073	2.050	2.024	1.992	1.954	1.908	1.851	1.776	1.678	1.615	1.608	1.601
2860	2.191	2.182	2.172	2.161	2.148	2.135	2.119	2.102	2.082	2.059	2.032	2.000	1.962	1.916	1.857	1.783	1.683	1.619	1.612	1.605
2880	2.201	2.191	2.181	2.170	2.158	2.144	2.129	2.111	2.091	2.068	2.040	2.008	1.970	1.923	1.864	1.789	1.688	1.624	1.617	1.610
2900	2.211	2.201	2.191	2.180	2.167	2.154	2.138	2.120	2.100	2.076	2.049	2.017	1.978	1.930	1.871	1.795	1.693	1.628	1.621	1.614
2920	2.221	2.211	2.201	2.190	2.177	2.163	2.147	2.129	2.109	2.085	2.057	2.025	1.985	1.938	1.878	1.801	1.698	1.633	1.626	1.618
2940	2.231	2.221	2.211	2.199	2.187	2.172	2.156	2.138	2.118	2.094	2.066	2.033	1.993	1.945	1.885	1.807	1.703	1.637	1.630	1.623
2960	2.241	2.231	2.221	2.209	2.196	2.182	2.166	2.147	2.127	2.102	2.074	2.041	2.001	1.952	1.891	1.813	1.708	1.642	1.635	1.627
2980	2.251	2.241	2.230	2.219	2.206	2.191	2.175	2.157	2.136	2.111	2.083	2.049	2.009	1.960	1.898	1.819	1.714	1.646	1.639	1.632
3000	2.261	2.251	2.240	2.229	2.216	2.201	2.184	2.166	2.145	2.120	2.091	2.057	2.017	1.967	1.905	1.825	1.719	1.651	1.644	1.636

SIGMA = 5266.0

V–V_c = 13

H	0.08	0.09	0.10	0.2	0.4	0.6	0.8	1.0	1.2	1.4	1.6	1.8	2.0	2.2	2.4	2.6	2.8	3.0	3.2	3.4
3100	1.658	1.666	1.674	1.745	1.856	1.939	2.004	2.056	2.099	2.134	2.164	2.190	2.212	2.232	2.249	2.264	2.278	2.290	2.301	2.311
3200	1.680	1.688	1.696	1.770	1.887	1.974	2.042	2.096	2.141	2.178	2.209	2.236	2.259	2.279	2.297	2.313	2.328	2.340	2.352	2.363
3300	1.702	1.711	1.719	1.797	1.918	2.009	2.080	2.136	2.183	2.221	2.254	2.282	2.306	2.327	2.346	2.363	2.378	2.391	2.403	2.414
3400	1.724	1.733	1.742	1.823	1.949	2.044	2.118	2.177	2.225	2.265	2.300	2.329	2.354	2.376	2.396	2.413	2.429	2.443	2.455	2.467
3500	1.747	1.756	1.765	1.849	1.981	2.080	2.156	2.218	2.268	2.310	2.345	2.376	2.402	2.425	2.446	2.464	2.480	2.494	2.508	2.520
3600	1.769	1.779	1.788	1.875	2.012	2.115	2.195	2.259	2.311	2.355	2.392	2.424	2.451	2.475	2.496	2.515	2.532	2.547	2.561	2.573
3700	1.791	1.801	1.811	1.902	2.044	2.151	2.234	2.301	2.355	2.400	2.439	2.472	2.500	2.525	2.547	2.567	2.584	2.600	2.614	2.627
3800	1.814	1.824	1.835	1.929	2.077	2.187	2.274	2.343	2.399	2.446	2.486	2.520	2.550	2.576	2.599	2.619	2.637	2.653	2.668	2.682
3900	1.836	1.847	1.858	1.955	2.109	2.224	2.314	2.385	2.444	2.493	2.534	2.569	2.600	2.627	2.651	2.672	2.690	2.707	2.723	2.737
4000	1.859	1.870	1.881	1.982	2.141	2.261	2.354	2.428	2.489	2.539	2.582	2.619	2.651	2.678	2.703	2.725	2.744	2.762	2.778	2.792
4100	1.881	1.893	1.904	2.009	2.174	2.298	2.394	2.471	2.534	2.586	2.631	2.669	2.702	2.731	2.756	2.779	2.799	2.817	2.834	2.849
4200	1.904	1.916	1.928	2.037	2.207	2.335	2.435	2.514	2.580	2.634	2.680	2.719	2.753	2.783	2.810	2.833	2.854	2.873	2.890	2.905
4300	1.926	1.939	1.951	2.064	2.240	2.373	2.476	2.558	2.626	2.682	2.729	2.770	2.805	2.836	2.864	2.888	2.910	2.929	2.947	2.963
4400	1.949	1.962	1.975	2.091	2.274	2.411	2.517	2.602	2.672	2.730	2.779	2.821	2.858	2.890	2.918	2.943	2.966	2.986	3.004	3.021
4500	1.972	1.985	1.999	2.119	2.307	2.449	2.559	2.647	2.719	2.779	2.830	2.873	2.911	2.944	2.973	2.999	3.022	3.043	3.062	3.079
4600	1.995	2.009	2.022	2.146	2.341	2.487	2.601	2.692	2.766	2.828	2.881	2.926	2.965	2.999	3.029	3.056	3.079	3.101	3.120	3.138
4700	2.018	2.032	2.046	2.174	2.375	2.526	2.643	2.737	2.814	2.878	2.932	2.978	3.019	3.054	3.085	3.112	3.137	3.159	3.179	3.198
4800	2.041	2.055	2.070	2.202	2.409	2.565	2.686	2.783	2.862	2.928	2.984	3.032	3.073	3.109	3.141	3.170	3.195	3.218	3.239	3.258
4900	2.064	2.079	2.094	2.230	2.444	2.604	2.729	2.829	2.910	2.978	3.036	3.085	3.128	3.165	3.198	3.228	3.254	3.278	3.299	3.319
5000	2.087	2.102	2.118	2.258	2.478	2.644	2.772	2.875	2.959	3.029	3.089	3.139	3.183	3.222	3.256	3.286	3.313	3.338	3.360	3.380
5100	2.110	2.126	2.142	2.286	2.513	2.683	2.816	2.922	3.008	3.081	3.142	3.194	3.239	3.279	3.314	3.345	3.373	3.398	3.421	3.442
5200	2.133	2.150	2.166	2.315	2.548	2.723	2.860	2.969	3.058	3.132	3.195	3.249	3.296	3.337	3.373	3.405	3.434	3.459	3.483	3.504
5300	2.156	2.173	2.190	2.343	2.583	2.764	2.904	3.016	3.108	3.184	3.249	3.305	3.353	3.395	3.432	3.465	3.494	3.521	3.545	3.567
5400	2.179	2.197	2.214	2.372	2.619	2.804	2.949	3.064	3.158	3.237	3.304	3.361	3.410	3.453	3.492	3.525	3.556	3.583	3.608	3.630
5500	2.203	2.221	2.239	2.400	2.655	2.845	2.993	3.112	3.209	3.290	3.358	3.417	3.468	3.512	3.552	3.587	3.618	3.646	3.671	3.694
5600	2.226	2.245	2.263	2.429	2.690	2.886	3.039	3.161	3.260	3.343	3.414	3.474	3.526	3.572	3.612	3.648	3.680	3.709	3.735	3.759
5700	2.249	2.269	2.287	2.458	2.726	2.928	3.084	3.209	3.312	3.397	3.470	3.531	3.585	3.632	3.674	3.710	3.743	3.773	3.800	3.824
5800	2.273	2.293	2.312	2.487	2.763	2.969	3.130	3.259	3.364	3.452	3.526	3.589	3.644	3.693	3.735	3.773	3.807	3.837	3.865	3.890
5900	2.296	2.317	2.336	2.516	2.799	3.011	3.176	3.308	3.416	3.506	3.582	3.648	3.704	3.754	3.797	3.836	3.871	3.902	3.930	3.956
6000	2.320	2.341	2.361	2.546	2.836	3.053	3.223	3.358	3.469	3.561	3.639	3.706	3.764	3.815	3.860	3.900	3.935	3.968	3.997	4.023
6100	2.344	2.365	2.386	2.575	2.873	3.096	3.270	3.408	3.522	3.617	3.697	3.766	3.825	3.877	3.923	3.964	4.001	4.034	4.063	4.090
6200	2.367	2.389	2.410	2.605	2.910	3.139	3.317	3.459	3.576	3.673	3.755	3.825	3.886	3.940	3.987	4.029	4.066	4.100	4.131	4.158
6300	2.391	2.413	2.435	2.634	2.947	3.182	3.364	3.510	3.630	3.729	3.813	3.886	3.948	4.003	4.051	4.094	4.132	4.167	4.198	4.227
6400	2.415	2.438	2.460	2.664	2.985	3.225	3.412	3.562	3.684	3.786	3.872	3.946	4.010	4.066	4.116	4.160	4.199	4.235	4.267	4.296
6500	2.439	2.462	2.485	2.694	3.022	3.269	3.460	3.613	3.739	3.843	3.932	4.007	4.073	4.130	4.181	4.226	4.266	4.303	4.336	4.365
6600	2.463	2.487	2.510	2.724	3.060	3.312	3.508	3.665	3.794	3.901	3.991	4.069	4.136	4.195	4.247	4.293	4.334	4.371	4.405	4.436
6700	2.486	2.511	2.535	2.754	3.098	3.356	3.557	3.718	3.849	3.959	4.051	4.131	4.200	4.260	4.313	4.360	4.403	4.441	4.475	4.506
6800	2.510	2.536	2.560	2.784	3.137	3.401	3.606	3.771	3.905	4.017	4.111	4.193	4.264	4.325	4.380	4.428	4.471	4.510	4.546	4.578
6900	2.535	2.560	2.585	2.815	3.175	3.445	3.656	3.824	3.961	4.076	4.173	4.256	4.328	4.391	4.447	4.497	4.541	4.581	4.617	4.649
7000	2.559	2.585	2.611	2.845	3.214	3.490	3.705	3.877	4.018	4.135	4.235	4.320	4.393	4.458	4.515	4.565	4.611	4.651	4.688	4.722
7100	2.583	2.610	2.636	2.876	3.253	3.536	3.755	3.931	4.075	4.195	4.297	4.384	4.459	4.525	4.583	4.635	4.681	4.723	4.761	4.795
7200	2.607	2.635	2.662	2.907	3.292	3.581	3.806	3.986	4.133	4.255	4.359	4.448	4.525	4.592	4.652	4.705	4.752	4.795	4.833	4.868

$V - V_c = 13$ SIGMA = 5266.0

H / x	3.4	3.2	3.0	2.8	2.6	2.4	2.2	2.0	1.8	1.6	1.4	1.2	1.0	0.8	0.6	0.4	0.2	0.10	0.09	0.08
7300	4.942	4.907	4.867	4.824	4.775	4.721	4.660	4.592	4.513	4.422	4.316	4.191	4.040	3.856	3.627	3.331	2.938	2.687	2.659	2.631
7400	5.017	4.980	4.940	4.896	4.846	4.791	4.729	4.659	4.578	4.485	4.377	4.249	4.095	3.907	3.673	3.371	2.969	2.713	2.684	2.656
7500	5.092	5.055	5.014	4.968	4.918	4.861	4.798	4.726	4.644	4.549	4.438	4.308	4.151	3.959	3.719	3.411	3.000	2.738	2.709	2.680
7600	5.168	5.130	5.088	5.041	4.990	4.932	4.867	4.794	4.710	4.613	4.500	4.367	4.206	4.010	3.766	3.451	3.031	2.764	2.734	2.704
7700	5.244	5.205	5.162	5.115	5.062	5.004	4.937	4.862	4.777	4.678	4.562	4.426	4.262	4.062	3.812	3.491	3.062	2.790	2.759	2.729
7800	5.321	5.281	5.238	5.189	5.135	5.075	5.008	4.931	4.844	4.743	4.625	4.486	4.319	4.115	3.859	3.531	3.094	2.815	2.785	2.753
7900	5.399	5.358	5.313	5.264	5.209	5.148	5.079	5.001	4.911	4.808	4.688	4.546	4.376	4.167	3.907	3.572	3.125	2.841	2.810	2.778
8000	5.476	5.435	5.389	5.339	5.283	5.221	5.150	5.071	4.979	4.874	4.752	4.607	4.433	4.220	3.954	3.613	3.157	2.867	2.835	2.803
8100	5.555	5.513	5.466	5.415	5.358	5.294	5.222	5.141	5.048	4.941	4.816	4.668	4.490	4.273	4.002	3.654	3.189	2.893	2.861	2.827
8200	5.634	5.591	5.544	5.491	5.433	5.368	5.295	5.212	5.117	5.008	4.880	4.729	4.548	4.327	4.050	3.695	3.221	2.919	2.886	2.852
8300	5.714	5.670	5.621	5.568	5.509	5.442	5.368	5.283	5.186	5.075	4.945	4.791	4.606	4.381	4.099	3.736	3.253	2.945	2.911	2.877
8400	5.795	5.749	5.700	5.645	5.585	5.517	5.441	5.355	5.256	5.143	5.010	4.853	4.665	4.435	4.147	3.778	3.285	2.971	2.937	2.902
8500	5.875	5.829	5.779	5.723	5.662	5.593	5.515	5.427	5.327	5.211	5.076	4.916	4.724	4.489	4.196	3.820	3.317	2.998	2.963	2.927
8600	5.956	5.909	5.858	5.802	5.739	5.669	5.590	5.500	5.398	5.279	5.142	4.979	4.783	4.544	4.246	3.862	3.350	3.024	2.988	2.952
8700	6.038	5.990	5.938	5.881	5.817	5.745	5.664	5.573	5.469	5.348	5.208	5.042	4.843	4.599	4.295	3.904	3.382	3.050	3.014	2.977
8800	6.120	6.072	6.019	5.960	5.895	5.822	5.740	5.647	5.541	5.418	5.275	5.106	4.903	4.655	4.345	3.946	3.415	3.077	3.040	3.002
8900	6.203	6.154	6.100	6.040	5.974	5.899	5.816	5.721	5.613	5.488	5.342	5.170	4.963	4.711	4.395	3.989	3.448	3.103	3.065	3.027
9000	6.287	6.237	6.182	6.121	6.053	5.977	5.892	5.796	5.686	5.558	5.410	5.235	5.024	4.767	4.445	4.032	3.481	3.130	3.091	3.052
9100	6.371	6.320	6.264	6.202	6.133	6.056	5.969	5.871	5.759	5.629	5.478	5.299	5.085	4.823	4.496	4.075	3.514	3.157	3.117	3.077
9200	6.456	6.404	6.347	6.283	6.213	6.135	6.047	5.947	5.832	5.700	5.547	5.365	5.147	4.880	4.547	4.118	3.547	3.183	3.143	3.103
9300	6.541	6.488	6.430	6.365	6.294	6.214	6.124	6.023	5.906	5.772	5.616	5.431	5.209	4.937	4.598	4.162	3.580	3.210	3.169	3.128
9400	6.627	6.573	6.514	6.448	6.375	6.294	6.203	6.099	5.981	5.844	5.685	5.497	5.271	4.995	4.649	4.205	3.614	3.237	3.196	3.153
9500	6.713	6.658	6.598	6.531	6.457	6.375	6.282	6.176	6.056	5.917	5.755	5.563	5.333	5.052	4.701	4.249	3.647	3.264	3.222	3.179
9600	6.800	6.744	6.683	6.615	6.540	6.456	6.361	6.254	6.131	5.990	5.825	5.630	5.396	5.110	4.753	4.293	3.681	3.291	3.248	3.204
9700	6.887	6.831	6.768	6.699	6.623	6.537	6.441	6.332	6.207	6.064	5.896	5.697	5.459	5.169	4.805	4.338	3.715	3.318	3.274	3.230
9800	6.975	6.918	6.854	6.784	6.706	6.619	6.521	6.411	6.284	6.137	5.967	5.765	5.523	5.227	4.858	4.382	3.748	3.345	3.301	3.255
9900	7.064	7.005	6.941	6.869	6.790	6.702	6.602	6.490	6.361	6.212	6.038	5.833	5.587	5.286	4.910	4.427	3.782	3.372	3.327	3.281
10000	7.153	7.093	7.028	6.955	6.875	6.785	6.684	6.569	6.438	6.287	6.110	5.902	5.652	5.346	4.963	4.472	3.816	3.399	3.354	3.307
10100	7.243	7.182	7.115	7.042	6.960	6.868	6.766	6.649	6.516	6.362	6.183	5.971	5.716	5.405	5.017	4.517	3.851	3.427	3.380	3.333
10200	7.333	7.271	7.204	7.129	7.045	6.953	6.848	6.729	6.594	6.438	6.255	6.040	5.781	5.465	5.070	4.562	3.885	3.454	3.407	3.358
10300	7.424	7.361	7.292	7.216	7.132	7.037	6.931	6.810	6.673	6.514	6.329	6.110	5.847	5.526	5.124	4.608	3.920	3.482	3.433	3.384
10400	7.515	7.451	7.381	7.304	7.218	7.122	7.014	6.892	6.752	6.590	6.402	6.180	5.913	5.586	5.178	4.654	3.954	3.509	3.460	3.410
10500	7.607	7.542	7.471	7.393	7.305	7.208	7.098	6.974	6.832	6.668	6.476	6.250	5.979	5.647	5.233	4.700	3.989	3.537	3.487	3.436
10600	7.699	7.634	7.561	7.482	7.393	7.294	7.182	7.056	6.912	6.745	6.551	6.321	6.045	5.708	5.287	4.746	4.024	3.564	3.514	3.462
10700	7.792	7.726	7.652	7.571	7.481	7.381	7.267	7.139	6.992	6.823	6.626	6.392	6.112	5.770	5.342	4.792	4.059	3.592	3.541	3.488
10800	7.886	7.818	7.744	7.661	7.570	7.468	7.353	7.222	7.073	6.901	6.701	6.464	6.179	5.832	5.397	4.839	4.094	3.620	3.568	3.514
10900	7.980	7.911	7.836	7.752	7.659	7.555	7.438	7.306	7.155	6.980	6.777	6.536	6.247	5.894	5.453	4.885	4.129	3.648	3.595	3.541
11000	8.075	8.005	7.929	7.843	7.749	7.643	7.525	7.390	7.237	7.059	6.853	6.608	6.315	5.957	5.508	4.932	4.164	3.675	3.622	3.567
11100	8.170	8.099	8.021	7.935	7.839	7.732	7.612	7.475	7.319	7.139	6.929	6.681	6.383	6.019	5.564	4.980	4.200	3.703	3.649	3.593
11200	8.266	8.194	8.115	8.027	7.930	7.821	7.699	7.560	7.402	7.219	7.006	6.754	6.452	6.083	5.621	5.027	4.235	3.731	3.676	3.619
11300	8.362	8.289	8.209	8.120	8.021	7.911	7.787	7.646	7.485	7.300	7.084	6.828	6.521	6.146	5.677	5.075	4.271	3.760	3.703	3.646
11400	8.459	8.385	8.303	8.213	8.113	8.001	7.875	7.732	7.569	7.381	7.161	6.902	6.590	6.210	5.734	5.122	4.307	3.788	3.731	3.672

$V-V_c = 13$

SIGMA = 5266.0

H	3.4	3.2	3.0	2.8	2.6	2.4	2.2	2.0	1.8	1.6	1.4	1.2	1.0	0.8	0.6	0.4	0.2	0.10	0.09	0.08
11500	8.557	8.481	8.398	8.307	8.205	8.092	7.964	7.819	7.654	7.463	7.240	6.976	6.660	6.274	5.791	5.170	4.343	3.816	3.758	3.699
11600	8.655	8.578	8.494	8.401	8.298	8.183	8.053	7.906	7.738	7.544	7.318	7.051	6.730	6.338	5.849	5.219	4.379	3.844	3.785	3.725
11700	8.753	8.676	8.590	8.496	8.392	8.275	8.143	7.994	7.823	7.627	7.397	7.126	6.801	6.403	5.906	5.267	4.415	3.873	3.813	3.752
11800	8.852	8.774	8.687	8.592	8.486	8.367	8.233	8.082	7.909	7.710	7.477	7.202	6.872	6.468	5.964	5.316	4.451	3.901	3.840	3.779
11900	8.952	8.872	8.785	8.688	8.580	8.460	8.324	8.171	7.995	7.793	7.557	7.278	6.943	6.534	6.022	5.364	4.488	3.929	3.868	3.805
12000	9.052	8.971	8.882	8.784	8.675	8.553	8.415	8.260	8.082	7.877	7.637	7.354	7.015	6.599	6.081	5.414	4.524	3.958	3.896	3.832
12100	9.153	9.071	8.981	8.881	8.770	8.647	8.507	8.349	8.169	7.961	7.718	7.431	7.086	6.666	6.139	5.463	4.561	3.987	3.923	3.859
12200	9.254	9.171	9.080	8.979	8.866	8.741	8.600	8.439	8.257	8.045	7.799	7.508	7.159	6.732	6.198	5.512	4.597	4.015	3.951	3.886
12300	9.356	9.272	9.179	9.077	8.963	8.836	8.692	8.530	8.345	8.130	7.881	7.586	7.231	6.799	6.258	5.562	4.634	4.044	3.979	3.913
12400	9.459	9.373	9.279	9.175	9.060	8.931	8.786	8.621	8.433	8.216	7.963	7.664	7.304	6.866	6.317	5.612	4.671	4.073	4.007	3.940
12500	9.562	9.475	9.380	9.275	9.157	9.027	8.879	8.713	8.522	8.302	8.045	7.742	7.378	6.933	6.377	5.662	4.708	4.102	4.035	3.967
12600	9.666	9.578	9.481	9.374	9.256	9.123	8.974	8.805	8.611	8.388	8.128	7.821	7.452	7.001	6.437	5.712	4.746	4.131	4.063	3.994
12700	9.770	9.681	9.583	9.474	9.354	9.226	9.068	8.897	8.701	8.475	8.211	7.900	7.526	7.069	6.497	5.763	4.783	4.160	4.091	4.021
12800	9.874	9.784	9.685	9.575	9.453	9.317	9.164	8.990	8.791	8.562	8.295	7.979	7.600	7.137	6.558	5.813	4.821	4.189	4.119	4.048
12900	9.980	9.888	9.788	9.676	9.553	9.415	9.259	9.083	8.882	8.650	8.379	8.059	7.675	7.205	6.619	5.864	4.858	4.218	4.147	4.076
13000	10.09	9.99	9.89	9.78	9.65	9.51	9.36	9.18	8.97	8.74	8.46	8.14	7.75	7.27	6.68	5.92	4.90	4.25	4.18	4.10
13100	10.19	10.10	9.99	9.88	9.75	9.61	9.45	9.27	9.07	8.83	8.55	8.22	7.83	7.34	6.74	5.97	4.93	4.28	4.20	4.13
13200	10.30	10.20	10.10	9.98	9.85	9.71	9.55	9.37	9.16	8.92	8.63	8.30	7.90	7.41	6.80	6.02	4.97	4.31	4.23	4.16
13300	10.41	10.31	10.20	10.09	9.96	9.81	9.65	9.46	9.25	9.01	8.72	8.38	7.98	7.48	6.86	6.07	5.01	4.34	4.26	4.19
13400	10.51	10.42	10.31	10.19	10.06	9.91	9.75	9.56	9.34	9.10	8.81	8.46	8.05	7.55	6.93	6.12	5.05	4.36	4.29	4.21
13500	10.62	10.52	10.42	10.30	10.16	10.01	9.85	9.65	9.44	9.19	8.89	8.55	8.13	7.62	6.99	6.17	5.09	4.39	4.32	4.24
13600	10.73	10.63	10.52	10.40	10.26	10.11	9.94	9.75	9.53	9.28	8.98	8.63	8.21	7.69	7.05	6.23	5.12	4.42	4.35	4.27
13700	10.84	10.74	10.63	10.51	10.37	10.22	10.04	9.85	9.63	9.37	9.07	8.71	8.29	7.77	7.12	6.28	5.16	4.45	4.38	4.30
13800	10.95	10.85	10.74	10.61	10.47	10.32	10.14	9.95	9.72	9.46	9.16	8.80	8.36	7.84	7.18	6.33	5.20	4.48	4.40	4.32
13900	11.06	10.96	10.84	10.72	10.58	10.42	10.24	10.04	9.82	9.55	9.24	8.88	8.44	7.91	7.24	6.38	5.24	4.51	4.43	4.35
14000	11.17	11.07	10.95	10.83	10.68	10.52	10.35	10.14	9.91	9.64	9.33	8.96	8.52	7.98	7.31	6.44	5.28	4.54	4.46	4.38
14100	11.29	11.18	11.06	10.93	10.79	10.63	10.45	10.24	10.01	9.74	9.42	9.05	8.60	8.05	7.37	6.49	5.32	4.57	4.49	4.41
14200	11.40	11.29	11.18	11.04	10.90	10.73	10.55	10.34	10.10	9.83	9.51	9.13	8.68	8.13	7.43	6.54	5.36	4.60	4.52	4.43
14300	11.51	11.40	11.28	11.15	11.00	10.84	10.65	10.44	10.20	9.92	9.60	9.22	8.76	8.20	7.50	6.60	5.40	4.63	4.55	4.46
14400	11.63	11.52	11.39	11.26	11.11	10.94	10.76	10.54	10.30	10.02	9.69	9.30	8.84	8.27	7.56	6.65	5.44	4.66	4.58	4.49
14500	11.74	11.63	11.51	11.37	11.22	11.05	10.86	10.64	10.40	10.11	9.78	9.39	8.92	8.35	7.63	6.71	5.48	4.69	4.61	4.52
14600	11.85	11.74	11.62	11.48	11.33	11.16	10.96	10.75	10.50	10.21	9.87	9.48	9.00	8.42	7.69	6.76	5.52	4.72	4.64	4.55
14700	11.97	11.86	11.73	11.59	11.44	11.26	11.07	10.85	10.60	10.30	9.97	9.56	9.08	8.50	7.76	6.82	5.56	4.75	4.67	4.58
14800	12.09	11.97	11.84	11.70	11.54	11.37	11.17	10.95	10.70	10.40	10.06	9.65	9.17	8.57	7.83	6.87	5.60	4.78	4.69	4.60
14900	12.20	12.09	11.96	11.81	11.66	11.48	11.28	11.05	10.80	10.50	10.15	9.74	9.25	8.65	7.89	6.93	5.64	4.81	4.72	4.63
15000	12.32	12.20	12.07	11.93	11.77	11.59	11.38	11.16	10.90	10.59	10.24	9.83	9.33	8.72	7.96	6.98	5.68	4.85	4.75	4.66

X

SIGMA = 4541.0

V-V_c = 14

H	0.08	0.09	0.10	0.2	0.4	0.6	0.8	1.0	1.2	1.4	1.6	1.8	2.0	2.2	2.4	2.6	2.8	3.0	3.2	3.4
0	1.000	1.000	1.000	1.000	1.000	1.000	1.000	1.000	1.000	1.000	1.000	1.000	1.000	1.000	1.000	1.000	1.000	1.000	1.000	1.000
20	1.005	1.005	1.005	1.005	1.006	1.006	1.006	1.007	1.007	1.007	1.007	1.007	1.007	1.007	1.008	1.008	1.008	1.008	1.008	1.008
40	1.009	1.010	1.010	1.010	1.011	1.012	1.013	1.013	1.014	1.014	1.014	1.015	1.015	1.015	1.015	1.015	1.015	1.015	1.016	1.016
60	1.014	1.014	1.014	1.015	1.017	1.018	1.019	1.020	1.021	1.021	1.021	1.022	1.022	1.022	1.023	1.023	1.023	1.023	1.023	1.024
80	1.019	1.019	1.019	1.021	1.023	1.024	1.026	1.027	1.027	1.028	1.029	1.029	1.030	1.030	1.030	1.031	1.031	1.031	1.031	1.031
100	1.024	1.024	1.024	1.026	1.028	1.030	1.032	1.033	1.034	1.035	1.036	1.036	1.037	1.037	1.038	1.038	1.039	1.039	1.039	1.039
120	1.028	1.029	1.029	1.031	1.034	1.037	1.038	1.040	1.041	1.042	1.043	1.044	1.045	1.045	1.046	1.046	1.047	1.047	1.047	1.047
140	1.033	1.033	1.034	1.036	1.040	1.043	1.045	1.047	1.048	1.049	1.050	1.051	1.052	1.053	1.053	1.054	1.054	1.055	1.055	1.055
160	1.038	1.038	1.039	1.041	1.046	1.049	1.051	1.053	1.055	1.057	1.058	1.059	1.060	1.060	1.061	1.062	1.062	1.063	1.063	1.063
180	1.043	1.043	1.043	1.047	1.051	1.055	1.058	1.060	1.062	1.064	1.065	1.066	1.067	1.068	1.069	1.069	1.070	1.071	1.071	1.071
200	1.047	1.048	1.048	1.052	1.057	1.061	1.064	1.067	1.069	1.071	1.072	1.074	1.075	1.076	1.077	1.077	1.078	1.079	1.079	1.079
220	1.052	1.053	1.053	1.057	1.063	1.067	1.071	1.074	1.076	1.078	1.080	1.081	1.082	1.083	1.084	1.085	1.086	1.087	1.087	1.088
240	1.057	1.057	1.058	1.062	1.069	1.074	1.078	1.081	1.083	1.085	1.087	1.089	1.090	1.091	1.092	1.093	1.094	1.095	1.095	1.096
260	1.062	1.062	1.063	1.067	1.075	1.080	1.084	1.088	1.090	1.093	1.095	1.096	1.098	1.099	1.100	1.101	1.102	1.103	1.103	1.104
280	1.067	1.067	1.068	1.073	1.080	1.086	1.091	1.094	1.097	1.100	1.102	1.104	1.105	1.107	1.108	1.109	1.110	1.111	1.112	1.112
300	1.071	1.072	1.072	1.078	1.086	1.092	1.097	1.101	1.104	1.107	1.109	1.111	1.113	1.114	1.116	1.117	1.118	1.119	1.120	1.120
320	1.076	1.077	1.077	1.083	1.092	1.099	1.104	1.108	1.112	1.114	1.117	1.119	1.121	1.122	1.124	1.125	1.126	1.127	1.128	1.129
340	1.081	1.082	1.082	1.088	1.098	1.105	1.111	1.115	1.119	1.122	1.124	1.127	1.129	1.130	1.132	1.133	1.134	1.135	1.136	1.137
360	1.086	1.086	1.087	1.092	1.104	1.111	1.117	1.122	1.126	1.129	1.132	1.134	1.136	1.138	1.140	1.141	1.142	1.144	1.144	1.145
380	1.090	1.091	1.092	1.099	1.110	1.118	1.124	1.129	1.133	1.137	1.139	1.142	1.144	1.146	1.148	1.149	1.151	1.152	1.153	1.154
400	1.095	1.096	1.097	1.104	1.115	1.124	1.131	1.136	1.140	1.144	1.147	1.150	1.152	1.154	1.156	1.157	1.159	1.160	1.161	1.162
420	1.100	1.101	1.102	1.109	1.121	1.130	1.137	1.143	1.148	1.151	1.155	1.157	1.160	1.162	1.164	1.165	1.167	1.168	1.169	1.171
440	1.105	1.106	1.107	1.115	1.127	1.137	1.144	1.150	1.155	1.159	1.162	1.165	1.168	1.170	1.172	1.174	1.175	1.177	1.178	1.179
460	1.110	1.111	1.111	1.120	1.133	1.143	1.151	1.157	1.162	1.166	1.170	1.173	1.176	1.178	1.180	1.182	1.184	1.185	1.186	1.188
480	1.114	1.115	1.116	1.125	1.139	1.150	1.158	1.164	1.169	1.174	1.178	1.181	1.184	1.186	1.188	1.190	1.192	1.193	1.195	1.196
500	1.119	1.120	1.121	1.130	1.145	1.156	1.164	1.171	1.177	1.181	1.185	1.189	1.192	1.194	1.196	1.198	1.200	1.202	1.203	1.205
520	1.124	1.125	1.126	1.136	1.151	1.162	1.171	1.178	1.184	1.189	1.193	1.197	1.200	1.202	1.205	1.207	1.209	1.210	1.212	1.213
540	1.129	1.130	1.131	1.141	1.157	1.169	1.178	1.185	1.191	1.197	1.201	1.204	1.208	1.210	1.213	1.215	1.217	1.219	1.220	1.222
560	1.134	1.135	1.136	1.146	1.163	1.175	1.185	1.193	1.199	1.204	1.209	1.212	1.216	1.219	1.221	1.223	1.225	1.227	1.229	1.230
580	1.138	1.140	1.141	1.152	1.169	1.182	1.192	1.200	1.206	1.212	1.216	1.220	1.224	1.227	1.229	1.232	1.234	1.236	1.237	1.239
600	1.143	1.144	1.146	1.157	1.175	1.188	1.199	1.207	1.214	1.219	1.224	1.228	1.232	1.235	1.238	1.240	1.242	1.244	1.246	1.248
620	1.148	1.149	1.151	1.162	1.181	1.195	1.206	1.214	1.221	1.227	1.232	1.236	1.240	1.243	1.246	1.249	1.251	1.253	1.255	1.256
640	1.153	1.154	1.156	1.168	1.187	1.201	1.212	1.221	1.229	1.235	1.240	1.244	1.248	1.251	1.254	1.257	1.259	1.262	1.263	1.265
660	1.158	1.159	1.160	1.173	1.193	1.208	1.219	1.229	1.236	1.242	1.248	1.252	1.256	1.260	1.263	1.266	1.268	1.270	1.272	1.274
680	1.163	1.164	1.165	1.178	1.199	1.214	1.226	1.236	1.244	1.250	1.256	1.260	1.265	1.268	1.271	1.274	1.277	1.279	1.281	1.283
700	1.167	1.169	1.170	1.184	1.205	1.221	1.233	1.243	1.251	1.258	1.264	1.269	1.273	1.276	1.280	1.283	1.285	1.288	1.290	1.292
720	1.172	1.174	1.175	1.189	1.211	1.227	1.240	1.250	1.259	1.266	1.272	1.277	1.281	1.285	1.288	1.291	1.294	1.296	1.299	1.301
740	1.177	1.179	1.180	1.195	1.217	1.234	1.247	1.258	1.266	1.274	1.280	1.285	1.289	1.293	1.297	1.300	1.303	1.305	1.307	1.309
760	1.182	1.183	1.185	1.200	1.223	1.241	1.254	1.265	1.274	1.281	1.288	1.293	1.298	1.302	1.305	1.308	1.311	1.314	1.316	1.318
780	1.187	1.188	1.190	1.205	1.229	1.247	1.261	1.272	1.282	1.289	1.296	1.301	1.306	1.310	1.314	1.317	1.320	1.323	1.325	1.327
800	1.192	1.193	1.195	1.211	1.235	1.254	1.268	1.280	1.289	1.297	1.304	1.309	1.314	1.319	1.322	1.326	1.329	1.332	1.334	1.336

x

SIGMA = 4541.0

$V - V_c = 14$

H	0.08	0.09	0.10	0.2	0.4	0.6	0.8	1.0	1.2	1.4	1.6	1.8	2.0	2.2	2.4	2.6	2.8	3.0	3.2	3.4
820	1.196	1.198	1.200	1.216	1.241	1.261	1.275	1.287	1.297	1.305	1.312	1.318	1.323	1.327	1.331	1.335	1.338	1.340	1.343	1.345
840	1.201	1.203	1.205	1.222	1.248	1.267	1.282	1.295	1.305	1.313	1.320	1.326	1.331	1.336	1.340	1.343	1.346	1.349	1.352	1.354
860	1.206	1.208	1.210	1.227	1.254	1.274	1.289	1.302	1.312	1.321	1.328	1.334	1.340	1.344	1.348	1.352	1.355	1.358	1.361	1.363
880	1.211	1.213	1.215	1.232	1.260	1.281	1.297	1.309	1.320	1.329	1.336	1.343	1.348	1.353	1.357	1.361	1.364	1.367	1.370	1.373
900	1.216	1.218	1.220	1.238	1.266	1.287	1.304	1.317	1.328	1.337	1.344	1.351	1.357	1.361	1.366	1.370	1.373	1.376	1.379	1.382
920	1.221	1.223	1.225	1.243	1.272	1.294	1.311	1.324	1.335	1.345	1.353	1.359	1.365	1.370	1.375	1.379	1.382	1.385	1.388	1.391
940	1.226	1.228	1.230	1.249	1.278	1.301	1.318	1.332	1.343	1.353	1.361	1.368	1.374	1.379	1.383	1.387	1.391	1.394	1.397	1.400
960	1.230	1.233	1.235	1.254	1.285	1.307	1.325	1.339	1.351	1.361	1.369	1.376	1.382	1.387	1.392	1.396	1.400	1.403	1.407	1.409
980	1.235	1.237	1.240	1.260	1.291	1.314	1.332	1.347	1.359	1.369	1.377	1.384	1.391	1.396	1.401	1.405	1.409	1.413	1.416	1.419
1000	1.240	1.242	1.245	1.265	1.297	1.321	1.340	1.355	1.367	1.377	1.386	1.393	1.399	1.405	1.410	1.414	1.418	1.422	1.425	1.428
1020	1.245	1.247	1.250	1.270	1.303	1.328	1.347	1.362	1.375	1.385	1.394	1.402	1.408	1.414	1.419	1.423	1.427	1.431	1.434	1.437
1040	1.250	1.252	1.255	1.276	1.309	1.335	1.354	1.370	1.383	1.393	1.402	1.410	1.417	1.423	1.428	1.432	1.436	1.440	1.443	1.447
1060	1.255	1.257	1.260	1.281	1.316	1.341	1.361	1.377	1.390	1.401	1.411	1.419	1.425	1.431	1.437	1.441	1.446	1.449	1.453	1.456
1080	1.260	1.262	1.265	1.287	1.322	1.348	1.369	1.385	1.398	1.410	1.419	1.427	1.434	1.440	1.446	1.450	1.455	1.459	1.462	1.465
1100	1.265	1.267	1.270	1.292	1.328	1.355	1.376	1.393	1.406	1.418	1.427	1.436	1.443	1.449	1.455	1.460	1.464	1.468	1.472	1.475
1120	1.269	1.272	1.275	1.298	1.334	1.362	1.383	1.400	1.414	1.426	1.436	1.444	1.452	1.458	1.464	1.469	1.473	1.477	1.481	1.484
1140	1.274	1.277	1.280	1.303	1.341	1.369	1.391	1.408	1.422	1.434	1.444	1.453	1.460	1.467	1.473	1.478	1.482	1.487	1.490	1.494
1160	1.279	1.282	1.285	1.309	1.347	1.376	1.398	1.416	1.430	1.443	1.453	1.462	1.469	1.476	1.482	1.487	1.492	1.496	1.500	1.503
1180	1.284	1.287	1.290	1.314	1.353	1.383	1.405	1.424	1.438	1.451	1.461	1.470	1.478	1.485	1.491	1.496	1.501	1.505	1.509	1.513
1200	1.289	1.292	1.295	1.320	1.360	1.390	1.413	1.431	1.446	1.459	1.470	1.479	1.487	1.494	1.500	1.506	1.510	1.515	1.519	1.522
1220	1.294	1.297	1.300	1.325	1.366	1.396	1.420	1.439	1.455	1.467	1.478	1.488	1.496	1.503	1.509	1.515	1.520	1.524	1.528	1.532
1240	1.299	1.302	1.305	1.331	1.372	1.403	1.428	1.447	1.463	1.476	1.487	1.497	1.505	1.512	1.518	1.524	1.529	1.534	1.538	1.542
1260	1.304	1.307	1.310	1.337	1.379	1.410	1.435	1.455	1.471	1.484	1.496	1.505	1.514	1.521	1.528	1.533	1.539	1.543	1.548	1.551
1280	1.309	1.312	1.315	1.342	1.385	1.417	1.442	1.463	1.479	1.493	1.504	1.514	1.523	1.530	1.537	1.543	1.548	1.553	1.557	1.561
1300	1.314	1.317	1.320	1.348	1.391	1.424	1.450	1.470	1.487	1.501	1.513	1.523	1.532	1.539	1.546	1.552	1.558	1.562	1.567	1.571
1320	1.318	1.322	1.325	1.353	1.398	1.431	1.457	1.478	1.495	1.510	1.522	1.532	1.541	1.549	1.556	1.562	1.567	1.572	1.577	1.581
1340	1.323	1.327	1.330	1.359	1.404	1.438	1.465	1.486	1.504	1.518	1.530	1.541	1.550	1.558	1.565	1.571	1.577	1.582	1.586	1.590
1360	1.328	1.332	1.335	1.364	1.411	1.445	1.472	1.494	1.512	1.527	1.539	1.550	1.559	1.567	1.574	1.581	1.586	1.591	1.596	1.600
1380	1.333	1.337	1.340	1.370	1.417	1.452	1.480	1.502	1.520	1.535	1.548	1.559	1.568	1.576	1.584	1.590	1.596	1.601	1.606	1.610
1400	1.338	1.342	1.345	1.376	1.424	1.460	1.488	1.510	1.528	1.544	1.557	1.568	1.577	1.586	1.593	1.600	1.606	1.611	1.616	1.620
1420	1.343	1.347	1.350	1.381	1.430	1.467	1.495	1.518	1.537	1.552	1.565	1.577	1.586	1.595	1.602	1.609	1.615	1.620	1.625	1.630
1440	1.348	1.352	1.355	1.387	1.436	1.474	1.503	1.526	1.545	1.561	1.574	1.586	1.596	1.604	1.612	1.619	1.625	1.630	1.635	1.640
1460	1.353	1.357	1.360	1.392	1.443	1.481	1.510	1.534	1.553	1.569	1.583	1.595	1.605	1.614	1.621	1.628	1.635	1.640	1.645	1.650
1480	1.358	1.362	1.365	1.398	1.449	1.488	1.518	1.542	1.562	1.578	1.592	1.604	1.614	1.623	1.631	1.638	1.644	1.650	1.655	1.660
1500	1.363	1.367	1.370	1.404	1.456	1.495	1.526	1.550	1.570	1.587	1.601	1.613	1.623	1.632	1.641	1.648	1.654	1.660	1.665	1.670
1520	1.368	1.372	1.375	1.409	1.462	1.502	1.533	1.558	1.578	1.595	1.610	1.622	1.633	1.642	1.650	1.657	1.664	1.670	1.675	1.680
1540	1.373	1.377	1.380	1.415	1.469	1.509	1.541	1.566	1.587	1.604	1.619	1.631	1.642	1.651	1.660	1.667	1.674	1.680	1.685	1.690
1560	1.378	1.382	1.385	1.420	1.475	1.517	1.549	1.574	1.595	1.613	1.628	1.640	1.651	1.661	1.669	1.677	1.684	1.690	1.695	1.700
1580	1.383	1.387	1.391	1.426	1.482	1.524	1.556	1.582	1.604	1.622	1.637	1.649	1.661	1.670	1.679	1.687	1.694	1.700	1.705	1.710
1600	1.388	1.392	1.396	1.432	1.488	1.531	1.564	1.591	1.612	1.630	1.646	1.659	1.670	1.680	1.689	1.696	1.703	1.710	1.715	1.721
1620	1.393	1.397	1.401	1.437	1.495	1.538	1.572	1.599	1.621	1.639	1.655	1.668	1.679	1.690	1.698	1.706	1.713	1.720	1.726	1.731
1640	1.398	1.402	1.406	1.443	1.502	1.545	1.580	1.607	1.629	1.648	1.664	1.677	1.689	1.699	1.708	1.716	1.723	1.730	1.736	1.741

x

$V - V_c = 14$ SIGMA = 4541.0

x	0.08	0.09	0.10	0.2	0.4	0.6	0.8	1.0	1.2	1.4	1.6	1.8	2.0	2.2	2.4	2.6	2.8	3.0	3.2	3.4
1660	1.403	1.407	1.411	1.449	1.508	1.553	1.587	1.615	1.638	1.657	1.673	1.686	1.698	1.709	1.718	1.726	1.733	1.740	1.746	1.751
1680	1.408	1.412	1.416	1.454	1.515	1.560	1.595	1.623	1.646	1.666	1.682	1.696	1.708	1.718	1.728	1.736	1.743	1.750	1.756	1.762
1700	1.412	1.417	1.421	1.460	1.521	1.567	1.603	1.632	1.655	1.675	1.691	1.705	1.717	1.728	1.738	1.746	1.753	1.760	1.766	1.772
1720	1.417	1.422	1.426	1.466	1.528	1.575	1.611	1.640	1.664	1.683	1.700	1.714	1.727	1.738	1.747	1.756	1.764	1.770	1.777	1.782
1740	1.422	1.427	1.431	1.472	1.535	1.582	1.619	1.648	1.672	1.692	1.709	1.724	1.737	1.748	1.757	1.766	1.774	1.781	1.787	1.793
1760	1.427	1.432	1.436	1.477	1.541	1.589	1.627	1.656	1.681	1.701	1.719	1.733	1.746	1.757	1.767	1.776	1.784	1.791	1.797	1.803
1780	1.432	1.437	1.442	1.483	1.548	1.597	1.634	1.665	1.690	1.710	1.728	1.743	1.756	1.767	1.777	1.786	1.794	1.801	1.808	1.814
1800	1.437	1.442	1.447	1.489	1.555	1.604	1.642	1.673	1.698	1.719	1.737	1.752	1.765	1.777	1.787	1.796	1.804	1.812	1.818	1.824
1820	1.442	1.447	1.452	1.494	1.561	1.611	1.650	1.682	1.707	1.728	1.746	1.762	1.775	1.787	1.797	1.806	1.814	1.822	1.829	1.835
1840	1.447	1.452	1.457	1.500	1.568	1.619	1.658	1.690	1.716	1.737	1.756	1.771	1.785	1.797	1.807	1.816	1.825	1.832	1.839	1.845
1860	1.452	1.457	1.462	1.506	1.575	1.626	1.666	1.698	1.725	1.746	1.765	1.781	1.795	1.807	1.817	1.827	1.835	1.843	1.850	1.856
1880	1.457	1.462	1.467	1.512	1.581	1.634	1.674	1.707	1.733	1.755	1.774	1.790	1.804	1.816	1.827	1.837	1.845	1.853	1.860	1.866
1900	1.462	1.467	1.472	1.517	1.588	1.641	1.682	1.715	1.742	1.765	1.784	1.800	1.814	1.826	1.837	1.847	1.856	1.864	1.871	1.877
1920	1.467	1.472	1.478	1.523	1.595	1.648	1.690	1.724	1.751	1.774	1.793	1.810	1.824	1.836	1.847	1.857	1.866	1.874	1.881	1.888
1940	1.472	1.478	1.483	1.529	1.601	1.656	1.698	1.732	1.760	1.783	1.802	1.819	1.834	1.846	1.858	1.868	1.876	1.885	1.892	1.898
1960	1.477	1.483	1.488	1.535	1.608	1.663	1.706	1.741	1.769	1.792	1.812	1.829	1.844	1.856	1.868	1.878	1.887	1.895	1.902	1.909
1980	1.482	1.488	1.493	1.540	1.615	1.671	1.714	1.749	1.778	1.801	1.821	1.839	1.853	1.867	1.878	1.888	1.897	1.906	1.913	1.920
2000	1.487	1.493	1.498	1.546	1.622	1.678	1.722	1.758	1.786	1.811	1.831	1.848	1.863	1.877	1.888	1.899	1.908	1.916	1.924	1.931
2020	1.492	1.498	1.503	1.552	1.628	1.686	1.730	1.766	1.795	1.820	1.840	1.858	1.873	1.887	1.899	1.909	1.918	1.927	1.935	1.941
2040	1.497	1.503	1.508	1.558	1.635	1.693	1.739	1.775	1.804	1.829	1.850	1.868	1.883	1.897	1.909	1.919	1.929	1.938	1.945	1.952
2060	1.502	1.508	1.514	1.564	1.642	1.701	1.747	1.783	1.813	1.838	1.859	1.878	1.893	1.907	1.919	1.930	1.940	1.948	1.956	1.963
2080	1.508	1.513	1.519	1.569	1.649	1.708	1.755	1.792	1.822	1.848	1.869	1.887	1.903	1.917	1.929	1.940	1.950	1.959	1.967	1.974
2100	1.513	1.518	1.524	1.575	1.656	1.716	1.763	1.801	1.831	1.857	1.879	1.897	1.913	1.927	1.940	1.951	1.961	1.970	1.978	1.985
2120	1.518	1.523	1.529	1.581	1.663	1.724	1.771	1.809	1.840	1.866	1.888	1.907	1.923	1.938	1.950	1.961	1.971	1.980	1.989	1.996
2140	1.523	1.529	1.534	1.587	1.669	1.731	1.779	1.818	1.849	1.876	1.898	1.917	1.933	1.948	1.961	1.972	1.982	1.991	2.000	2.007
2160	1.528	1.534	1.539	1.593	1.676	1.739	1.788	1.827	1.859	1.885	1.908	1.927	1.944	1.958	1.971	1.983	1.993	2.002	2.010	2.018
2180	1.533	1.539	1.545	1.598	1.683	1.747	1.796	1.835	1.868	1.895	1.917	1.937	1.954	1.969	1.982	1.993	2.004	2.013	2.021	2.029
2200	1.538	1.544	1.550	1.604	1.690	1.754	1.804	1.844	1.877	1.904	1.927	1.947	1.964	1.979	1.992	2.004	2.015	2.024	2.032	2.040
2220	1.543	1.549	1.555	1.610	1.697	1.762	1.812	1.853	1.886	1.913	1.937	1.957	1.974	1.989	2.003	2.015	2.025	2.035	2.043	2.051
2240	1.548	1.554	1.560	1.616	1.704	1.770	1.821	1.862	1.895	1.923	1.947	1.967	1.984	2.000	2.013	2.025	2.036	2.046	2.055	2.062
2260	1.553	1.559	1.565	1.622	1.711	1.777	1.829	1.870	1.904	1.932	1.956	1.977	1.995	2.010	2.024	2.036	2.047	2.057	2.066	2.074
2280	1.558	1.564	1.571	1.628	1.718	1.785	1.837	1.879	1.913	1.942	1.966	1.987	2.005	2.021	2.034	2.047	2.058	2.068	2.077	2.085
2300	1.563	1.569	1.576	1.634	1.725	1.793	1.846	1.888	1.923	1.952	1.976	1.997	2.015	2.031	2.045	2.058	2.069	2.079	2.088	2.096
2320	1.568	1.575	1.581	1.645	1.731	1.800	1.854	1.897	1.932	1.961	1.986	2.007	2.026	2.042	2.056	2.069	2.080	2.090	2.099	2.107
2340	1.573	1.580	1.586	1.651	1.738	1.808	1.862	1.906	1.941	1.971	1.996	2.017	2.036	2.052	2.066	2.079	2.091	2.101	2.110	2.119
2360	1.578	1.585	1.592	1.657	1.745	1.816	1.871	1.915	1.951	1.980	2.006	2.027	2.046	2.063	2.077	2.090	2.102	2.112	2.121	2.130
2380	1.583	1.590	1.597	1.663	1.752	1.824	1.879	1.924	1.960	1.990	2.016	2.038	2.057	2.073	2.088	2.101	2.113	2.123	2.133	2.141
2400	1.588	1.595	1.602	1.669	1.759	1.831	1.888	1.932	1.969	2.000	2.026	2.048	2.067	2.084	2.099	2.112	2.124	2.134	2.144	2.153
2420	1.593	1.600	1.607	1.675	1.766	1.839	1.896	1.941	1.979	2.009	2.036	2.058	2.078	2.095	2.110	2.123	2.135	2.146	2.155	2.164
2440	1.599	1.606	1.612	1.681	1.773	1.847	1.904	1.950	1.988	2.019	2.046	2.068	2.088	2.105	2.120	2.134	2.146	2.157	2.167	2.176
2460	1.604	1.611	1.618	1.687	1.780	1.855	1.913	1.959	1.997	2.029	2.056	2.079	2.099	2.116	2.131	2.145	2.157	2.168	2.178	2.187
2480	1.609	1.616	1.623	1.687	1.787	1.863	1.921	1.968	2.007	2.039	2.066	2.089	2.109	2.127	2.142	2.156	2.168	2.179	2.189	2.199

SIGMA = 4541.0

v-v_c = 14

H	3.4	3.2	3.0	2.8	2.6	2.4	2.2	2.0	1.8	1.6	1.4	1.2	1.0	0.8	0.6	0.4	0.2	0.10	0.09	0.08
x																				
2500	2.210	2.201	2.191	2.180	2.167	2.153	2.137	2.120	2.099	2.076	2.048	2.016	1.977	1.930	1.871	1.794	1.693	1.628	1.621	1.614
2520	2.222	2.212	2.202	2.191	2.178	2.164	2.148	2.130	2.110	2.086	2.058	2.026	1.986	1.938	1.879	1.801	1.699	1.633	1.626	1.619
2540	2.233	2.224	2.214	2.202	2.189	2.175	2.159	2.141	2.120	2.096	2.068	2.035	1.995	1.947	1.886	1.809	1.705	1.639	1.631	1.624
2560	2.245	2.235	2.225	2.213	2.200	2.186	2.170	2.151	2.130	2.106	2.078	2.045	2.005	1.956	1.894	1.816	1.711	1.644	1.637	1.629
2580	2.257	2.247	2.236	2.225	2.212	2.197	2.181	2.162	2.141	2.116	2.088	2.054	2.014	1.964	1.902	1.823	1.717	1.649	1.642	1.634
2600	2.268	2.259	2.248	2.236	2.223	2.208	2.192	2.173	2.151	2.127	2.098	2.064	2.023	1.973	1.910	1.830	1.723	1.654	1.647	1.639
2620	2.280	2.270	2.259	2.247	2.234	2.219	2.202	2.184	2.162	2.137	2.108	2.073	2.032	1.981	1.918	1.837	1.729	1.660	1.652	1.644
2640	2.292	2.282	2.271	2.259	2.245	2.230	2.213	2.194	2.172	2.147	2.118	2.083	2.041	1.990	1.926	1.844	1.735	1.665	1.657	1.649
2660	2.304	2.294	2.282	2.270	2.257	2.241	2.224	2.205	2.183	2.157	2.128	2.092	2.050	1.999	1.934	1.851	1.741	1.670	1.662	1.655
2680	2.315	2.305	2.294	2.282	2.268	2.253	2.235	2.216	2.193	2.168	2.138	2.102	2.059	2.007	1.942	1.858	1.747	1.675	1.668	1.660
2700	2.327	2.317	2.306	2.293	2.279	2.264	2.246	2.227	2.204	2.178	2.148	2.112	2.069	2.016	1.950	1.865	1.753	1.681	1.673	1.665
2720	2.339	2.329	2.317	2.305	2.291	2.275	2.257	2.238	2.215	2.188	2.158	2.121	2.078	2.025	1.958	1.873	1.759	1.686	1.678	1.670
2740	2.351	2.341	2.329	2.316	2.302	2.286	2.269	2.248	2.225	2.199	2.168	2.131	2.087	2.033	1.966	1.880	1.765	1.691	1.683	1.675
2760	2.363	2.352	2.341	2.328	2.314	2.298	2.280	2.259	2.236	2.209	2.178	2.141	2.096	2.042	1.974	1.887	1.771	1.697	1.688	1.680
2780	2.375	2.364	2.352	2.339	2.325	2.309	2.291	2.270	2.247	2.220	2.188	2.151	2.106	2.051	1.982	1.894	1.777	1.702	1.694	1.685
2800	2.387	2.376	2.364	2.351	2.337	2.320	2.302	2.281	2.257	2.230	2.198	2.160	2.115	2.060	1.990	1.901	1.783	1.707	1.699	1.690
2820	2.399	2.388	2.376	2.363	2.348	2.332	2.313	2.292	2.268	2.240	2.208	2.170	2.124	2.068	1.999	1.909	1.789	1.713	1.704	1.696
2840	2.411	2.400	2.388	2.374	2.360	2.343	2.324	2.303	2.279	2.251	2.218	2.180	2.134	2.077	2.007	1.916	1.795	1.718	1.709	1.701
2860	2.423	2.412	2.400	2.386	2.371	2.354	2.336	2.314	2.290	2.262	2.229	2.190	2.143	2.086	2.015	1.923	1.801	1.723	1.715	1.706
2880	2.435	2.424	2.412	2.398	2.383	2.366	2.347	2.325	2.301	2.272	2.239	2.200	2.152	2.095	2.023	1.930	1.807	1.728	1.720	1.711
2900	2.447	2.436	2.423	2.410	2.394	2.377	2.358	2.336	2.311	2.283	2.249	2.209	2.162	2.104	2.031	1.938	1.813	1.734	1.725	1.716
2920	2.459	2.448	2.435	2.422	2.406	2.389	2.369	2.347	2.322	2.293	2.259	2.219	2.171	2.113	2.039	1.945	1.819	1.739	1.730	1.721
2940	2.472	2.460	2.447	2.433	2.418	2.400	2.381	2.359	2.333	2.304	2.270	2.229	2.181	2.121	2.047	1.952	1.825	1.744	1.736	1.726
2960	2.484	2.472	2.459	2.445	2.429	2.412	2.392	2.370	2.344	2.314	2.280	2.239	2.190	2.130	2.056	1.959	1.831	1.750	1.741	1.732
2980	2.496	2.484	2.471	2.457	2.441	2.423	2.403	2.381	2.355	2.325	2.290	2.249	2.200	2.139	2.064	1.967	1.837	1.755	1.746	1.737
3000	2.508	2.497	2.483	2.469	2.453	2.435	2.415	2.392	2.366	2.336	2.301	2.259	2.209	2.148	2.072	1.974	1.843	1.760	1.751	1.742

$V-V_c = 14$ SIGMA = 4541.0

H	0.08	0.09	0.10	0.2	0.4	0.6	0.8	1.0	1.2	1.4	1.6	1.8	2.0	2.2	2.4	2.6	2.8	3.0	3.2	3.4
X																				
3100	1.768	1.778	1.787	1.874	2.011	2.113	2.193	2.257	2.309	2.353	2.390	2.421	2.448	2.472	2.494	2.512	2.529	2.544	2.558	2.570
3200	1.794	1.804	1.814	1.905	2.048	2.155	2.239	2.305	2.360	2.405	2.444	2.477	2.506	2.531	2.553	2.572	2.590	2.606	2.620	2.633
3300	1.820	1.830	1.841	1.936	2.085	2.197	2.284	2.354	2.411	2.459	2.499	2.533	2.563	2.589	2.612	2.633	2.651	2.668	2.683	2.696
3400	1.846	1.857	1.868	1.967	2.123	2.240	2.331	2.403	2.463	2.513	2.554	2.590	2.622	2.649	2.673	2.694	2.714	2.731	2.746	2.760
3500	1.872	1.883	1.895	1.998	2.161	2.283	2.377	2.453	2.515	2.567	2.611	2.648	2.681	2.709	2.734	2.756	2.776	2.794	2.811	2.825
3600	1.898	1.910	1.922	2.030	2.199	2.326	2.424	2.503	2.568	2.622	2.667	2.706	2.740	2.770	2.796	2.819	2.840	2.859	2.876	2.891
3700	1.924	1.937	1.949	2.061	2.237	2.369	2.472	2.554	2.621	2.677	2.725	2.765	2.801	2.831	2.859	2.883	2.904	2.924	2.941	2.957
3800	1.951	1.964	1.977	2.093	2.276	2.413	2.520	2.605	2.675	2.733	2.783	2.825	2.862	2.894	2.922	2.947	2.969	2.990	3.008	3.025
3900	1.977	1.991	2.004	2.125	2.315	2.458	2.568	2.657	2.730	2.790	2.841	2.885	2.923	2.956	2.986	3.012	3.035	3.056	3.075	3.092
4000	2.004	2.018	2.031	2.157	2.354	2.502	2.617	2.709	2.785	2.847	2.900	2.946	2.985	3.020	3.050	3.077	3.102	3.123	3.143	3.161
4100	2.030	2.045	2.059	2.189	2.394	2.547	2.666	2.762	2.840	2.905	2.960	3.007	3.048	3.084	3.116	3.144	3.169	3.191	3.212	3.230
4200	2.057	2.072	2.087	2.222	2.434	2.593	2.716	2.815	2.896	2.963	3.021	3.069	3.112	3.149	3.182	3.211	3.237	3.260	3.281	3.301
4300	2.083	2.099	2.115	2.254	2.474	2.638	2.766	2.869	2.953	3.022	3.081	3.132	3.176	3.214	3.248	3.278	3.305	3.330	3.352	3.372
4400	2.110	2.126	2.142	2.287	2.514	2.684	2.817	2.923	3.010	3.082	3.143	3.195	3.241	3.281	3.316	3.347	3.375	3.400	3.423	3.443
4500	2.137	2.154	2.170	2.320	2.555	2.731	2.868	2.977	3.067	3.142	3.205	3.259	3.306	3.347	3.384	3.416	3.445	3.471	3.494	3.516
4600	2.164	2.181	2.198	2.353	2.596	2.778	2.919	3.033	3.125	3.202	3.268	3.324	3.372	3.415	3.452	3.486	3.516	3.542	3.567	3.589
4700	2.191	2.209	2.226	2.386	2.637	2.825	2.971	3.088	3.184	3.264	3.331	3.389	3.439	3.483	3.522	3.556	3.587	3.615	3.640	3.663
4800	2.218	2.237	2.255	2.419	2.678	2.872	3.023	3.144	3.243	3.325	3.395	3.455	3.507	3.552	3.592	3.627	3.659	3.688	3.714	3.737
4900	2.245	2.264	2.283	2.453	2.720	2.920	3.076	3.201	3.303	3.388	3.460	3.521	3.575	3.621	3.663	3.699	3.732	3.762	3.788	3.813
5000	2.272	2.292	2.311	2.487	2.762	2.969	3.129	3.258	3.363	3.451	3.525	3.588	3.643	3.692	3.734	3.772	3.806	3.836	3.864	3.889
5100	2.300	2.320	2.340	2.521	2.804	3.017	3.183	3.315	3.424	3.514	3.590	3.656	3.713	3.762	3.806	3.845	3.880	3.911	3.940	3.966
5200	2.327	2.348	2.368	2.555	2.847	3.066	3.237	3.373	3.485	3.578	3.657	3.724	3.783	3.834	3.879	3.919	3.955	3.987	4.017	4.043
5300	2.355	2.376	2.397	2.589	2.890	3.116	3.291	3.432	3.547	3.643	3.724	3.793	3.853	3.906	3.953	3.994	4.031	4.064	4.094	4.122
5400	2.382	2.404	2.426	2.623	2.933	3.165	3.346	3.491	3.609	3.708	3.791	3.863	3.925	3.979	4.027	4.069	4.105	4.142	4.173	4.201
5500	2.410	2.432	2.455	2.658	2.976	3.215	3.401	3.550	3.672	3.773	3.859	3.933	3.997	4.052	4.102	4.145	4.185	4.220	4.252	4.281
5600	2.437	2.461	2.484	2.692	3.020	3.266	3.457	3.610	3.735	3.840	3.928	4.004	4.069	4.127	4.177	4.222	4.262	4.299	4.332	4.361
5700	2.465	2.489	2.513	2.727	3.064	3.317	3.513	3.671	3.799	3.907	3.997	4.075	4.142	4.201	4.253	4.300	4.341	4.378	4.412	4.443
5800	2.493	2.517	2.542	2.762	3.108	3.368	3.570	3.732	3.864	3.974	4.067	4.147	4.216	4.277	4.330	4.378	4.420	4.459	4.493	4.525
5900	2.521	2.546	2.571	2.797	3.153	3.420	3.627	3.793	3.929	4.042	4.138	4.220	4.291	4.353	4.408	4.457	4.501	4.540	4.575	4.608
6000	2.548	2.575	2.600	2.832	3.198	3.471	3.684	3.855	3.994	4.110	4.209	4.293	4.366	4.430	4.486	4.536	4.581	4.622	4.658	4.691
6100	2.576	2.603	2.629	2.868	3.243	3.524	3.742	3.917	4.060	4.180	4.280	4.367	4.442	4.507	4.565	4.617	4.663	4.704	4.742	4.776
6200	2.605	2.632	2.659	2.904	3.288	3.576	3.801	3.980	4.127	4.249	4.353	4.441	4.518	4.586	4.645	4.698	4.745	4.787	4.826	4.861
6300	2.633	2.661	2.688	2.939	3.334	3.629	3.859	4.043	4.194	4.319	4.426	4.517	4.595	4.664	4.725	4.779	4.828	4.871	4.911	4.947
6400	2.661	2.690	2.718	2.975	3.380	3.683	3.919	4.107	4.262	4.390	4.499	4.592	4.673	4.744	4.806	4.862	4.912	4.956	4.997	5.033
6500	2.689	2.719	2.748	3.011	3.426	3.737	3.978	4.172	4.330	4.462	4.573	4.669	4.752	4.824	4.888	4.945	4.996	5.042	5.083	5.121
6600	2.718	2.748	2.778	3.048	3.472	3.791	4.038	4.236	4.398	4.534	4.648	4.746	4.831	4.905	4.971	5.029	5.081	5.128	5.170	5.209
6700	2.746	2.777	2.807	3.084	3.519	3.845	4.099	4.302	4.468	4.606	4.723	4.823	4.910	4.986	5.054	5.113	5.167	5.215	5.258	5.298
6800	2.774	2.806	2.837	3.121	3.566	3.900	4.160	4.367	4.537	4.679	4.799	4.902	4.991	5.069	5.137	5.198	5.253	5.302	5.347	5.387
6900	2.803	2.836	2.868	3.158	3.613	3.955	4.221	4.434	4.608	4.753	4.875	4.981	5.072	5.151	5.222	5.284	5.340	5.391	5.436	5.478
7000	2.832	2.865	2.898	3.194	3.661	4.011	4.283	4.502	4.678	4.827	4.952	5.060	5.153	5.235	5.307	5.371	5.428	5.480	5.526	5.569
7100	2.860	2.894	2.928	3.232	3.709	4.067	4.345	4.568	4.750	4.902	5.030	5.140	5.236	5.319	5.393	5.458	5.517	5.570	5.617	5.661
7200	2.889	2.924	2.958	3.269	3.757	4.123	4.408	4.635	4.822	4.977	5.108	5.221	5.319	5.404	5.479	5.546	5.606	5.660	5.709	5.753

$V - V_c = 14$ SIGMA = 4541.0

H	0.08	0.09	0.10	0.2	0.4	0.6	0.8	1.0	1.2	1.4	1.6	1.8	2.0	2.2	2.4	2.6	2.8	3.0	3.2	3.4
7300	2.918	2.954	2.989	3.306	3.805	4.180	4.471	4.704	4.894	5.053	5.187	5.302	5.402	5.489	5.567	5.635	5.696	5.751	5.801	5.847
7400	2.947	2.983	3.019	3.344	3.854	4.237	4.534	4.772	4.967	5.129	5.267	5.384	5.486	5.576	5.654	5.724	5.787	5.843	5.894	5.941
7500	2.976	3.013	3.050	3.382	3.903	4.294	4.598	4.841	5.040	5.206	5.347	5.467	5.571	5.662	5.743	5.815	5.879	5.936	5.988	6.036
7600	3.005	3.043	3.080	3.419	3.952	4.352	4.662	4.911	5.114	5.284	5.427	5.550	5.657	5.750	5.832	5.905	5.971	6.030	6.083	6.131
7700	3.034	3.073	3.111	3.457	4.002	4.410	4.727	4.981	5.189	5.362	5.509	5.634	5.743	5.838	5.922	5.997	6.064	6.124	6.178	6.228
7800	3.063	3.103	3.142	3.496	4.051	4.468	4.792	5.052	5.264	5.441	5.590	5.719	5.830	5.927	6.013	6.089	6.157	6.219	6.274	6.325
7900	3.093	3.133	3.173	3.534	4.101	4.527	4.858	5.123	5.339	5.520	5.673	5.804	5.917	6.017	6.104	6.182	6.252	6.314	6.371	6.423
8000	3.122	3.163	3.204	3.573	4.152	4.586	4.924	5.194	5.416	5.600	5.756	5.889	6.005	6.107	6.196	6.276	6.347	6.411	6.469	6.521
8100	3.152	3.194	3.235	3.611	4.202	4.646	4.991	5.267	5.492	5.680	5.839	5.976	6.094	6.198	6.289	6.370	6.443	6.508	6.567	6.621
8200	3.181	3.224	3.266	3.650	4.253	4.706	5.058	5.339	5.569	5.761	5.924	6.063	6.183	6.289	6.382	6.465	6.539	6.606	6.666	6.721
8300	3.211	3.255	3.298	3.689	4.305	4.766	5.125	5.412	5.647	5.843	6.008	6.150	6.274	6.381	6.476	6.561	6.636	6.704	6.766	6.822
8400	3.240	3.285	3.329	3.728	4.356	4.827	5.193	5.486	5.725	5.925	6.094	6.239	6.364	6.474	6.571	6.657	6.734	6.804	6.866	6.923
8500	3.270	3.316	3.361	3.768	4.408	4.888	5.261	5.560	5.804	6.008	6.180	6.328	6.456	6.568	6.666	6.754	6.833	6.904	6.968	7.026
8600	3.300	3.346	3.392	3.807	4.460	4.949	5.330	5.634	5.883	6.091	6.266	6.417	6.548	6.662	6.762	6.852	6.932	7.004	7.070	7.129
8700	3.330	3.377	3.424	3.847	4.512	5.011	5.399	5.709	5.963	6.175	6.354	6.507	6.640	6.757	6.859	6.951	7.032	7.106	7.172	7.233
8800	3.360	3.408	3.455	3.887	4.565	5.073	5.468	5.785	6.043	6.259	6.442	6.598	6.733	6.852	6.955	7.050	7.133	7.208	7.276	7.337
8900	3.390	3.439	3.487	3.927	4.617	5.135	5.538	5.861	6.124	6.344	6.530	6.689	6.827	6.948	7.055	7.150	7.234	7.311	7.380	7.443
9000	3.420	3.470	3.519	3.967	4.671	5.198	5.609	5.937	6.206	6.429	6.619	6.781	6.922	7.045	7.154	7.250	7.337	7.414	7.485	7.549
9100	3.450	3.501	3.551	4.007	4.724	5.261	5.679	6.014	6.288	6.516	6.708	6.874	7.017	7.143	7.253	7.352	7.440	7.519	7.591	7.656
9200	3.480	3.532	3.583	4.048	4.778	5.325	5.751	6.091	6.370	6.602	6.799	6.967	7.113	7.241	7.353	7.454	7.543	7.624	7.697	7.763
9300	3.510	3.563	3.615	4.088	4.832	5.389	5.822	6.169	6.453	6.689	6.889	7.061	7.210	7.340	7.454	7.556	7.648	7.730	7.804	7.872
9400	3.541	3.595	3.648	4.129	4.886	5.453	5.894	6.248	6.536	6.777	6.981	7.155	7.307	7.439	7.556	7.660	7.753	7.836	7.912	7.981
9500	3.571	3.626	3.680	4.170	4.940	5.518	5.967	6.326	6.620	6.865	7.073	7.251	7.405	7.539	7.658	7.764	7.858	7.944	8.021	8.091
9600	3.602	3.658	3.713	4.211	4.995	5.583	6.040	6.406	6.705	6.954	7.165	7.346	7.503	7.640	7.761	7.869	7.965	8.052	8.130	8.201
9700	3.632	3.689	3.745	4.253	5.050	5.648	6.113	6.486	6.790	7.044	7.259	7.443	7.602	7.742	7.865	7.974	8.072	8.160	8.240	8.313
9800	3.663	3.721	3.778	4.294	5.105	5.714	6.187	6.566	6.876	7.134	7.352	7.540	7.702	7.844	7.969	8.080	8.180	8.270	8.351	8.425
9900	3.694	3.753	3.810	4.336	5.161	5.780	6.262	6.647	6.962	7.224	7.447	7.637	7.802	7.947	8.074	8.187	8.289	8.380	8.463	8.538
10000	3.725	3.784	3.843	4.377	5.217	5.847	6.336	6.728	7.049	7.316	7.542	7.735	7.903	8.050	8.180	8.295	8.398	8.491	8.575	8.651
10100	3.755	3.816	3.876	4.419	5.273	5.913	6.411	6.810	7.136	7.407	7.637	7.834	8.005	8.154	8.286	8.403	8.508	8.603	8.688	8.766
10200	3.786	3.848	3.909	4.461	5.330	5.981	6.487	6.892	7.223	7.500	7.733	7.934	8.107	8.259	8.393	8.512	8.619	8.715	8.802	8.881
10300	3.817	3.880	3.942	4.504	5.386	6.048	6.563	6.975	7.312	7.593	7.830	8.034	8.210	8.365	8.501	8.622	8.730	8.828	8.916	8.996
10400	3.848	3.912	3.975	4.546	5.443	6.116	6.639	7.058	7.400	7.686	7.927	8.134	8.314	8.471	8.609	8.733	8.843	8.942	9.032	9.113
10500	3.880	3.945	4.009	4.589	5.501	6.184	6.716	7.142	7.490	7.780	8.025	8.236	8.418	8.578	8.719	8.844	8.955	9.056	9.148	9.230
10600	3.911	3.977	4.042	4.631	5.558	6.253	6.793	7.226	7.580	7.874	8.124	8.338	8.523	8.685	8.828	8.955	9.069	9.172	9.264	9.349
10700	3.942	4.009	4.075	4.674	5.616	6.322	6.871	7.311	7.670	7.970	8.223	8.440	8.629	8.793	8.939	9.068	9.184	9.288	9.382	9.467
10800	3.974	4.042	4.109	4.717	5.674	6.391	6.949	7.396	7.761	8.065	8.323	8.544	8.735	8.902	9.050	9.181	9.299	9.404	9.500	9.587
10900	4.005	4.074	4.142	4.761	5.732	6.461	7.028	7.481	7.852	8.162	8.423	8.647	8.842	9.012	9.162	9.295	9.414	9.522	9.619	9.707
11000	4.036	4.107	4.176	4.804	5.791	6.531	7.107	7.568	7.944	8.258	8.524	8.752	8.949	9.122	9.274	9.410	9.531	9.640	9.739	9.828
11100	4.068	4.140	4.210	4.848	5.850	6.602	7.186	7.654	8.037	8.356	8.626	8.857	9.057	9.233	9.388	9.525	9.648	9.759	9.859	9.950
11200	4.10	4.17	4.24	4.89	5.91	6.67	7.27	7.74	8.13	8.45	8.73	8.96	9.17	9.34	9.50	9.64	9.77	9.88	9.98	10.07
11300	4.13	4.21	4.28	4.94	5.97	6.74	7.35	7.83	8.22	8.55	8.83	9.07	9.28	9.46	9.62	9.76	9.88	10.00	10.10	10.20
11400	4.16	4.24	4.31	4.98	6.03	6.82	7.43	7.92	8.32	8.65	8.93	9.18	9.39	9.57	9.73	9.88	10.00	10.12	10.23	10.32

x

SIGMA = 4541.0

$V-V_c = 14$

H \ x	0.08	0.09	0.10	0.2	0.4	0.6	0.8	1.0	1.2	1.4	1.6	1.8	2.0	2.2	2.4	2.6	2.8	3.0	3.2	3.4
11500	4.20	4.27	4.35	5.02	6.09	6.89	7.51	8.01	8.41	8.75	9.04	9.28	9.50	9.68	9.85	9.99	10.12	10.24	10.35	10.45
11600	4.23	4.30	4.38	5.07	6.15	6.96	7.59	8.09	8.51	8.85	9.14	9.39	9.61	9.80	9.96	10.11	10.25	10.36	10.47	10.57
11730	4.26	4.34	4.41	5.11	6.21	7.03	7.67	8.18	8.60	8.95	9.25	9.50	9.72	9.91	10.08	10.23	10.37	10.49	10.60	10.70
11830	4.29	4.37	4.45	5.16	6.27	7.11	7.75	8.27	8.70	9.05	9.35	9.61	9.83	10.03	10.20	10.35	10.49	10.61	10.72	10.82
11900	4.32	4.40	4.48	5.20	6.33	7.18	7.84	8.36	8.80	9.16	9.46	9.72	9.95	10.14	10.32	10.47	10.61	10.74	10.85	10.95
12000	4.36	4.44	4.52	5.25	6.39	7.25	7.92	8.46	8.89	9.26	9.57	9.83	10.06	10.26	10.44	10.59	10.74	10.86	10.98	11.08
12100	4.39	4.47	4.55	5.29	6.45	7.33	8.00	8.55	8.99	9.36	9.67	9.94	10.17	10.38	10.56	10.72	10.86	10.99	11.10	11.21
12200	4.42	4.50	4.59	5.34	6.52	7.40	8.09	8.64	9.09	9.46	9.78	10.05	10.29	10.50	10.68	10.84	10.98	11.12	11.23	11.34
12300	4.45	4.54	4.62	5.38	6.58	7.48	8.17	8.73	9.19	9.57	9.89	10.17	10.41	10.61	10.80	10.96	11.11	11.24	11.36	11.47
12400	4.49	4.57	4.66	5.43	6.64	7.55	8.26	8.82	9.29	9.67	10.00	10.28	10.52	10.73	10.92	11.09	11.24	11.37	11.49	11.60
12500	4.52	4.61	4.69	5.47	6.70	7.63	8.34	8.92	9.39	9.78	10.11	10.39	10.64	10.85	11.04	11.21	11.36	11.50	11.62	11.74
12600	4.55	4.64	4.73	5.52	6.77	7.70	8.43	9.01	9.49	9.88	10.22	10.51	10.76	10.98	11.17	11.34	11.49	11.63	11.75	11.87
12700	4.58	4.67	4.76	5.57	6.83	7.78	8.52	9.11	9.59	9.99	10.33	10.62	10.88	11.10	11.29	11.47	11.62	11.76	11.89	12.00
12800	4.62	4.71	4.80	5.61	6.89	7.86	8.60	9.20	9.69	10.10	10.44	10.74	10.99	11.22	11.42	11.59	11.75	11.89	12.02	12.14
12900	4.65	4.74	4.83	5.66	6.96	7.93	8.69	9.30	9.79	10.21	10.56	10.85	11.11	11.34	11.54	11.72	11.88	12.02	12.15	12.27
13000	4.68	4.78	4.87	5.71	7.02	8.01	8.78	9.39	9.89	10.31	10.67	10.97	11.24	11.47	11.67	11.85	12.01	12.16	12.29	12.41
13100	4.71	4.81	4.90	5.75	7.09	8.09	8.87	9.49	10.00	10.42	10.78	11.09	11.36	11.59	11.80	11.98	12.14	12.29	12.42	12.54
13200	4.75	4.84	4.94	5.80	7.15	8.17	8.95	9.59	10.10	10.53	10.90	11.21	11.48	11.72	11.92	12.11	12.27	12.42	12.56	12.68
13300	4.78	4.88	4.97	5.85	7.22	8.24	9.04	9.68	10.21	10.64	11.01	11.33	11.60	11.84	12.05	12.24	12.41	12.56	12.70	12.82
13400	4.81	4.91	5.01	5.89	7.28	8.32	9.13	9.78	10.31	10.75	11.13	11.45	11.72	11.97	12.18	12.37	12.54	12.69	12.83	12.96
13500	4.85	4.95	5.05	5.94	7.35	8.40	9.22	9.88	10.42	10.86	11.24	11.57	11.85	12.09	12.31	12.50	12.68	12.83	12.97	13.10
13600	4.88	4.98	5.08	5.99	7.41	8.48	9.31	9.98	10.52	10.97	11.36	11.69	11.97	12.22	12.44	12.64	12.81	12.97	13.11	13.24
13700	4.91	5.02	5.12	6.04	7.48	8.56	9.40	10.08	10.63	11.09	11.47	11.81	12.10	12.35	12.57	12.77	12.95	13.11	13.25	13.38
13800	4.95	5.05	5.15	6.08	7.55	8.64	9.49	10.18	10.73	11.20	11.59	11.93	12.22	12.48	12.70	12.90	13.08	13.24	13.39	13.52
13900	4.98	5.09	5.19	6.13	7.61	8.72	9.59	10.28	10.84	11.31	11.71	12.05	12.35	12.61	12.84	13.04	13.22	13.38	13.53	13.67
14000	5.02	5.12	5.23	6.18	7.68	8.80	9.68	10.38	10.95	11.43	11.83	12.18	12.48	12.74	12.97	13.17	13.36	13.52	13.67	13.81
14100	5.05	5.16	5.26	6.23	7.75	8.88	9.77	10.48	11.06	11.54	11.95	12.30	12.60	12.87	13.10	13.31	13.50	13.66	13.82	13.95
14200	5.08	5.19	5.30	6.28	7.81	8.97	9.86	10.58	11.17	11.66	12.07	12.42	12.73	13.00	13.24	13.45	13.64	13.81	13.96	14.10
14300	5.12	5.23	5.34	6.33	7.88	9.05	9.96	10.68	11.28	11.77	12.19	12.55	12.86	13.13	13.37	13.59	13.78	13.95	14.11	14.25
14400	5.15	5.26	5.37	6.38	7.95	9.13	10.05	10.78	11.39	11.89	12.31	12.67	12.99	13.26	13.51	13.72	13.92	14.09	14.25	14.39
14500	5.18	5.30	5.41	6.42	8.02	9.21	10.14	10.89	11.50	12.00	12.43	12.80	13.12	13.40	13.64	13.86	14.06	14.24	14.39	14.54
14600	5.22	5.33	5.45	6.47	8.09	9.30	10.24	10.99	11.61	12.12	12.56	12.93	13.25	13.53	13.78	14.00	14.20	14.38	14.54	14.69
14700	5.25	5.37	5.48	6.52	8.16	9.38	10.33	11.10	11.72	12.24	12.68	13.05	13.38	13.67	13.92	14.14	14.34	14.52	14.69	14.84
14800	5.29	5.41	5.52	6.57	8.23	9.46	10.43	11.20	11.83	12.36	12.80	13.18	13.51	13.80	14.06	14.28	14.48	14.67	14.84	14.99
14900	5.32	5.44	5.56	6.62	8.29	9.55	10.52	11.31	11.94	12.48	12.93	13.31	13.65	13.94	14.20	14.43	14.63	14.82	14.98	15.14
15000	5.36	5.48	5.60	6.67	8.36	9.63	10.62	11.41	12.06	12.60	13.05	13.44	13.78	14.08	14.34	14.57	14.78	14.96	15.13	15.29

SIGMA = 3956.0

V−V$_c$ = 15

H X	0.08	0.09	0.10	0.2	0.4	0.6	0.8	1.0	1.2	1.4	1.6	1.8	2.0	2.2	2.4	2.6	2.8	3.0	3.2	3.4
0	1.000	1.000	1.000	1.000	1.000	1.000	1.000	1.000	1.000	1.000	1.000	1.000	1.000	1.000	1.000	1.000	1.000	1.000	1.000	1.000
20	1.005	1.005	1.006	1.006	1.007	1.007	1.007	1.008	1.008	1.008	1.008	1.008	1.008	1.009	1.009	1.009	1.009	1.009	1.009	1.009
40	1.011	1.011	1.011	1.012	1.013	1.014	1.015	1.015	1.016	1.016	1.016	1.017	1.017	1.017	1.017	1.017	1.018	1.018	1.018	1.018
60	1.016	1.016	1.017	1.018	1.020	1.021	1.022	1.023	1.024	1.024	1.025	1.025	1.025	1.026	1.026	1.026	1.027	1.027	1.027	1.027
80	1.022	1.022	1.022	1.024	1.026	1.028	1.029	1.031	1.031	1.032	1.033	1.034	1.034	1.034	1.035	1.035	1.036	1.036	1.036	1.036
100	1.027	1.027	1.028	1.030	1.033	1.035	1.037	1.038	1.039	1.040	1.041	1.042	1.043	1.043	1.044	1.044	1.044	1.045	1.045	1.045
120	1.033	1.033	1.033	1.036	1.039	1.042	1.044	1.046	1.047	1.049	1.050	1.050	1.051	1.051	1.053	1.053	1.053	1.054	1.054	1.054
140	1.038	1.038	1.039	1.041	1.046	1.049	1.052	1.054	1.055	1.057	1.058	1.059	1.060	1.061	1.061	1.062	1.062	1.063	1.063	1.064
160	1.044	1.044	1.044	1.047	1.052	1.056	1.059	1.061	1.063	1.065	1.066	1.067	1.068	1.069	1.070	1.071	1.071	1.072	1.073	1.073
180	1.049	1.049	1.050	1.053	1.059	1.063	1.067	1.069	1.071	1.073	1.075	1.076	1.077	1.078	1.079	1.080	1.081	1.081	1.082	1.082
200	1.054	1.055	1.055	1.059	1.066	1.070	1.074	1.077	1.080	1.082	1.083	1.085	1.086	1.087	1.088	1.089	1.090	1.090	1.091	1.092
220	1.060	1.060	1.061	1.065	1.072	1.078	1.082	1.085	1.088	1.090	1.092	1.093	1.095	1.096	1.097	1.098	1.099	1.100	1.100	1.101
240	1.065	1.066	1.067	1.071	1.079	1.085	1.089	1.093	1.096	1.098	1.100	1.102	1.104	1.105	1.106	1.107	1.108	1.109	1.110	1.110
260	1.071	1.072	1.072	1.077	1.086	1.092	1.097	1.101	1.104	1.107	1.109	1.111	1.112	1.114	1.115	1.116	1.117	1.118	1.119	1.120
280	1.076	1.077	1.078	1.083	1.092	1.099	1.104	1.109	1.112	1.115	1.117	1.119	1.121	1.123	1.124	1.125	1.127	1.127	1.128	1.129
300	1.082	1.083	1.083	1.089	1.099	1.106	1.112	1.117	1.120	1.123	1.126	1.128	1.130	1.132	1.133	1.135	1.136	1.137	1.138	1.139
320	1.087	1.088	1.089	1.095	1.106	1.114	1.120	1.125	1.129	1.132	1.135	1.137	1.139	1.141	1.143	1.144	1.145	1.146	1.148	1.148
340	1.093	1.094	1.094	1.102	1.113	1.121	1.127	1.133	1.137	1.140	1.143	1.146	1.148	1.150	1.152	1.153	1.155	1.156	1.157	1.158
360	1.098	1.099	1.100	1.108	1.119	1.128	1.135	1.141	1.145	1.149	1.152	1.155	1.157	1.159	1.161	1.163	1.164	1.165	1.167	1.168
380	1.104	1.105	1.106	1.114	1.126	1.136	1.143	1.149	1.153	1.157	1.161	1.164	1.166	1.168	1.170	1.172	1.174	1.175	1.176	1.177
400	1.109	1.110	1.111	1.120	1.133	1.143	1.151	1.157	1.162	1.166	1.170	1.173	1.175	1.178	1.180	1.182	1.183	1.185	1.186	1.187
420	1.115	1.116	1.117	1.126	1.140	1.150	1.158	1.165	1.170	1.175	1.178	1.182	1.184	1.187	1.189	1.191	1.193	1.194	1.196	1.197
440	1.120	1.121	1.122	1.132	1.147	1.158	1.166	1.173	1.179	1.183	1.187	1.191	1.194	1.196	1.198	1.200	1.202	1.204	1.205	1.207
460	1.126	1.127	1.128	1.138	1.153	1.165	1.174	1.181	1.187	1.192	1.196	1.200	1.203	1.206	1.208	1.210	1.212	1.214	1.215	1.217
480	1.131	1.133	1.134	1.144	1.160	1.172	1.182	1.189	1.196	1.201	1.205	1.209	1.212	1.215	1.217	1.220	1.222	1.223	1.225	1.226
500	1.137	1.138	1.139	1.150	1.167	1.180	1.190	1.198	1.204	1.209	1.214	1.218	1.221	1.224	1.227	1.229	1.231	1.233	1.235	1.236
520	1.142	1.144	1.145	1.156	1.174	1.187	1.198	1.206	1.213	1.218	1.223	1.227	1.231	1.234	1.236	1.239	1.241	1.243	1.245	1.246
540	1.148	1.149	1.151	1.162	1.181	1.195	1.205	1.214	1.221	1.227	1.232	1.236	1.240	1.243	1.246	1.248	1.251	1.253	1.255	1.256
560	1.154	1.155	1.156	1.168	1.188	1.202	1.213	1.222	1.230	1.236	1.241	1.245	1.249	1.253	1.256	1.258	1.261	1.263	1.265	1.266
580	1.159	1.160	1.162	1.175	1.195	1.210	1.221	1.230	1.238	1.245	1.250	1.255	1.259	1.262	1.265	1.268	1.270	1.273	1.275	1.277
600	1.165	1.166	1.168	1.181	1.202	1.217	1.229	1.239	1.247	1.254	1.259	1.264	1.268	1.272	1.275	1.278	1.280	1.283	1.285	1.287
620	1.170	1.172	1.173	1.187	1.209	1.225	1.237	1.247	1.256	1.262	1.268	1.273	1.278	1.281	1.285	1.288	1.290	1.293	1.295	1.297
640	1.176	1.177	1.179	1.193	1.216	1.232	1.245	1.256	1.264	1.271	1.277	1.282	1.287	1.291	1.294	1.298	1.300	1.303	1.305	1.307
660	1.181	1.183	1.185	1.199	1.222	1.240	1.253	1.264	1.273	1.280	1.287	1.292	1.297	1.301	1.304	1.307	1.310	1.313	1.315	1.317
680	1.187	1.189	1.190	1.205	1.229	1.247	1.261	1.273	1.282	1.289	1.296	1.301	1.306	1.310	1.314	1.317	1.320	1.323	1.325	1.328
700	1.192	1.194	1.196	1.212	1.236	1.255	1.270	1.281	1.291	1.298	1.305	1.311	1.316	1.320	1.324	1.327	1.330	1.333	1.336	1.338
720	1.198	1.200	1.202	1.218	1.243	1.263	1.278	1.290	1.299	1.307	1.314	1.320	1.325	1.330	1.334	1.337	1.341	1.343	1.346	1.348
740	1.204	1.205	1.207	1.224	1.250	1.270	1.286	1.298	1.308	1.317	1.324	1.330	1.335	1.340	1.344	1.347	1.351	1.354	1.356	1.359
760	1.209	1.211	1.213	1.230	1.258	1.278	1.294	1.307	1.317	1.326	1.333	1.339	1.345	1.350	1.354	1.358	1.361	1.364	1.367	1.369
780	1.215	1.217	1.219	1.237	1.265	1.286	1.302	1.315	1.326	1.335	1.342	1.349	1.355	1.359	1.364	1.368	1.371	1.374	1.377	1.380
800	1.220	1.222	1.224	1.243	1.272	1.293	1.310	1.324	1.335	1.344	1.352	1.359	1.364	1.369	1.374	1.378	1.381	1.385	1.387	1.390

V-V$_c$ = 15   SIGMA =   3956.0

H / x	3.4	3.2	3.0	2.8	2.6	2.4	2.2	2.0	1.8	1.6	1.4	1.2	1.0	0.8	0.6	0.4	0.2	0.10	0.09	0.08
820	1.401	1.398	1.395	1.392	1.388	1.384	1.379	1.374	1.368	1.361	1.353	1.344	1.332	1.318	1.301	1.279	1.249	1.230	1.228	1.226
840	1.411	1.408	1.405	1.402	1.398	1.394	1.389	1.384	1.378	1.371	1.362	1.353	1.341	1.327	1.309	1.286	1.255	1.236	1.234	1.231
860	1.422	1.419	1.416	1.412	1.409	1.404	1.399	1.394	1.388	1.380	1.372	1.362	1.350	1.335	1.317	1.293	1.261	1.241	1.239	1.237
880	1.433	1.430	1.426	1.423	1.419	1.414	1.409	1.404	1.397	1.390	1.381	1.371	1.358	1.343	1.324	1.300	1.268	1.247	1.245	1.243
900	1.443	1.440	1.437	1.433	1.429	1.425	1.419	1.414	1.407	1.399	1.390	1.380	1.367	1.352	1.332	1.307	1.274	1.253	1.251	1.248
920	1.454	1.451	1.448	1.444	1.440	1.435	1.430	1.424	1.417	1.409	1.400	1.389	1.376	1.360	1.340	1.314	1.280	1.259	1.256	1.254
940	1.465	1.462	1.458	1.454	1.450	1.445	1.440	1.434	1.427	1.419	1.409	1.398	1.385	1.368	1.348	1.322	1.287	1.264	1.262	1.259
960	1.476	1.472	1.469	1.465	1.460	1.456	1.450	1.444	1.437	1.428	1.419	1.407	1.393	1.377	1.356	1.329	1.293	1.270	1.268	1.265
980	1.487	1.483	1.480	1.475	1.471	1.466	1.460	1.454	1.446	1.438	1.428	1.416	1.402	1.385	1.364	1.336	1.299	1.276	1.273	1.271
1000	1.497	1.494	1.490	1.486	1.481	1.476	1.470	1.464	1.456	1.448	1.438	1.426	1.411	1.394	1.372	1.343	1.306	1.282	1.279	1.276
1020	1.508	1.505	1.501	1.497	1.492	1.487	1.481	1.474	1.466	1.457	1.447	1.435	1.420	1.402	1.379	1.350	1.312	1.287	1.285	1.282
1040	1.519	1.516	1.512	1.508	1.503	1.497	1.491	1.484	1.476	1.467	1.457	1.444	1.429	1.410	1.387	1.358	1.318	1.293	1.290	1.287
1060	1.530	1.527	1.523	1.518	1.513	1.508	1.502	1.494	1.486	1.477	1.466	1.453	1.438	1.419	1.395	1.365	1.325	1.299	1.296	1.293
1080	1.542	1.538	1.534	1.529	1.524	1.518	1.512	1.505	1.496	1.487	1.476	1.463	1.447	1.427	1.403	1.372	1.331	1.305	1.302	1.299
1100	1.553	1.549	1.545	1.540	1.535	1.529	1.522	1.515	1.507	1.497	1.485	1.472	1.456	1.436	1.411	1.380	1.337	1.310	1.307	1.304
1120	1.564	1.560	1.556	1.551	1.545	1.540	1.533	1.525	1.517	1.507	1.495	1.481	1.465	1.445	1.419	1.387	1.344	1.316	1.313	1.310
1140	1.575	1.571	1.567	1.562	1.556	1.550	1.543	1.536	1.527	1.517	1.505	1.491	1.474	1.453	1.427	1.394	1.350	1.322	1.319	1.316
1160	1.586	1.582	1.578	1.573	1.567	1.561	1.554	1.546	1.537	1.527	1.514	1.500	1.483	1.462	1.435	1.402	1.356	1.328	1.325	1.321
1180	1.598	1.593	1.589	1.584	1.578	1.572	1.565	1.556	1.547	1.537	1.524	1.510	1.492	1.470	1.444	1.409	1.363	1.333	1.330	1.327
1200	1.609	1.605	1.600	1.595	1.589	1.582	1.575	1.567	1.557	1.547	1.534	1.519	1.501	1.479	1.452	1.416	1.369	1.339	1.336	1.333
1220	1.620	1.616	1.611	1.606	1.600	1.593	1.586	1.577	1.568	1.557	1.544	1.528	1.510	1.488	1.460	1.424	1.376	1.345	1.342	1.338
1240	1.632	1.627	1.622	1.617	1.611	1.604	1.596	1.588	1.578	1.567	1.554	1.538	1.519	1.496	1.468	1.431	1.382	1.351	1.347	1.344
1260	1.643	1.638	1.633	1.628	1.622	1.615	1.607	1.598	1.588	1.577	1.563	1.548	1.528	1.505	1.476	1.438	1.388	1.357	1.353	1.350
1280	1.654	1.650	1.645	1.639	1.633	1.626	1.618	1.609	1.599	1.587	1.573	1.557	1.538	1.514	1.484	1.446	1.395	1.362	1.359	1.355
1300	1.666	1.661	1.656	1.650	1.644	1.637	1.629	1.620	1.609	1.597	1.583	1.567	1.547	1.523	1.492	1.453	1.401	1.368	1.365	1.361
1320	1.678	1.673	1.667	1.662	1.655	1.648	1.640	1.630	1.620	1.608	1.593	1.576	1.556	1.531	1.501	1.461	1.408	1.374	1.370	1.367
1340	1.689	1.684	1.679	1.673	1.666	1.659	1.650	1.641	1.630	1.618	1.603	1.586	1.565	1.540	1.509	1.468	1.414	1.380	1.376	1.372
1360	1.701	1.696	1.690	1.684	1.677	1.670	1.661	1.652	1.641	1.628	1.613	1.596	1.575	1.549	1.517	1.476	1.421	1.386	1.382	1.378
1380	1.712	1.707	1.702	1.696	1.689	1.681	1.672	1.663	1.651	1.638	1.623	1.605	1.584	1.558	1.525	1.483	1.427	1.392	1.388	1.384
1400	1.724	1.719	1.713	1.707	1.700	1.692	1.683	1.673	1.662	1.649	1.633	1.615	1.593	1.567	1.534	1.491	1.434	1.397	1.393	1.389
1420	1.736	1.731	1.725	1.718	1.711	1.703	1.694	1.684	1.673	1.659	1.643	1.625	1.603	1.576	1.542	1.498	1.440	1.403	1.399	1.395
1440	1.748	1.742	1.736	1.730	1.723	1.714	1.705	1.695	1.683	1.670	1.654	1.635	1.612	1.585	1.550	1.506	1.447	1.409	1.405	1.401
1460	1.759	1.754	1.748	1.741	1.734	1.726	1.716	1.706	1.694	1.680	1.664	1.645	1.622	1.594	1.559	1.513	1.453	1.415	1.411	1.406
1480	1.771	1.766	1.760	1.753	1.745	1.737	1.728	1.717	1.705	1.690	1.674	1.655	1.631	1.603	1.567	1.521	1.460	1.421	1.417	1.412
1500	1.783	1.778	1.771	1.764	1.757	1.748	1.739	1.728	1.715	1.701	1.684	1.664	1.641	1.612	1.575	1.529	1.466	1.427	1.422	1.418
1520	1.795	1.789	1.783	1.776	1.768	1.760	1.750	1.739	1.726	1.712	1.694	1.674	1.650	1.621	1.584	1.536	1.473	1.433	1.428	1.424
1540	1.807	1.801	1.795	1.788	1.780	1.771	1.761	1.750	1.737	1.722	1.705	1.684	1.660	1.630	1.592	1.544	1.479	1.438	1.434	1.429
1560	1.819	1.813	1.807	1.799	1.791	1.782	1.772	1.761	1.748	1.733	1.715	1.694	1.669	1.639	1.601	1.551	1.486	1.444	1.440	1.435
1580	1.831	1.825	1.819	1.811	1.803	1.794	1.784	1.772	1.759	1.743	1.725	1.704	1.679	1.648	1.609	1.559	1.493	1.450	1.446	1.441
1600	1.843	1.837	1.830	1.823	1.815	1.805	1.795	1.783	1.770	1.754	1.736	1.714	1.688	1.657	1.617	1.567	1.499	1.456	1.451	1.447
1620	1.856	1.849	1.842	1.835	1.826	1.817	1.806	1.794	1.781	1.765	1.746	1.724	1.698	1.666	1.626	1.574	1.506	1.462	1.457	1.452
1640	1.868	1.861	1.854	1.847	1.838	1.829	1.818	1.806	1.792	1.775	1.757	1.734	1.708	1.675	1.634	1.582	1.512	1.468	1.463	1.458

SIGMA = 3956.0

$V-V_c = 15$

H x	0.08	0.09	0.10	0.2	0.4	0.6	0.8	1.0	1.2	1.4	1.6	1.8	2.0	2.2	2.4	2.6	2.8	3.0	3.2	3.4
1660	1.464	1.469	1.474	1.519	1.590	1.643	1.684	1.717	1.745	1.767	1.786	1.803	1.817	1.829	1.840	1.850	1.859	1.866	1.873	1.880
1680	1.469	1.475	1.480	1.526	1.598	1.652	1.694	1.727	1.755	1.778	1.797	1.814	1.828	1.841	1.852	1.862	1.870	1.878	1.886	1.892
1700	1.475	1.480	1.486	1.532	1.605	1.660	1.703	1.737	1.765	1.788	1.808	1.825	1.839	1.852	1.863	1.873	1.882	1.891	1.898	1.904
1720	1.481	1.486	1.491	1.539	1.613	1.669	1.712	1.747	1.775	1.799	1.819	1.836	1.851	1.864	1.875	1.885	1.894	1.903	1.910	1.917
1740	1.487	1.492	1.497	1.545	1.621	1.677	1.721	1.756	1.785	1.809	1.830	1.847	1.862	1.875	1.887	1.897	1.906	1.915	1.922	1.929
1760	1.493	1.498	1.503	1.552	1.629	1.686	1.731	1.766	1.796	1.820	1.840	1.858	1.873	1.887	1.899	1.909	1.919	1.927	1.935	1.942
1780	1.498	1.504	1.509	1.559	1.636	1.695	1.740	1.776	1.806	1.831	1.851	1.869	1.885	1.898	1.910	1.921	1.931	1.939	1.947	1.954
1800	1.504	1.510	1.515	1.565	1.644	1.703	1.749	1.786	1.816	1.841	1.862	1.881	1.896	1.910	1.922	1.933	1.943	1.952	1.959	1.967
1820	1.510	1.516	1.521	1.572	1.652	1.712	1.759	1.796	1.826	1.852	1.873	1.892	1.908	1.922	1.934	1.945	1.955	1.964	1.972	1.979
1840	1.516	1.521	1.527	1.579	1.660	1.721	1.768	1.806	1.837	1.863	1.884	1.903	1.919	1.934	1.946	1.957	1.967	1.976	1.984	1.992
1860	1.521	1.527	1.533	1.585	1.668	1.729	1.777	1.816	1.847	1.873	1.896	1.915	1.931	1.945	1.958	1.969	1.980	1.989	1.997	2.004
1880	1.527	1.533	1.539	1.592	1.676	1.738	1.787	1.826	1.858	1.884	1.907	1.926	1.943	1.957	1.970	1.982	1.992	2.001	2.009	2.017
1900	1.533	1.539	1.545	1.599	1.683	1.747	1.796	1.836	1.868	1.895	1.918	1.937	1.954	1.969	1.982	1.994	2.004	2.013	2.022	2.030
1920	1.539	1.545	1.551	1.605	1.691	1.756	1.806	1.846	1.879	1.906	1.929	1.949	1.966	1.981	1.994	2.006	2.017	2.026	2.035	2.042
1940	1.545	1.551	1.557	1.612	1.699	1.764	1.815	1.856	1.889	1.917	1.940	1.960	1.978	1.993	2.006	2.018	2.029	2.039	2.047	2.055
1960	1.550	1.557	1.563	1.619	1.707	1.773	1.825	1.866	1.900	1.928	1.951	1.972	1.989	2.005	2.018	2.031	2.041	2.051	2.060	2.068
1980	1.556	1.563	1.569	1.626	1.715	1.782	1.834	1.876	1.910	1.939	1.963	1.983	2.001	2.017	2.031	2.043	2.054	2.064	2.073	2.081
2000	1.562	1.568	1.575	1.632	1.723	1.791	1.844	1.886	1.921	1.950	1.974	1.995	2.013	2.029	2.043	2.055	2.066	2.076	2.085	2.094
2020	1.568	1.574	1.581	1.639	1.731	1.800	1.853	1.896	1.931	1.961	1.985	2.006	2.025	2.041	2.055	2.068	2.079	2.089	2.098	2.107
2040	1.574	1.580	1.587	1.646	1.739	1.809	1.863	1.906	1.942	1.972	1.997	2.018	2.037	2.053	2.067	2.080	2.092	2.102	2.111	2.120
2060	1.579	1.586	1.593	1.653	1.747	1.818	1.873	1.917	1.953	1.983	2.008	2.030	2.049	2.065	2.080	2.093	2.104	2.115	2.124	2.133
2080	1.585	1.592	1.599	1.659	1.755	1.827	1.882	1.927	1.963	1.994	2.019	2.042	2.061	2.077	2.092	2.105	2.117	2.127	2.137	2.146
2100	1.591	1.598	1.605	1.666	1.763	1.836	1.892	1.937	1.974	2.005	2.031	2.053	2.073	2.090	2.104	2.118	2.130	2.140	2.150	2.159
2120	1.597	1.604	1.611	1.673	1.771	1.845	1.902	1.947	1.985	2.016	2.042	2.065	2.085	2.102	2.117	2.130	2.142	2.153	2.163	2.172
2140	1.603	1.610	1.617	1.680	1.779	1.854	1.911	1.958	1.996	2.027	2.054	2.077	2.097	2.114	2.129	2.143	2.155	2.166	2.176	2.185
2160	1.609	1.616	1.623	1.687	1.787	1.863	1.921	1.968	2.006	2.038	2.065	2.089	2.109	2.126	2.142	2.156	2.168	2.179	2.189	2.198
2180	1.614	1.622	1.629	1.694	1.795	1.872	1.931	1.978	2.017	2.050	2.077	2.101	2.121	2.139	2.154	2.168	2.181	2.192	2.202	2.212
2200	1.620	1.628	1.635	1.700	1.803	1.881	1.941	1.989	2.028	2.061	2.089	2.112	2.133	2.151	2.167	2.181	2.194	2.205	2.215	2.225
2220	1.626	1.634	1.641	1.707	1.811	1.890	1.951	1.999	2.039	2.072	2.100	2.124	2.145	2.163	2.180	2.194	2.207	2.218	2.229	2.238
2240	1.632	1.639	1.647	1.714	1.820	1.899	1.960	2.010	2.050	2.084	2.112	2.136	2.157	2.176	2.192	2.207	2.220	2.231	2.242	2.252
2260	1.638	1.645	1.653	1.721	1.828	1.908	1.970	2.020	2.061	2.095	2.124	2.148	2.170	2.188	2.205	2.220	2.233	2.245	2.255	2.265
2280	1.644	1.651	1.659	1.728	1.836	1.917	1.980	2.031	2.072	2.106	2.135	2.160	2.182	2.201	2.218	2.232	2.246	2.258	2.269	2.278
2300	1.650	1.657	1.665	1.735	1.844	1.926	1.990	2.041	2.083	2.118	2.147	2.172	2.194	2.213	2.230	2.245	2.259	2.271	2.282	2.292
2320	1.655	1.663	1.671	1.742	1.852	1.935	2.000	2.052	2.094	2.129	2.159	2.185	2.207	2.226	2.243	2.258	2.272	2.284	2.295	2.305
2340	1.661	1.669	1.677	1.748	1.860	1.945	2.010	2.062	2.105	2.141	2.171	2.197	2.219	2.239	2.256	2.271	2.285	2.298	2.309	2.319
2360	1.667	1.675	1.683	1.755	1.869	1.954	2.020	2.073	2.116	2.152	2.183	2.209	2.232	2.251	2.269	2.284	2.298	2.311	2.322	2.333
2380	1.673	1.681	1.689	1.762	1.877	1.963	2.030	2.083	2.127	2.164	2.195	2.221	2.244	2.264	2.282	2.298	2.312	2.324	2.336	2.346
2400	1.679	1.687	1.695	1.769	1.885	1.972	2.040	2.094	2.138	2.175	2.207	2.233	2.256	2.277	2.295	2.311	2.325	2.338	2.349	2.360
2420	1.685	1.693	1.701	1.776	1.893	1.981	2.050	2.105	2.150	2.187	2.218	2.246	2.269	2.290	2.308	2.324	2.338	2.351	2.363	2.374
2440	1.691	1.699	1.707	1.783	1.902	1.991	2.060	2.115	2.161	2.198	2.230	2.258	2.282	2.302	2.321	2.337	2.352	2.365	2.377	2.387
2460	1.697	1.705	1.714	1.790	1.910	2.000	2.070	2.126	2.172	2.210	2.242	2.270	2.294	2.315	2.334	2.350	2.365	2.378	2.390	2.401
2480	1.702	1.711	1.720	1.797	1.918	2.009	2.080	2.137	2.183	2.222	2.255	2.283	2.307	2.328	2.347	2.363	2.378	2.392	2.404	2.415

V-V$_c$ = 15   SIGMA = 3956.0

H \ x	3.4	3.2	3.0	2.8	2.6	2.4	2.2	2.0	1.8	1.6	1.4	1.2	1.0	0.8	0.6	0.4	0.2	0.10	0.09	0.08
2500	2.429	2.418	2.405	2.392	2.377	2.360	2.341	2.319	2.295	2.267	2.234	2.194	2.148	2.090	2.019	1.927	1.804	1.726	1.717	1.708
2520	2.443	2.432	2.419	2.405	2.390	2.373	2.354	2.332	2.307	2.279	2.245	2.206	2.158	2.100	2.028	1.935	1.811	1.732	1.723	1.714
2540	2.457	2.445	2.433	2.419	2.404	2.386	2.367	2.345	2.320	2.291	2.257	2.217	2.169	2.111	2.037	1.943	1.818	1.738	1.729	1.720
2560	2.471	2.459	2.447	2.433	2.417	2.400	2.380	2.358	2.332	2.303	2.269	2.229	2.180	2.121	2.047	1.952	1.825	1.744	1.735	1.726
2580	2.485	2.473	2.460	2.446	2.430	2.413	2.393	2.371	2.345	2.315	2.281	2.240	2.191	2.131	2.056	1.960	1.832	1.750	1.741	1.732
2600	2.499	2.487	2.474	2.460	2.444	2.426	2.406	2.383	2.357	2.327	2.293	2.251	2.202	2.141	2.066	1.968	1.839	1.756	1.747	1.738
2620	2.513	2.501	2.488	2.473	2.457	2.439	2.419	2.396	2.370	2.340	2.304	2.263	2.213	2.152	2.075	1.977	1.846	1.762	1.753	1.744
2640	2.527	2.515	2.502	2.487	2.471	2.453	2.432	2.409	2.383	2.352	2.316	2.274	2.224	2.162	2.085	1.985	1.853	1.768	1.759	1.750
2660	2.541	2.529	2.516	2.501	2.485	2.466	2.445	2.422	2.395	2.364	2.328	2.286	2.235	2.172	2.094	1.994	1.860	1.775	1.765	1.756
2680	2.556	2.543	2.530	2.515	2.498	2.480	2.459	2.435	2.408	2.377	2.340	2.297	2.246	2.183	2.104	2.002	1.867	1.781	1.771	1.762
2700	2.570	2.557	2.544	2.529	2.512	2.493	2.472	2.448	2.421	2.389	2.352	2.309	2.257	2.193	2.113	2.011	1.874	1.787	1.777	1.768
2720	2.584	2.572	2.558	2.543	2.526	2.507	2.485	2.461	2.433	2.402	2.364	2.320	2.268	2.203	2.123	2.019	1.881	1.793	1.783	1.774
2740	2.599	2.586	2.572	2.556	2.539	2.520	2.499	2.474	2.446	2.414	2.376	2.332	2.279	2.214	2.132	2.028	1.888	1.799	1.789	1.779
2760	2.613	2.600	2.586	2.570	2.553	2.534	2.512	2.487	2.459	2.427	2.389	2.344	2.290	2.224	2.142	2.036	1.895	1.805	1.795	1.785
2780	2.627	2.614	2.600	2.584	2.567	2.547	2.525	2.500	2.472	2.439	2.401	2.355	2.301	2.235	2.151	2.045	1.902	1.812	1.802	1.791
2800	2.642	2.629	2.614	2.598	2.581	2.561	2.539	2.514	2.485	2.452	2.413	2.367	2.312	2.245	2.161	2.053	1.909	1.818	1.808	1.797
2820	2.656	2.643	2.629	2.613	2.595	2.575	2.552	2.527	2.498	2.464	2.425	2.379	2.323	2.256	2.171	2.062	1.916	1.824	1.814	1.803
2840	2.671	2.658	2.643	2.627	2.609	2.588	2.566	2.540	2.511	2.477	2.437	2.391	2.335	2.266	2.180	2.070	1.923	1.830	1.820	1.809
2860	2.685	2.672	2.657	2.641	2.623	2.602	2.579	2.553	2.524	2.489	2.450	2.402	2.346	2.277	2.190	2.079	1.931	1.836	1.826	1.815
2880	2.700	2.686	2.672	2.655	2.637	2.616	2.593	2.567	2.537	2.502	2.462	2.414	2.357	2.287	2.200	2.087	1.938	1.842	1.832	1.821
2900	2.715	2.701	2.686	2.669	2.651	2.630	2.606	2.580	2.550	2.515	2.474	2.426	2.368	2.298	2.209	2.096	1.945	1.849	1.838	1.827
2920	2.729	2.716	2.700	2.683	2.665	2.644	2.620	2.593	2.563	2.528	2.487	2.438	2.380	2.308	2.219	2.105	1.952	1.855	1.844	1.833
2940	2.744	2.730	2.715	2.698	2.679	2.658	2.634	2.607	2.576	2.540	2.499	2.450	2.391	2.319	2.229	2.113	1.959	1.861	1.850	1.839
2960	2.759	2.745	2.729	2.712	2.693	2.672	2.648	2.620	2.589	2.553	2.511	2.462	2.402	2.330	2.239	2.122	1.966	1.867	1.856	1.845
2980	2.774	2.760	2.744	2.726	2.707	2.686	2.661	2.634	2.602	2.566	2.524	2.474	2.414	2.340	2.249	2.131	1.973	1.873	1.862	1.851
3000	2.789	2.774	2.758	2.741	2.721	2.700	2.675	2.647	2.616	2.579	2.536	2.486	2.425	2.351	2.258	2.139	1.981	1.880	1.868	1.857

SIGMA = 3956.0

V-V_c = 15

H x	3.4	3.2	3.0	2.8	2.6	2.4	2.2	2.0	1.8	1.6	1.4	1.2	1.0	0.8	0.6	0.4	0.2	0.10	0.09	0.08
3100	2.864	2.849	2.832	2.813	2.793	2.770	2.745	2.715	2.682	2.644	2.599	2.546	2.482	2.405	2.308	2.183	2.017	1.911	1.899	1.887
3200	2.940	2.924	2.906	2.887	2.866	2.842	2.815	2.784	2.750	2.709	2.662	2.607	2.541	2.459	2.358	2.227	2.053	1.942	1.930	1.917
3300	3.016	3.000	2.982	2.962	2.939	2.914	2.886	2.854	2.818	2.776	2.727	2.669	2.599	2.514	2.408	2.271	2.089	1.973	1.961	1.948
3400	3.094	3.077	3.058	3.037	3.014	2.988	2.958	2.925	2.887	2.843	2.792	2.731	2.659	2.570	2.459	2.316	2.126	2.005	1.991	1.976
3500	3.173	3.155	3.135	3.113	3.089	3.062	3.031	2.996	2.957	2.911	2.857	2.794	2.718	2.626	2.510	2.361	2.163	2.036	2.022	2.008
3600	3.253	3.234	3.214	3.191	3.165	3.137	3.105	3.069	3.027	2.980	2.924	2.858	2.779	2.683	2.562	2.407	2.200	2.068	2.054	2.039
3700	3.334	3.314	3.293	3.269	3.243	3.213	3.180	3.142	3.099	3.049	2.991	2.923	2.840	2.740	2.614	2.452	2.237	2.100	2.085	2.069
3800	3.416	3.395	3.373	3.348	3.321	3.290	3.255	3.216	3.171	3.119	3.059	2.988	2.902	2.798	2.667	2.499	2.274	2.132	2.116	2.100
3900	3.499	3.477	3.454	3.428	3.400	3.368	3.332	3.291	3.244	3.191	3.128	3.054	2.965	2.856	2.720	2.545	2.312	2.164	2.147	2.131
4000	3.582	3.560	3.536	3.509	3.480	3.447	3.409	3.367	3.318	3.263	3.197	3.120	3.028	2.915	2.774	2.592	2.350	2.196	2.179	2.162
4100	3.667	3.644	3.619	3.592	3.561	3.526	3.487	3.443	3.393	3.335	3.268	3.188	3.092	2.975	2.828	2.639	2.388	2.228	2.211	2.193
4200	3.753	3.729	3.703	3.675	3.643	3.607	3.567	3.521	3.469	3.409	3.339	3.256	3.156	3.034	2.882	2.687	2.426	2.261	2.242	2.224
4300	3.840	3.815	3.788	3.758	3.725	3.688	3.647	3.599	3.545	3.483	3.410	3.324	3.221	3.095	2.938	2.735	2.465	2.293	2.274	2.255
4400	3.928	3.902	3.874	3.843	3.809	3.771	3.727	3.678	3.622	3.558	3.483	3.394	3.287	3.156	2.993	2.783	2.504	2.326	2.306	2.286
4500	4.016	3.990	3.961	3.929	3.894	3.854	3.809	3.758	3.701	3.634	3.556	3.464	3.353	3.218	3.049	2.832	2.543	2.359	2.338	2.318
4600	4.106	4.079	4.049	4.016	3.979	3.938	3.892	3.839	3.779	3.710	3.630	3.535	3.420	3.281	3.106	2.881	2.582	2.391	2.370	2.349
4700	4.197	4.169	4.138	4.104	4.066	4.023	3.975	3.921	3.859	3.788	3.704	3.606	3.487	3.343	3.163	2.931	2.621	2.424	2.403	2.381
4800	4.289	4.259	4.228	4.192	4.153	4.109	4.060	4.004	3.940	3.866	3.780	3.678	3.556	3.407	3.220	2.981	2.661	2.457	2.435	2.412
4900	4.381	4.351	4.318	4.282	4.241	4.196	4.145	4.087	4.021	3.945	3.856	3.751	3.625	3.471	3.278	3.031	2.701	2.491	2.468	2.444
5000	4.475	4.444	4.410	4.372	4.331	4.284	4.231	4.171	4.103	4.025	3.933	3.825	3.695	3.536	3.337	3.081	2.741	2.524	2.500	2.476
5100	4.570	4.538	4.503	4.464	4.421	4.372	4.318	4.257	4.186	4.105	4.011	3.899	3.765	3.601	3.396	3.132	2.781	2.557	2.533	2.508
5200	4.665	4.632	4.596	4.556	4.512	4.462	4.406	4.343	4.270	4.187	4.089	3.974	3.836	3.667	3.455	3.184	2.822	2.591	2.566	2.540
5300	4.762	4.728	4.691	4.649	4.604	4.552	4.495	4.429	4.355	4.269	4.168	4.050	3.907	3.733	3.515	3.235	2.862	2.625	2.599	2.572
5400	4.860	4.824	4.786	4.744	4.697	4.644	4.584	4.517	4.440	4.352	4.248	4.126	3.979	3.800	3.576	3.287	2.903	2.656	2.632	2.604
5500	4.958	4.922	4.883	4.839	4.790	4.736	4.675	4.606	4.527	4.435	4.329	4.203	4.052	3.867	3.636	3.340	2.944	2.692	2.665	2.636
5600	5.058	5.021	4.980	4.935	4.885	4.829	4.766	4.695	4.614	4.520	4.410	4.281	4.125	3.935	3.698	3.393	2.985	2.726	2.698	2.669
5700	5.158	5.120	5.079	5.032	4.981	4.923	4.859	4.785	4.702	4.605	4.492	4.359	4.199	4.004	3.760	3.446	3.027	2.761	2.731	2.701
5800	5.260	5.221	5.178	5.130	5.077	5.018	4.952	4.877	4.790	4.691	4.575	4.438	4.274	4.073	3.822	3.499	3.069	2.795	2.765	2.734
5900	5.363	5.322	5.278	5.229	5.175	5.114	5.046	4.969	4.880	4.778	4.659	4.518	4.349	4.143	3.885	3.553	3.111	2.829	2.798	2.767
6000	5.466	5.425	5.379	5.329	5.273	5.211	5.141	5.061	4.970	4.866	4.743	4.599	4.425	4.213	3.948	3.607	3.153	2.864	2.832	2.799
6100	5.571	5.528	5.482	5.430	5.373	5.309	5.237	5.155	5.062	4.954	4.828	4.680	4.502	4.284	4.012	3.662	3.195	2.898	2.866	2.832
6200	5.676	5.633	5.585	5.532	5.473	5.407	5.333	5.250	5.154	5.043	4.914	4.762	4.579	4.355	4.076	3.717	3.238	2.933	2.899	2.865
6300	5.783	5.738	5.689	5.635	5.575	5.507	5.431	5.345	5.247	5.133	5.001	4.844	4.657	4.427	4.141	3.772	3.281	2.968	2.933	2.898
6400	5.890	5.845	5.794	5.738	5.676	5.607	5.529	5.441	5.340	5.224	5.088	4.928	4.735	4.500	4.206	3.828	3.324	3.003	2.967	2.932
6500	5.999	5.952	5.900	5.843	5.780	5.709	5.629	5.538	5.435	5.316	5.176	5.012	4.814	4.573	4.272	3.884	3.367	3.038	3.002	2.965
6600	6.108	6.060	6.007	5.949	5.885	5.811	5.729	5.636	5.530	5.408	5.265	5.097	4.894	4.647	4.338	3.940	3.410	3.073	3.036	2.998
6700	6.219	6.169	6.115	6.055	5.988	5.914	5.830	5.735	5.626	5.501	5.355	5.182	4.975	4.721	4.404	3.997	3.454	3.108	3.070	3.032
6800	6.330	6.280	6.224	6.163	6.094	6.018	5.932	5.835	5.723	5.595	5.445	5.268	5.056	4.796	4.471	4.054	3.498	3.144	3.105	3.065
6900	6.443	6.391	6.334	6.271	6.201	6.123	6.035	5.935	5.821	5.690	5.536	5.355	5.137	4.871	4.539	4.110	3.542	3.179	3.139	3.099
7000	6.556	6.503	6.445	6.380	6.309	6.229	6.139	6.036	5.920	5.785	5.628	5.442	5.220	4.947	4.607	4.170	3.586	3.215	3.174	3.132
7100	6.671	6.616	6.557	6.491	6.417	6.335	6.243	6.139	6.019	5.881	5.721	5.531	5.303	5.024	4.676	4.228	3.631	3.251	3.209	3.166
7200	6.786	6.730	6.669	6.602	6.527	6.443	6.349	6.242	6.119	5.978	5.814	5.620	5.386	5.101	4.745	4.286	3.675	3.287	3.244	3.200

$V - V_c = 15$ SIGMA = 3956.0

H \ x	3.4	3.2	3.0	2.8	2.6	2.4	2.2	2.0	1.8	1.6	1.4	1.2	1.0	0.8	0.6	0.4	0.2	0.10	0.09	0.08
7300	6.902	6.846	6.783	6.714	6.637	6.551	6.455	6.346	6.221	6.076	5.908	5.709	5.471	5.179	4.814	4.345	3.720	3.323	3.279	3.234
7400	7.020	6.962	6.898	6.827	6.749	6.662	6.562	6.450	6.322	6.175	6.003	5.799	5.555	5.257	4.884	4.405	3.766	3.359	3.314	3.268
7500	7.138	7.079	7.013	6.941	6.861	6.771	6.670	6.556	6.425	6.274	6.098	5.890	5.641	5.336	4.955	4.464	3.811	3.395	3.349	3.303
7600	7.258	7.197	7.130	7.056	6.974	6.882	6.779	6.662	6.529	6.375	6.195	5.982	5.727	5.415	5.026	4.525	3.856	3.431	3.384	3.337
7700	7.378	7.316	7.248	7.172	7.088	6.995	6.889	6.770	6.633	6.476	6.292	6.075	5.814	5.495	5.097	4.585	3.902	3.468	3.420	3.371
7800	7.499	7.436	7.366	7.289	7.203	7.108	7.000	6.878	6.738	6.577	6.390	6.168	5.901	5.576	5.169	4.646	3.948	3.504	3.455	3.406
7900	7.622	7.557	7.486	7.407	7.319	7.222	7.112	6.987	6.844	6.680	6.488	6.261	5.989	5.657	5.241	4.707	3.994	3.541	3.491	3.443
8000	7.745	7.679	7.606	7.526	7.436	7.336	7.224	7.097	6.951	6.783	6.587	6.356	6.078	5.739	5.314	4.768	4.041	3.578	3.527	3.475
8100	7.869	7.802	7.727	7.645	7.554	7.452	7.337	7.207	7.059	6.887	6.687	6.451	6.167	5.821	5.387	4.830	4.088	3.615	3.563	3.510
8200	7.995	7.926	7.850	7.766	7.673	7.569	7.452	7.319	7.167	6.992	6.788	6.547	6.257	5.904	5.461	4.893	4.134	3.652	3.599	3.545
8300	8.121	8.050	7.973	7.888	7.793	7.686	7.567	7.431	7.277	7.098	6.890	6.644	6.348	5.987	5.536	4.955	4.181	3.689	3.635	3.580
8400	8.248	8.176	8.097	8.010	7.913	7.805	7.683	7.545	7.387	7.205	6.992	6.741	6.439	6.071	5.610	5.018	4.229	3.726	3.671	3.615
8500	8.376	8.303	8.223	8.134	8.035	7.924	7.800	7.659	7.498	7.312	7.095	6.839	6.531	6.155	5.686	5.082	4.276	3.764	3.707	3.650
8600	8.506	8.431	8.349	8.258	8.157	8.044	7.918	7.774	7.610	7.420	7.199	6.937	6.624	6.240	5.761	5.145	4.324	3.801	3.744	3.685
8700	8.636	8.560	8.476	8.383	8.280	8.166	8.036	7.890	7.723	7.529	7.303	7.037	6.717	6.326	5.838	5.209	4.372	3.839	3.780	3.720
8800	8.767	8.689	8.604	8.510	8.405	8.288	8.156	8.006	7.836	7.638	7.409	7.137	6.811	6.412	5.914	5.274	4.420	3.877	3.817	3.756
8900	8.899	8.820	8.733	8.637	8.530	8.411	8.276	8.124	7.950	7.749	7.515	7.238	6.905	6.499	5.991	5.339	4.468	3.914	3.853	3.791
9000	9.032	8.952	8.863	8.765	8.656	8.534	8.397	8.242	8.065	7.860	7.621	7.339	7.000	6.586	6.069	5.404	4.517	3.952	3.890	3.827
9100	9.167	9.084	8.994	8.894	8.783	8.659	8.520	8.361	8.181	7.972	7.729	7.441	7.096	6.674	6.147	5.469	4.566	3.990	3.927	3.863
9200	9.302	9.218	9.126	9.024	8.911	8.785	8.643	8.482	8.297	8.085	7.837	7.544	7.193	6.763	6.226	5.535	4.615	4.029	3.964	3.898
9300	9.438	9.353	9.259	9.155	9.040	8.911	8.767	8.602	8.415	8.198	7.946	7.648	7.290	6.852	6.305	5.602	4.664	4.067	4.001	3.934
9400	9.575	9.488	9.393	9.287	9.170	9.039	8.891	8.724	8.533	8.313	8.056	7.752	7.387	6.942	6.384	5.668	4.713	4.105	4.039	3.970
9500	9.713	9.625	9.528	9.420	9.301	9.167	9.017	8.847	8.652	8.428	8.166	7.857	7.486	7.032	6.464	5.735	4.763	4.144	4.076	4.006
9600	9.852	9.762	9.663	9.554	9.432	9.296	9.144	8.970	8.772	8.544	8.277	7.962	7.584	7.122	6.545	5.802	4.813	4.183	4.113	4.043
9700	9.992	9.901	9.800	9.689	9.565	9.427	9.271	9.095	8.893	8.661	8.389	8.069	7.684	7.214	6.626	5.870	4.863	4.221	4.151	4.079
9800	10.13	10.04	9.94	9.82	9.70	9.56	9.40	9.22	9.01	8.78	8.50	8.18	7.78	7.31	6.71	5.94	4.91	4.26	4.19	4.12
9900	10.28	10.18	10.08	9.96	9.83	9.69	9.53	9.35	9.14	8.90	8.62	8.28	7.89	7.40	6.79	6.01	4.96	4.30	4.23	4.15
10000	10.42	10.32	10.22	10.10	9.97	9.82	9.66	9.47	9.26	9.02	8.73	8.39	7.99	7.49	6.87	6.08	5.01	4.34	4.26	4.19
10100	10.56	10.46	10.36	10.24	10.10	9.96	9.79	9.60	9.38	9.14	8.84	8.50	8.09	7.58	6.95	6.14	5.06	4.38	4.30	4.23
10200	10.71	10.61	10.50	10.38	10.24	10.09	9.92	9.73	9.51	9.26	8.96	8.61	8.19	7.68	7.04	6.21	5.12	4.42	4.34	4.26
10300	10.85	10.75	10.64	10.52	10.38	10.23	10.05	9.86	9.64	9.38	9.08	8.72	8.29	7.77	7.12	6.28	5.17	4.46	4.38	4.30
10400	11.00	10.90	10.78	10.66	10.52	10.36	10.19	9.99	9.76	9.50	9.19	8.83	8.40	7.87	7.21	6.35	5.22	4.50	4.42	4.34
10500	11.15	11.04	10.93	10.80	10.66	10.50	10.32	10.12	9.89	9.62	9.31	8.94	8.50	7.96	7.29	6.43	5.27	4.54	4.46	4.37
10600	11.30	11.19	11.07	10.94	10.80	10.64	10.46	10.25	10.02	9.75	9.43	9.06	8.61	8.06	7.38	6.50	5.32	4.58	4.49	4.41
10700	11.45	11.34	11.22	11.09	10.94	10.78	10.59	10.39	10.15	9.87	9.55	9.17	8.71	8.16	7.46	6.57	5.37	4.62	4.53	4.45
10800	11.60	11.49	11.37	11.23	11.08	10.92	10.73	10.52	10.28	10.00	9.67	9.28	8.82	8.26	7.55	6.64	5.43	4.66	4.57	4.48
10900	11.75	11.64	11.52	11.38	11.23	11.06	10.87	10.65	10.41	10.12	9.79	9.40	8.93	8.35	7.64	6.71	5.48	4.70	4.61	4.52
11000	11.90	11.79	11.66	11.53	11.37	11.20	11.01	10.79	10.54	10.25	9.91	9.51	9.04	8.45	7.72	6.78	5.53	4.74	4.65	4.56
11100	12.06	11.94	11.81	11.67	11.52	11.34	11.15	10.93	10.67	10.38	10.04	9.63	9.15	8.55	7.81	6.86	5.59	4.78	4.69	4.60
11200	12.21	12.10	11.97	11.82	11.66	11.49	11.29	11.06	10.80	10.51	10.16	9.75	9.25	8.65	7.90	6.93	5.64	4.82	4.73	4.63
11300	12.37	12.25	12.12	11.97	11.81	11.63	11.43	11.20	10.94	10.64	10.28	9.86	9.36	8.75	7.99	7.00	5.69	4.86	4.77	4.67
11400	12.53	12.40	12.27	12.12	11.96	11.78	11.57	11.34	11.07	10.77	10.41	9.98	9.47	8.85	8.08	7.08	5.75	4.90	4.81	4.71

SIGMA = 3956.0

$V-V_c = 15$

H	0.08	0.09	0.10	0.2	0.4	0.6	0.8	1.0	1.2	1.4	1.6	1.8	2.0	2.2	2.4	2.6	2.8	3.0	3.2	3.4
11500	4.75	4.84	4.94	5.80	7.15	8.17	8.95	9.59	10.10	10.53	10.90	11.21	11.48	11.72	11.92	12.11	12.28	12.43	12.56	12.68
11600	4.79	4.88	4.98	5.85	7.23	8.26	9.06	9.70	10.22	10.66	11.03	11.34	11.62	11.86	12.07	12.26	12.43	12.58	12.72	12.84
11700	4.82	4.92	5.02	5.91	7.30	8.35	9.16	9.81	10.34	10.79	11.16	11.48	11.76	12.00	12.22	12.41	12.58	12.74	12.88	13.00
11800	4.86	4.96	5.06	5.96	7.38	8.44	9.26	9.92	10.46	10.91	11.29	11.62	11.90	12.15	12.37	12.56	12.74	12.89	13.03	13.16
11900	4.90	5.00	5.10	6.02	7.45	8.53	9.37	10.04	10.58	11.04	11.43	11.76	12.05	12.30	12.52	12.72	12.89	13.05	13.19	13.32
12000	4.94	5.04	5.15	6.07	7.53	8.62	9.47	10.15	10.71	11.17	11.56	11.90	12.19	12.44	12.67	12.87	13.05	13.21	13.36	13.49
12100	4.98	5.08	5.19	6.13	7.61	8.71	9.58	10.27	10.83	11.30	11.70	12.04	12.33	12.59	12.82	13.02	13.21	13.37	13.52	13.65
12200	5.02	5.12	5.23	6.18	7.68	8.81	9.68	10.38	10.95	11.43	11.83	12.18	12.48	12.74	12.97	13.18	13.36	13.53	13.68	13.82
12300	5.06	5.16	5.27	6.24	7.76	8.90	9.79	10.50	11.08	11.56	11.97	12.32	12.63	12.89	13.13	13.34	13.52	13.69	13.84	13.98
12400	5.09	5.20	5.31	6.29	7.84	8.99	9.89	10.61	11.20	11.69	12.11	12.47	12.77	13.04	13.28	13.49	13.68	13.85	14.01	14.15
12500	5.13	5.25	5.35	6.35	7.92	9.09	10.00	10.73	11.33	11.83	12.25	12.61	12.92	13.20	13.44	13.65	13.84	14.02	14.17	14.32
12600	5.17	5.29	5.40	6.41	7.99	9.18	10.11	10.85	11.46	11.96	12.39	12.75	13.07	13.35	13.59	13.81	14.01	14.18	14.34	14.49
12700	5.21	5.33	5.44	6.46	8.07	9.28	10.22	10.97	11.58	12.09	12.53	12.90	13.22	13.50	13.75	13.97	14.17	14.35	14.51	14.65
12800	5.25	5.37	5.48	6.52	8.15	9.37	10.33	11.09	11.71	12.23	12.67	13.05	13.37	13.66	13.91	14.13	14.33	14.51	14.68	14.83
12900	5.29	5.41	5.52	6.58	8.23	9.47	10.44	11.21	11.84	12.37	12.81	13.19	13.52	13.81	14.07	14.30	14.50	14.68	14.85	15.00
13000	5.33	5.45	5.57	6.63	8.31	9.57	10.55	11.33	11.97	12.50	12.95	13.34	13.68	13.97	14.23	14.46	14.66	14.85	15.02	15.17
13100	5.37	5.49	5.61	6.69	8.39	9.67	10.66	11.45	12.10	12.64	13.10	13.49	13.83	14.13	14.39	14.62	14.83	15.02	15.19	15.34
13200	5.41	5.53	5.65	6.75	8.47	9.76	10.77	11.57	12.23	12.78	13.24	13.64	13.98	14.29	14.55	14.79	15.00	15.19	15.36	15.52
13300	5.45	5.57	5.70	6.81	8.55	9.86	10.88	11.69	12.36	12.92	13.39	13.79	14.14	14.44	14.71	14.95	15.17	15.36	15.54	15.69
13400	5.49	5.61	5.74	6.86	8.63	9.96	10.99	11.82	12.49	13.06	13.53	13.94	14.29	14.60	14.86	15.12	15.34	15.53	15.71	15.87
13500	5.53	5.66	5.78	6.92	8.71	10.06	11.10	11.94	12.63	13.20	13.68	14.09	14.45	14.76	15.04	15.29	15.51	15.71	15.89	16.05
13600	5.57	5.70	5.82	6.98	8.80	10.16	11.22	12.07	12.76	13.34	13.83	14.25	14.61	14.93	15.21	15.46	15.68	15.88	16.06	16.23
13700	5.61	5.74	5.87	7.04	8.88	10.26	11.33	12.19	12.89	13.48	13.97	14.40	14.77	15.09	15.37	15.63	15.85	16.06	16.24	16.41
13800	5.65	5.78	5.91	7.10	8.96	10.36	11.45	12.32	13.03	13.62	14.12	14.55	14.93	15.25	15.54	15.80	16.03	16.23	16.42	16.59
13900	5.69	5.82	5.96	7.16	9.04	10.46	11.56	12.44	13.16	13.76	14.27	14.71	15.09	15.42	15.71	15.97	16.20	16.41	16.60	16.77
14000	5.73	5.87	6.00	7.22	9.13	10.56	11.68	12.57	13.30	13.91	14.42	14.87	15.25	15.58	15.88	16.14	16.37	16.59	16.78	16.95
14100	5.77	5.91	6.04	7.28	9.21	10.66	11.79	12.70	13.44	14.05	14.58	15.02	15.41	15.75	16.05	16.31	16.55	16.76	16.96	17.13
14200	5.81	5.95	6.09	7.34	9.30	10.77	11.91	12.83	13.58	14.20	14.73	15.18	15.57	15.92	16.22	16.49	16.73	16.94	17.14	17.32
14300	5.85	5.99	6.13	7.39	9.38	10.87	12.03	12.96	13.71	14.35	14.88	15.34	15.74	16.08	16.39	16.66	16.91	17.13	17.32	17.53
14400	5.89	6.03	6.18	7.46	9.47	10.97	12.15	13.09	13.85	14.49	15.03	15.50	15.90	16.25	16.56	16.84	17.09	17.31	17.51	17.69
14500	5.93	6.08	6.22	7.52	9.55	11.08	12.27	13.22	13.99	14.64	15.19	15.66	16.07	16.42	16.74	17.02	17.27	17.49	17.69	17.88
14600	5.97	6.12	6.26	7.58	9.64	11.18	12.38	13.35	14.13	14.79	15.34	15.82	16.23	16.59	16.91	17.19	17.45	17.67	17.88	18.07
14700	6.01	6.16	6.31	7.64	9.72	11.29	12.50	13.48	14.27	14.94	15.50	15.98	16.40	16.76	17.09	17.37	17.63	17.86	18.07	18.26
14800	6.06	6.21	6.35	7.70	9.81	11.39	12.62	13.61	14.42	15.09	15.66	16.14	16.57	16.94	17.26	17.55	17.81	18.04	18.26	18.45
14900	6.10	6.25	6.40	7.76	9.90	11.50	12.75	13.74	14.56	15.24	15.81	16.31	16.73	17.11	17.44	17.73	17.99	18.23	18.44	18.64
15000	6.14	6.29	6.44	7.82	9.98	11.60	12.87	13.88	14.70	15.39	15.97	16.47	16.90	17.28	17.62	17.91	18.18	18.42	18.63	18.83

x

$V - V_c = 16$ SIGMA = 3477.0

H x	3.4	3.2	3.0	2.8	2.6	2.4	2.2	2.0	1.8	1.6	1.4	1.2	1.0	0.8	0.6	0.4	0.2	0.10	0.09	0.08
0	1.000	1.000	1.000	1.000	1.000	1.000	1.000	1.000	1.000	1.000	1.000	1.000	1.000	1.000	1.000	1.000	1.000	1.000	1.000	1.000
20	1.010	1.010	1.010	1.010	1.010	1.010	1.010	1.010	1.009	1.009	1.009	1.009	1.009	1.008	1.008	1.007	1.007	1.006	1.006	1.006
40	1.020	1.020	1.020	1.020	1.020	1.020	1.020	1.019	1.019	1.019	1.018	1.018	1.017	1.017	1.016	1.015	1.013	1.013	1.012	1.012
60	1.031	1.031	1.030	1.030	1.030	1.030	1.029	1.029	1.029	1.028	1.027	1.027	1.026	1.025	1.024	1.022	1.020	1.019	1.019	1.019
80	1.041	1.041	1.041	1.040	1.040	1.040	1.039	1.039	1.038	1.037	1.037	1.036	1.035	1.033	1.032	1.030	1.027	1.025	1.025	1.025
100	1.052	1.051	1.051	1.051	1.050	1.050	1.049	1.048	1.048	1.047	1.046	1.045	1.044	1.042	1.040	1.037	1.034	1.031	1.031	1.031
120	1.062	1.062	1.061	1.061	1.060	1.060	1.059	1.058	1.057	1.056	1.055	1.054	1.052	1.050	1.048	1.045	1.040	1.038	1.037	1.037
140	1.073	1.072	1.072	1.071	1.071	1.070	1.069	1.068	1.067	1.066	1.065	1.063	1.061	1.059	1.056	1.052	1.047	1.044	1.044	1.043
160	1.083	1.083	1.082	1.081	1.081	1.080	1.079	1.078	1.077	1.076	1.074	1.072	1.070	1.067	1.064	1.060	1.054	1.050	1.050	1.050
180	1.094	1.093	1.092	1.092	1.091	1.090	1.089	1.088	1.087	1.085	1.084	1.081	1.079	1.076	1.072	1.067	1.061	1.057	1.056	1.056
200	1.105	1.104	1.103	1.102	1.101	1.100	1.099	1.098	1.097	1.095	1.093	1.091	1.088	1.085	1.080	1.075	1.068	1.063	1.063	1.062
220	1.115	1.115	1.114	1.113	1.112	1.111	1.110	1.108	1.107	1.105	1.103	1.100	1.097	1.093	1.089	1.082	1.074	1.069	1.069	1.068
240	1.126	1.125	1.124	1.123	1.122	1.121	1.120	1.118	1.116	1.114	1.112	1.109	1.106	1.102	1.097	1.090	1.081	1.076	1.075	1.074
260	1.137	1.136	1.135	1.134	1.133	1.132	1.130	1.128	1.126	1.124	1.122	1.119	1.115	1.110	1.105	1.098	1.088	1.082	1.081	1.081
280	1.148	1.147	1.146	1.145	1.143	1.142	1.140	1.139	1.136	1.134	1.131	1.128	1.124	1.119	1.113	1.105	1.095	1.088	1.088	1.087
300	1.159	1.158	1.156	1.155	1.154	1.152	1.151	1.149	1.147	1.144	1.141	1.137	1.133	1.128	1.121	1.113	1.102	1.095	1.094	1.093
320	1.170	1.169	1.167	1.166	1.165	1.163	1.161	1.159	1.157	1.154	1.151	1.147	1.142	1.137	1.130	1.121	1.109	1.101	1.100	1.099
340	1.181	1.180	1.178	1.177	1.175	1.174	1.172	1.169	1.167	1.164	1.160	1.156	1.151	1.145	1.138	1.128	1.116	1.108	1.107	1.106
360	1.192	1.191	1.189	1.188	1.186	1.184	1.182	1.180	1.177	1.174	1.170	1.166	1.161	1.154	1.146	1.136	1.123	1.114	1.113	1.112
380	1.203	1.202	1.200	1.199	1.197	1.195	1.193	1.190	1.187	1.184	1.180	1.175	1.170	1.163	1.155	1.144	1.129	1.120	1.119	1.118
400	1.214	1.213	1.211	1.210	1.208	1.206	1.203	1.201	1.198	1.194	1.190	1.185	1.179	1.172	1.163	1.152	1.136	1.127	1.126	1.125
420	1.225	1.224	1.222	1.221	1.219	1.216	1.214	1.211	1.208	1.204	1.200	1.195	1.188	1.181	1.172	1.159	1.143	1.133	1.132	1.131
440	1.237	1.235	1.233	1.232	1.230	1.227	1.225	1.222	1.218	1.214	1.210	1.204	1.198	1.190	1.180	1.167	1.150	1.140	1.138	1.137
460	1.248	1.246	1.245	1.243	1.240	1.238	1.235	1.232	1.229	1.224	1.220	1.214	1.207	1.199	1.188	1.175	1.157	1.146	1.145	1.143
480	1.259	1.258	1.256	1.254	1.252	1.249	1.246	1.243	1.239	1.235	1.230	1.224	1.217	1.208	1.197	1.183	1.164	1.152	1.151	1.150
500	1.271	1.269	1.267	1.265	1.263	1.260	1.257	1.253	1.250	1.245	1.240	1.234	1.226	1.217	1.205	1.191	1.171	1.159	1.157	1.156
520	1.282	1.281	1.278	1.276	1.274	1.271	1.268	1.264	1.260	1.255	1.250	1.243	1.236	1.226	1.214	1.199	1.178	1.165	1.164	1.162
540	1.294	1.292	1.290	1.288	1.285	1.282	1.279	1.275	1.271	1.266	1.260	1.253	1.245	1.235	1.223	1.207	1.185	1.172	1.170	1.169
560	1.306	1.304	1.301	1.299	1.296	1.293	1.290	1.286	1.281	1.276	1.270	1.263	1.255	1.244	1.231	1.214	1.192	1.178	1.176	1.175
580	1.317	1.315	1.313	1.310	1.307	1.304	1.301	1.297	1.292	1.287	1.280	1.273	1.264	1.253	1.240	1.222	1.199	1.185	1.183	1.181
600	1.329	1.327	1.324	1.322	1.319	1.315	1.312	1.307	1.303	1.297	1.291	1.283	1.274	1.262	1.248	1.230	1.206	1.191	1.189	1.188
620	1.341	1.338	1.336	1.333	1.330	1.327	1.323	1.318	1.313	1.308	1.301	1.293	1.283	1.272	1.257	1.238	1.213	1.197	1.196	1.194
640	1.352	1.350	1.348	1.345	1.341	1.338	1.334	1.329	1.324	1.318	1.311	1.303	1.293	1.281	1.266	1.246	1.220	1.204	1.202	1.200
660	1.364	1.362	1.359	1.356	1.353	1.349	1.345	1.340	1.335	1.329	1.322	1.313	1.303	1.290	1.275	1.254	1.227	1.210	1.208	1.207
680	1.376	1.374	1.371	1.368	1.364	1.361	1.356	1.351	1.346	1.339	1.332	1.323	1.312	1.299	1.283	1.262	1.235	1.217	1.215	1.213
700	1.388	1.386	1.383	1.380	1.376	1.372	1.368	1.363	1.357	1.350	1.342	1.333	1.322	1.309	1.292	1.270	1.242	1.223	1.221	1.219
720	1.400	1.398	1.395	1.391	1.388	1.384	1.379	1.374	1.368	1.361	1.353	1.343	1.332	1.318	1.301	1.278	1.249	1.230	1.228	1.226
740	1.412	1.409	1.406	1.403	1.399	1.395	1.390	1.385	1.379	1.372	1.363	1.354	1.342	1.328	1.310	1.287	1.256	1.236	1.234	1.232
760	1.424	1.422	1.418	1.415	1.411	1.407	1.402	1.396	1.390	1.382	1.374	1.364	1.352	1.337	1.318	1.295	1.263	1.243	1.241	1.238
780	1.437	1.434	1.430	1.427	1.423	1.418	1.413	1.407	1.401	1.393	1.385	1.374	1.362	1.346	1.327	1.303	1.270	1.249	1.247	1.245
800	1.449	1.446	1.442	1.439	1.434	1.430	1.425	1.419	1.412	1.404	1.395	1.384	1.372	1.356	1.336	1.311	1.277	1.256	1.253	1.251

SIGMA = 3477.0

$V-V_c = 16$

H / x	0.08	0.09	0.10	0.2	0.4	0.6	0.8	1.0	1.2	1.4	1.6	1.8	2.0	2.2	2.4	2.6	2.8	3.0	3.2	3.4
820	1.257	1.260	1.262	1.284	1.319	1.345	1.365	1.382	1.395	1.406	1.415	1.423	1.430	1.436	1.442	1.446	1.451	1.454	1.458	1.461
840	1.264	1.266	1.269	1.292	1.327	1.354	1.375	1.392	1.405	1.417	1.426	1.434	1.442	1.448	1.453	1.458	1.463	1.467	1.470	1.473
860	1.270	1.273	1.275	1.299	1.335	1.363	1.384	1.402	1.416	1.427	1.437	1.446	1.453	1.459	1.465	1.470	1.475	1.479	1.482	1.486
880	1.277	1.279	1.282	1.306	1.344	1.372	1.394	1.412	1.426	1.438	1.448	1.457	1.465	1.471	1.477	1.482	1.487	1.491	1.495	1.498
900	1.283	1.286	1.288	1.313	1.352	1.381	1.404	1.422	1.437	1.449	1.459	1.468	1.476	1.483	1.489	1.494	1.499	1.503	1.507	1.511
920	1.289	1.292	1.295	1.320	1.360	1.390	1.413	1.432	1.447	1.460	1.471	1.480	1.488	1.495	1.501	1.506	1.511	1.516	1.520	1.523
940	1.296	1.299	1.302	1.328	1.368	1.399	1.423	1.442	1.458	1.471	1.482	1.491	1.499	1.506	1.513	1.518	1.523	1.528	1.532	1.536
960	1.302	1.305	1.308	1.335	1.377	1.408	1.433	1.452	1.468	1.482	1.493	1.503	1.511	1.518	1.525	1.530	1.536	1.540	1.545	1.548
980	1.309	1.312	1.315	1.342	1.385	1.417	1.442	1.462	1.479	1.493	1.504	1.514	1.523	1.530	1.537	1.543	1.548	1.553	1.557	1.561
1000	1.315	1.318	1.321	1.349	1.393	1.426	1.452	1.473	1.490	1.504	1.515	1.526	1.534	1.542	1.549	1.555	1.560	1.565	1.570	1.574
1020	1.321	1.325	1.328	1.357	1.402	1.436	1.462	1.483	1.500	1.515	1.527	1.537	1.546	1.554	1.561	1.567	1.573	1.578	1.582	1.587
1040	1.328	1.331	1.334	1.364	1.410	1.445	1.472	1.493	1.511	1.526	1.538	1.549	1.558	1.566	1.573	1.580	1.585	1.591	1.595	1.599
1060	1.334	1.338	1.341	1.371	1.419	1.454	1.482	1.504	1.522	1.537	1.550	1.561	1.570	1.578	1.586	1.592	1.598	1.603	1.608	1.612
1080	1.341	1.344	1.348	1.378	1.427	1.463	1.492	1.514	1.533	1.548	1.561	1.572	1.582	1.590	1.598	1.605	1.611	1.616	1.621	1.625
1100	1.347	1.351	1.354	1.386	1.435	1.473	1.501	1.525	1.544	1.559	1.573	1.584	1.594	1.603	1.610	1.617	1.623	1.629	1.634	1.638
1120	1.354	1.357	1.361	1.393	1.444	1.482	1.511	1.535	1.554	1.571	1.584	1.596	1.606	1.615	1.623	1.630	1.636	1.642	1.647	1.651
1140	1.360	1.364	1.367	1.400	1.452	1.491	1.521	1.546	1.565	1.582	1.596	1.608	1.618	1.627	1.635	1.642	1.649	1.654	1.660	1.664
1160	1.367	1.370	1.374	1.408	1.461	1.500	1.531	1.556	1.576	1.593	1.607	1.620	1.630	1.640	1.648	1.655	1.661	1.667	1.673	1.677
1180	1.373	1.377	1.381	1.415	1.469	1.510	1.541	1.567	1.587	1.605	1.619	1.632	1.642	1.652	1.660	1.668	1.674	1.680	1.686	1.691
1200	1.380	1.383	1.387	1.422	1.478	1.519	1.551	1.577	1.598	1.616	1.631	1.644	1.655	1.664	1.673	1.680	1.687	1.693	1.699	1.704
1220	1.386	1.390	1.394	1.430	1.486	1.529	1.562	1.588	1.609	1.627	1.643	1.656	1.667	1.677	1.685	1.693	1.700	1.706	1.712	1.717
1240	1.392	1.397	1.401	1.437	1.495	1.538	1.572	1.599	1.621	1.639	1.654	1.668	1.679	1.689	1.698	1.706	1.713	1.719	1.725	1.730
1260	1.399	1.403	1.407	1.445	1.503	1.548	1.582	1.609	1.632	1.650	1.666	1.680	1.692	1.702	1.711	1.719	1.726	1.733	1.739	1.744
1280	1.405	1.410	1.414	1.452	1.512	1.557	1.592	1.620	1.643	1.662	1.678	1.692	1.704	1.714	1.724	1.732	1.739	1.746	1.752	1.757
1300	1.412	1.416	1.421	1.459	1.521	1.567	1.602	1.631	1.654	1.674	1.690	1.704	1.716	1.727	1.736	1.745	1.752	1.759	1.765	1.771
1320	1.418	1.423	1.427	1.467	1.529	1.576	1.612	1.642	1.665	1.685	1.702	1.716	1.729	1.740	1.749	1.758	1.766	1.772	1.779	1.784
1340	1.425	1.429	1.434	1.474	1.538	1.586	1.623	1.652	1.677	1.697	1.714	1.729	1.741	1.752	1.762	1.771	1.779	1.786	1.792	1.798
1360	1.431	1.436	1.441	1.482	1.547	1.595	1.633	1.663	1.688	1.709	1.726	1.741	1.754	1.765	1.775	1.784	1.792	1.799	1.806	1.812
1380	1.438	1.443	1.447	1.489	1.555	1.605	1.643	1.674	1.699	1.720	1.738	1.754	1.767	1.778	1.788	1.797	1.805	1.813	1.819	1.825
1400	1.444	1.449	1.454	1.497	1.564	1.614	1.654	1.685	1.711	1.732	1.750	1.766	1.779	1.791	1.801	1.811	1.819	1.826	1.833	1.839
1420	1.451	1.456	1.461	1.504	1.573	1.624	1.664	1.696	1.722	1.744	1.762	1.778	1.792	1.804	1.814	1.824	1.832	1.840	1.847	1.853
1440	1.458	1.463	1.467	1.512	1.581	1.634	1.674	1.707	1.734	1.756	1.775	1.791	1.805	1.817	1.828	1.837	1.846	1.853	1.860	1.867
1460	1.464	1.469	1.474	1.519	1.590	1.643	1.685	1.718	1.745	1.768	1.787	1.803	1.817	1.830	1.841	1.851	1.859	1.867	1.874	1.881
1480	1.471	1.476	1.481	1.527	1.599	1.653	1.695	1.729	1.757	1.780	1.799	1.816	1.830	1.843	1.854	1.864	1.873	1.881	1.888	1.895
1500	1.477	1.482	1.488	1.534	1.608	1.663	1.706	1.740	1.768	1.792	1.811	1.828	1.843	1.856	1.867	1.877	1.886	1.895	1.902	1.909
1520	1.484	1.489	1.494	1.542	1.617	1.673	1.716	1.751	1.780	1.804	1.824	1.841	1.856	1.869	1.881	1.891	1.900	1.908	1.916	1.923
1540	1.490	1.496	1.501	1.549	1.626	1.683	1.727	1.762	1.791	1.816	1.836	1.854	1.869	1.882	1.894	1.904	1.914	1.922	1.930	1.937
1560	1.497	1.502	1.508	1.557	1.634	1.692	1.738	1.774	1.803	1.828	1.849	1.866	1.882	1.896	1.907	1.918	1.928	1.936	1.944	1.951
1580	1.503	1.509	1.514	1.565	1.643	1.702	1.748	1.785	1.815	1.840	1.861	1.879	1.895	1.909	1.921	1.932	1.941	1.950	1.958	1.965
1600	1.510	1.516	1.521	1.572	1.652	1.712	1.759	1.796	1.827	1.852	1.874	1.892	1.908	1.921	1.934	1.945	1.955	1.964	1.972	1.979
1620	1.517	1.522	1.528	1.580	1.661	1.722	1.769	1.807	1.838	1.864	1.886	1.905	1.921	1.935	1.948	1.959	1.969	1.978	1.986	1.994
1640	1.523	1.529	1.535	1.587	1.670	1.732	1.780	1.819	1.850	1.877	1.899	1.918	1.934	1.949	1.962	1.973	1.983	1.992	2.001	2.008

SIGMA = 3477.0

$V - V_c = 16$

H \ x	3.4	3.2	3.0	2.8	2.6	2.4	2.2	2.0	1.8	1.6	1.4	1.2	1.0	0.8	0.6	0.4	0.2	0.10	0.09	0.08
1660	2.022	2.015	2.006	1.997	1.987	1.975	1.962	1.948	1.931	1.911	1.889	1.862	1.830	1.791	1.742	1.679	1.595	1.542	1.536	1.530
1680	2.037	2.029	2.021	2.011	2.001	1.989	1.976	1.961	1.944	1.924	1.901	1.874	1.841	1.802	1.752	1.688	1.603	1.548	1.542	1.536
1700	2.051	2.044	2.035	2.025	2.015	2.003	1.989	1.974	1.957	1.937	1.914	1.886	1.853	1.812	1.762	1.697	1.610	1.555	1.549	1.543
1720	2.066	2.058	2.049	2.039	2.029	2.017	2.003	1.988	1.970	1.950	1.926	1.898	1.864	1.823	1.772	1.706	1.618	1.562	1.556	1.549
1740	2.081	2.073	2.064	2.054	2.043	2.030	2.017	2.001	1.983	1.963	1.938	1.910	1.876	1.834	1.782	1.715	1.626	1.569	1.562	1.556
1760	2.095	2.087	2.078	2.068	2.057	2.044	2.030	2.014	1.996	1.975	1.951	1.922	1.887	1.845	1.792	1.724	1.633	1.575	1.569	1.563
1780	2.110	2.102	2.092	2.082	2.071	2.058	2.044	2.028	2.010	1.988	1.963	1.934	1.899	1.856	1.802	1.733	1.641	1.582	1.576	1.569
1800	2.125	2.116	2.107	2.097	2.085	2.072	2.058	2.041	2.023	2.001	1.976	1.946	1.911	1.867	1.812	1.742	1.649	1.589	1.583	1.576
1820	2.140	2.131	2.122	2.111	2.099	2.086	2.072	2.055	2.036	2.014	1.989	1.958	1.922	1.878	1.822	1.751	1.656	1.596	1.589	1.583
1840	2.155	2.146	2.136	2.125	2.114	2.100	2.086	2.069	2.049	2.027	2.001	1.971	1.934	1.889	1.833	1.760	1.664	1.603	1.596	1.589
1860	2.169	2.161	2.151	2.140	2.128	2.115	2.099	2.082	2.063	2.040	2.014	1.983	1.945	1.900	1.843	1.770	1.672	1.610	1.603	1.596
1880	2.184	2.175	2.165	2.155	2.142	2.129	2.113	2.096	2.076	2.053	2.027	1.995	1.957	1.911	1.853	1.779	1.680	1.616	1.609	1.602
1900	2.199	2.190	2.180	2.169	2.157	2.143	2.127	2.110	2.090	2.066	2.039	2.007	1.969	1.922	1.863	1.788	1.687	1.623	1.616	1.609
1920	2.215	2.205	2.195	2.184	2.171	2.157	2.141	2.124	2.103	2.080	2.052	2.020	1.981	1.933	1.874	1.797	1.695	1.630	1.623	1.616
1940	2.230	2.220	2.210	2.198	2.186	2.172	2.156	2.137	2.117	2.093	2.065	2.032	1.993	1.944	1.884	1.806	1.703	1.637	1.630	1.622
1960	2.245	2.235	2.225	2.213	2.200	2.186	2.170	2.151	2.130	2.106	2.078	2.045	2.004	1.955	1.894	1.816	1.711	1.644	1.636	1.629
1980	2.260	2.250	2.240	2.228	2.215	2.200	2.184	2.165	2.144	2.119	2.091	2.057	2.016	1.967	1.905	1.825	1.718	1.651	1.643	1.636
2000	2.275	2.266	2.255	2.243	2.230	2.215	2.198	2.179	2.158	2.133	2.104	2.069	2.028	1.978	1.915	1.834	1.726	1.658	1.650	1.642
2020	2.291	2.281	2.270	2.258	2.244	2.229	2.212	2.193	2.171	2.146	2.117	2.082	2.040	1.989	1.925	1.843	1.734	1.664	1.657	1.649
2040	2.306	2.296	2.285	2.273	2.259	2.244	2.227	2.207	2.185	2.160	2.130	2.095	2.052	2.000	1.936	1.853	1.742	1.671	1.664	1.656
2060	2.322	2.311	2.300	2.288	2.274	2.258	2.241	2.221	2.199	2.173	2.143	2.107	2.064	2.012	1.946	1.862	1.750	1.678	1.670	1.662
2080	2.337	2.327	2.315	2.303	2.289	2.273	2.256	2.236	2.213	2.187	2.156	2.120	2.076	2.023	1.957	1.871	1.758	1.685	1.677	1.669
2100	2.353	2.342	2.331	2.318	2.304	2.288	2.270	2.250	2.227	2.200	2.169	2.132	2.088	2.035	1.967	1.881	1.765	1.692	1.684	1.676
2120	2.368	2.358	2.346	2.333	2.319	2.303	2.284	2.264	2.241	2.214	2.182	2.145	2.100	2.046	1.978	1.890	1.773	1.699	1.691	1.682
2140	2.384	2.373	2.361	2.348	2.334	2.317	2.299	2.278	2.255	2.227	2.195	2.158	2.113	2.057	1.988	1.900	1.781	1.706	1.698	1.689
2160	2.399	2.389	2.377	2.363	2.349	2.332	2.314	2.293	2.269	2.241	2.209	2.171	2.125	2.069	1.999	1.909	1.789	1.713	1.704	1.696
2180	2.415	2.404	2.392	2.379	2.364	2.347	2.328	2.307	2.283	2.255	2.222	2.183	2.137	2.080	2.010	1.918	1.797	1.720	1.711	1.703
2200	2.431	2.420	2.408	2.394	2.379	2.362	2.343	2.321	2.297	2.268	2.235	2.196	2.149	2.092	2.020	1.928	1.805	1.727	1.718	1.709
2220	2.447	2.436	2.423	2.409	2.394	2.377	2.358	2.336	2.311	2.282	2.249	2.209	2.162	2.103	2.031	1.937	1.813	1.734	1.725	1.716
2240	2.463	2.451	2.439	2.425	2.409	2.392	2.372	2.350	2.325	2.296	2.262	2.222	2.174	2.115	2.041	1.947	1.821	1.741	1.732	1.723
2260	2.479	2.467	2.454	2.440	2.425	2.407	2.387	2.365	2.339	2.310	2.276	2.235	2.186	2.127	2.052	1.956	1.829	1.747	1.739	1.729
2280	2.495	2.483	2.470	2.456	2.440	2.422	2.402	2.380	2.354	2.324	2.289	2.248	2.199	2.138	2.063	1.966	1.837	1.754	1.745	1.736
2300	2.511	2.499	2.486	2.471	2.455	2.437	2.417	2.394	2.368	2.338	2.303	2.261	2.211	2.150	2.074	1.976	1.845	1.761	1.752	1.743
2320	2.527	2.515	2.502	2.487	2.471	2.453	2.432	2.409	2.382	2.352	2.316	2.274	2.223	2.162	2.084	1.985	1.853	1.768	1.759	1.750
2340	2.543	2.531	2.517	2.503	2.486	2.468	2.447	2.424	2.397	2.366	2.330	2.287	2.236	2.173	2.095	1.995	1.861	1.775	1.766	1.756
2360	2.559	2.547	2.533	2.518	2.502	2.483	2.462	2.438	2.411	2.380	2.343	2.300	2.248	2.185	2.106	2.004	1.869	1.782	1.773	1.763
2380	2.575	2.563	2.549	2.534	2.517	2.498	2.477	2.453	2.426	2.394	2.357	2.313	2.261	2.197	2.117	2.014	1.877	1.789	1.780	1.770
2400	2.592	2.579	2.565	2.550	2.533	2.514	2.492	2.468	2.440	2.408	2.371	2.327	2.274	2.209	2.128	2.024	1.885	1.796	1.787	1.777
2420	2.608	2.595	2.581	2.566	2.549	2.529	2.508	2.483	2.455	2.422	2.385	2.340	2.286	2.221	2.139	2.033	1.893	1.803	1.793	1.783
2440	2.625	2.612	2.597	2.582	2.564	2.545	2.523	2.498	2.469	2.437	2.398	2.353	2.299	2.233	2.150	2.043	1.901	1.810	1.800	1.790
2460	2.641	2.628	2.614	2.598	2.580	2.560	2.538	2.513	2.484	2.451	2.412	2.366	2.312	2.244	2.161	2.053	1.909	1.817	1.807	1.797
2480	2.658	2.644	2.630	2.614	2.596	2.576	2.553	2.528	2.499	2.465	2.426	2.380	2.324	2.256	2.172	2.062	1.917	1.824	1.814	1.804

SIGMA = 3477.0

V−Vc = 16

H \ x	3.4	3.2	3.0	2.8	2.6	2.4	2.2	2.0	1.8	1.6	1.4	1.2	1.0	0.8	0.6	0.4	0.2	0.10	0.09	0.08
2500	2.674	2.661	2.646	2.630	2.612	2.591	2.569	2.543	2.514	2.480	2.440	2.393	2.337	2.268	2.183	2.072	1.925	1.831	1.821	1.811
2520	2.691	2.677	2.662	2.646	2.628	2.607	2.584	2.558	2.528	2.494	2.454	2.407	2.350	2.280	2.194	2.082	1.933	1.838	1.828	1.817
2540	2.707	2.694	2.679	2.662	2.644	2.623	2.600	2.573	2.543	2.508	2.468	2.420	2.363	2.292	2.205	2.092	1.941	1.845	1.835	1.824
2560	2.724	2.710	2.695	2.678	2.660	2.639	2.615	2.589	2.558	2.523	2.482	2.434	2.375	2.304	2.216	2.102	1.949	1.852	1.842	1.831
2580	2.741	2.727	2.711	2.694	2.676	2.654	2.631	2.604	2.573	2.537	2.496	2.447	2.388	2.317	2.227	2.111	1.957	1.860	1.849	1.838
2600	2.758	2.744	2.728	2.711	2.692	2.670	2.646	2.619	2.588	2.552	2.510	2.461	2.401	2.329	2.238	2.121	1.966	1.867	1.856	1.845
2620	2.775	2.760	2.745	2.727	2.708	2.686	2.662	2.634	2.603	2.567	2.524	2.474	2.414	2.341	2.249	2.131	1.974	1.874	1.863	1.851
2640	2.791	2.777	2.761	2.744	2.724	2.702	2.678	2.650	2.618	2.581	2.538	2.488	2.427	2.353	2.260	2.141	1.982	1.881	1.870	1.858
2660	2.808	2.794	2.778	2.760	2.740	2.718	2.693	2.665	2.633	2.596	2.553	2.502	2.440	2.365	2.271	2.151	1.990	1.888	1.877	1.865
2680	2.825	2.811	2.794	2.776	2.757	2.734	2.709	2.681	2.648	2.611	2.567	2.515	2.453	2.377	2.283	2.161	1.998	1.895	1.883	1.872
2700	2.843	2.828	2.811	2.793	2.773	2.750	2.725	2.696	2.663	2.625	2.581	2.529	2.466	2.390	2.294	2.171	2.006	1.902	1.890	1.879
2720	2.860	2.845	2.828	2.810	2.789	2.766	2.741	2.712	2.679	2.640	2.596	2.543	2.479	2.402	2.305	2.181	2.015	1.909	1.897	1.886
2740	2.877	2.862	2.845	2.826	2.806	2.783	2.757	2.727	2.694	2.655	2.610	2.557	2.493	2.414	2.316	2.191	2.023	1.916	1.904	1.892
2760	2.894	2.879	2.862	2.843	2.822	2.799	2.773	2.743	2.709	2.670	2.624	2.570	2.506	2.427	2.328	2.201	2.031	1.923	1.911	1.899
2780	2.911	2.896	2.879	2.860	2.839	2.815	2.789	2.759	2.724	2.685	2.639	2.584	2.519	2.439	2.339	2.211	2.039	1.930	1.918	1.906
2800	2.929	2.913	2.896	2.877	2.855	2.831	2.805	2.774	2.740	2.700	2.653	2.598	2.532	2.451	2.350	2.221	2.048	1.937	1.925	1.913
2820	2.946	2.930	2.913	2.893	2.872	2.848	2.821	2.790	2.755	2.715	2.668	2.612	2.545	2.464	2.362	2.231	2.056	1.945	1.932	1.920
2840	2.963	2.947	2.930	2.910	2.889	2.864	2.837	2.806	2.771	2.730	2.682	2.626	2.559	2.476	2.373	2.241	2.064	1.952	1.939	1.927
2860	2.981	2.965	2.947	2.927	2.905	2.881	2.853	2.822	2.786	2.745	2.697	2.640	2.572	2.489	2.385	2.251	2.072	1.959	1.946	1.934
2880	2.999	2.982	2.964	2.944	2.922	2.897	2.869	2.838	2.802	2.760	2.712	2.654	2.585	2.501	2.396	2.261	2.081	1.966	1.953	1.940
2900	3.016	3.000	2.981	2.961	2.939	2.914	2.886	2.854	2.817	2.775	2.726	2.668	2.599	2.514	2.408	2.271	2.089	1.973	1.960	1.947
2920	3.034	3.017	2.999	2.978	2.956	2.930	2.902	2.870	2.833	2.791	2.741	2.683	2.612	2.527	2.419	2.281	2.097	1.980	1.967	1.954
2940	3.051	3.035	3.016	2.995	2.973	2.947	2.918	2.886	2.849	2.806	2.756	2.697	2.626	2.539	2.431	2.291	2.106	1.987	1.974	1.961
2960	3.069	3.052	3.033	3.013	2.990	2.964	2.935	2.902	2.864	2.821	2.771	2.711	2.639	2.552	2.442	2.302	2.114	1.995	1.981	1.968
2980	3.087	3.070	3.051	3.030	3.007	2.981	2.951	2.918	2.880	2.837	2.786	2.725	2.653	2.564	2.454	2.312	2.122	2.002	1.988	1.975
3000	3.105	3.087	3.068	3.047	3.024	2.997	2.968	2.934	2.896	2.852	2.800	2.739	2.666	2.577	2.466	2.322	2.131	2.009	1.996	1.982

SIGMA = 3477.0

$V-V_c = 16$

H	3.4	3.2	3.0	2.8	2.6	2.4	2.2	2.0	1.8	1.6	1.4	1.2	1.0	0.8	0.6	0.4	0.2	0.10	0.09	0.08
3100	3.195	3.177	3.156	3.134	3.110	3.082	3.051	3.016	2.976	2.929	2.875	2.811	2.735	2.641	2.524	2.373	2.173	2.045	2.031	2.016
3200	3.286	3.267	3.246	3.223	3.197	3.168	3.135	3.099	3.056	3.008	2.951	2.884	2.804	2.706	2.583	2.425	2.215	2.081	2.066	2.051
3300	3.379	3.359	3.336	3.312	3.285	3.255	3.221	3.182	3.138	3.087	3.028	2.958	2.874	2.771	2.643	2.478	2.257	2.117	2.102	2.086
3400	3.472	3.451	3.428	3.403	3.375	3.343	3.308	3.267	3.221	3.168	3.106	3.033	2.945	2.837	2.703	2.530	2.300	2.154	2.138	2.121
3500	3.567	3.546	3.522	3.495	3.465	3.432	3.395	3.353	3.305	3.250	3.185	3.108	3.017	2.904	2.764	2.584	2.343	2.190	2.173	2.156
3600	3.664	3.641	3.616	3.588	3.557	3.523	3.484	3.440	3.390	3.332	3.265	3.185	3.089	2.972	2.826	2.637	2.387	2.227	2.209	2.191
3700	3.761	3.738	3.712	3.683	3.651	3.615	3.574	3.528	3.476	3.416	3.345	3.262	3.162	3.040	2.888	2.692	2.430	2.264	2.245	2.227
3800	3.860	3.836	3.808	3.778	3.745	3.707	3.665	3.618	3.563	3.500	3.427	3.341	3.237	3.109	2.951	2.746	2.474	2.301	2.282	2.262
3900	3.961	3.935	3.906	3.875	3.840	3.801	3.758	3.708	3.652	3.586	3.510	3.420	3.312	3.179	3.014	2.802	2.518	2.338	2.318	2.298
4000	4.062	4.035	4.006	3.973	3.937	3.897	3.851	3.800	3.741	3.673	3.594	3.500	3.387	3.250	3.078	2.857	2.563	2.375	2.355	2.334
4100	4.165	4.137	4.106	4.073	4.035	3.993	3.946	3.892	3.831	3.760	3.678	3.581	3.464	3.321	3.143	2.913	2.607	2.413	2.391	2.370
4200	4.269	4.240	4.208	4.173	4.134	4.091	4.042	3.986	3.922	3.849	3.764	3.663	3.541	3.393	3.208	2.970	2.652	2.450	2.428	2.405
4300	4.374	4.344	4.311	4.275	4.234	4.189	4.138	4.081	4.015	3.939	3.850	3.745	3.620	3.466	3.275	3.027	2.698	2.488	2.465	2.442
4400	4.481	4.450	4.416	4.378	4.336	4.289	4.236	4.177	4.108	4.030	3.938	3.829	3.699	3.540	3.341	3.085	2.743	2.526	2.502	2.478
4500	4.589	4.556	4.521	4.482	4.439	4.390	4.336	4.274	4.203	4.121	4.026	3.914	3.779	3.614	3.408	3.143	2.789	2.564	2.539	2.514
4600	4.698	4.665	4.628	4.587	4.543	4.492	4.436	4.372	4.299	4.214	4.116	3.999	3.860	3.689	3.475	3.201	2.835	2.602	2.577	2.551
4700	4.808	4.774	4.736	4.694	4.648	4.596	4.537	4.471	4.395	4.308	4.206	4.086	3.941	3.765	3.544	3.260	2.882	2.641	2.614	2.587
4800	4.920	4.884	4.845	4.802	4.754	4.700	4.640	4.571	4.493	4.403	4.297	4.173	4.024	3.841	3.613	3.319	2.928	2.679	2.652	2.624
4900	5.033	4.996	4.956	4.911	4.861	4.806	4.744	4.673	4.592	4.499	4.390	4.261	4.107	3.918	3.682	3.379	2.975	2.718	2.690	2.661
5000	5.147	5.109	5.067	5.021	4.970	4.913	4.848	4.775	4.692	4.596	4.483	4.350	4.191	3.996	3.753	3.440	3.022	2.757	2.728	2.698
5100	5.263	5.224	5.180	5.133	5.080	5.021	4.954	4.879	4.793	4.693	4.577	4.440	4.276	4.075	3.824	3.501	3.070	2.796	2.766	2.735
5200	5.380	5.339	5.295	5.246	5.191	5.130	5.061	4.984	4.895	4.792	4.673	4.531	4.362	4.154	3.895	3.562	3.118	2.835	2.804	2.772
5300	5.498	5.456	5.410	5.360	5.303	5.240	5.170	5.090	4.998	4.892	4.769	4.623	4.448	4.234	3.967	3.624	3.166	2.874	2.842	2.809
5400	5.617	5.574	5.527	5.475	5.417	5.352	5.279	5.196	5.102	4.993	4.866	4.716	4.536	4.315	4.040	3.686	3.214	2.914	2.880	2.847
5500	5.738	5.693	5.645	5.591	5.531	5.465	5.390	5.304	5.207	5.095	4.964	4.809	4.624	4.397	4.113	3.749	3.262	2.953	2.919	2.884
5600	5.860	5.814	5.764	5.709	5.647	5.579	5.501	5.414	5.314	5.198	5.063	4.904	4.713	4.479	4.187	3.812	3.311	2.993	2.958	2.922
5700	5.983	5.936	5.884	5.828	5.764	5.694	5.614	5.524	5.421	5.302	5.163	4.999	4.803	4.562	4.262	3.876	3.360	3.033	2.997	2.960
5800	6.107	6.059	6.006	5.948	5.882	5.810	5.728	5.635	5.529	5.407	5.264	5.096	4.893	4.646	4.337	3.940	3.410	3.073	3.036	2.998
5900	6.233	6.184	6.129	6.069	6.002	5.927	5.843	5.748	5.639	5.513	5.366	5.193	4.985	4.731	4.413	4.004	3.460	3.113	3.075	3.036
6000	6.360	6.309	6.253	6.191	6.123	6.046	5.959	5.861	5.749	5.620	5.469	5.291	5.077	4.816	4.489	4.069	3.510	3.153	3.114	3.074
6100	6.488	6.436	6.379	6.315	6.244	6.165	6.077	5.976	5.861	5.728	5.573	5.390	5.171	4.902	4.566	4.135	3.560	3.194	3.153	3.112
6200	6.618	6.564	6.505	6.440	6.367	6.286	6.195	6.092	5.973	5.837	5.678	5.490	5.265	4.989	4.644	4.201	3.610	3.234	3.193	3.151
6300	6.749	6.694	6.633	6.566	6.492	6.408	6.315	6.209	6.087	5.947	5.784	5.591	5.359	5.076	4.722	4.268	3.661	3.275	3.233	3.189
6400	6.881	6.824	6.762	6.693	6.617	6.532	6.436	6.326	6.202	6.058	5.891	5.693	5.455	5.165	4.801	4.335	3.712	3.316	3.272	3.228
6500	7.014	6.956	6.893	6.822	6.744	6.656	6.557	6.446	6.318	6.170	5.999	5.795	5.552	5.254	4.881	4.402	3.763	3.357	3.312	3.267
6600	7.149	7.090	7.024	6.952	6.871	6.781	6.680	6.566	6.435	6.284	6.107	5.899	5.649	5.343	4.961	4.470	3.815	3.398	3.352	3.306
6700	7.285	7.224	7.157	7.083	7.000	6.908	6.805	6.687	6.553	6.398	6.217	6.003	5.747	5.434	5.042	4.538	3.867	3.440	3.393	3.345
6800	7.422	7.360	7.291	7.215	7.131	7.036	6.930	6.809	6.672	6.513	6.328	6.109	5.846	5.525	5.123	4.607	3.919	3.481	3.433	3.384
6900	7.561	7.497	7.426	7.348	7.262	7.165	7.056	6.933	6.792	6.629	6.439	6.215	5.946	5.617	5.205	4.677	3.972	3.523	3.473	3.423
7000	7.701	7.635	7.563	7.483	7.394	7.295	7.184	7.057	6.913	6.746	6.552	6.322	6.046	5.709	5.288	4.746	4.024	3.565	3.514	3.463
7100	7.842	7.775	7.701	7.619	7.528	7.427	7.313	7.183	7.035	6.865	6.665	6.430	6.148	5.803	5.371	4.817	4.077	3.607	3.555	3.502
7200	7.984	7.916	7.840	7.756	7.663	7.559	7.442	7.310	7.159	6.984	6.780	6.539	6.250	5.897	5.455	4.888	4.131	3.649	3.596	3.542

x

V–V$_c$ = 16   SIGMA = 3477.0

H \ x	3.4	3.2	3.0	2.8	2.6	2.4	2.2	2.0	1.8	1.6	1.4	1.2	1.0	0.8	0.6	0.4	0.2	0.10	0.09	0.08
7300	8.128	8.058	7.980	7.894	7.799	7.693	7.573	7.438	7.283	7.104	6.896	6.649	6.353	5.992	5.540	4.959	4.184	3.691	3.637	3.582
7400	8.273	8.201	8.122	8.034	7.937	7.828	7.706	7.567	7.408	7.225	7.012	6.760	6.457	6.087	5.625	5.031	4.238	3.734	3.678	3.621
7500	8.419	8.345	8.264	8.175	8.075	7.964	7.839	7.697	7.535	7.348	7.129	6.871	6.562	6.184	5.711	5.103	4.292	3.776	3.719	3.661
7600	8.567	8.491	8.408	8.317	8.215	8.101	7.973	7.828	7.662	7.471	7.248	6.984	6.668	6.281	5.797	5.175	4.346	3.819	3.761	3.702
7700	8.715	8.638	8.554	8.460	8.356	8.240	8.109	7.960	7.791	7.595	7.367	7.098	6.774	6.378	5.884	5.248	4.401	3.862	3.802	3.742
7800	8.865	8.787	8.700	8.604	8.498	8.379	8.245	8.094	7.921	7.721	7.488	7.212	6.881	6.477	5.972	5.322	4.456	3.905	3.844	3.782
7900	9.017	8.936	8.848	8.750	8.641	8.520	8.383	8.228	8.051	7.847	7.609	7.327	6.989	6.576	6.060	5.396	4.511	3.948	3.886	3.823
8000	9.169	9.087	8.997	8.897	8.786	8.662	8.522	8.364	8.183	7.974	7.731	7.443	7.098	6.676	6.149	5.471	4.567	3.991	3.928	3.863
8100	9.323	9.239	9.147	9.045	8.932	8.805	8.662	8.501	8.316	8.103	7.854	7.560	7.208	6.777	6.238	5.546	4.622	4.035	3.970	3.904
8200	9.479	9.393	9.298	9.194	9.078	8.949	8.803	8.638	8.450	8.232	7.978	7.678	7.318	6.878	6.328	5.621	4.678	4.078	4.012	3.945
8300	9.635	9.547	9.451	9.345	9.227	9.094	8.946	8.777	8.585	8.363	8.104	7.797	7.430	6.981	6.419	5.697	4.735	4.122	4.055	3.986
8400	9.793	9.703	9.605	9.497	9.376	9.241	9.089	8.917	8.721	8.494	8.230	7.917	7.542	7.084	6.510	5.774	4.791	4.166	4.097	4.027
8500	9.952	9.861	9.760	9.649	9.526	9.389	9.234	9.059	8.858	8.627	8.357	8.038	7.655	7.187	6.602	5.851	4.848	4.210	4.140	4.068
8600	10.10	10.02	9.92	9.80	9.68	9.54	9.38	9.20	9.01	8.76	8.48	8.16	7.77	7.29	6.70	5.93	4.91	4.25	4.18	4.11
8700	10.27	10.18	10.07	9.96	9.83	9.69	9.53	9.34	9.14	8.89	8.61	8.28	7.88	7.40	6.79	6.01	4.96	4.30	4.23	4.15
8800	10.44	10.34	10.23	10.12	9.99	9.84	9.67	9.49	9.28	9.03	8.74	8.41	8.00	7.50	6.88	6.08	5.02	4.34	4.27	4.19
8900	10.60	10.50	10.39	10.27	10.14	9.99	9.82	9.63	9.42	9.17	8.87	8.53	8.12	7.61	6.98	6.16	5.08	4.39	4.31	4.23
9000	10.77	10.67	10.55	10.43	10.30	10.14	9.97	9.78	9.56	9.30	9.01	8.65	8.23	7.72	7.07	6.24	5.14	4.43	4.36	4.28
9100	10.93	10.83	10.72	10.59	10.45	10.30	10.13	9.93	9.70	9.44	9.14	8.78	8.35	7.82	7.17	6.32	5.20	4.48	4.40	4.32
9200	11.10	11.00	10.88	10.75	10.61	10.45	10.28	10.08	9.85	9.58	9.27	8.91	8.47	7.93	7.26	6.40	5.25	4.52	4.44	4.36
9300	11.27	11.16	11.05	10.92	10.77	10.61	10.43	10.23	9.99	9.72	9.41	9.04	8.59	8.04	7.36	6.48	5.31	4.57	4.49	4.40
9400	11.44	11.33	11.21	11.08	10.93	10.77	10.59	10.38	10.14	9.86	9.54	9.16	8.71	8.15	7.46	6.56	5.37	4.61	4.53	4.45
9500	11.61	11.50	11.38	11.25	11.10	10.93	10.74	10.53	10.29	10.01	9.68	9.29	8.83	8.26	7.56	6.65	5.43	4.66	4.57	4.49
9600	11.79	11.67	11.55	11.41	11.26	11.10	10.90	10.68	10.44	10.15	9.82	9.43	8.95	8.38	7.66	6.73	5.49	4.71	4.62	4.53
9700	11.96	11.85	11.72	11.58	11.43	11.26	11.06	10.84	10.59	10.30	9.96	9.56	9.08	8.49	7.75	6.81	5.55	4.75	4.66	4.57
9800	12.14	12.02	11.89	11.75	11.59	11.42	11.22	10.99	10.74	10.44	10.10	9.69	9.20	8.60	7.85	6.89	5.61	4.80	4.71	4.62
9900	12.31	12.19	12.06	11.92	11.76	11.59	11.38	11.15	10.89	10.59	10.24	9.82	9.32	8.72	7.96	6.98	5.67	4.84	4.75	4.66
10000	12.49	12.37	12.24	12.09	11.93	11.74	11.54	11.31	11.04	10.74	10.38	9.96	9.45	8.83	8.06	7.06	5.73	4.89	4.80	4.70
10100	12.67	12.55	12.41	12.26	12.10	11.91	11.70	11.47	11.20	10.88	10.52	10.10	9.58	8.95	8.16	7.15	5.80	4.94	4.84	4.75
10200	12.85	12.73	12.59	12.44	12.27	12.10	11.87	11.63	11.35	11.04	10.66	10.23	9.70	9.06	8.26	7.23	5.86	4.98	4.89	4.79
10300	13.03	12.91	12.77	12.61	12.44	12.27	12.03	11.79	11.51	11.19	10.81	10.36	9.83	9.18	8.36	7.32	5.92	5.03	4.93	4.83
10400	13.22	13.09	12.94	12.79	12.61	12.42	12.20	11.95	11.67	11.34	10.95	10.50	9.96	9.30	8.47	7.40	5.98	5.08	4.98	4.88
10500	13.40	13.27	13.12	12.96	12.79	12.61	12.37	12.11	11.82	11.51	11.10	10.64	10.09	9.42	8.57	7.49	6.04	5.12	5.02	4.92
10600	13.59	13.45	13.31	13.14	12.96	12.76	12.53	12.28	11.98	11.67	11.25	10.78	10.22	9.53	8.68	7.58	6.11	5.17	5.07	4.96
10700	13.77	13.64	13.49	13.32	13.14	12.93	12.70	12.44	12.14	11.82	11.40	10.92	10.35	9.65	8.78	7.66	6.17	5.22	5.11	5.01
10800	13.96	13.82	13.67	13.50	13.32	13.11	12.87	12.61	12.31	11.98	11.55	11.06	10.48	9.77	8.89	7.75	6.23	5.27	5.16	5.05
10900	14.15	14.01	13.86	13.69	13.50	13.28	13.05	12.78	12.47	12.14	11.70	11.20	10.62	9.90	9.00	7.84	6.30	5.31	5.21	5.10
11000	14.34	14.20	14.05	13.87	13.69	13.46	13.22	12.95	12.63	12.27	11.85	11.35	10.75	10.02	9.10	7.93	6.36	5.36	5.25	5.14
11100	14.53	14.39	14.24	14.05	13.87	13.66	13.39	13.12	12.80	12.43	12.00	11.49	10.88	10.14	9.21	8.02	6.42	5.41	5.30	5.18
11200	14.73	14.58	14.42	14.24	14.04	13.82	13.57	13.29	12.96	12.59	12.15	11.64	11.02	10.26	9.32	8.11	6.49	5.46	5.34	5.23
11300	14.92	14.77	14.61	14.43	14.23	14.00	13.75	13.46	13.13	12.75	12.31	11.78	11.16	10.39	9.43	8.20	6.55	5.51	5.39	5.27
11400	15.12	14.97	14.80	14.62	14.41	14.18	13.92	13.63	13.30	12.91	12.46	11.93	11.29	10.51	9.54	8.29	6.62	5.55	5.44	5.32

$V - V_c = 16$　　　　　SIGMA = 3477.0

H \ x	0.08	0.09	0.10	0.2	0.4	0.6	0.8	1.0	1.2	1.4	1.6	1.8	2.0	2.2	2.4	2.6	2.8	3.0	3.2	3.4
11500	5.36	5.48	5.60	6.68	8.38	9.65	10.64	11.43	12.08	12.62	13.07	13.47	13.81	14.10	14.36	14.60	14.80	14.99	15.16	15.32
11600	5.41	5.53	5.65	6.75	8.47	9.76	10.77	11.57	12.23	12.78	13.24	13.64	13.98	14.28	14.55	14.78	15.00	15.19	15.36	15.51
11700	5.45	5.58	5.70	6.81	8.56	9.87	10.89	11.71	12.38	12.93	13.40	13.81	14.16	14.46	14.73	14.97	15.19	15.38	15.56	15.71
11800	5.50	5.62	5.75	6.88	8.65	9.99	11.02	11.85	12.53	13.09	13.57	13.98	14.33	14.65	14.92	15.16	15.38	15.58	15.75	15.92
11900	5.54	5.67	5.80	6.95	8.75	10.10	11.15	11.99	12.68	13.25	13.74	14.15	14.51	14.83	15.11	15.35	15.58	15.77	15.95	16.12
12000	5.59	5.72	5.85	7.01	8.84	10.21	11.28	12.13	12.83	13.41	13.91	14.33	14.69	15.01	15.30	15.55	15.77	15.97	16.16	16.32
12100	5.63	5.77	5.90	7.08	8.93	10.33	11.41	12.28	12.98	13.57	14.07	14.50	14.87	15.20	15.49	15.74	15.97	16.17	16.36	16.53
12200	5.68	5.82	5.95	7.15	9.03	10.44	11.54	12.42	13.14	13.74	14.24	14.68	15.06	15.39	15.68	15.93	16.17	16.37	16.56	16.73
12300	5.73	5.86	6.00	7.21	9.12	10.56	11.67	12.56	13.29	13.90	14.42	14.86	15.24	15.57	15.87	16.13	16.37	16.58	16.77	16.94
12400	5.77	5.91	6.05	7.28	9.22	10.67	11.80	12.71	13.45	14.07	14.59	15.04	15.42	15.76	16.06	16.33	16.57	16.78	16.97	17.15
12500	5.82	5.96	6.10	7.35	9.31	10.79	11.94	12.85	13.61	14.23	14.76	15.21	15.61	15.95	16.26	16.53	16.77	16.98	17.18	17.36
12600	5.86	6.01	6.15	7.42	9.41	10.91	12.07	13.00	13.76	14.40	14.94	15.40	15.79	16.14	16.45	16.73	16.97	17.19	17.39	17.57
12700	5.91	6.06	6.20	7.48	9.51	11.03	12.21	13.15	13.92	14.57	15.11	15.58	15.98	16.34	16.65	16.93	17.17	17.40	17.60	17.78
12800	5.96	6.10	6.25	7.55	9.61	11.14	12.34	13.30	14.08	14.73	15.29	15.76	16.16	16.53	16.85	17.13	17.38	17.60	17.81	18.00
12900	6.00	6.15	6.30	7.62	9.70	11.26	12.48	13.45	14.24	14.90	15.46	15.94	16.36	16.72	17.05	17.33	17.59	17.82	18.02	18.21
13000	6.05	6.20	6.35	7.69	9.80	11.38	12.61	13.60	14.40	15.07	15.64	16.13	16.55	16.92	17.25	17.54	17.79	18.03	18.24	18.43
13100	6.10	6.25	6.40	7.76	9.90	11.50	12.75	13.75	14.57	15.25	15.82	16.31	16.74	17.12	17.45	17.74	18.00	18.24	18.45	18.65
13200	6.15	6.30	6.45	7.83	10.00	11.62	12.89	13.90	14.73	15.42	16.00	16.50	16.94	17.31	17.65	17.95	18.21	18.45	18.67	18.87
13300	6.19	6.35	6.50	7.90	10.10	11.75	13.03	14.05	14.89	15.59	16.18	16.69	17.13	17.51	17.85	18.16	18.42	18.67	18.89	19.09
13400	6.24	6.40	6.55	7.97	10.20	11.87	13.17	14.21	15.06	15.77	16.36	16.88	17.32	17.71	18.06	18.36	18.64	18.89	19.11	19.31
13500	6.29	6.45	6.61	8.04	10.30	11.99	13.31	14.36	15.22	15.94	16.55	17.07	17.52	17.92	18.26	18.57	18.85	19.10	19.33	19.53
13600	6.33	6.50	6.66	8.11	10.40	12.12	13.45	14.52	15.39	16.12	16.73	17.26	17.71	18.12	18.47	18.79	19.07	19.32	19.55	19.76
13700	6.38	6.55	6.71	8.18	10.50	12.24	13.59	14.67	15.56	16.29	16.92	17.45	17.92	18.32	18.68	19.00	19.28	19.54	19.77	19.98
13800	6.43	6.60	6.76	8.26	10.60	12.36	13.73	14.83	15.73	16.47	17.10	17.65	18.12	18.53	18.89	19.21	19.50	19.76	19.98	20.21
13900	6.48	6.65	6.81	8.33	10.71	12.49	13.88	14.99	15.90	16.65	17.29	17.84	18.32	18.73	19.10	19.43	19.72	19.98	20.22	20.44
14000	6.53	6.70	6.87	8.40	10.81	12.62	14.02	15.15	16.07	16.83	17.48	18.04	18.52	18.94	19.31	19.64	19.94	20.21	20.45	20.67
14100	6.57	6.75	6.92	8.47	10.91	12.74	14.17	15.31	16.24	17.01	17.67	18.23	18.72	19.15	19.53	19.86	20.16	20.43	20.67	20.90
14200	6.62	6.80	6.97	8.54	11.02	12.87	14.31	15.47	16.41	17.20	17.86	18.43	18.93	19.36	19.74	20.08	20.38	20.66	20.90	21.13
14300	6.67	6.85	7.02	8.62	11.12	13.00	14.46	15.63	16.58	17.38	18.05	18.63	19.13	19.57	19.96	20.30	20.61	20.88	21.13	21.36
14400	6.72	6.90	7.08	8.69	11.23	13.13	14.61	15.79	16.76	17.56	18.25	18.83	19.34	19.78	20.17	20.52	20.83	21.11	21.37	21.60
14500	6.77	6.95	7.13	8.76	11.33	13.26	14.75	15.95	16.93	17.75	18.44	19.03	19.54	19.99	20.39	20.74	21.06	21.34	21.60	21.83
14600	6.82	7.00	7.18	8.84	11.44	13.39	14.90	16.11	17.11	17.93	18.63	19.23	19.75	20.21	20.61	20.97	21.28	21.57	21.83	22.07
14700	6.86	7.05	7.24	8.91	11.54	13.52	15.05	16.28	17.28	18.12	18.83	19.44	19.96	20.42	20.83	21.19	21.51	21.80	22.07	22.31
14800	6.91	7.10	7.29	8.99	11.65	13.65	15.20	16.44	17.46	18.31	19.03	19.64	20.17	20.64	21.05	21.42	21.74	22.04	22.30	22.55
14900	6.96	7.16	7.34	9.06	11.76	13.78	15.35	16.61	17.64	18.50	19.22	19.85	20.38	20.86	21.27	21.64	21.97	22.27	22.54	22.79
15000	7.01	7.21	7.40	9.13	11.86	13.91	15.50	16.78	17.82	18.69	19.42	20.05	20.60	21.08	21.50	21.87	22.21	22.51	22.78	23.03

SIGMA = 3080.0

$V-V_c = 17$

H x	0.08	0.09	0.10	0.2	0.4	0.6	0.8	1.0	1.2	1.4	1.6	1.8	2.0	2.2	2.4	2.6	2.8	3.0	3.2	3.4
0	1.000	1.000	1.000	1.000	1.000	1.000	1.000	1.000	1.000	1.000	1.000	1.000	1.000	1.000	1.000	1.000	1.000	1.000	1.000	1.000
20	1.007	1.007	1.007	1.008	1.008	1.009	1.009	1.010	1.010	1.010	1.011	1.011	1.011	1.011	1.011	1.011	1.011	1.011	1.011	1.012
40	1.014	1.014	1.014	1.015	1.017	1.018	1.019	1.020	1.020	1.021	1.021	1.021	1.022	1.022	1.022	1.022	1.023	1.023	1.023	1.023
60	1.021	1.021	1.021	1.023	1.025	1.027	1.028	1.029	1.030	1.031	1.032	1.032	1.033	1.033	1.034	1.034	1.034	1.035	1.035	1.035
80	1.028	1.028	1.028	1.030	1.034	1.036	1.038	1.039	1.041	1.042	1.042	1.043	1.044	1.044	1.045	1.045	1.046	1.046	1.046	1.047
100	1.035	1.035	1.036	1.038	1.042	1.045	1.047	1.049	1.051	1.052	1.053	1.054	1.055	1.056	1.056	1.057	1.057	1.058	1.058	1.058
120	1.042	1.042	1.043	1.046	1.051	1.054	1.057	1.059	1.061	1.063	1.064	1.065	1.066	1.067	1.068	1.068	1.069	1.069	1.070	1.070
140	1.049	1.049	1.050	1.053	1.059	1.063	1.067	1.069	1.071	1.073	1.075	1.076	1.077	1.078	1.079	1.080	1.080	1.081	1.082	1.082
160	1.056	1.056	1.057	1.061	1.068	1.072	1.076	1.079	1.082	1.084	1.086	1.087	1.088	1.090	1.091	1.091	1.092	1.093	1.094	1.094
180	1.063	1.064	1.064	1.069	1.076	1.082	1.086	1.089	1.092	1.095	1.097	1.098	1.100	1.101	1.103	1.103	1.104	1.105	1.106	1.106
200	1.070	1.071	1.071	1.076	1.085	1.091	1.096	1.100	1.103	1.105	1.107	1.109	1.111	1.112	1.114	1.115	1.116	1.117	1.118	1.118
220	1.077	1.078	1.076	1.084	1.093	1.100	1.105	1.110	1.113	1.116	1.119	1.121	1.122	1.124	1.125	1.127	1.128	1.129	1.130	1.131
240	1.084	1.085	1.086	1.092	1.102	1.109	1.115	1.120	1.124	1.127	1.130	1.132	1.134	1.136	1.137	1.139	1.141	1.141	1.142	1.143
260	1.091	1.092	1.093	1.100	1.111	1.119	1.125	1.130	1.134	1.138	1.141	1.143	1.145	1.147	1.149	1.151	1.152	1.153	1.154	1.155
280	1.098	1.099	1.100	1.107	1.119	1.128	1.135	1.140	1.145	1.149	1.152	1.155	1.157	1.159	1.161	1.163	1.164	1.165	1.166	1.168
300	1.105	1.106	1.107	1.115	1.128	1.137	1.145	1.151	1.156	1.160	1.163	1.166	1.169	1.171	1.173	1.175	1.176	1.178	1.179	1.180
320	1.112	1.113	1.114	1.123	1.137	1.147	1.155	1.161	1.166	1.171	1.174	1.178	1.180	1.183	1.185	1.187	1.188	1.190	1.191	1.193
340	1.119	1.121	1.122	1.131	1.145	1.156	1.165	1.172	1.177	1.182	1.186	1.189	1.192	1.195	1.197	1.199	1.201	1.202	1.204	1.205
360	1.127	1.128	1.129	1.139	1.154	1.166	1.175	1.182	1.188	1.193	1.197	1.201	1.204	1.207	1.209	1.211	1.213	1.215	1.216	1.218
380	1.134	1.135	1.136	1.146	1.163	1.175	1.185	1.193	1.199	1.204	1.209	1.212	1.216	1.219	1.221	1.223	1.226	1.227	1.229	1.230
400	1.141	1.142	1.143	1.154	1.172	1.185	1.195	1.203	1.210	1.215	1.220	1.224	1.228	1.231	1.233	1.236	1.238	1.240	1.242	1.243
420	1.148	1.149	1.150	1.162	1.181	1.194	1.205	1.214	1.221	1.227	1.232	1.236	1.240	1.243	1.246	1.248	1.251	1.253	1.254	1.256
440	1.155	1.156	1.158	1.170	1.190	1.204	1.215	1.224	1.232	1.238	1.243	1.248	1.252	1.255	1.258	1.261	1.263	1.265	1.267	1.269
460	1.162	1.164	1.165	1.178	1.198	1.214	1.226	1.235	1.243	1.249	1.255	1.260	1.264	1.267	1.271	1.273	1.276	1.278	1.280	1.282
480	1.169	1.171	1.172	1.186	1.207	1.223	1.236	1.246	1.254	1.261	1.267	1.272	1.276	1.280	1.283	1.286	1.289	1.291	1.293	1.295
500	1.176	1.178	1.179	1.194	1.216	1.233	1.246	1.257	1.265	1.272	1.278	1.284	1.288	1.292	1.296	1.299	1.301	1.304	1.306	1.308
520	1.183	1.185	1.187	1.202	1.225	1.243	1.257	1.267	1.276	1.284	1.290	1.296	1.300	1.304	1.308	1.311	1.314	1.317	1.319	1.321
540	1.191	1.192	1.194	1.210	1.234	1.253	1.267	1.278	1.288	1.296	1.302	1.308	1.313	1.317	1.321	1.324	1.327	1.330	1.332	1.335
560	1.198	1.200	1.201	1.218	1.243	1.262	1.277	1.289	1.299	1.307	1.314	1.320	1.325	1.330	1.333	1.337	1.340	1.343	1.346	1.348
580	1.205	1.207	1.209	1.226	1.252	1.272	1.288	1.300	1.310	1.319	1.326	1.332	1.337	1.342	1.346	1.350	1.353	1.356	1.359	1.361
600	1.212	1.214	1.216	1.234	1.261	1.282	1.298	1.311	1.322	1.331	1.338	1.344	1.350	1.355	1.359	1.363	1.366	1.369	1.372	1.375
620	1.219	1.221	1.223	1.242	1.270	1.292	1.309	1.322	1.333	1.342	1.350	1.357	1.363	1.368	1.372	1.376	1.379	1.383	1.386	1.388
640	1.226	1.229	1.231	1.250	1.279	1.302	1.319	1.333	1.345	1.354	1.362	1.369	1.375	1.380	1.385	1.389	1.393	1.396	1.399	1.402
660	1.234	1.236	1.238	1.258	1.289	1.312	1.330	1.344	1.356	1.366	1.374	1.382	1.388	1.393	1.398	1.402	1.406	1.409	1.413	1.415
680	1.241	1.243	1.245	1.266	1.298	1.322	1.341	1.356	1.368	1.378	1.387	1.394	1.400	1.406	1.411	1.415	1.419	1.423	1.426	1.429
700	1.248	1.250	1.253	1.274	1.307	1.332	1.351	1.367	1.379	1.390	1.399	1.407	1.413	1.419	1.424	1.429	1.433	1.436	1.440	1.443
720	1.255	1.258	1.260	1.282	1.316	1.342	1.362	1.378	1.391	1.402	1.411	1.419	1.426	1.432	1.437	1.442	1.446	1.450	1.454	1.457
740	1.262	1.265	1.267	1.290	1.325	1.352	1.373	1.389	1.403	1.414	1.424	1.432	1.439	1.445	1.451	1.455	1.460	1.464	1.467	1.471
760	1.270	1.272	1.275	1.298	1.335	1.362	1.383	1.401	1.415	1.426	1.436	1.445	1.452	1.458	1.464	1.469	1.473	1.477	1.481	1.484
780	1.277	1.279	1.282	1.306	1.344	1.372	1.394	1.412	1.426	1.438	1.449	1.457	1.465	1.471	1.477	1.482	1.487	1.491	1.495	1.498
800	1.284	1.287	1.289	1.314	1.353	1.382	1.405	1.423	1.438	1.451	1.461	1.470	1.478	1.485	1.491	1.496	1.501	1.505	1.509	1.513

SIGMA = 3080.0

V–V$_c$ = 17

H	3.4	3.2	3.0	2.8	2.6	2.4	2.2	2.0	1.8	1.6	1.4	1.2	1.0	0.8	0.6	0.4	0.2	0.10	0.09	0.08
820	1.527	1.523	1.519	1.515	1.510	1.504	1.498	1.491	1.483	1.474	1.463	1.450	1.435	1.416	1.393	1.363	1.322	1.297	1.294	1.291
840	1.541	1.537	1.533	1.528	1.523	1.518	1.511	1.504	1.496	1.486	1.475	1.462	1.446	1.427	1.403	1.372	1.331	1.304	1.301	1.298
860	1.555	1.551	1.547	1.542	1.537	1.531	1.525	1.517	1.509	1.499	1.488	1.474	1.458	1.438	1.413	1.381	1.339	1.312	1.309	1.306
880	1.570	1.566	1.561	1.556	1.551	1.545	1.538	1.531	1.522	1.512	1.500	1.486	1.469	1.449	1.423	1.391	1.347	1.319	1.316	1.313
900	1.584	1.580	1.575	1.570	1.565	1.559	1.552	1.544	1.535	1.525	1.512	1.498	1.481	1.460	1.434	1.400	1.355	1.327	1.323	1.320
920	1.598	1.594	1.590	1.585	1.579	1.573	1.565	1.557	1.548	1.537	1.525	1.510	1.493	1.471	1.444	1.410	1.363	1.334	1.331	1.327
940	1.613	1.609	1.604	1.599	1.593	1.586	1.579	1.571	1.561	1.550	1.538	1.522	1.504	1.482	1.455	1.419	1.372	1.341	1.338	1.335
960	1.628	1.623	1.618	1.613	1.607	1.600	1.593	1.584	1.575	1.563	1.550	1.535	1.516	1.493	1.465	1.428	1.380	1.349	1.345	1.342
980	1.642	1.638	1.633	1.627	1.621	1.614	1.607	1.598	1.588	1.576	1.563	1.547	1.528	1.505	1.475	1.438	1.388	1.356	1.353	1.349
1000	1.657	1.652	1.647	1.642	1.635	1.628	1.620	1.611	1.601	1.589	1.576	1.559	1.540	1.516	1.486	1.448	1.396	1.364	1.360	1.357
1020	1.672	1.667	1.662	1.656	1.650	1.642	1.634	1.625	1.615	1.602	1.588	1.572	1.552	1.527	1.496	1.457	1.405	1.371	1.368	1.364
1040	1.687	1.682	1.676	1.670	1.664	1.656	1.648	1.639	1.628	1.616	1.601	1.584	1.564	1.538	1.507	1.467	1.413	1.379	1.375	1.371
1060	1.702	1.697	1.691	1.685	1.678	1.671	1.662	1.653	1.642	1.629	1.614	1.596	1.575	1.550	1.518	1.476	1.421	1.386	1.382	1.378
1080	1.717	1.711	1.706	1.700	1.693	1.685	1.676	1.666	1.655	1.642	1.627	1.609	1.587	1.561	1.528	1.486	1.430	1.394	1.390	1.386
1100	1.732	1.726	1.721	1.714	1.707	1.699	1.690	1.680	1.669	1.655	1.640	1.622	1.599	1.573	1.539	1.496	1.438	1.401	1.397	1.393
1120	1.747	1.741	1.736	1.729	1.722	1.714	1.705	1.694	1.682	1.669	1.653	1.634	1.612	1.584	1.550	1.505	1.446	1.409	1.405	1.400
1140	1.762	1.757	1.750	1.744	1.736	1.728	1.719	1.708	1.696	1.682	1.666	1.647	1.624	1.596	1.560	1.515	1.455	1.416	1.412	1.408
1160	1.777	1.772	1.765	1.759	1.751	1.743	1.733	1.722	1.710	1.696	1.679	1.659	1.636	1.607	1.571	1.525	1.463	1.424	1.419	1.415
1180	1.793	1.787	1.780	1.774	1.766	1.757	1.747	1.736	1.724	1.709	1.692	1.672	1.648	1.619	1.582	1.535	1.471	1.431	1.427	1.422
1200	1.808	1.802	1.796	1.789	1.781	1.772	1.762	1.751	1.738	1.723	1.705	1.685	1.660	1.630	1.593	1.544	1.480	1.439	1.434	1.430
1220	1.823	1.817	1.811	1.804	1.795	1.786	1.776	1.765	1.752	1.736	1.719	1.698	1.673	1.642	1.603	1.554	1.488	1.446	1.442	1.437
1240	1.839	1.833	1.826	1.819	1.810	1.801	1.791	1.779	1.766	1.750	1.732	1.711	1.685	1.654	1.614	1.564	1.497	1.454	1.449	1.444
1260	1.855	1.848	1.841	1.834	1.825	1.816	1.805	1.793	1.780	1.764	1.745	1.724	1.697	1.665	1.625	1.574	1.505	1.461	1.457	1.452
1280	1.870	1.864	1.857	1.849	1.840	1.831	1.820	1.808	1.794	1.778	1.759	1.736	1.710	1.677	1.636	1.584	1.514	1.469	1.464	1.459
1300	1.886	1.879	1.872	1.864	1.856	1.846	1.835	1.822	1.808	1.791	1.772	1.749	1.722	1.689	1.647	1.594	1.522	1.477	1.472	1.467
1320	1.902	1.895	1.888	1.880	1.871	1.861	1.849	1.837	1.822	1.805	1.786	1.763	1.735	1.701	1.658	1.603	1.531	1.484	1.479	1.474
1340	1.918	1.911	1.903	1.895	1.886	1.876	1.864	1.851	1.836	1.819	1.799	1.776	1.747	1.713	1.669	1.613	1.539	1.492	1.487	1.481
1360	1.933	1.927	1.919	1.911	1.901	1.891	1.879	1.866	1.851	1.833	1.813	1.789	1.760	1.724	1.680	1.623	1.548	1.499	1.494	1.489
1380	1.949	1.942	1.935	1.926	1.917	1.906	1.894	1.881	1.865	1.847	1.827	1.802	1.772	1.736	1.691	1.633	1.556	1.507	1.502	1.496
1400	1.965	1.958	1.950	1.942	1.932	1.921	1.909	1.895	1.880	1.861	1.840	1.815	1.785	1.748	1.702	1.643	1.565	1.515	1.509	1.504
1420	1.982	1.974	1.966	1.957	1.948	1.937	1.924	1.910	1.894	1.876	1.854	1.828	1.798	1.760	1.714	1.653	1.573	1.522	1.517	1.511
1440	1.998	1.990	1.982	1.973	1.963	1.952	1.939	1.925	1.909	1.890	1.868	1.842	1.811	1.772	1.725	1.664	1.582	1.530	1.524	1.518
1460	2.014	2.006	1.998	1.989	1.979	1.967	1.954	1.940	1.923	1.904	1.882	1.855	1.823	1.785	1.736	1.674	1.590	1.538	1.532	1.526
1480	2.030	2.023	2.014	2.005	1.994	1.983	1.970	1.955	1.938	1.918	1.896	1.869	1.836	1.797	1.747	1.684	1.599	1.545	1.539	1.533
1500	2.047	2.039	2.030	2.021	2.010	1.998	1.985	1.970	1.953	1.933	1.909	1.882	1.849	1.809	1.759	1.694	1.608	1.553	1.547	1.541
1520	2.063	2.055	2.046	2.037	2.026	2.014	2.000	1.985	1.967	1.947	1.923	1.896	1.862	1.821	1.770	1.704	1.616	1.561	1.554	1.548
1540	2.080	2.071	2.063	2.053	2.042	2.029	2.016	2.000	1.982	1.962	1.937	1.909	1.875	1.833	1.781	1.714	1.625	1.568	1.562	1.556
1560	2.096	2.088	2.079	2.069	2.058	2.045	2.031	2.015	1.997	1.976	1.952	1.923	1.888	1.846	1.793	1.725	1.634	1.576	1.569	1.563
1580	2.113	2.104	2.095	2.085	2.074	2.061	2.047	2.030	2.012	1.991	1.966	1.936	1.901	1.858	1.804	1.735	1.642	1.584	1.577	1.570
1600	2.129	2.121	2.111	2.101	2.090	2.077	2.062	2.046	2.027	2.005	1.980	1.950	1.914	1.870	1.815	1.745	1.651	1.591	1.585	1.578
1620	2.146	2.137	2.128	2.117	2.106	2.093	2.078	2.061	2.042	2.020	1.994	1.964	1.927	1.883	1.827	1.755	1.660	1.599	1.592	1.585
1640	2.163	2.154	2.144	2.134	2.122	2.108	2.093	2.076	2.057	2.035	2.008	1.978	1.940	1.895	1.838	1.766	1.668	1.607	1.600	1.593

x

SIGMA = 3080.0

$V-V_c = 17$

x \ H	3.4	3.2	3.0	2.8	2.6	2.4	2.2	2.0	1.8	1.6	1.4	1.2	1.0	0.8	0.6	0.4	0.2	0.10	0.09	0.08
1660	2.180	2.171	2.161	2.150	2.138	2.124	2.109	2.092	2.072	2.049	2.023	1.991	1.954	1.906	1.850	1.776	1.677	1.614	1.607	1.600
1680	2.197	2.188	2.178	2.167	2.154	2.140	2.125	2.107	2.087	2.064	2.037	2.005	1.967	1.920	1.862	1.786	1.686	1.622	1.615	1.606
1700	2.214	2.205	2.194	2.183	2.171	2.157	2.141	2.123	2.103	2.079	2.052	2.019	1.980	1.933	1.873	1.797	1.695	1.630	1.623	1.615
1720	2.231	2.222	2.211	2.200	2.187	2.173	2.157	2.139	2.118	2.094	2.066	2.033	1.994	1.945	1.885	1.807	1.703	1.638	1.630	1.623
1740	2.248	2.239	2.228	2.216	2.203	2.189	2.173	2.154	2.133	2.109	2.081	2.047	2.007	1.958	1.896	1.818	1.712	1.645	1.638	1.630
1760	2.265	2.256	2.245	2.233	2.220	2.205	2.189	2.170	2.149	2.124	2.095	2.061	2.020	1.971	1.908	1.828	1.721	1.653	1.646	1.638
1780	2.283	2.273	2.262	2.250	2.237	2.222	2.205	2.186	2.164	2.139	2.110	2.075	2.034	1.983	1.920	1.838	1.730	1.661	1.653	1.646
1800	2.300	2.290	2.279	2.267	2.253	2.238	2.221	2.202	2.180	2.154	2.125	2.089	2.047	1.996	1.932	1.849	1.739	1.669	1.661	1.653
1820	2.317	2.307	2.296	2.284	2.270	2.254	2.237	2.218	2.195	2.169	2.139	2.104	2.061	2.009	1.943	1.860	1.746	1.676	1.669	1.661
1840	2.335	2.324	2.313	2.301	2.287	2.271	2.253	2.234	2.211	2.185	2.154	2.118	2.075	2.022	1.955	1.870	1.756	1.684	1.676	1.668
1860	2.352	2.342	2.330	2.318	2.303	2.288	2.270	2.250	2.227	2.200	2.169	2.132	2.088	2.034	1.967	1.881	1.765	1.692	1.684	1.676
1880	2.370	2.359	2.348	2.335	2.320	2.304	2.286	2.266	2.242	2.215	2.184	2.147	2.102	2.047	1.979	1.891	1.774	1.700	1.692	1.683
1900	2.388	2.377	2.365	2.352	2.337	2.321	2.303	2.282	2.258	2.231	2.199	2.161	2.116	2.060	1.991	1.902	1.783	1.708	1.699	1.691
1920	2.405	2.394	2.382	2.369	2.354	2.338	2.319	2.298	2.274	2.246	2.214	2.175	2.129	2.073	2.003	1.913	1.792	1.715	1.707	1.698
1940	2.423	2.412	2.400	2.386	2.371	2.355	2.336	2.314	2.290	2.262	2.229	2.190	2.143	2.086	2.015	1.923	1.801	1.723	1.715	1.706
1960	2.441	2.430	2.417	2.404	2.388	2.371	2.352	2.331	2.306	2.277	2.244	2.204	2.157	2.099	2.027	1.934	1.810	1.731	1.722	1.713
1980	2.459	2.448	2.435	2.421	2.406	2.388	2.369	2.347	2.322	2.293	2.259	2.219	2.171	2.112	2.039	1.945	1.819	1.739	1.730	1.721
2000	2.477	2.465	2.453	2.439	2.423	2.405	2.386	2.363	2.338	2.308	2.274	2.234	2.185	2.125	2.051	1.955	1.828	1.747	1.738	1.729
2020	2.495	2.483	2.470	2.456	2.440	2.422	2.402	2.380	2.354	2.324	2.289	2.248	2.199	2.138	2.063	1.966	1.837	1.755	1.746	1.736
2040	2.513	2.501	2.488	2.474	2.458	2.440	2.419	2.396	2.370	2.340	2.305	2.263	2.213	2.152	2.075	1.977	1.846	1.762	1.753	1.744
2060	2.531	2.519	2.506	2.491	2.475	2.457	2.436	2.413	2.386	2.356	2.320	2.278	2.227	2.165	2.087	1.988	1.855	1.770	1.761	1.752
2080	2.550	2.537	2.524	2.509	2.492	2.474	2.453	2.430	2.403	2.372	2.335	2.292	2.241	2.178	2.100	1.999	1.864	1.778	1.769	1.759
2100	2.568	2.555	2.542	2.527	2.510	2.491	2.470	2.446	2.419	2.387	2.351	2.307	2.255	2.191	2.112	2.009	1.873	1.786	1.776	1.767
2120	2.586	2.574	2.560	2.545	2.526	2.509	2.487	2.463	2.435	2.403	2.366	2.322	2.269	2.205	2.124	2.020	1.882	1.794	1.784	1.774
2140	2.605	2.592	2.578	2.562	2.545	2.526	2.504	2.480	2.452	2.419	2.382	2.337	2.284	2.218	2.136	2.031	1.891	1.802	1.792	1.782
2160	2.623	2.610	2.596	2.580	2.563	2.543	2.522	2.497	2.468	2.435	2.397	2.352	2.298	2.232	2.149	2.042	1.900	1.810	1.800	1.790
2180	2.642	2.629	2.614	2.598	2.581	2.561	2.539	2.514	2.485	2.452	2.413	2.367	2.312	2.245	2.161	2.053	1.909	1.818	1.808	1.797
2200	2.660	2.647	2.633	2.617	2.599	2.579	2.556	2.531	2.501	2.468	2.429	2.382	2.327	2.259	2.173	2.064	1.918	1.826	1.815	1.805
2220	2.679	2.666	2.651	2.635	2.617	2.596	2.573	2.548	2.518	2.484	2.444	2.397	2.341	2.272	2.186	2.075	1.927	1.834	1.823	1.813
2240	2.698	2.684	2.669	2.653	2.635	2.614	2.591	2.565	2.535	2.500	2.460	2.412	2.355	2.286	2.198	2.086	1.937	1.841	1.831	1.821
2260	2.717	2.703	2.688	2.671	2.653	2.632	2.608	2.582	2.552	2.517	2.476	2.428	2.370	2.299	2.211	2.097	1.946	1.849	1.839	1.828
2280	2.736	2.722	2.706	2.689	2.671	2.650	2.626	2.599	2.568	2.533	2.492	2.443	2.384	2.313	2.223	2.108	1.955	1.857	1.847	1.836
2300	2.755	2.741	2.725	2.708	2.689	2.668	2.644	2.616	2.585	2.549	2.508	2.458	2.399	2.326	2.236	2.119	1.964	1.865	1.854	1.843
2320	2.774	2.759	2.744	2.726	2.707	2.685	2.661	2.634	2.602	2.566	2.524	2.474	2.414	2.340	2.248	2.131	1.973	1.873	1.862	1.851
2340	2.793	2.778	2.762	2.745	2.725	2.703	2.679	2.651	2.619	2.582	2.540	2.489	2.428	2.354	2.261	2.142	1.983	1.881	1.870	1.859
2360	2.812	2.797	2.781	2.763	2.744	2.722	2.697	2.668	2.636	2.599	2.556	2.504	2.443	2.368	2.274	2.153	1.992	1.889	1.878	1.866
2380	2.831	2.816	2.800	2.782	2.762	2.740	2.714	2.686	2.653	2.616	2.572	2.520	2.458	2.382	2.286	2.164	2.001	1.897	1.886	1.874
2400	2.851	2.836	2.819	2.801	2.781	2.758	2.732	2.703	2.670	2.632	2.588	2.535	2.472	2.395	2.299	2.175	2.010	1.905	1.894	1.882
2420	2.870	2.855	2.838	2.820	2.799	2.776	2.750	2.721	2.688	2.649	2.604	2.551	2.487	2.409	2.312	2.187	2.020	1.913	1.902	1.890
2440	2.889	2.874	2.857	2.838	2.818	2.794	2.768	2.739	2.705	2.666	2.620	2.567	2.502	2.423	2.325	2.198	2.029	1.921	1.909	1.897
2460	2.909	2.893	2.876	2.857	2.836	2.813	2.786	2.756	2.722	2.683	2.637	2.582	2.517	2.437	2.337	2.209	2.038	1.929	1.917	1.905
2480	2.928	2.913	2.895	2.876	2.855	2.831	2.804	2.774	2.740	2.700	2.653	2.598	2.532	2.451	2.350	2.220	2.047	1.937	1.925	1.913

146

$V - V_c = 17$ SIGMA = 3083.0

x	H 0.08	0.09	0.10	0.2	0.4	0.6	0.8	1.0	1.2	1.4	1.6	1.8	2.0	2.2	2.4	2.6	2.8	3.0	3.2	3.4
2500	1.921	1.933	1.945	2.057	2.232	2.363	2.465	2.547	2.614	2.669	2.717	2.757	2.792	2.823	2.850	2.874	2.895	2.915	2.932	2.948
2520	1.928	1.941	1.953	2.066	2.243	2.376	2.479	2.562	2.630	2.686	2.734	2.774	2.810	2.841	2.868	2.893	2.914	2.934	2.952	2.968
2540	1.936	1.949	1.961	2.075	2.255	2.389	2.493	2.577	2.645	2.702	2.751	2.792	2.828	2.859	2.887	2.911	2.933	2.953	2.971	2.987
2560	1.944	1.957	1.970	2.085	2.266	2.402	2.508	2.592	2.661	2.719	2.768	2.810	2.846	2.878	2.906	2.930	2.953	2.973	2.991	3.007
2580	1.952	1.965	1.978	2.094	2.277	2.415	2.522	2.607	2.677	2.736	2.785	2.827	2.864	2.896	2.924	2.949	2.972	2.992	3.010	3.027
2600	1.959	1.973	1.986	2.104	2.289	2.428	2.536	2.623	2.693	2.752	2.802	2.845	2.882	2.914	2.943	2.968	2.991	3.012	3.030	3.047
2620	1.967	1.981	1.994	2.113	2.300	2.441	2.550	2.638	2.709	2.769	2.819	2.863	2.900	2.933	2.962	2.988	3.011	3.031	3.050	3.067
2640	1.975	1.989	2.002	2.122	2.312	2.454	2.565	2.653	2.725	2.786	2.837	2.880	2.918	2.952	2.981	3.007	3.030	3.051	3.070	3.087
2660	1.983	1.997	2.010	2.132	2.323	2.467	2.579	2.668	2.742	2.803	2.854	2.898	2.937	2.970	3.000	3.026	3.050	3.071	3.090	3.107
2680	1.991	2.004	2.018	2.141	2.335	2.480	2.593	2.684	2.758	2.819	2.872	2.916	2.955	2.989	3.019	3.045	3.069	3.091	3.110	3.128
2700	1.998	2.012	2.026	2.151	2.347	2.494	2.608	2.699	2.774	2.836	2.889	2.934	2.973	3.008	3.038	3.065	3.089	3.110	3.130	3.148
2720	2.006	2.020	2.034	2.160	2.358	2.507	2.622	2.715	2.790	2.853	2.907	2.952	2.992	3.026	3.057	3.084	3.108	3.130	3.150	3.168
2740	2.014	2.028	2.042	2.170	2.370	2.520	2.637	2.730	2.807	2.870	2.924	2.970	3.010	3.045	3.076	3.104	3.128	3.150	3.170	3.189
2760	2.022	2.036	2.051	2.179	2.382	2.533	2.651	2.746	2.823	2.887	2.942	2.988	3.029	3.064	3.095	3.123	3.148	3.170	3.191	3.209
2780	2.030	2.044	2.059	2.189	2.393	2.547	2.666	2.761	2.839	2.904	2.959	3.007	3.047	3.083	3.115	3.143	3.168	3.191	3.211	3.230
2800	2.038	2.052	2.067	2.198	2.405	2.560	2.680	2.777	2.856	2.921	2.977	3.025	3.066	3.102	3.134	3.163	3.188	3.211	3.231	3.250
2820	2.046	2.060	2.075	2.208	2.417	2.573	2.695	2.793	2.872	2.939	2.995	3.043	3.085	3.121	3.154	3.182	3.208	3.231	3.252	3.271
2840	2.053	2.068	2.083	2.217	2.428	2.587	2.710	2.808	2.889	2.956	3.013	3.061	3.104	3.141	3.173	3.202	3.228	3.251	3.272	3.292
2860	2.061	2.076	2.091	2.227	2.440	2.600	2.724	2.824	2.905	2.973	3.031	3.080	3.122	3.160	3.193	3.222	3.248	3.272	3.293	3.312
2880	2.069	2.084	2.100	2.237	2.452	2.614	2.739	2.840	2.922	2.991	3.049	3.098	3.141	3.179	3.212	3.242	3.268	3.292	3.314	3.333
2900	2.077	2.093	2.108	2.246	2.464	2.627	2.754	2.856	2.939	3.008	3.067	3.117	3.160	3.198	3.232	3.262	3.289	3.313	3.334	3.354
2920	2.085	2.101	2.116	2.256	2.476	2.641	2.769	2.871	2.955	3.025	3.085	3.135	3.179	3.218	3.252	3.282	3.309	3.333	3.355	3.375
2940	2.093	2.109	2.124	2.265	2.488	2.654	2.784	2.887	2.972	3.043	3.103	3.154	3.198	3.237	3.272	3.302	3.329	3.354	3.376	3.396
2960	2.101	2.117	2.132	2.275	2.500	2.668	2.799	2.903	2.989	3.060	3.121	3.173	3.217	3.257	3.291	3.322	3.350	3.375	3.397	3.417
2980	2.109	2.125	2.141	2.285	2.511	2.681	2.814	2.919	3.006	3.078	3.139	3.191	3.237	3.276	3.311	3.342	3.370	3.395	3.418	3.439
3000	2.116	2.133	2.149	2.294	2.523	2.695	2.829	2.935	3.023	3.096	3.157	3.210	3.256	3.296	3.331	3.363	3.391	3.416	3.439	3.460

$V-V_c = 17$ SIGMA = 3080.0

x	0.08	0.09	0.10	0.2	0.4	0.6	0.8	1.0	1.2	1.4	1.6	1.8	2.0	2.2	2.4	2.6	2.8	3.0	3.2	3.4
3100	2.156	2.173	2.190	2.343	2.583	2.764	2.904	3.016	3.108	3.185	3.249	3.305	3.353	3.395	3.432	3.465	3.495	3.521	3.545	3.567
3200	2.196	2.214	2.232	2.392	2.644	2.833	2.980	3.098	3.194	3.275	3.343	3.401	3.451	3.495	3.534	3.569	3.600	3.628	3.653	3.676
3300	2.236	2.255	2.273	2.441	2.706	2.904	3.058	3.181	3.282	3.366	3.437	3.498	3.551	3.597	3.638	3.674	3.707	3.736	3.762	3.786
3400	2.276	2.296	2.315	2.491	2.767	2.975	3.136	3.265	3.371	3.459	3.533	3.597	3.651	3.701	3.743	3.781	3.815	3.846	3.873	3.899
3500	2.316	2.337	2.357	2.541	2.830	3.047	3.215	3.350	3.461	3.553	3.630	3.697	3.755	3.805	3.850	3.890	3.925	3.957	3.986	4.012
3600	2.357	2.378	2.399	2.591	2.893	3.119	3.295	3.436	3.552	3.648	3.729	3.798	3.859	3.912	3.958	4.000	4.037	4.070	4.100	4.128
3700	2.397	2.420	2.442	2.642	2.957	3.193	3.377	3.524	3.644	3.744	3.829	3.901	3.964	4.019	4.068	4.111	4.150	4.185	4.216	4.245
3800	2.438	2.461	2.484	2.693	3.021	3.267	3.459	3.612	3.737	3.841	3.930	4.005	4.071	4.128	4.179	4.224	4.264	4.301	4.334	4.363
3900	2.479	2.503	2.527	2.744	3.086	3.342	3.542	3.701	3.831	3.940	4.032	4.111	4.179	4.239	4.292	4.339	4.381	4.418	4.453	4.484
4000	2.520	2.545	2.570	2.796	3.152	3.418	3.626	3.791	3.927	4.040	4.136	4.218	4.289	4.351	4.406	4.455	4.498	4.538	4.573	4.606
4100	2.561	2.587	2.613	2.848	3.218	3.495	3.710	3.883	4.024	4.141	4.241	4.326	4.400	4.465	4.522	4.572	4.618	4.659	4.695	4.729
4200	2.602	2.630	2.657	2.901	3.285	3.572	3.796	3.975	4.122	4.244	4.347	4.436	4.512	4.580	4.639	4.691	4.739	4.781	4.819	4.854
4300	2.644	2.672	2.700	2.954	3.352	3.651	3.883	4.069	4.221	4.347	4.455	4.547	4.626	4.696	4.757	4.812	4.861	4.905	4.945	4.981
4400	2.686	2.715	2.744	3.007	3.420	3.730	3.971	4.163	4.321	4.452	4.564	4.659	4.741	4.814	4.878	4.934	4.985	5.031	5.072	5.109
4500	2.727	2.758	2.788	3.060	3.488	3.809	4.059	4.259	4.422	4.559	4.674	4.773	4.858	4.933	4.999	5.058	5.110	5.158	5.201	5.240
4600	2.769	2.801	2.832	3.114	3.558	3.890	4.149	4.356	4.525	4.666	4.785	4.888	4.976	5.054	5.122	5.183	5.238	5.287	5.331	5.371
4700	2.811	2.844	2.876	3.168	3.627	3.971	4.239	4.453	4.628	4.774	4.898	5.004	5.096	5.176	5.247	5.310	5.366	5.417	5.463	5.505
4800	2.854	2.888	2.921	3.223	3.698	4.054	4.331	4.552	4.733	4.884	5.012	5.122	5.217	5.300	5.373	5.438	5.496	5.549	5.596	5.639
4900	2.896	2.931	2.966	3.278	3.769	4.137	4.423	4.652	4.839	4.995	5.127	5.241	5.339	5.425	5.500	5.568	5.628	5.682	5.731	5.776
5000	2.939	2.975	3.011	3.333	3.840	4.220	4.516	4.753	4.946	5.108	5.241	5.361	5.463	5.551	5.630	5.699	5.761	5.817	5.866	5.914
5100	2.982	3.019	3.056	3.389	3.912	4.305	4.610	4.855	5.055	5.221	5.362	5.483	5.588	5.679	5.760	5.832	5.896	5.954	6.006	6.054
5200	3.025	3.063	3.101	3.445	3.985	4.390	4.706	4.958	5.164	5.336	5.481	5.606	5.714	5.809	5.892	5.966	6.033	6.092	6.146	6.195
5300	3.068	3.107	3.146	3.501	4.058	4.476	4.802	5.062	5.275	5.452	5.602	5.731	5.842	5.940	6.026	6.102	6.171	6.232	6.288	6.339
5400	3.111	3.152	3.192	3.558	4.132	4.563	4.899	5.167	5.386	5.569	5.724	5.856	5.971	6.072	6.161	6.240	6.310	6.374	6.431	6.483
5500	3.154	3.196	3.238	3.615	4.207	4.651	4.997	5.273	5.499	5.687	5.847	5.984	6.102	6.206	6.297	6.378	6.451	6.517	6.576	6.630
5600	3.198	3.241	3.284	3.672	4.282	4.740	5.096	5.380	5.613	5.807	5.971	6.112	6.234	6.341	6.435	6.519	6.594	6.661	6.722	6.778
5700	3.241	3.286	3.330	3.730	4.358	4.829	5.195	5.488	5.728	5.928	6.097	6.242	6.368	6.478	6.575	6.661	6.738	6.807	6.870	6.927
5800	3.285	3.331	3.377	3.788	4.434	4.919	5.296	5.598	5.845	6.050	6.224	6.373	6.503	6.616	6.716	6.804	6.884	6.955	7.020	7.078
5900	3.329	3.377	3.423	3.846	4.511	5.010	5.398	5.708	5.962	6.174	6.353	6.506	6.639	6.755	6.858	6.949	7.031	7.104	7.171	7.231
6000	3.373	3.422	3.470	3.905	4.589	5.102	5.500	5.820	6.081	6.298	6.482	6.640	6.777	6.896	7.002	7.096	7.180	7.255	7.324	7.386
6100	3.418	3.468	3.517	3.964	4.667	5.194	5.604	5.932	6.200	6.424	6.613	6.777	6.916	7.039	7.147	7.244	7.330	7.408	7.478	7.542
6200	3.462	3.514	3.564	4.024	4.746	5.287	5.709	6.046	6.321	6.551	6.745	6.916	7.056	7.183	7.294	7.393	7.482	7.562	7.634	7.700
6300	3.507	3.560	3.612	4.084	4.825	5.381	5.814	6.160	6.443	6.679	6.879	7.056	7.198	7.328	7.443	7.544	7.635	7.717	7.792	7.859
6400	3.552	3.606	3.659	4.144	4.905	5.476	5.920	6.276	6.566	6.809	7.014	7.189	7.342	7.475	7.593	7.697	7.791	7.875	7.951	8.020
6500	3.597	3.652	3.707	4.204	4.986	5.572	6.028	6.392	6.691	6.939	7.150	7.330	7.486	7.623	7.744	7.851	7.947	8.033	8.112	8.183
6600	3.642	3.699	3.755	4.265	5.067	5.668	6.136	6.510	6.816	7.071	7.287	7.472	7.633	7.773	7.897	8.007	8.105	8.194	8.274	8.347
6700	3.687	3.746	3.803	4.327	5.149	5.766	6.245	6.629	6.943	7.205	7.426	7.616	7.780	7.924	8.051	8.164	8.265	8.356	8.438	8.513
6800	3.732	3.793	3.852	4.388	5.231	5.864	6.355	6.749	7.071	7.339	7.566	7.761	7.929	8.077	8.207	8.323	8.426	8.519	8.604	8.680
6900	3.778	3.840	3.900	4.450	5.314	5.962	6.466	6.870	7.200	7.475	7.707	7.907	8.080	8.231	8.364	8.483	8.589	8.685	8.771	8.850
7000	3.824	3.887	3.949	4.512	5.398	6.062	6.579	6.992	7.330	7.612	7.850	8.054	8.231	8.386	8.523	8.645	8.753	8.851	8.940	9.020
7100	3.870	3.934	3.998	4.575	5.482	6.162	6.691	7.115	7.461	7.750	7.994	8.203	8.385	8.543	8.683	8.808	8.919	9.020	9.110	9.193
7200	3.916	3.982	4.047	4.638	5.567	6.264	6.805	7.239	7.593	7.889	8.139	8.353	8.539	8.702	8.845	8.973	9.087	9.189	9.282	9.367

SIGMA = 3080.0

V−V$_c$ = 17

H	3.4	3.2	3.0	2.8	2.6	2.4	2.2	2.0	1.8	1.6	1.4	1.2	1.0	0.8	0.6	0.4	0.2	0.10	0.09	0.08
7300	9.542	9.456	9.361	9.256	9.139	9.008	8.862	8.695	8.505	8.286	8.030	7.727	7.364	6.920	6.365	5.652	4.701	4.096	4.030	3.962
7400	9.720	9.631	9.534	9.426	9.307	9.173	9.023	8.853	8.658	8.433	8.171	7.862	7.490	7.036	6.468	5.738	4.765	4.146	4.078	4.008
7500	9.899	9.808	9.709	9.598	9.476	9.339	9.186	9.011	8.812	8.583	8.314	7.998	7.617	7.153	6.572	5.825	4.829	4.195	4.126	4.055
7600	10.08	9.99	9.88	9.77	9.65	9.51	9.35	9.17	8.97	8.73	8.46	8.13	7.75	7.27	6.68	5.91	4.89	4.25	4.17	4.10
7700	10.26	10.17	10.06	9.95	9.82	9.68	9.52	9.33	9.12	8.88	8.60	8.27	7.88	7.39	6.78	6.00	4.96	4.30	4.22	4.15
7800	10.45	10.35	10.24	10.12	9.99	9.85	9.68	9.50	9.28	9.04	8.75	8.41	8.01	7.51	6.89	6.09	5.02	4.35	4.27	4.20
7900	10.63	10.53	10.42	10.30	10.17	10.02	9.85	9.66	9.44	9.19	8.90	8.55	8.14	7.63	6.99	6.18	5.09	4.40	4.32	4.24
8000	10.82	10.72	10.61	10.48	10.35	10.19	10.02	9.83	9.60	9.35	9.05	8.69	8.27	7.75	7.10	6.27	5.15	4.45	4.37	4.29
8100	11.01	10.90	10.79	10.66	10.52	10.35	10.19	9.99	9.77	9.50	9.20	8.84	8.40	7.87	7.21	6.36	5.22	4.50	4.42	4.34
8200	11.20	11.09	10.98	10.85	10.70	10.54	10.37	10.16	9.93	9.66	9.35	8.98	8.54	8.00	7.32	6.45	5.29	4.55	4.47	4.38
8300	11.39	11.28	11.16	11.03	10.89	10.72	10.54	10.33	10.10	9.82	9.50	9.13	8.67	8.12	7.43	6.54	5.35	4.60	4.52	4.43
8400	11.58	11.47	11.35	11.22	11.07	10.90	10.72	10.50	10.26	9.98	9.66	9.27	8.81	8.25	7.54	6.63	5.42	4.65	4.57	4.48
8500	11.78	11.67	11.54	11.41	11.25	11.08	10.90	10.68	10.43	10.14	9.81	9.42	8.95	8.37	7.65	6.72	5.49	4.70	4.62	4.53
8600	11.97	11.86	11.73	11.59	11.44	11.27	11.07	10.85	10.60	10.31	9.97	9.57	9.09	8.50	7.76	6.82	5.56	4.75	4.67	4.58
8700	12.17	12.06	11.93	11.79	11.63	11.45	11.25	11.03	10.77	10.47	10.13	9.72	9.23	8.63	7.88	6.91	5.63	4.81	4.72	4.62
8800	12.37	12.25	12.12	11.98	11.82	11.64	11.43	11.20	10.94	10.64	10.29	9.87	9.37	8.76	7.99	7.01	5.69	4.86	4.77	4.67
8900	12.57	12.45	12.32	12.17	12.01	11.82	11.60	11.38	11.11	10.81	10.45	10.02	9.51	8.88	8.10	7.10	5.76	4.91	4.82	4.72
9000	12.78	12.65	12.52	12.37	12.20	12.01	11.80	11.56	11.29	10.97	10.61	10.17	9.65	9.02	8.22	7.20	5.83	4.96	4.87	4.77
9100	12.98	12.86	12.72	12.56	12.39	12.20	11.99	11.74	11.47	11.14	10.77	10.33	9.80	9.15	8.34	7.29	5.90	5.02	4.92	4.82
9200	13.19	13.06	12.92	12.76	12.59	12.39	12.17	11.93	11.64	11.32	10.93	10.48	9.94	9.28	8.45	7.39	5.97	5.07	4.97	4.87
9300	13.40	13.27	13.13	12.96	12.78	12.59	12.36	12.11	11.82	11.49	11.10	10.64	10.09	9.41	8.57	7.49	6.04	5.12	5.02	4.92
9400	13.61	13.47	13.33	13.16	12.98	12.78	12.55	12.30	12.00	11.66	11.27	10.80	10.24	9.55	8.69	7.59	6.11	5.18	5.07	4.97
9500	13.82	13.68	13.53	13.37	13.18	12.98	12.75	12.48	12.18	11.84	11.43	10.96	10.38	9.68	8.81	7.68	6.18	5.23	5.12	5.02
9600	14.03	13.89	13.74	13.57	13.38	13.17	12.94	12.67	12.37	12.01	11.60	11.12	10.53	9.82	8.93	7.78	6.26	5.28	5.18	5.07
9700	14.25	14.11	13.95	13.78	13.59	13.37	13.13	12.86	12.55	12.19	11.77	11.28	10.68	9.96	9.05	7.88	6.33	5.34	5.23	5.12
9800	14.46	14.32	14.16	13.99	13.79	13.57	13.33	13.05	12.74	12.37	11.94	11.44	10.83	10.10	9.17	7.98	6.40	5.39	5.28	5.17
9900	14.68	14.53	14.37	14.20	14.00	13.78	13.53	13.24	12.92	12.55	12.12	11.60	10.99	10.23	9.29	8.08	6.47	5.45	5.33	5.22
10000	14.90	14.75	14.59	14.41	14.20	13.98	13.73	13.44	13.11	12.73	12.29	11.77	11.14	10.37	9.42	8.19	6.54	5.50	5.39	5.27
10100	15.12	14.97	14.80	14.62	14.41	14.18	13.93	13.63	13.30	12.91	12.46	11.93	11.30	10.52	9.54	8.29	6.62	5.55	5.44	5.32
10200	15.35	15.19	15.02	14.83	14.62	14.39	14.13	13.83	13.49	13.10	12.64	12.10	11.45	10.66	9.67	8.39	6.69	5.61	5.49	5.37
10300	15.57	15.41	15.24	15.05	14.84	14.60	14.34	14.03	13.68	13.28	12.82	12.27	11.61	10.80	9.79	8.49	6.77	5.66	5.54	5.42
10400	15.80	15.64	15.46	15.27	15.05	14.81	14.54	14.23	13.88	13.47	13.00	12.44	11.77	10.94	9.92	8.60	6.84	5.72	5.60	5.47
10500	16.02	15.86	15.68	15.48	15.26	15.02	14.74	14.43	14.07	13.66	13.18	12.61	11.92	11.09	10.05	8.70	6.91	5.78	5.65	5.52
10600	16.25	16.09	15.91	15.70	15.48	15.23	14.95	14.63	14.27	13.85	13.36	12.78	12.08	11.24	10.17	8.81	6.99	5.83	5.70	5.57
10700	16.48	16.32	16.13	15.93	15.70	15.45	15.16	14.84	14.47	14.04	13.54	12.95	12.25	11.38	10.30	8.91	7.06	5.89	5.76	5.63
10800	16.72	16.55	16.36	16.15	15.92	15.66	15.37	15.04	14.66	14.23	13.72	13.13	12.41	11.53	10.43	9.02	7.14	5.94	5.81	5.68
10900	16.95	16.78	16.59	16.38	16.14	15.88	15.58	15.25	14.87	14.42	13.91	13.30	12.57	11.68	10.56	9.13	7.22	6.00	5.87	5.73
11000	17.19	17.01	16.82	16.60	16.36	16.10	15.80	15.46	15.07	14.62	14.10	13.48	12.73	11.83	10.69	9.24	7.29	6.06	5.92	5.78
11100	17.42	17.25	17.05	16.83	16.59	16.32	16.01	15.67	15.27	14.81	14.28	13.65	12.90	11.98	10.83	9.34	7.37	6.11	5.97	5.83
11200	17.66	17.48	17.28	17.06	16.81	16.54	16.23	15.88	15.47	15.01	14.46	13.83	13.07	12.13	10.96	9.45	7.45	6.17	6.03	5.89
11300	17.90	17.72	17.52	17.29	17.04	16.76	16.45	16.09	15.67	15.21	14.66	14.01	13.23	12.28	11.09	9.56	7.52	6.23	6.08	5.94
11400	18.15	17.96	17.75	17.52	17.27	16.98	16.66	16.30	15.89	15.41	14.85	14.19	13.40	12.44	11.23	9.67	7.60	6.28	6.14	5.99

x

$$V - V_c = 17 \qquad \text{SIGMA} = 3080.0$$

H	3.4	3.2	3.0	2.8	2.6	2.4	2.2	2.0	1.8	1.6	1.4	1.2	1.0	0.8	0.6	0.4	0.2	0.10	0.09	0.08
11500	18.39	18.20	17.99	17.76	17.50	17.21	16.89	16.52	16.10	15.61	15.04	14.37	13.57	12.59	11.36	9.78	7.68	6.34	6.19	6.04
11600	18.64	18.44	18.23	17.99	17.73	17.44	17.11	16.73	16.31	15.81	15.24	14.56	13.74	12.74	11.50	9.90	7.76	6.40	6.25	6.10
11700	18.88	18.69	18.47	18.23	17.96	17.67	17.33	16.95	16.52	16.02	15.43	14.74	13.91	12.90	11.63	10.01	7.84	6.46	6.30	6.15
11800	19.13	18.93	18.71	18.47	18.20	17.90	17.56	17.17	16.73	16.22	15.63	14.93	14.09	13.06	11.77	10.12	7.92	6.51	6.36	6.20
11900	19.38	19.18	18.96	18.71	18.44	18.13	17.78	17.39	16.94	16.43	15.83	15.11	14.26	13.22	11.91	10.23	8.00	6.57	6.42	6.26
12000	19.64	19.43	19.20	18.95	18.67	18.36	18.01	17.61	17.16	16.64	16.02	15.30	14.43	13.37	12.05	10.35	8.08	6.63	6.47	6.31
12100	19.89	19.68	19.45	19.20	18.91	18.60	18.24	17.84	17.38	16.84	16.22	15.49	14.61	13.53	12.19	10.46	8.16	6.69	6.53	6.36
12200	20.15	19.93	19.70	19.44	19.15	18.83	18.47	18.06	17.59	17.05	16.42	15.68	14.79	13.69	12.33	10.58	8.24	6.75	6.58	6.42
12300	20.40	20.19	19.95	19.69	19.40	19.07	18.70	18.29	17.81	17.27	16.63	15.87	14.96	13.86	12.47	10.69	8.32	6.81	6.64	6.47
12400	20.66	20.44	20.20	19.94	19.64	19.31	18.94	18.52	18.03	17.48	16.83	16.06	15.14	14.02	12.61	10.81	8.40	6.87	6.70	6.52
12500	20.92	20.70	20.46	20.19	19.89	19.55	19.17	18.74	18.26	17.69	17.03	16.26	15.32	14.18	12.76	10.92	8.48	6.92	6.75	6.58
12600	21.18	20.96	20.71	20.44	20.13	19.79	19.41	18.98	18.48	17.91	17.24	16.45	15.50	14.35	12.90	11.04	8.56	6.98	6.81	6.63
12700	21.45	21.22	20.97	20.69	20.38	20.04	19.65	19.21	18.70	18.12	17.45	16.65	15.69	14.51	13.05	11.16	8.64	7.04	6.87	6.69
12800	21.71	21.48	21.23	20.94	20.63	20.28	19.89	19.44	18.93	18.34	17.65	16.84	15.87	14.68	13.19	11.28	8.73	7.10	6.93	6.74
12900	21.98	21.74	21.49	21.20	20.88	20.53	20.13	19.68	19.16	18.56	17.86	17.04	16.05	14.85	13.34	11.40	8.81	7.16	6.98	6.80
13000	22.25	22.01	21.75	21.46	21.14	20.78	20.37	19.91	19.39	18.78	18.07	17.24	16.24	15.01	13.48	11.52	8.89	7.22	7.04	6.85
13100	22.52	22.28	22.01	21.72	21.39	21.03	20.61	20.15	19.62	19.00	18.29	17.44	16.42	15.18	13.63	11.64	8.98	7.28	7.10	6.91
13200	22.79	22.55	22.28	21.98	21.65	21.29	20.86	20.38	19.85	19.23	18.50	17.64	16.61	15.35	13.78	11.76	9.06	7.35	7.16	6.96
13300	23.06	22.82	22.54	22.24	21.90	21.53	21.11	20.63	20.08	19.45	18.71	17.84	16.80	15.52	13.93	11.88	9.15	7.41	7.21	7.02
13400	23.34	23.09	22.81	22.50	22.16	21.78	21.35	20.87	20.32	19.68	18.93	18.05	16.99	15.70	14.08	12.00	9.23	7.47	7.27	7.08
13500	23.62	23.36	23.08	22.77	22.42	22.04	21.60	21.11	20.55	19.90	19.15	18.25	17.18	15.87	14.23	12.12	9.32	7.53	7.33	7.13
13600	23.89	23.63	23.35	23.04	22.69	22.30	21.86	21.36	20.79	20.13	19.36	18.46	17.37	16.04	14.38	12.25	9.40	7.59	7.39	7.19
13700	24.17	23.91	23.62	23.30	22.95	22.55	22.11	21.60	21.03	20.36	19.58	18.67	17.56	16.22	14.54	12.37	9.49	7.65	7.45	7.24
13800	24.46	24.19	23.90	23.57	23.22	22.81	22.36	21.86	21.27	20.59	19.80	18.87	17.76	16.39	14.69	12.50	9.57	7.71	7.51	7.30
13900	24.74	24.47	24.17	23.85	23.48	23.08	22.62	22.10	21.51	20.82	20.03	19.08	17.95	16.57	14.84	12.62	9.66	7.77	7.57	7.36
14000	25.02	24.75	24.45	24.12	23.75	23.34	22.87	22.35	21.75	21.06	20.25	19.29	18.15	16.75	15.00	12.75	9.75	7.84	7.63	7.41
14100	25.31	25.03	24.73	24.39	24.02	23.60	23.13	22.60	21.99	21.29	20.47	19.51	18.35	16.93	15.15	12.87	9.83	7.90	7.69	7.47
14200	25.60	25.32	25.01	24.67	24.29	23.87	23.39	22.85	22.24	21.53	20.70	19.72	18.54	17.11	15.31	13.00	9.92	7.96	7.75	7.53
14300	25.89	25.60	25.29	24.95	24.56	24.14	23.65	23.11	22.49	21.77	20.93	19.93	18.74	17.29	15.47	13.13	10.01	8.02	7.81	7.58
14400	26.18	25.89	25.58	25.23	24.84	24.41	23.92	23.36	22.73	22.00	21.15	20.15	18.94	17.47	15.63	13.26	10.10	8.09	7.87	7.64
14500	26.47	26.18	25.86	25.51	25.11	24.68	24.18	23.62	22.98	22.24	21.38	20.36	19.14	17.65	15.78	13.39	10.19	8.15	7.93	7.70
14600	26.77	26.47	26.15	25.79	25.39	24.95	24.45	23.88	23.23	22.49	21.61	20.58	19.35	17.83	15.94	13.51	10.28	8.21	7.99	7.76
14700	27.06	26.76	26.44	26.07	25.67	25.23	24.71	24.13	23.48	22.73	21.84	20.80	19.55	18.02	16.10	13.64	10.36	8.28	8.05	7.81
14800	27.36	27.06	26.73	26.36	25.95	25.50	24.98	24.40	23.74	22.97	22.08	21.02	19.75	18.20	16.27	13.78	10.45	8.34	8.11	7.87
14900	27.66	27.35	27.02	26.65	26.23	25.77	25.25	24.66	23.99	23.22	22.31	21.24	19.96	18.39	16.43	13.91	10.54	8.40	8.17	7.93
15000	27.96	27.65	27.31	26.94	26.52	26.05	25.52	24.93	24.25	23.46	22.55	21.46	20.16	18.58	16.59	14.04	10.63	8.47	8.23	7.99

x

$V-V_c = 18$ SIGMA = 2747.0

X / H	0.08	0.09	0.10	0.2	0.4	0.6	0.8	1.0	1.2	1.4	1.6	1.8	2.0	2.2	2.4	2.6	2.8	3.0	3.2	3.4
C	1.000	1.000	1.000	1.000	1.000	1.000	1.000	1.000	1.000	1.000	1.000	1.000	1.000	1.000	1.000	1.000	1.000	1.000	1.000	1.000
20	1.008	1.008	1.008	1.009	1.009	1.010	1.011	1.011	1.011	1.012	1.012	1.012	1.012	1.012	1.013	1.013	1.013	1.013	1.013	1.013
40	1.016	1.016	1.016	1.017	1.019	1.020	1.021	1.022	1.023	1.023	1.024	1.024	1.024	1.025	1.025	1.025	1.025	1.026	1.026	1.026
60	1.023	1.024	1.024	1.026	1.028	1.030	1.032	1.033	1.034	1.035	1.036	1.036	1.037	1.037	1.038	1.038	1.038	1.039	1.039	1.039
80	1.031	1.032	1.032	1.034	1.038	1.042	1.042	1.044	1.045	1.047	1.048	1.048	1.049	1.050	1.050	1.051	1.051	1.052	1.052	1.052
100	1.039	1.040	1.040	1.043	1.047	1.051	1.053	1.055	1.057	1.058	1.060	1.061	1.062	1.062	1.063	1.064	1.064	1.065	1.065	1.066
120	1.047	1.047	1.048	1.051	1.057	1.061	1.064	1.066	1.069	1.070	1.072	1.073	1.074	1.075	1.076	1.077	1.077	1.078	1.078	1.079
140	1.055	1.055	1.056	1.060	1.066	1.071	1.075	1.078	1.080	1.082	1.084	1.085	1.087	1.088	1.089	1.090	1.090	1.091	1.092	1.092
160	1.063	1.063	1.064	1.069	1.076	1.081	1.086	1.089	1.092	1.094	1.096	1.098	1.099	1.101	1.102	1.103	1.104	1.104	1.105	1.106
180	1.071	1.071	1.072	1.077	1.085	1.092	1.097	1.100	1.104	1.106	1.108	1.110	1.112	1.114	1.115	1.116	1.117	1.118	1.119	1.119
200	1.079	1.079	1.080	1.086	1.095	1.102	1.108	1.112	1.115	1.118	1.121	1.123	1.125	1.127	1.128	1.129	1.130	1.131	1.132	1.133
220	1.086	1.087	1.088	1.095	1.105	1.113	1.119	1.123	1.127	1.131	1.133	1.136	1.138	1.140	1.141	1.143	1.144	1.145	1.146	1.147
240	1.094	1.095	1.096	1.103	1.115	1.123	1.130	1.135	1.139	1.143	1.146	1.148	1.151	1.153	1.154	1.156	1.157	1.159	1.160	1.161
260	1.102	1.103	1.104	1.112	1.124	1.134	1.141	1.146	1.151	1.155	1.158	1.161	1.164	1.166	1.168	1.169	1.171	1.172	1.174	1.175
280	1.110	1.111	1.112	1.121	1.134	1.144	1.152	1.158	1.163	1.167	1.171	1.174	1.177	1.179	1.181	1.183	1.185	1.186	1.188	1.189
300	1.118	1.119	1.120	1.129	1.144	1.155	1.163	1.170	1.175	1.180	1.184	1.187	1.190	1.192	1.195	1.197	1.198	1.200	1.202	1.203
320	1.126	1.127	1.128	1.138	1.154	1.165	1.174	1.182	1.187	1.192	1.197	1.200	1.203	1.206	1.208	1.210	1.212	1.214	1.216	1.217
340	1.134	1.135	1.136	1.147	1.164	1.174	1.186	1.193	1.200	1.205	1.209	1.213	1.216	1.219	1.222	1.224	1.226	1.228	1.230	1.231
360	1.142	1.143	1.145	1.156	1.173	1.187	1.197	1.205	1.212	1.218	1.222	1.226	1.230	1.233	1.236	1.238	1.240	1.242	1.244	1.246
380	1.150	1.151	1.153	1.165	1.183	1.197	1.208	1.217	1.224	1.230	1.235	1.240	1.243	1.247	1.249	1.252	1.254	1.256	1.258	1.260
400	1.158	1.159	1.161	1.173	1.193	1.208	1.220	1.229	1.237	1.243	1.248	1.253	1.257	1.260	1.263	1.266	1.269	1.271	1.273	1.275
420	1.166	1.167	1.169	1.182	1.203	1.219	1.231	1.241	1.249	1.256	1.261	1.266	1.270	1.274	1.277	1.280	1.283	1.285	1.287	1.289
440	1.174	1.176	1.177	1.191	1.213	1.230	1.243	1.253	1.262	1.269	1.275	1.280	1.284	1.288	1.291	1.294	1.297	1.300	1.302	1.304
460	1.182	1.184	1.185	1.200	1.223	1.241	1.254	1.265	1.274	1.281	1.288	1.293	1.298	1.302	1.305	1.309	1.312	1.314	1.316	1.319
480	1.190	1.192	1.193	1.209	1.233	1.252	1.266	1.277	1.287	1.294	1.301	1.307	1.312	1.316	1.320	1.323	1.326	1.329	1.331	1.333
500	1.198	1.200	1.202	1.218	1.243	1.263	1.278	1.290	1.299	1.308	1.314	1.320	1.325	1.330	1.334	1.337	1.341	1.343	1.346	1.348
520	1.206	1.208	1.210	1.227	1.254	1.274	1.289	1.302	1.312	1.321	1.328	1.334	1.339	1.344	1.348	1.352	1.355	1.358	1.361	1.363
540	1.214	1.216	1.218	1.236	1.264	1.285	1.301	1.314	1.325	1.334	1.341	1.348	1.353	1.358	1.363	1.366	1.370	1.373	1.376	1.378
560	1.222	1.224	1.226	1.245	1.274	1.296	1.313	1.327	1.338	1.347	1.355	1.362	1.367	1.373	1.377	1.381	1.385	1.388	1.391	1.393
580	1.230	1.232	1.234	1.254	1.284	1.307	1.325	1.339	1.351	1.360	1.369	1.376	1.382	1.387	1.392	1.396	1.400	1.403	1.406	1.409
600	1.238	1.240	1.243	1.263	1.294	1.318	1.337	1.351	1.364	1.374	1.382	1.390	1.396	1.401	1.406	1.411	1.415	1.418	1.421	1.424
620	1.246	1.249	1.251	1.272	1.305	1.329	1.349	1.364	1.377	1.387	1.396	1.404	1.410	1.416	1.421	1.425	1.430	1.433	1.436	1.439
640	1.254	1.257	1.259	1.281	1.315	1.341	1.361	1.377	1.390	1.401	1.410	1.418	1.424	1.430	1.436	1.440	1.445	1.448	1.452	1.455
660	1.262	1.265	1.267	1.290	1.325	1.352	1.373	1.390	1.403	1.414	1.424	1.432	1.439	1.445	1.451	1.455	1.460	1.464	1.467	1.471
680	1.270	1.273	1.276	1.299	1.336	1.363	1.385	1.403	1.416	1.428	1.438	1.446	1.453	1.460	1.466	1.471	1.475	1.479	1.483	1.486
700	1.279	1.281	1.284	1.308	1.346	1.375	1.397	1.416	1.429	1.441	1.452	1.460	1.468	1.475	1.481	1.486	1.490	1.495	1.498	1.502
720	1.287	1.289	1.292	1.317	1.357	1.386	1.409	1.428	1.443	1.455	1.466	1.475	1.483	1.490	1.496	1.501	1.506	1.510	1.514	1.518
740	1.295	1.298	1.300	1.326	1.367	1.398	1.421	1.440	1.456	1.469	1.480	1.489	1.497	1.504	1.511	1.516	1.521	1.526	1.530	1.534
760	1.303	1.306	1.309	1.336	1.378	1.409	1.434	1.453	1.469	1.483	1.494	1.504	1.512	1.519	1.526	1.532	1.537	1.542	1.546	1.550
780	1.311	1.314	1.317	1.345	1.388	1.421	1.446	1.466	1.483	1.497	1.508	1.518	1.527	1.535	1.541	1.547	1.553	1.557	1.562	1.566
800	1.319	1.322	1.325	1.354	1.399	1.432	1.458	1.479	1.496	1.511	1.523	1.533	1.542	1.550	1.557	1.563	1.568	1.573	1.578	1.582

SIGMA = 2747.0

V-V_c = 18

H x	0.08	0.09	0.10	0.2	0.4	0.6	0.8	1.0	1.2	1.4	1.6	1.8	2.0	2.2	2.4	2.6	2.8	3.0	3.2	3.4
820	1.327	1.331	1.334	1.363	1.409	1.444	1.471	1.492	1.510	1.525	1.537	1.548	1.557	1.565	1.572	1.578	1.584	1.589	1.594	1.598
840	1.335	1.339	1.342	1.372	1.420	1.456	1.483	1.505	1.524	1.539	1.552	1.562	1.572	1.580	1.588	1.594	1.600	1.605	1.610	1.614
860	1.344	1.347	1.350	1.382	1.431	1.467	1.496	1.519	1.537	1.553	1.566	1.577	1.587	1.596	1.603	1.610	1.616	1.621	1.626	1.631
880	1.352	1.355	1.359	1.391	1.441	1.479	1.508	1.532	1.551	1.567	1.581	1.592	1.602	1.611	1.619	1.626	1.632	1.638	1.643	1.647
900	1.360	1.364	1.367	1.400	1.452	1.491	1.521	1.545	1.565	1.581	1.595	1.607	1.618	1.627	1.635	1.642	1.648	1.654	1.659	1.664
920	1.368	1.372	1.376	1.409	1.463	1.503	1.534	1.558	1.579	1.596	1.610	1.622	1.633	1.642	1.650	1.658	1.664	1.670	1.676	1.680
940	1.376	1.380	1.384	1.419	1.473	1.514	1.546	1.572	1.593	1.610	1.625	1.637	1.648	1.658	1.666	1.674	1.681	1.687	1.692	1.697
960	1.384	1.388	1.392	1.428	1.484	1.526	1.559	1.585	1.607	1.625	1.640	1.653	1.664	1.674	1.682	1.690	1.697	1.703	1.709	1.714
980	1.393	1.397	1.401	1.437	1.495	1.538	1.572	1.599	1.621	1.639	1.655	1.668	1.679	1.690	1.698	1.706	1.713	1.720	1.726	1.731
1000	1.401	1.405	1.409	1.447	1.506	1.550	1.585	1.612	1.635	1.654	1.670	1.683	1.695	1.705	1.715	1.723	1.730	1.736	1.742	1.748
1020	1.409	1.413	1.418	1.456	1.517	1.562	1.598	1.626	1.649	1.668	1.685	1.699	1.711	1.721	1.731	1.739	1.747	1.753	1.759	1.765
1040	1.417	1.422	1.426	1.466	1.528	1.574	1.611	1.640	1.663	1.683	1.700	1.714	1.727	1.737	1.747	1.756	1.763	1.770	1.776	1.782
1060	1.425	1.430	1.434	1.475	1.539	1.586	1.624	1.653	1.678	1.698	1.715	1.730	1.742	1.754	1.763	1.772	1.780	1.787	1.793	1.799
1080	1.434	1.438	1.443	1.484	1.550	1.599	1.637	1.667	1.692	1.713	1.730	1.745	1.758	1.770	1.780	1.789	1.797	1.804	1.810	1.816
1100	1.442	1.447	1.451	1.494	1.561	1.611	1.650	1.681	1.706	1.728	1.746	1.761	1.774	1.786	1.796	1.805	1.814	1.821	1.828	1.834
1120	1.450	1.455	1.460	1.503	1.572	1.623	1.663	1.695	1.721	1.743	1.761	1.777	1.790	1.802	1.813	1.822	1.831	1.838	1.845	1.851
1140	1.458	1.463	1.468	1.513	1.583	1.635	1.676	1.709	1.735	1.758	1.776	1.792	1.806	1.819	1.830	1.839	1.848	1.855	1.862	1.869
1160	1.467	1.472	1.477	1.522	1.594	1.648	1.689	1.723	1.750	1.773	1.792	1.808	1.823	1.835	1.846	1.856	1.865	1.873	1.880	1.886
1180	1.475	1.480	1.485	1.532	1.605	1.660	1.702	1.737	1.765	1.788	1.807	1.824	1.839	1.852	1.863	1.873	1.883	1.890	1.897	1.904
1200	1.483	1.489	1.494	1.541	1.616	1.672	1.716	1.751	1.779	1.803	1.823	1.840	1.855	1.868	1.880	1.890	1.899	1.908	1.915	1.922
1220	1.492	1.497	1.502	1.551	1.627	1.685	1.729	1.765	1.794	1.818	1.839	1.856	1.872	1.885	1.897	1.907	1.917	1.925	1.933	1.940
1240	1.500	1.505	1.511	1.561	1.639	1.697	1.743	1.779	1.809	1.834	1.855	1.873	1.888	1.902	1.914	1.925	1.934	1.943	1.951	1.958
1260	1.508	1.514	1.520	1.570	1.650	1.710	1.756	1.793	1.824	1.849	1.870	1.889	1.905	1.919	1.931	1.942	1.952	1.960	1.968	1.976
1280	1.517	1.522	1.528	1.580	1.661	1.722	1.770	1.808	1.839	1.864	1.886	1.905	1.921	1.936	1.948	1.959	1.969	1.978	1.986	1.994
1300	1.525	1.531	1.537	1.589	1.672	1.735	1.783	1.822	1.854	1.880	1.902	1.921	1.938	1.953	1.965	1.977	1.987	1.996	2.004	2.012
1320	1.533	1.539	1.545	1.599	1.684	1.747	1.797	1.836	1.869	1.896	1.918	1.938	1.955	1.970	1.983	1.994	2.005	2.014	2.023	2.030
1340	1.542	1.548	1.554	1.609	1.695	1.760	1.810	1.851	1.884	1.911	1.934	1.954	1.972	1.987	2.000	2.012	2.023	2.032	2.041	2.049
1360	1.550	1.556	1.562	1.618	1.707	1.773	1.824	1.865	1.899	1.927	1.951	1.971	1.989	2.004	2.018	2.030	2.040	2.050	2.059	2.067
1380	1.558	1.565	1.571	1.628	1.718	1.785	1.838	1.880	1.914	1.943	1.967	1.988	2.006	2.021	2.035	2.047	2.058	2.068	2.077	2.086
1400	1.567	1.573	1.580	1.638	1.729	1.798	1.852	1.894	1.929	1.958	1.983	2.004	2.023	2.039	2.053	2.065	2.077	2.086	2.096	2.104
1420	1.575	1.582	1.588	1.648	1.741	1.811	1.865	1.909	1.945	1.974	1.999	2.021	2.040	2.056	2.070	2.083	2.095	2.105	2.114	2.123
1440	1.583	1.590	1.597	1.657	1.752	1.824	1.879	1.924	1.960	1.990	2.016	2.038	2.057	2.074	2.088	2.101	2.113	2.123	2.133	2.142
1460	1.592	1.599	1.605	1.667	1.764	1.837	1.893	1.938	1.975	2.006	2.032	2.055	2.074	2.091	2.106	2.119	2.131	2.142	2.152	2.160
1480	1.600	1.607	1.614	1.677	1.776	1.850	1.907	1.953	1.991	2.022	2.049	2.072	2.091	2.109	2.124	2.138	2.150	2.161	2.170	2.179
1500	1.609	1.616	1.623	1.687	1.787	1.863	1.921	1.968	2.007	2.039	2.066	2.089	2.109	2.126	2.142	2.156	2.168	2.179	2.189	2.198
1520	1.617	1.624	1.631	1.697	1.799	1.876	1.935	1.983	2.022	2.055	2.082	2.106	2.126	2.144	2.160	2.174	2.187	2.198	2.208	2.217
1540	1.625	1.633	1.640	1.706	1.811	1.889	1.949	1.998	2.038	2.071	2.099	2.123	2.144	2.162	2.178	2.192	2.205	2.217	2.227	2.237
1560	1.634	1.641	1.649	1.716	1.822	1.902	1.964	2.013	2.054	2.087	2.116	2.140	2.161	2.180	2.196	2.211	2.224	2.236	2.246	2.256
158C	1.642	1.650	1.658	1.726	1.834	1.915	1.978	2.028	2.069	2.104	2.133	2.158	2.179	2.198	2.215	2.230	2.243	2.255	2.265	2.275
1600	1.651	1.659	1.666	1.736	1.846	1.928	1.992	2.043	2.085	2.120	2.150	2.175	2.197	2.216	2.233	2.248	2.262	2.274	2.285	2.295
1620	1.659	1.667	1.675	1.746	1.858	1.941	2.006	2.058	2.101	2.137	2.167	2.192	2.215	2.234	2.252	2.267	2.281	2.293	2.304	2.314
1640	1.668	1.676	1.684	1.756	1.869	1.955	2.021	2.074	2.117	2.153	2.184	2.210	2.233	2.253	2.270	2.286	2.300	2.312	2.323	2.334

SIGMA = 2747.0

V-V_c = 18

H \ x	3.4	3.2	3.0	2.8	2.6	2.4	2.2	2.0	1.8	1.6	1.4	1.2	1.0	0.8	0.6	0.4	0.2	0.10	0.09	0.08
1660	2.353	2.343	2.331	2.319	2.304	2.289	2.271	2.251	2.228	2.201	2.170	2.133	2.089	2.035	1.968	1.881	1.766	1.692	1.684	1.676
1680	2.373	2.363	2.351	2.338	2.323	2.307	2.289	2.269	2.245	2.218	2.187	2.149	2.104	2.050	1.981	1.893	1.776	1.701	1.693	1.685
1700	2.393	2.382	2.370	2.357	2.342	2.326	2.308	2.287	2.263	2.235	2.203	2.165	2.120	2.064	1.995	1.905	1.786	1.710	1.702	1.693
1720	2.413	2.402	2.390	2.376	2.361	2.345	2.326	2.305	2.281	2.253	2.220	2.182	2.135	2.079	2.008	1.917	1.796	1.719	1.710	1.702
1740	2.433	2.422	2.409	2.396	2.381	2.364	2.345	2.323	2.299	2.270	2.237	2.198	2.151	2.093	2.021	1.929	1.806	1.727	1.719	1.710
1760	2.453	2.442	2.429	2.415	2.400	2.383	2.363	2.341	2.316	2.288	2.254	2.214	2.166	2.108	2.035	1.941	1.816	1.736	1.727	1.719
1780	2.473	2.462	2.449	2.435	2.419	2.402	2.382	2.360	2.334	2.305	2.271	2.230	2.182	2.123	2.048	1.953	1.826	1.745	1.736	1.727
1800	2.493	2.482	2.469	2.454	2.439	2.421	2.401	2.378	2.353	2.323	2.288	2.247	2.198	2.137	2.062	1.965	1.836	1.754	1.745	1.736
1820	2.514	2.502	2.489	2.474	2.458	2.440	2.420	2.397	2.371	2.340	2.305	2.263	2.213	2.152	2.076	1.977	1.846	1.763	1.753	1.744
1840	2.534	2.522	2.509	2.494	2.478	2.459	2.439	2.415	2.389	2.358	2.322	2.280	2.229	2.167	2.089	1.989	1.856	1.772	1.762	1.753
1860	2.555	2.542	2.529	2.514	2.497	2.479	2.458	2.434	2.407	2.376	2.340	2.297	2.245	2.182	2.103	2.002	1.866	1.780	1.771	1.761
1880	2.575	2.563	2.549	2.534	2.517	2.498	2.477	2.453	2.425	2.394	2.357	2.313	2.261	2.197	2.117	2.014	1.877	1.789	1.780	1.770
1900	2.596	2.583	2.569	2.554	2.537	2.518	2.496	2.472	2.444	2.412	2.374	2.330	2.277	2.212	2.130	2.026	1.887	1.798	1.788	1.778
1920	2.617	2.604	2.590	2.574	2.557	2.537	2.515	2.491	2.462	2.430	2.392	2.347	2.293	2.227	2.144	2.038	1.897	1.807	1.797	1.787
1940	2.637	2.624	2.610	2.594	2.576	2.557	2.535	2.510	2.481	2.448	2.409	2.363	2.309	2.242	2.158	2.051	1.907	1.816	1.806	1.795
1960	2.658	2.645	2.630	2.614	2.596	2.577	2.554	2.529	2.499	2.466	2.427	2.380	2.325	2.257	2.172	2.063	1.917	1.825	1.814	1.804
1980	2.679	2.666	2.651	2.635	2.617	2.596	2.574	2.548	2.518	2.484	2.444	2.397	2.341	2.272	2.186	2.075	1.928	1.834	1.823	1.813
2000	2.700	2.687	2.672	2.655	2.637	2.617	2.593	2.567	2.537	2.502	2.462	2.414	2.357	2.287	2.200	2.088	1.938	1.842	1.832	1.821
2020	2.721	2.708	2.692	2.676	2.657	2.636	2.613	2.586	2.556	2.521	2.480	2.431	2.373	2.302	2.214	2.100	1.948	1.851	1.841	1.830
2040	2.743	2.729	2.713	2.696	2.677	2.656	2.632	2.605	2.575	2.539	2.498	2.449	2.390	2.318	2.228	2.112	1.958	1.860	1.849	1.838
2060	2.764	2.750	2.734	2.717	2.698	2.676	2.652	2.625	2.594	2.557	2.515	2.466	2.406	2.333	2.242	2.125	1.969	1.869	1.858	1.847
2080	2.785	2.771	2.755	2.738	2.718	2.696	2.672	2.644	2.613	2.576	2.533	2.483	2.422	2.349	2.256	2.137	1.979	1.878	1.867	1.856
2100	2.807	2.792	2.776	2.758	2.739	2.717	2.692	2.664	2.632	2.595	2.551	2.500	2.439	2.364	2.270	2.150	1.989	1.887	1.876	1.864
2120	2.828	2.814	2.797	2.779	2.759	2.737	2.712	2.683	2.651	2.613	2.569	2.518	2.455	2.379	2.285	2.162	2.000	1.896	1.885	1.873
2140	2.850	2.835	2.818	2.800	2.780	2.757	2.732	2.703	2.670	2.632	2.587	2.535	2.472	2.395	2.299	2.175	2.010	1.905	1.893	1.882
2160	2.872	2.856	2.840	2.821	2.801	2.778	2.752	2.723	2.689	2.651	2.606	2.552	2.489	2.411	2.313	2.188	2.020	1.914	1.902	1.890
2180	2.893	2.878	2.861	2.842	2.822	2.798	2.772	2.743	2.709	2.670	2.624	2.570	2.505	2.426	2.327	2.200	2.031	1.923	1.911	1.899
2200	2.915	2.900	2.883	2.864	2.843	2.819	2.792	2.762	2.728	2.688	2.642	2.588	2.522	2.442	2.342	2.213	2.041	1.932	1.920	1.908
2220	2.937	2.922	2.904	2.885	2.864	2.840	2.813	2.782	2.748	2.707	2.661	2.605	2.539	2.458	2.356	2.226	2.052	1.941	1.929	1.916
2240	2.959	2.943	2.926	2.906	2.885	2.860	2.833	2.802	2.767	2.726	2.679	2.623	2.556	2.473	2.371	2.238	2.062	1.950	1.938	1.925
2260	2.981	2.965	2.947	2.928	2.906	2.881	2.854	2.822	2.787	2.746	2.697	2.641	2.573	2.489	2.385	2.251	2.073	1.959	1.947	1.934
2280	3.004	2.987	2.969	2.949	2.927	2.902	2.874	2.843	2.806	2.765	2.716	2.658	2.589	2.505	2.400	2.264	2.083	1.968	1.955	1.943
2300	3.026	3.009	2.991	2.971	2.948	2.923	2.895	2.863	2.826	2.784	2.735	2.676	2.606	2.521	2.414	2.277	2.094	1.977	1.964	1.951
2320	3.048	3.031	3.013	2.992	2.970	2.944	2.916	2.883	2.846	2.803	2.753	2.694	2.623	2.537	2.429	2.290	2.104	1.986	1.973	1.960
2340	3.071	3.054	3.035	3.014	2.991	2.965	2.936	2.903	2.866	2.823	2.772	2.712	2.641	2.553	2.443	2.303	2.115	1.995	1.982	1.969
2360	3.093	3.076	3.057	3.036	3.013	2.987	2.957	2.924	2.886	2.842	2.791	2.730	2.658	2.569	2.458	2.315	2.125	2.004	1.991	1.977
2380	3.116	3.098	3.079	3.058	3.034	3.008	2.978	2.944	2.906	2.862	2.810	2.748	2.675	2.585	2.473	2.328	2.136	2.013	2.000	1.986
2400	3.139	3.121	3.101	3.080	3.056	3.029	2.999	2.965	2.926	2.881	2.829	2.767	2.692	2.601	2.488	2.341	2.147	2.022	2.009	1.995
2420	3.161	3.143	3.124	3.102	3.078	3.051	3.020	2.986	2.946	2.901	2.848	2.785	2.709	2.617	2.502	2.354	2.157	2.032	2.018	2.004
2440	3.184	3.166	3.146	3.124	3.100	3.072	3.041	3.006	2.966	2.920	2.867	2.803	2.727	2.634	2.517	2.367	2.168	2.041	2.027	2.012
2460	3.207	3.189	3.169	3.146	3.121	3.094	3.063	3.027	2.987	2.940	2.886	2.821	2.744	2.650	2.532	2.381	2.178	2.050	2.036	2.021
2480	3.230	3.212	3.191	3.169	3.143	3.115	3.084	3.048	3.007	2.960	2.905	2.840	2.762	2.666	2.547	2.394	2.189	2.059	2.045	2.030

$V - V_c = 18$ SIGMA = 2747.0

H \ X	3.4	3.2	3.0	2.8	2.6	2.4	2.2	2.0	1.8	1.6	1.4	1.2	1.0	0.8	0.6	0.4	0.2	0.10	0.09	0.08
2500	3.253	3.235	3.214	3.191	3.166	3.137	3.105	3.069	3.028	2.980	2.924	2.858	2.779	2.683	2.562	2.407	2.200	2.068	2.054	2.039
2520	3.277	3.257	3.237	3.213	3.188	3.159	3.127	3.090	3.048	3.000	2.943	2.877	2.797	2.699	2.577	2.420	2.211	2.077	2.063	2.048
2540	3.300	3.281	3.259	3.236	3.210	3.181	3.148	3.111	3.069	3.020	2.963	2.895	2.814	2.716	2.592	2.433	2.221	2.086	2.072	2.056
2560	3.323	3.304	3.282	3.259	3.232	3.203	3.170	3.132	3.089	3.040	2.982	2.914	2.832	2.732	2.607	2.446	2.232	2.096	2.081	2.065
2580	3.347	3.327	3.305	3.281	3.255	3.225	3.191	3.153	3.110	3.060	3.002	2.933	2.850	2.749	2.622	2.460	2.243	2.105	2.090	2.074
2600	3.370	3.350	3.328	3.304	3.277	3.247	3.213	3.175	3.131	3.080	3.021	2.951	2.868	2.765	2.637	2.473	2.254	2.114	2.099	2.083
2620	3.394	3.374	3.351	3.327	3.300	3.269	3.235	3.196	3.152	3.101	3.041	2.970	2.885	2.782	2.653	2.486	2.264	2.123	2.108	2.092
2640	3.417	3.397	3.375	3.350	3.322	3.291	3.257	3.217	3.173	3.121	3.060	2.989	2.903	2.799	2.668	2.500	2.275	2.132	2.117	2.101
2660	3.441	3.421	3.398	3.373	3.345	3.314	3.279	3.239	3.194	3.141	3.080	3.008	2.921	2.815	2.683	2.513	2.286	2.142	2.126	2.110
2680	3.465	3.444	3.421	3.396	3.368	3.336	3.301	3.261	3.215	3.162	3.100	3.027	2.939	2.832	2.698	2.526	2.297	2.151	2.135	2.118
2700	3.489	3.468	3.445	3.419	3.390	3.359	3.323	3.282	3.236	3.182	3.120	3.046	2.957	2.849	2.714	2.540	2.308	2.160	2.144	2.127
2720	3.513	3.492	3.468	3.442	3.413	3.381	3.345	3.304	3.257	3.203	3.140	3.065	2.975	2.866	2.729	2.553	2.319	2.169	2.153	2.136
2740	3.537	3.515	3.492	3.466	3.436	3.404	3.367	3.326	3.278	3.224	3.160	3.084	2.994	2.883	2.745	2.567	2.330	2.179	2.162	2.145
2760	3.561	3.539	3.515	3.489	3.459	3.427	3.390	3.348	3.300	3.244	3.180	3.103	3.012	2.900	2.760	2.580	2.340	2.188	2.171	2.154
2780	3.585	3.563	3.539	3.512	3.483	3.449	3.412	3.369	3.321	3.265	3.200	3.123	3.030	2.917	2.776	2.594	2.351	2.197	2.180	2.163
2800	3.610	3.587	3.563	3.536	3.506	3.472	3.434	3.391	3.342	3.286	3.220	3.142	3.048	2.934	2.791	2.607	2.362	2.206	2.189	2.172
2820	3.634	3.612	3.587	3.560	3.529	3.495	3.457	3.414	3.364	3.307	3.240	3.161	3.067	2.951	2.807	2.621	2.373	2.216	2.198	2.181
2840	3.659	3.636	3.611	3.583	3.552	3.518	3.479	3.436	3.386	3.328	3.260	3.181	3.085	2.968	2.822	2.635	2.384	2.225	2.207	2.190
2860	3.683	3.660	3.635	3.607	3.576	3.541	3.502	3.458	3.407	3.349	3.281	3.200	3.104	2.986	2.838	2.648	2.395	2.234	2.217	2.199
2880	3.708	3.685	3.659	3.631	3.599	3.564	3.525	3.480	3.429	3.370	3.301	3.220	3.122	3.003	2.854	2.662	2.406	2.244	2.226	2.207
2900	3.733	3.709	3.683	3.655	3.623	3.588	3.548	3.502	3.451	3.391	3.322	3.239	3.141	3.020	2.870	2.676	2.417	2.253	2.235	2.216
2920	3.757	3.734	3.708	3.679	3.647	3.611	3.571	3.525	3.473	3.412	3.342	3.259	3.159	3.038	2.885	2.690	2.428	2.262	2.244	2.225
2940	3.782	3.758	3.732	3.703	3.670	3.634	3.594	3.547	3.495	3.434	3.363	3.279	3.178	3.055	2.901	2.703	2.440	2.272	2.253	2.234
2960	3.807	3.783	3.757	3.727	3.694	3.658	3.617	3.570	3.517	3.455	3.383	3.299	3.197	3.072	2.917	2.717	2.451	2.281	2.262	2.243
2980	3.832	3.808	3.781	3.751	3.718	3.681	3.640	3.593	3.539	3.477	3.404	3.318	3.216	3.090	2.933	2.731	2.462	2.290	2.272	2.252
3000	3.858	3.833	3.806	3.776	3.742	3.705	3.663	3.615	3.561	3.498	3.425	3.338	3.234	3.108	2.949	2.745	2.473	2.300	2.281	2.261

$V-V_c = 18$ SIGMA = 2747.0

H x	3.4	3.2	3.0	2.8	2.6	2.4	2.2	2.0	1.8	1.6	1.4	1.2	1.0	0.8	0.6	0.4	0.2	0.10	0.09	0.08
3100	3.985	3.959	3.930	3.898	3.863	3.824	3.780	3.730	3.673	3.607	3.530	3.439	3.330	3.196	3.029	2.815	2.529	2.347	2.327	2.306
3200	4.114	4.086	4.056	4.023	3.986	3.945	3.899	3.846	3.786	3.717	3.636	3.540	3.426	3.286	3.111	2.885	2.585	2.394	2.373	2.352
3300	4.245	4.216	4.185	4.150	4.111	4.068	4.019	3.964	3.901	3.829	3.744	3.644	3.524	3.377	3.193	2.957	2.642	2.442	2.420	2.397
3400	4.378	4.348	4.315	4.279	4.238	4.193	4.142	4.084	4.018	3.942	3.853	3.748	3.623	3.469	3.276	3.029	2.699	2.490	2.466	2.443
3500	4.513	4.482	4.447	4.409	4.367	4.319	4.266	4.206	4.137	4.057	3.964	3.855	3.723	3.562	3.361	3.102	2.757	2.538	2.513	2.489
3600	4.650	4.618	4.582	4.542	4.497	4.448	4.392	4.329	4.257	4.174	4.077	3.962	3.825	3.656	3.446	3.176	2.815	2.586	2.561	2.535
3700	4.790	4.755	4.718	4.676	4.630	4.578	4.520	4.454	4.379	4.292	4.191	4.071	3.927	3.752	3.532	3.250	2.874	2.634	2.608	2.581
3800	4.931	4.895	4.856	4.813	4.764	4.711	4.650	4.581	4.503	4.412	4.307	4.182	4.032	3.849	3.620	3.325	2.933	2.683	2.656	2.628
3900	5.074	5.037	4.996	4.951	4.901	4.845	4.782	4.710	4.628	4.534	4.424	4.294	4.137	3.947	3.708	3.401	2.992	2.732	2.703	2.674
4000	5.220	5.181	5.138	5.091	5.039	4.981	4.915	4.840	4.755	4.657	4.542	4.407	4.244	4.046	3.797	3.478	3.052	2.781	2.751	2.721
4100	5.367	5.327	5.283	5.234	5.179	5.119	5.050	4.973	4.884	4.782	4.663	4.522	4.353	4.146	3.888	3.555	3.113	2.831	2.800	2.768
4200	5.517	5.475	5.429	5.376	5.321	5.258	5.187	5.107	5.015	4.908	4.786	4.638	4.462	4.247	3.979	3.634	3.173	2.880	2.848	2.815
4300	5.668	5.625	5.577	5.524	5.466	5.400	5.326	5.242	5.147	5.037	4.908	4.756	4.573	4.350	4.071	3.713	3.235	2.930	2.897	2.863
4400	5.822	5.777	5.727	5.672	5.611	5.543	5.467	5.380	5.281	5.166	5.033	4.875	4.685	4.454	4.165	3.792	3.296	2.981	2.946	2.910
4500	5.978	5.931	5.879	5.822	5.759	5.689	5.611	5.519	5.416	5.298	5.159	4.995	4.799	4.559	4.259	3.873	3.358	3.031	2.995	2.958
4600	6.135	6.087	6.034	5.975	5.909	5.836	5.754	5.660	5.554	5.431	5.287	5.117	4.914	4.665	4.354	3.954	3.421	3.082	3.044	3.006
4700	6.295	6.245	6.190	6.129	6.061	5.985	5.900	5.803	5.693	5.565	5.417	5.241	5.030	4.772	4.450	4.036	3.484	3.133	3.094	3.055
4800	6.457	6.405	6.348	6.285	6.214	6.136	6.048	5.948	5.833	5.702	5.548	5.366	5.152	4.881	4.548	4.119	3.547	3.184	3.144	3.103
4900	6.621	6.567	6.508	6.443	6.370	6.289	6.198	6.094	5.976	5.840	5.680	5.492	5.267	4.991	4.646	4.203	3.611	3.235	3.194	3.152
5000	6.787	6.731	6.670	6.603	6.527	6.444	6.349	6.242	6.120	5.979	5.815	5.620	5.387	5.102	4.745	4.287	3.676	3.287	3.244	3.200
5100	6.955	6.897	6.834	6.764	6.687	6.600	6.503	6.392	6.266	6.120	5.950	5.749	5.508	5.214	4.845	4.372	3.740	3.339	3.294	3.249
5200	7.125	7.065	7.000	6.928	6.848	6.759	6.658	6.544	6.413	6.263	6.087	5.880	5.631	5.327	4.947	4.458	3.806	3.391	3.345	3.299
5300	7.297	7.236	7.168	7.094	7.011	6.919	6.815	6.697	6.563	6.407	6.226	6.012	5.755	5.441	5.049	4.544	3.871	3.443	3.396	3.348
5400	7.471	7.408	7.338	7.262	7.176	7.081	6.974	6.852	6.714	6.554	6.367	6.146	5.881	5.557	5.152	4.632	3.937	3.496	3.447	3.398
5500	7.647	7.582	7.510	7.431	7.343	7.245	7.135	7.009	6.866	6.701	6.509	6.281	6.008	5.674	5.256	4.720	4.004	3.549	3.498	3.447
5600	7.825	7.758	7.684	7.603	7.512	7.411	7.297	7.168	7.021	6.851	6.652	6.417	6.136	5.792	5.361	4.808	4.071	3.602	3.550	3.497
5700	8.005	7.936	7.860	7.776	7.683	7.579	7.462	7.329	7.177	7.002	6.797	6.555	6.265	5.911	5.468	4.898	4.138	3.655	3.602	3.548
5800	8.188	8.117	8.038	7.952	7.856	7.749	7.628	7.491	7.335	7.154	6.944	6.695	6.396	6.031	5.575	4.988	4.206	3.709	3.654	3.598
5900	8.372	8.299	8.218	8.129	8.031	7.920	7.796	7.655	7.494	7.308	7.092	6.836	6.528	6.153	5.683	5.079	4.275	3.762	3.706	3.649
6000	8.558	8.483	8.400	8.309	8.207	8.094	7.966	7.821	7.655	7.464	7.241	6.978	6.662	6.275	5.792	5.171	4.343	3.816	3.758	3.699
6100	8.747	8.670	8.584	8.490	8.386	8.269	8.137	7.988	7.818	7.622	7.392	7.122	6.796	6.399	5.902	5.264	4.413	3.871	3.811	3.750
6200	8.937	8.858	8.770	8.674	8.566	8.446	8.311	8.158	7.983	7.781	7.545	7.267	6.933	6.524	6.014	5.357	4.482	3.925	3.864	3.802
6300	9.130	9.048	8.958	8.859	8.748	8.625	8.486	8.329	8.149	7.941	7.699	7.413	7.070	6.650	6.126	5.451	4.552	3.980	3.917	3.853
6400	9.325	9.241	9.148	9.046	8.933	8.806	8.663	8.502	8.317	8.104	7.855	7.561	7.209	6.778	6.239	5.546	4.623	4.035	3.970	3.904
6500	9.521	9.435	9.340	9.235	9.119	8.989	8.842	8.676	8.487	8.268	8.013	7.711	7.349	6.906	6.353	5.642	4.694	4.090	4.024	3.956
6600	9.720	9.631	9.534	9.426	9.307	9.173	9.023	8.853	8.658	8.434	8.171	7.862	7.490	7.036	6.468	5.738	4.765	4.146	4.078	4.008
6700	9.921	9.830	9.730	9.620	9.497	9.360	9.206	9.031	8.831	8.601	8.332	8.014	7.633	7.167	6.584	5.836	4.837	4.202	4.132	4.060
6800	10.12	10.03	9.93	9.81	9.69	9.55	9.39	9.21	9.01	8.77	8.49	8.17	7.78	7.30	6.70	5.93	4.91	4.26	4.19	4.11
6900	10.33	10.23	10.13	10.01	9.88	9.74	9.58	9.39	9.18	8.94	8.66	8.32	7.92	7.43	6.82	6.03	4.98	4.31	4.24	4.17
7000	10.54	10.44	10.33	10.21	10.08	9.93	9.76	9.58	9.36	9.11	8.82	8.48	8.07	7.57	6.94	6.13	5.06	4.37	4.29	4.22
7100	10.74	10.64	10.53	10.41	10.28	10.12	9.95	9.76	9.54	9.29	8.99	8.64	8.22	7.70	7.06	6.23	5.13	4.43	4.35	4.27
7200	10.95	10.85	10.74	10.61	10.48	10.32	10.15	9.95	9.72	9.46	9.16	8.80	8.37	7.84	7.18	6.33	5.20	4.48	4.40	4.32

$V - V_c = 18$ SIGMA = 2747.0

H \ x	0.08	0.09	0.10	0.2	0.4	0.6	0.8	1.0	1.2	1.4	1.6	1.8	2.0	2.2	2.4	2.6	2.8	3.0	3.2	3.4
7300	4.38	4.46	4.54	5.28	6.43	7.30	7.98	8.52	8.96	9.33	9.64	9.91	10.14	10.34	10.52	10.68	10.82	10.95	11.06	11.17
7400	4.43	4.52	4.60	5.35	6.54	7.43	8.12	8.67	9.12	9.50	9.82	10.09	10.33	10.53	10.72	10.88	11.03	11.16	11.28	11.38
7500	4.46	4.57	4.66	5.43	6.64	7.55	8.26	8.82	9.29	9.67	10.00	10.28	10.52	10.73	10.92	11.09	11.23	11.37	11.49	11.60
7600	4.54	4.63	4.71	5.50	6.74	7.67	8.40	8.98	9.45	9.85	10.18	10.47	10.71	10.93	11.12	11.29	11.45	11.58	11.71	11.82
7700	4.59	4.68	4.77	5.58	6.85	7.80	8.54	9.13	9.62	10.02	10.36	10.66	10.91	11.13	11.33	11.50	11.66	11.80	11.93	12.04
7800	4.65	4.74	4.83	5.66	6.95	7.93	8.68	9.29	9.79	10.20	10.55	10.85	11.11	11.33	11.53	11.71	11.87	12.02	12.15	12.26
7900	4.70	4.80	4.89	5.73	7.06	8.06	8.83	9.45	9.96	10.38	10.74	11.04	11.31	11.54	11.74	11.93	12.09	12.24	12.37	12.49
8000	4.76	4.85	4.95	5.81	7.17	8.18	8.98	9.61	10.13	10.56	10.92	11.24	11.51	11.75	11.95	12.14	12.31	12.46	12.59	12.72
8100	4.81	4.91	5.01	5.89	7.28	8.31	9.12	9.77	10.30	10.74	11.11	11.43	11.71	11.95	12.16	12.36	12.53	12.68	12.82	12.95
8200	4.87	4.97	5.07	5.97	7.38	8.45	9.27	9.93	10.47	10.92	11.31	11.63	11.92	12.16	12.38	12.58	12.75	12.91	13.05	13.18
8300	4.92	5.02	5.13	6.05	7.49	8.58	9.42	10.10	10.65	11.11	11.50	11.83	12.12	12.38	12.60	12.80	12.97	13.13	13.28	13.41
8400	4.98	5.08	5.19	6.13	7.60	8.71	9.57	10.26	10.83	11.30	11.69	12.03	12.33	12.59	12.82	13.02	13.20	13.36	13.51	13.65
8500	5.03	5.14	5.25	6.21	7.71	8.85	9.72	10.43	11.00	11.48	11.89	12.24	12.54	12.80	13.04	13.24	13.43	13.60	13.75	13.88
8600	5.09	5.20	5.31	6.29	7.83	8.98	9.88	10.60	11.18	11.67	12.09	12.44	12.75	13.02	13.26	13.47	13.66	13.83	13.98	14.12
8700	5.14	5.26	5.37	6.37	7.94	9.12	10.03	10.77	11.37	11.87	12.29	12.65	12.97	13.24	13.48	13.70	13.89	14.07	14.22	14.37
8800	5.20	5.32	5.43	6.45	8.06	9.25	10.19	10.94	11.55	12.06	12.49	12.86	13.18	13.46	13.71	13.93	14.13	14.30	14.46	14.61
8900	5.26	5.37	5.49	6.53	8.16	9.39	10.35	11.11	11.73	12.25	12.69	13.07	13.40	13.68	13.94	14.16	14.36	14.54	14.71	14.85
9000	5.31	5.43	5.55	6.61	8.28	9.53	10.50	11.28	11.92	12.45	12.90	13.28	13.62	13.91	14.17	14.39	14.60	14.78	14.95	15.10
9100	5.37	5.49	5.61	6.69	8.39	9.67	10.66	11.46	12.11	12.65	13.10	13.50	13.84	14.13	14.40	14.63	14.84	15.03	15.20	15.35
9200	5.43	5.55	5.67	6.78	8.51	9.81	10.82	11.63	12.29	12.85	13.31	13.71	14.06	14.36	14.63	14.87	15.08	15.27	15.45	15.60
9300	5.49	5.61	5.74	6.86	8.63	9.95	10.98	11.81	12.48	13.05	13.52	13.93	14.28	14.59	14.87	15.11	15.33	15.52	15.70	15.86
9400	5.54	5.67	5.80	6.94	8.75	10.10	11.15	11.99	12.68	13.25	13.73	14.15	14.51	14.82	15.10	15.35	15.57	15.77	15.95	16.11
9500	5.60	5.73	5.86	7.03	8.86	10.24	11.31	12.17	12.87	13.45	13.95	14.37	14.74	15.06	15.34	15.59	15.82	16.02	16.21	16.37
9600	5.66	5.79	5.92	7.11	8.98	10.39	11.48	12.35	13.06	13.66	14.16	14.59	14.97	15.29	15.58	15.84	16.07	16.28	16.46	16.63
9700	5.72	5.85	5.99	7.20	9.10	10.53	11.64	12.53	13.26	13.86	14.38	14.82	15.20	15.53	15.83	16.09	16.32	16.53	16.72	16.89
9800	5.77	5.91	6.05	7.28	9.22	10.66	11.81	12.71	13.46	14.07	14.60	15.04	15.43	15.77	16.07	16.34	16.57	16.79	16.98	17.16
9900	5.83	5.97	6.11	7.37	9.34	10.83	11.98	12.90	13.66	14.28	14.81	15.26	15.67	16.01	16.32	16.63	16.84	17.05	17.25	17.43
10000	5.89	6.04	6.18	7.46	9.47	10.97	12.15	13.09	13.85	14.49	15.04	15.50	15.90	16.25	16.56	16.84	17.09	17.31	17.51	17.69
10100	5.95	6.10	6.24	7.54	9.59	11.12	12.32	13.27	14.06	14.71	15.26	15.73	16.14	16.50	16.81	17.10	17.35	17.57	17.78	17.96
10200	6.01	6.16	6.30	7.63	9.71	11.28	12.49	13.46	14.26	14.92	15.48	15.96	16.38	16.74	17.07	17.35	17.61	17.84	18.05	18.24
10300	6.07	6.22	6.37	7.72	9.84	11.43	12.66	13.65	14.46	15.14	15.71	16.20	16.62	16.99	17.32	17.61	17.87	18.11	18.32	18.51
10400	6.13	6.28	6.43	7.81	9.96	11.58	12.84	13.84	14.67	15.35	15.94	16.43	16.87	17.24	17.58	17.87	18.14	18.38	18.59	18.79
10500	6.19	6.34	6.50	7.89	10.09	11.73	13.01	14.04	14.88	15.57	16.17	16.67	17.11	17.49	17.83	18.13	18.40	18.65	18.87	19.07
10600	6.25	6.41	6.56	7.98	10.22	11.89	13.19	14.23	15.09	15.80	16.40	16.91	17.36	17.75	18.09	18.40	18.67	18.92	19.14	19.35
10700	6.31	6.47	6.63	8.07	10.34	12.05	13.37	14.43	15.30	16.02	16.63	17.15	17.61	18.00	18.35	18.67	18.94	19.20	19.42	19.63
10800	6.37	6.53	6.69	8.16	10.47	12.20	13.55	14.63	15.51	16.24	16.86	17.40	17.86	18.26	18.62	18.93	19.22	19.47	19.70	19.91
10900	6.43	6.60	6.76	8.25	10.60	12.36	13.73	14.82	15.72	16.47	17.10	17.64	18.11	18.52	18.88	19.20	19.49	19.75	19.99	20.20
11000	6.49	6.66	6.83	8.34	10.73	12.52	13.91	15.02	15.93	16.69	17.34	17.89	18.36	18.78	19.15	19.48	19.77	20.03	20.27	20.49
11100	6.55	6.72	6.89	8.44	10.86	12.68	14.09	15.23	16.15	16.93	17.58	18.13	18.62	19.04	19.42	19.75	20.05	20.33	20.56	20.78
11200	6.61	6.79	6.96	8.53	10.99	12.84	14.28	15.43	16.37	17.15	17.82	18.38	18.88	19.31	19.69	20.03	20.33	20.60	20.85	21.07
11300	6.67	6.85	7.03	8.62	11.12	13.00	14.46	15.63	16.59	17.38	18.06	18.64	19.14	19.58	19.96	20.31	20.61	20.89	21.14	21.37
11460	6.73	6.91	7.09	8.71	11.26	13.16	14.65	15.84	16.81	17.62	18.30	18.89	19.40	19.84	20.24	20.59	20.90	21.18	21.43	21.66

SIGMA = 2747.0

$V-V_c = 18$

H	3.4	3.2	3.0	2.8	2.6	2.4	2.2	2.0	1.8	1.6	1.4	1.2	1.0	0.8	0.6	0.4	0.2	0.10	0.09	0.08
11500	21.96	21.73	21.47	21.18	20.87	20.51	20.11	19.66	19.14	18.55	17.85	17.03	16.04	14.84	13.33	11.39	8.81	7.16	6.98	6.79
11600	22.27	22.03	21.76	21.47	21.15	20.79	20.39	19.93	19.40	18.79	18.09	17.25	16.25	15.02	13.49	11.52	8.90	7.23	7.04	6.86
11700	22.57	22.33	22.06	21.76	21.44	21.07	20.66	20.19	19.66	19.04	18.33	17.48	16.46	15.21	13.66	11.66	8.99	7.30	7.11	6.92
11800	22.87	22.63	22.36	22.06	21.72	21.35	20.93	20.46	19.92	19.29	18.57	17.70	16.67	15.41	13.83	11.79	9.09	7.36	7.17	6.98
11900	23.18	22.93	22.66	22.35	22.01	21.64	21.21	20.73	20.18	19.55	18.81	17.93	16.88	15.60	13.99	11.93	9.18	7.43	7.24	7.04
12000	23.49	23.24	22.96	22.65	22.31	21.92	21.49	21.00	20.44	19.80	19.05	18.16	17.09	15.79	14.16	12.07	9.28	7.50	7.30	7.11
12100	23.80	23.54	23.26	22.95	22.60	22.21	21.77	21.28	20.71	20.06	19.29	18.39	17.31	15.99	14.33	12.21	9.37	7.57	7.37	7.17
12200	24.11	23.85	23.57	23.25	22.89	22.50	22.06	21.55	20.98	20.31	19.54	18.62	17.52	16.18	14.50	12.35	9.47	7.64	7.44	7.23
12300	24.43	24.16	23.87	23.55	23.19	22.79	22.34	21.83	21.24	20.57	19.78	18.86	17.74	16.38	14.68	12.49	9.57	7.71	7.50	7.30
12400	24.75	24.48	24.18	23.85	23.49	23.08	22.63	22.11	21.51	20.83	20.03	19.09	17.96	16.58	14.85	12.63	9.66	7.78	7.57	7.36
12500	25.07	24.79	24.49	24.16	23.79	23.38	22.91	22.39	21.79	21.09	20.28	19.33	18.18	16.78	15.02	12.77	9.76	7.85	7.64	7.42
12600	25.39	25.11	24.81	24.47	24.10	23.68	23.20	22.67	22.06	21.36	20.54	19.56	18.40	16.98	15.20	12.91	9.86	7.92	7.70	7.49
12700	25.71	25.43	25.12	24.78	24.40	23.97	23.50	22.95	22.34	21.62	20.79	19.80	18.62	17.18	15.37	13.05	9.96	7.99	7.77	7.55
12800	26.04	25.75	25.44	25.09	24.71	24.27	23.79	23.24	22.61	21.89	21.04	20.04	18.85	17.38	15.55	13.19	10.05	8.06	7.84	7.61
12900	26.37	26.08	25.76	25.41	25.02	24.58	24.09	23.53	22.89	22.16	21.30	20.29	19.07	17.58	15.73	13.34	10.15	8.13	7.90	7.68
13000	26.70	26.40	26.08	25.72	25.33	24.88	24.38	23.82	23.17	22.43	21.56	20.53	19.30	17.79	15.91	13.48	10.25	8.20	7.97	7.74
13100	27.03	26.73	26.40	26.04	25.64	25.19	24.68	24.11	23.45	22.70	21.82	20.77	19.52	18.00	16.09	13.63	10.35	8.27	8.04	7.81
13200	27.36	27.06	26.73	26.36	25.95	25.50	24.98	24.40	23.74	22.97	22.08	21.02	19.75	18.20	16.27	13.78	10.45	8.34	8.11	7.87
13300	27.70	27.39	27.05	26.68	26.27	25.81	25.29	24.70	24.02	23.25	22.34	21.27	19.98	18.41	16.45	13.92	10.56	8.41	8.18	7.94
13400	28.03	27.72	27.38	27.01	26.59	26.12	25.59	24.99	24.31	23.52	22.60	21.52	20.21	18.62	16.63	14.07	10.66	8.48	8.25	8.00
13500	28.37	28.06	27.71	27.33	26.91	26.43	25.90	25.29	24.60	23.80	22.87	21.77	20.45	18.83	16.81	14.22	10.76	8.56	8.31	8.07
13600	28.72	28.37	28.05	27.66	27.23	26.75	26.21	25.59	24.89	24.08	23.14	22.02	20.68	19.04	17.00	14.37	10.86	8.63	8.38	8.13
13700	29.06	28.74	28.38	27.99	27.55	27.06	26.52	25.89	25.18	24.36	23.41	22.27	20.92	19.26	17.18	14.52	10.96	8.70	8.45	8.20
13800	29.41	29.08	28.72	28.32	27.88	27.38	26.83	26.20	25.48	24.65	23.68	22.53	21.15	19.47	17.37	14.67	11.07	8.77	8.52	8.27
13900	29.76	29.42	29.06	28.65	28.21	27.71	27.14	26.50	25.78	24.93	23.96	22.79	21.39	19.69	17.56	14.82	11.17	8.85	8.59	8.33
14000	30.11	29.76	29.40	28.99	28.54	28.03	27.46	26.81	26.07	25.22	24.22	23.04	21.63	19.91	17.75	14.97	11.27	8.92	8.66	8.40
14100	30.46	30.11	29.74	29.33	28.87	28.35	27.77	27.12	26.37	25.50	24.50	23.30	21.87	20.12	17.94	15.13	11.38	8.99	8.73	8.46
14200	30.81	30.47	30.09	29.67	29.20	28.68	28.09	27.43	26.67	25.79	24.77	23.56	22.11	20.34	18.13	15.28	11.48	9.07	8.80	8.53
14300	31.17	30.82	30.43	30.01	29.54	29.01	28.42	27.74	26.97	26.09	25.05	23.83	22.36	20.56	18.32	15.44	11.59	9.14	8.87	8.60
14400	31.53	31.17	30.78	30.35	29.87	29.34	28.74	28.06	27.28	26.38	25.33	24.09	22.60	20.78	18.51	15.59	11.70	9.22	8.94	8.67
14500	31.89	31.53	31.13	30.70	30.21	29.67	29.06	28.37	27.58	26.67	25.61	24.36	22.85	21.01	18.71	15.75	11.80	9.29	9.01	8.73
14600	32.25	31.89	31.49	31.05	30.55	30.01	29.39	28.69	27.89	26.97	25.89	24.62	23.10	21.23	18.90	15.90	11.91	9.37	9.09	8.80
14700	32.61	32.25	31.84	31.39	30.90	30.34	29.72	29.01	28.20	27.27	26.18	24.89	23.35	21.46	19.10	16.06	12.02	9.44	9.16	8.87
14800	32.98	32.61	32.20	31.75	31.24	30.68	30.05	29.33	28.51	27.57	26.46	25.16	23.60	21.68	19.29	16.22	12.12	9.52	9.23	6.94
14900	33.35	32.97	32.56	32.10	31.59	31.02	30.38	29.65	28.82	27.87	26.75	25.43	23.85	21.91	19.49	16.38	12.23	9.59	9.30	9.01
15000	33.72	33.34	32.92	32.45	31.94	31.36	30.71	29.98	29.14	28.17	27.04	25.70	24.10	22.14	19.69	16.54	12.34	9.67	9.37	9.07

SIGMA = 2465.0

$V-V_c = 19$

H	3.4	3.2	3.0	2.8	2.6	2.4	2.2	2.0	1.8	1.6	1.4	1.2	1.0	0.8	0.6	0.4	0.2	0.10	0.09	0.08
0	1.000	1.000	1.000	1.000	1.000	1.000	1.000	1.000	1.000	1.000	1.000	1.000	1.000	1.000	1.000	1.000	1.000	1.000	1.000	1.000
20	1.014	1.014	1.014	1.014	1.014	1.014	1.014	1.014	1.013	1.013	1.013	1.013	1.012	1.012	1.011	1.010	1.009	1.009	1.009	1.009
40	1.029	1.029	1.029	1.028	1.028	1.028	1.028	1.027	1.027	1.026	1.026	1.025	1.024	1.024	1.022	1.021	1.019	1.018	1.018	1.017
60	1.044	1.043	1.043	1.043	1.042	1.042	1.041	1.041	1.040	1.040	1.039	1.038	1.037	1.035	1.034	1.031	1.028	1.027	1.026	1.026
80	1.058	1.058	1.058	1.057	1.057	1.056	1.055	1.055	1.054	1.053	1.052	1.051	1.049	1.047	1.045	1.042	1.038	1.036	1.035	1.035
100	1.073	1.073	1.072	1.072	1.071	1.070	1.070	1.069	1.068	1.067	1.065	1.064	1.062	1.059	1.056	1.053	1.048	1.044	1.044	1.044
120	1.088	1.088	1.087	1.086	1.086	1.085	1.084	1.083	1.082	1.080	1.078	1.077	1.074	1.071	1.068	1.063	1.057	1.053	1.053	1.052
140	1.103	1.103	1.102	1.101	1.100	1.099	1.098	1.097	1.095	1.094	1.092	1.090	1.087	1.083	1.079	1.074	1.067	1.062	1.062	1.061
160	1.118	1.118	1.117	1.116	1.115	1.114	1.112	1.111	1.109	1.107	1.105	1.103	1.099	1.096	1.091	1.085	1.076	1.071	1.071	1.070
180	1.134	1.133	1.132	1.131	1.130	1.128	1.127	1.125	1.123	1.121	1.119	1.116	1.112	1.108	1.102	1.095	1.086	1.080	1.079	1.079
200	1.149	1.148	1.147	1.146	1.144	1.143	1.141	1.140	1.138	1.135	1.132	1.129	1.125	1.120	1.114	1.106	1.096	1.089	1.088	1.088
220	1.164	1.163	1.162	1.161	1.159	1.158	1.156	1.154	1.152	1.149	1.146	1.142	1.138	1.132	1.126	1.117	1.105	1.098	1.097	1.096
240	1.180	1.179	1.177	1.176	1.175	1.173	1.171	1.169	1.166	1.163	1.160	1.156	1.151	1.145	1.137	1.128	1.115	1.107	1.106	1.105
260	1.196	1.194	1.193	1.191	1.190	1.188	1.186	1.183	1.180	1.177	1.173	1.169	1.164	1.157	1.149	1.139	1.125	1.116	1.115	1.114
280	1.211	1.210	1.208	1.207	1.205	1.203	1.201	1.198	1.195	1.191	1.187	1.183	1.177	1.170	1.161	1.150	1.135	1.125	1.124	1.124
300	1.227	1.226	1.224	1.222	1.220	1.218	1.216	1.213	1.209	1.206	1.201	1.196	1.190	1.182	1.173	1.161	1.144	1.134	1.133	1.132
320	1.243	1.242	1.240	1.238	1.236	1.233	1.231	1.228	1.224	1.220	1.215	1.210	1.203	1.195	1.185	1.172	1.154	1.143	1.142	1.141
340	1.259	1.258	1.256	1.254	1.251	1.249	1.246	1.243	1.239	1.235	1.229	1.224	1.216	1.208	1.197	1.183	1.164	1.152	1.151	1.150
360	1.275	1.274	1.272	1.269	1.267	1.264	1.261	1.258	1.254	1.249	1.244	1.237	1.230	1.220	1.209	1.194	1.174	1.161	1.160	1.158
380	1.292	1.290	1.288	1.285	1.283	1.280	1.276	1.273	1.269	1.264	1.258	1.251	1.243	1.233	1.221	1.205	1.184	1.170	1.169	1.167
400	1.308	1.306	1.304	1.301	1.298	1.295	1.292	1.288	1.284	1.278	1.272	1.265	1.257	1.246	1.233	1.216	1.194	1.179	1.178	1.176
420	1.324	1.322	1.320	1.317	1.314	1.311	1.307	1.303	1.299	1.293	1.287	1.279	1.270	1.259	1.245	1.227	1.204	1.189	1.187	1.185
440	1.341	1.339	1.336	1.334	1.330	1.327	1.323	1.319	1.314	1.308	1.301	1.293	1.284	1.272	1.257	1.239	1.214	1.198	1.196	1.194
460	1.358	1.355	1.353	1.350	1.347	1.343	1.339	1.334	1.329	1.323	1.316	1.307	1.297	1.285	1.270	1.250	1.224	1.207	1.205	1.203
480	1.374	1.372	1.369	1.366	1.363	1.359	1.355	1.350	1.344	1.338	1.330	1.322	1.311	1.298	1.282	1.261	1.234	1.216	1.214	1.212
500	1.391	1.389	1.386	1.383	1.379	1.375	1.371	1.365	1.360	1.353	1.345	1.336	1.325	1.311	1.294	1.273	1.244	1.225	1.223	1.221
520	1.408	1.406	1.403	1.399	1.395	1.391	1.387	1.381	1.375	1.368	1.360	1.350	1.339	1.324	1.307	1.284	1.254	1.234	1.232	1.230
540	1.425	1.423	1.419	1.416	1.412	1.406	1.403	1.397	1.391	1.383	1.375	1.365	1.353	1.338	1.319	1.295	1.264	1.243	1.241	1.239
560	1.443	1.440	1.436	1.433	1.429	1.424	1.419	1.413	1.406	1.399	1.390	1.379	1.367	1.351	1.332	1.307	1.274	1.253	1.250	1.248
580	1.460	1.457	1.453	1.449	1.445	1.440	1.435	1.429	1.422	1.414	1.405	1.394	1.381	1.364	1.344	1.318	1.284	1.262	1.259	1.257
600	1.477	1.474	1.470	1.466	1.462	1.457	1.451	1.445	1.438	1.430	1.420	1.408	1.395	1.378	1.357	1.330	1.294	1.271	1.268	1.266
620	1.495	1.491	1.488	1.483	1.479	1.474	1.468	1.461	1.454	1.445	1.435	1.423	1.409	1.391	1.370	1.341	1.304	1.280	1.278	1.275
640	1.512	1.509	1.505	1.501	1.496	1.490	1.484	1.478	1.470	1.461	1.450	1.438	1.423	1.405	1.382	1.353	1.314	1.289	1.287	1.284
660	1.530	1.526	1.522	1.518	1.513	1.507	1.501	1.494	1.486	1.477	1.466	1.453	1.437	1.419	1.395	1.365	1.324	1.299	1.296	1.293
680	1.548	1.544	1.540	1.535	1.530	1.524	1.518	1.511	1.502	1.492	1.481	1.468	1.452	1.432	1.408	1.376	1.335	1.308	1.305	1.302
700	1.566	1.562	1.557	1.553	1.547	1.541	1.535	1.527	1.518	1.508	1.497	1.483	1.466	1.446	1.421	1.388	1.345	1.317	1.314	1.311
720	1.584	1.580	1.575	1.570	1.565	1.558	1.552	1.544	1.535	1.524	1.512	1.498	1.481	1.460	1.434	1.400	1.355	1.326	1.323	1.320
740	1.602	1.598	1.593	1.588	1.582	1.576	1.569	1.560	1.551	1.540	1.528	1.513	1.495	1.474	1.447	1.412	1.365	1.336	1.332	1.329
760	1.620	1.616	1.611	1.606	1.600	1.593	1.586	1.577	1.568	1.557	1.544	1.528	1.510	1.488	1.460	1.424	1.376	1.345	1.342	1.338
780	1.638	1.634	1.629	1.623	1.617	1.610	1.603	1.594	1.584	1.573	1.559	1.544	1.525	1.502	1.473	1.435	1.386	1.354	1.351	1.347
800	1.657	1.652	1.647	1.641	1.635	1.628	1.620	1.611	1.601	1.589	1.575	1.559	1.539	1.516	1.486	1.447	1.396	1.364	1.360	1.356

x

SIGMA =   2465.0

V-V$_c$ = 19

H / x	0.08	0.09	0.10	0.2	0.4	0.6	0.8	1.0	1.2	1.4	1.6	1.8	2.0	2.2	2.4	2.6	2.8	3.0	3.2	3.4
820	1.365	1.369	1.373	1.407	1.459	1.499	1.530	1.554	1.574	1.591	1.605	1.618	1.628	1.637	1.646	1.653	1.659	1.665	1.670	1.675
840	1.375	1.378	1.382	1.417	1.471	1.512	1.544	1.569	1.590	1.607	1.622	1.634	1.645	1.655	1.663	1.671	1.677	1.683	1.689	1.694
860	1.384	1.388	1.392	1.427	1.483	1.525	1.558	1.584	1.606	1.623	1.638	1.651	1.663	1.672	1.681	1.689	1.696	1.702	1.707	1.713
880	1.393	1.397	1.401	1.438	1.495	1.539	1.572	1.599	1.621	1.640	1.655	1.668	1.680	1.690	1.699	1.707	1.714	1.720	1.726	1.731
900	1.402	1.406	1.410	1.448	1.508	1.552	1.587	1.614	1.637	1.656	1.672	1.686	1.697	1.708	1.717	1.725	1.732	1.739	1.745	1.750
920	1.411	1.416	1.420	1.459	1.520	1.565	1.601	1.629	1.653	1.672	1.689	1.703	1.715	1.726	1.735	1.743	1.751	1.758	1.764	1.769
940	1.420	1.425	1.429	1.469	1.532	1.579	1.615	1.645	1.669	1.689	1.705	1.720	1.733	1.743	1.753	1.762	1.769	1.776	1.783	1.788
960	1.430	1.434	1.439	1.480	1.544	1.592	1.630	1.660	1.685	1.705	1.722	1.737	1.750	1.761	1.771	1.780	1.788	1.795	1.802	1.808
980	1.439	1.443	1.448	1.490	1.556	1.606	1.645	1.675	1.701	1.722	1.739	1.755	1.768	1.780	1.790	1.799	1.807	1.814	1.821	1.827
1000	1.448	1.453	1.458	1.501	1.569	1.620	1.659	1.691	1.717	1.738	1.757	1.772	1.786	1.798	1.808	1.818	1.826	1.833	1.840	1.846
1020	1.457	1.462	1.467	1.511	1.581	1.633	1.674	1.706	1.733	1.755	1.774	1.790	1.804	1.816	1.827	1.836	1.845	1.853	1.860	1.866
1040	1.466	1.471	1.476	1.522	1.593	1.647	1.689	1.722	1.749	1.772	1.791	1.808	1.822	1.834	1.845	1.855	1.864	1.872	1.879	1.885
1060	1.476	1.481	1.486	1.533	1.606	1.661	1.703	1.737	1.765	1.789	1.808	1.825	1.840	1.853	1.864	1.874	1.883	1.891	1.899	1.905
1080	1.485	1.490	1.495	1.543	1.618	1.674	1.718	1.753	1.782	1.806	1.826	1.843	1.858	1.871	1.883	1.893	1.902	1.911	1.918	1.925
1100	1.494	1.500	1.505	1.554	1.631	1.688	1.733	1.769	1.798	1.823	1.843	1.861	1.877	1.890	1.902	1.912	1.922	1.930	1.938	1.945
1120	1.503	1.509	1.514	1.564	1.643	1.702	1.748	1.785	1.815	1.840	1.861	1.879	1.895	1.909	1.921	1.932	1.941	1.950	1.958	1.965
1140	1.513	1.518	1.524	1.575	1.656	1.716	1.763	1.801	1.831	1.857	1.879	1.897	1.913	1.927	1.940	1.951	1.961	1.970	1.978	1.985
1160	1.522	1.528	1.534	1.586	1.668	1.730	1.778	1.817	1.848	1.874	1.896	1.915	1.932	1.946	1.959	1.970	1.981	1.990	1.998	2.005
1180	1.531	1.537	1.543	1.597	1.681	1.744	1.793	1.833	1.865	1.892	1.914	1.934	1.951	1.965	1.978	1.990	2.000	2.010	2.018	2.026
1200	1.540	1.547	1.553	1.607	1.694	1.758	1.809	1.849	1.882	1.909	1.932	1.952	1.969	1.984	1.998	2.010	2.020	2.030	2.038	2.046
1220	1.550	1.556	1.562	1.618	1.706	1.772	1.824	1.865	1.899	1.927	1.950	1.971	1.988	2.004	2.017	2.029	2.040	2.050	2.059	2.067
1240	1.559	1.565	1.572	1.629	1.719	1.787	1.839	1.881	1.915	1.944	1.968	1.989	2.007	2.023	2.037	2.049	2.060	2.070	2.079	2.087
1260	1.568	1.575	1.581	1.640	1.732	1.801	1.854	1.897	1.932	1.962	1.987	2.008	2.026	2.042	2.056	2.069	2.080	2.090	2.100	2.108
1280	1.577	1.584	1.591	1.651	1.745	1.815	1.870	1.914	1.950	1.979	2.005	2.026	2.045	2.062	2.076	2.089	2.101	2.111	2.121	2.129
1300	1.586	1.594	1.601	1.662	1.758	1.829	1.885	1.930	1.967	1.997	2.023	2.045	2.064	2.081	2.096	2.109	2.121	2.132	2.141	2.150
1320	1.596	1.603	1.610	1.673	1.770	1.844	1.901	1.947	1.984	2.015	2.041	2.064	2.084	2.101	2.116	2.129	2.141	2.152	2.162	2.171
1340	1.606	1.613	1.620	1.683	1.783	1.858	1.917	1.963	2.001	2.033	2.060	2.083	2.103	2.121	2.136	2.150	2.162	2.173	2.183	2.192
1360	1.615	1.622	1.630	1.694	1.796	1.873	1.932	1.980	2.019	2.051	2.079	2.102	2.122	2.140	2.156	2.170	2.183	2.194	2.204	2.213
1380	1.625	1.632	1.639	1.705	1.809	1.887	1.948	1.996	2.036	2.069	2.097	2.121	2.142	2.160	2.176	2.191	2.203	2.215	2.225	2.235
1400	1.634	1.641	1.649	1.716	1.822	1.902	1.964	2.013	2.054	2.087	2.116	2.140	2.162	2.180	2.197	2.211	2.224	2.236	2.246	2.256
1420	1.643	1.651	1.659	1.727	1.835	1.917	1.980	2.030	2.071	2.106	2.135	2.160	2.181	2.200	2.217	2.232	2.245	2.257	2.268	2.278
1440	1.653	1.661	1.668	1.738	1.849	1.931	1.995	2.047	2.089	2.124	2.154	2.179	2.201	2.220	2.237	2.253	2.266	2.278	2.289	2.299
1460	1.662	1.670	1.678	1.749	1.862	1.946	2.011	2.064	2.107	2.142	2.173	2.199	2.221	2.241	2.258	2.273	2.287	2.300	2.311	2.321
1480	1.672	1.680	1.688	1.761	1.875	1.961	2.027	2.081	2.125	2.161	2.192	2.218	2.241	2.261	2.279	2.294	2.308	2.321	2.333	2.343
1500	1.681	1.689	1.698	1.772	1.888	1.976	2.044	2.098	2.142	2.179	2.211	2.238	2.261	2.281	2.299	2.315	2.330	2.343	2.354	2.365
1520	1.690	1.699	1.707	1.783	1.901	1.990	2.060	2.115	2.160	2.198	2.230	2.257	2.281	2.302	2.320	2.337	2.351	2.364	2.376	2.387
1540	1.700	1.709	1.717	1.794	1.915	2.005	2.076	2.132	2.178	2.217	2.249	2.277	2.301	2.323	2.341	2.358	2.373	2.386	2.398	2.409
1560	1.709	1.718	1.727	1.805	1.928	2.020	2.092	2.150	2.197	2.236	2.269	2.297	2.322	2.343	2.362	2.379	2.394	2.408	2.420	2.431
1580	1.719	1.728	1.737	1.816	1.941	2.035	2.108	2.167	2.215	2.255	2.288	2.317	2.342	2.364	2.383	2.401	2.416	2.430	2.442	2.454
1600	1.728	1.737	1.746	1.827	1.955	2.050	2.125	2.184	2.233	2.273	2.308	2.337	2.363	2.385	2.405	2.422	2.438	2.452	2.465	2.476
1620	1.738	1.747	1.756	1.839	1.968	2.066	2.141	2.202	2.251	2.293	2.327	2.357	2.383	2.406	2.426	2.444	2.460	2.474	2.487	2.499
1640	1.747	1.757	1.766	1.850	1.982	2.081	2.158	2.219	2.270	2.312	2.347	2.378	2.404	2.427	2.447	2.466	2.482	2.496	2.509	2.521

SIGMA = 2465.0

$V - V_c = 19$

H \ x	0.08	0.09	0.10	0.2	0.4	0.6	0.8	1.0	1.2	1.4	1.6	1.8	2.0	2.2	2.4	2.6	2.8	3.0	3.2	3.4
1660	1.757	1.766	1.776	1.861	1.995	2.096	2.174	2.237	2.288	2.331	2.367	2.398	2.425	2.448	2.469	2.487	2.504	2.519	2.532	2.544
1680	1.766	1.776	1.786	1.873	2.009	2.111	2.191	2.255	2.307	2.350	2.387	2.418	2.446	2.469	2.491	2.509	2.526	2.541	2.555	2.567
1700	1.776	1.786	1.796	1.884	2.023	2.127	2.208	2.272	2.325	2.369	2.407	2.439	2.467	2.491	2.512	2.531	2.548	2.564	2.577	2.590
1720	1.786	1.796	1.805	1.895	2.036	2.142	2.224	2.290	2.344	2.389	2.427	2.459	2.488	2.512	2.534	2.553	2.571	2.586	2.600	2.613
1740	1.795	1.805	1.815	1.907	2.050	2.157	2.241	2.308	2.363	2.408	2.447	2.480	2.509	2.534	2.556	2.576	2.593	2.609	2.623	2.636
1760	1.805	1.815	1.825	1.918	2.064	2.173	2.258	2.326	2.382	2.428	2.467	2.501	2.530	2.555	2.578	2.598	2.616	2.632	2.646	2.660
1780	1.814	1.825	1.835	1.929	2.077	2.188	2.275	2.344	2.400	2.448	2.487	2.522	2.551	2.577	2.600	2.620	2.638	2.655	2.670	2.683
1800	1.824	1.835	1.845	1.941	2.091	2.204	2.292	2.362	2.419	2.467	2.508	2.542	2.573	2.599	2.622	2.643	2.661	2.678	2.693	2.707
1820	1.833	1.844	1.855	1.952	2.105	2.220	2.309	2.380	2.438	2.487	2.528	2.563	2.594	2.621	2.644	2.665	2.684	2.701	2.716	2.730
1840	1.843	1.854	1.865	1.964	2.119	2.235	2.326	2.398	2.458	2.507	2.549	2.585	2.616	2.643	2.667	2.688	2.707	2.724	2.740	2.754
1860	1.853	1.864	1.875	1.975	2.133	2.251	2.343	2.417	2.477	2.527	2.569	2.606	2.637	2.665	2.689	2.711	2.730	2.748	2.763	2.778
1880	1.862	1.874	1.885	1.987	2.147	2.267	2.363	2.435	2.496	2.547	2.590	2.627	2.659	2.687	2.712	2.734	2.753	2.771	2.787	2.801
1900	1.872	1.883	1.895	1.998	2.161	2.283	2.377	2.453	2.515	2.567	2.611	2.648	2.681	2.709	2.734	2.757	2.777	2.794	2.811	2.825
1920	1.882	1.893	1.905	2.010	2.175	2.299	2.395	2.472	2.535	2.587	2.632	2.670	2.703	2.732	2.757	2.780	2.800	2.818	2.835	2.850
1940	1.891	1.903	1.915	2.021	2.189	2.314	2.412	2.490	2.554	2.607	2.653	2.691	2.725	2.754	2.780	2.803	2.823	2.842	2.859	2.874
1960	1.901	1.913	1.925	2.033	2.203	2.330	2.430	2.509	2.574	2.628	2.674	2.713	2.747	2.776	2.803	2.826	2.847	2.866	2.883	2.898
1980	1.911	1.923	1.935	2.045	2.217	2.346	2.447	2.527	2.593	2.648	2.695	2.734	2.769	2.799	2.826	2.849	2.871	2.890	2.907	2.923
2000	1.920	1.933	1.945	2.056	2.231	2.362	2.465	2.546	2.613	2.669	2.716	2.756	2.791	2.822	2.849	2.872	2.894	2.914	2.931	2.947
2020	1.930	1.943	1.955	2.068	2.245	2.379	2.482	2.565	2.633	2.689	2.737	2.778	2.813	2.845	2.872	2.896	2.918	2.938	2.955	2.972
2040	1.940	1.952	1.965	2.080	2.260	2.395	2.500	2.584	2.653	2.710	2.758	2.800	2.836	2.867	2.895	2.920	2.942	2.962	2.980	2.996
2060	1.949	1.962	1.975	2.091	2.274	2.411	2.518	2.603	2.672	2.731	2.780	2.822	2.858	2.890	2.919	2.944	2.966	2.986	3.005	3.021
2080	1.959	1.972	1.985	2.103	2.288	2.427	2.535	2.622	2.692	2.751	2.801	2.844	2.881	2.913	2.942	2.967	2.990	3.011	3.029	3.046
2100	1.969	1.982	1.995	2.115	2.303	2.444	2.553	2.641	2.712	2.772	2.823	2.866	2.904	2.937	2.966	2.991	3.014	3.035	3.054	3.071
2120	1.979	1.992	2.005	2.127	2.317	2.460	2.571	2.660	2.733	2.793	2.844	2.888	2.927	2.960	2.989	3.015	3.039	3.060	3.079	3.096
2140	1.988	2.002	2.016	2.138	2.332	2.476	2.589	2.679	2.753	2.814	2.866	2.911	2.949	2.983	3.013	3.039	3.063	3.085	3.104	3.121
2160	1.998	2.012	2.026	2.150	2.346	2.493	2.607	2.698	2.773	2.835	2.888	2.933	2.972	3.007	3.037	3.064	3.088	3.109	3.129	3.147
2180	2.008	2.022	2.036	2.162	2.361	2.509	2.625	2.718	2.793	2.857	2.910	2.956	2.995	3.030	3.061	3.088	3.112	3.134	3.154	3.172
2200	2.018	2.032	2.046	2.174	2.375	2.526	2.643	2.737	2.814	2.878	2.932	2.978	3.019	3.054	3.085	3.112	3.137	3.159	3.179	3.198
2220	2.027	2.042	2.056	2.186	2.390	2.542	2.661	2.756	2.834	2.899	2.954	3.001	3.042	3.077	3.109	3.137	3.162	3.184	3.205	3.223
2240	2.037	2.052	2.066	2.198	2.404	2.559	2.680	2.776	2.855	2.921	2.976	3.024	3.065	3.101	3.133	3.161	3.187	3.210	3.230	3.249
2260	2.047	2.062	2.077	2.210	2.419	2.576	2.698	2.796	2.875	2.942	2.998	3.047	3.088	3.125	3.157	3.186	3.212	3.235	3.256	3.275
2280	2.057	2.072	2.087	2.222	2.434	2.593	2.716	2.815	2.896	2.964	3.020	3.070	3.112	3.149	3.182	3.211	3.237	3.260	3.282	3.301
2300	2.067	2.082	2.097	2.234	2.448	2.609	2.735	2.835	2.917	2.985	3.043	3.093	3.136	3.173	3.206	3.236	3.262	3.286	3.307	3.327
2320	2.077	2.092	2.107	2.246	2.463	2.626	2.753	2.855	2.938	3.007	3.065	3.116	3.159	3.197	3.231	3.261	3.287	3.311	3.333	3.353
2340	2.086	2.102	2.118	2.258	2.478	2.643	2.772	2.875	2.959	3.029	3.088	3.139	3.183	3.221	3.255	3.286	3.313	3.337	3.359	3.379
2360	2.096	2.112	2.128	2.270	2.493	2.660	2.790	2.894	2.980	3.051	3.111	3.162	3.207	3.246	3.280	3.311	3.338	3.363	3.385	3.406
2380	2.106	2.122	2.138	2.282	2.508	2.677	2.809	2.914	3.001	3.073	3.133	3.185	3.231	3.270	3.305	3.336	3.364	3.389	3.411	3.432
2400	2.116	2.132	2.146	2.294	2.523	2.694	2.828	2.934	3.022	3.095	3.156	3.209	3.255	3.295	3.330	3.361	3.390	3.415	3.438	3.458
2420	2.126	2.142	2.159	2.306	2.538	2.711	2.846	2.955	3.043	3.117	3.179	3.232	3.279	3.319	3.355	3.387	3.415	3.441	3.464	3.485
2440	2.136	2.152	2.169	2.318	2.553	2.726	2.865	2.975	3.064	3.139	3.202	3.256	3.303	3.344	3.380	3.412	3.441	3.467	3.491	3.512
2460	2.146	2.163	2.179	2.330	2.568	2.746	2.884	2.995	3.086	3.161	3.225	3.280	3.327	3.369	3.405	3.438	3.467	3.493	3.517	3.539
2480	2.156	2.173	2.196	2.342	2.583	2.763	2.903	3.015	3.107	3.183	3.248	3.304	3.352	3.394	3.431	3.464	3.493	3.520	3.544	3.566

SIGMA = 2465.0

V−V$_c$ = 19

X

H	3.4	3.2	3.0	2.8	2.6	2.4	2.2	2.0	1.8	1.6	1.4	1.2	1.0	0.8	0.6	0.4	0.2	0.10	0.09	0.08
2500	3.593	3.571	3.546	3.519	3.490	3.456	3.419	3.376	3.327	3.271	3.206	3.128	3.036	2.922	2.780	2.598	2.355	2.200	2.183	2.166
2520	3.620	3.598	3.573	3.546	3.515	3.482	3.444	3.401	3.351	3.295	3.228	3.150	3.056	2.941	2.798	2.613	2.367	2.210	2.193	2.175
2540	3.647	3.624	3.600	3.572	3.541	3.507	3.469	3.425	3.375	3.318	3.251	3.172	3.077	2.960	2.815	2.628	2.379	2.221	2.203	2.185
2560	3.674	3.652	3.626	3.599	3.568	3.533	3.494	3.450	3.400	3.341	3.274	3.193	3.097	2.979	2.832	2.643	2.391	2.231	2.213	2.195
2580	3.702	3.679	3.653	3.625	3.594	3.559	3.519	3.475	3.424	3.365	3.296	3.215	3.118	2.999	2.850	2.659	2.404	2.241	2.224	2.205
2600	3.729	3.706	3.680	3.652	3.620	3.585	3.545	3.500	3.448	3.388	3.319	3.237	3.138	3.018	2.868	2.674	2.416	2.252	2.234	2.215
2620	3.757	3.733	3.707	3.678	3.646	3.611	3.570	3.525	3.472	3.412	3.342	3.259	3.159	3.037	2.885	2.689	2.428	2.262	2.244	2.225
2640	3.785	3.761	3.735	3.705	3.673	3.637	3.596	3.550	3.497	3.436	3.365	3.281	3.180	3.057	2.903	2.705	2.441	2.273	2.254	2.235
2660	3.813	3.789	3.762	3.732	3.699	3.663	3.622	3.575	3.521	3.460	3.388	3.303	3.201	3.076	2.920	2.720	2.453	2.283	2.264	2.245
2680	3.841	3.816	3.789	3.759	3.726	3.689	3.647	3.600	3.546	3.484	3.411	3.325	3.222	3.096	2.938	2.736	2.465	2.294	2.275	2.255
2700	3.869	3.844	3.817	3.786	3.753	3.715	3.673	3.625	3.571	3.508	3.434	3.347	3.243	3.115	2.956	2.751	2.478	2.304	2.285	2.265
2720	3.897	3.872	3.844	3.814	3.780	3.742	3.699	3.651	3.596	3.532	3.457	3.369	3.264	3.135	2.974	2.767	2.490	2.314	2.295	2.275
2740	3.925	3.900	3.872	3.841	3.807	3.768	3.725	3.676	3.620	3.556	3.481	3.392	3.285	3.155	2.992	2.782	2.503	2.325	2.305	2.285
2760	3.954	3.928	3.900	3.868	3.834	3.795	3.751	3.702	3.645	3.580	3.504	3.414	3.306	3.174	3.010	2.798	2.515	2.335	2.316	2.295
2780	3.982	3.956	3.928	3.896	3.861	3.822	3.778	3.728	3.670	3.605	3.528	3.437	3.328	3.194	3.028	2.813	2.528	2.346	2.326	2.306
2800	4.011	3.984	3.956	3.924	3.888	3.848	3.804	3.753	3.696	3.629	3.551	3.459	3.349	3.214	3.046	2.829	2.540	2.356	2.336	2.316
2820	4.039	4.013	3.984	3.951	3.915	3.875	3.830	3.779	3.721	3.653	3.575	3.482	3.370	3.234	3.064	2.845	2.553	2.367	2.347	2.326
2840	4.068	4.041	4.012	3.979	3.943	3.902	3.857	3.805	3.746	3.678	3.599	3.505	3.392	3.254	3.082	2.861	2.565	2.378	2.357	2.336
2860	4.097	4.070	4.040	4.007	3.970	3.929	3.883	3.831	3.772	3.703	3.622	3.527	3.413	3.274	3.100	2.876	2.578	2.388	2.367	2.346
2880	4.126	4.099	4.068	4.035	3.998	3.957	3.910	3.857	3.797	3.727	3.646	3.550	3.435	3.294	3.118	2.892	2.591	2.399	2.378	2.356
2900	4.155	4.127	4.097	4.063	4.026	3.984	3.937	3.884	3.823	3.752	3.670	3.573	3.457	3.314	3.137	2.908	2.603	2.409	2.388	2.366
2920	4.184	4.156	4.125	4.091	4.054	4.011	3.964	3.910	3.848	3.777	3.694	3.596	3.478	3.335	3.155	2.924	2.616	2.420	2.398	2.376
2940	4.214	4.185	4.154	4.120	4.081	4.039	3.991	3.936	3.874	3.802	3.718	3.619	3.500	3.355	3.173	2.940	2.629	2.430	2.409	2.386
2960	4.243	4.214	4.183	4.148	4.109	4.066	4.018	3.963	3.900	3.827	3.742	3.642	3.522	3.375	3.192	2.956	2.641	2.441	2.419	2.397
2980	4.272	4.244	4.212	4.177	4.138	4.094	4.045	3.989	3.926	3.852	3.767	3.666	3.544	3.396	3.210	2.972	2.654	2.452	2.429	2.407
3000	4.302	4.273	4.241	4.205	4.166	4.122	4.072	4.016	3.952	3.877	3.791	3.689	3.566	3.416	3.229	2.988	2.667	2.462	2.440	2.417

SIGMA = 2465.0

$V - V_c = 19$

H	0.08	0.09	0.10	0.2	0.4	0.6	0.8	1.0	1.2	1.4	1.6	1.8	2.0	2.2	2.4	2.6	2.8	3.0	3.2	3.4
3100	2.468	2.492	2.516	2.731	3.069	3.322	3.519	3.677	3.806	3.914	4.005	4.083	4.150	4.210	4.262	4.308	4.350	4.387	4.421	4.452
3200	2.519	2.545	2.569	2.795	3.151	3.417	3.624	3.790	3.926	4.039	4.134	4.216	4.287	4.349	4.404	4.453	4.499	4.536	4.571	4.604
3300	2.571	2.597	2.623	2.861	3.233	3.513	3.730	3.904	4.047	4.165	4.265	4.352	4.426	4.491	4.549	4.600	4.646	4.687	4.724	4.758
3400	2.622	2.650	2.678	2.926	3.317	3.610	3.838	4.020	4.169	4.294	4.399	4.489	4.567	4.636	4.696	4.750	4.797	4.841	4.880	4.915
3500	2.674	2.704	2.732	2.993	3.402	3.708	3.947	4.138	4.294	4.424	4.534	4.629	4.710	4.782	4.845	4.901	4.952	4.997	5.038	5.075
3600	2.727	2.757	2.787	3.059	3.487	3.808	4.057	4.257	4.420	4.557	4.672	4.770	4.856	4.931	4.997	5.056	5.108	5.155	5.198	5.237
3700	2.779	2.811	2.842	3.127	3.574	3.909	4.169	4.378	4.549	4.691	4.811	4.914	5.004	5.082	5.151	5.212	5.267	5.317	5.361	5.402
3800	2.832	2.865	2.898	3.195	3.661	4.011	4.283	4.501	4.679	4.827	4.953	5.060	5.154	5.235	5.307	5.371	5.429	5.480	5.527	5.569
3900	2.885	2.919	2.954	3.263	3.749	4.114	4.398	4.625	4.811	4.965	5.096	5.208	5.306	5.391	5.466	5.533	5.592	5.646	5.695	5.739
4000	2.938	2.974	3.010	3.332	3.839	4.219	4.514	4.751	4.944	5.105	5.242	5.359	5.460	5.549	5.627	5.696	5.759	5.815	5.865	5.911
4100	2.991	3.029	3.066	3.402	3.929	4.324	4.632	4.878	5.080	5.247	5.389	5.511	5.616	5.709	5.790	5.863	5.927	5.986	6.038	6.086
4200	3.045	3.084	3.123	3.472	4.020	4.431	4.751	5.007	5.217	5.391	5.539	5.665	5.775	5.871	5.956	6.031	6.098	6.159	6.214	6.264
4300	3.099	3.140	3.180	3.542	4.112	4.540	4.872	5.138	5.356	5.537	5.691	5.822	5.936	6.036	6.124	6.202	6.272	6.335	6.392	6.444
4400	3.153	3.195	3.237	3.614	4.205	4.649	4.994	5.271	5.497	5.685	5.844	5.981	6.099	6.203	6.294	6.375	6.446	6.513	6.573	6.626
4500	3.208	3.251	3.294	3.685	4.299	4.760	5.118	5.405	5.639	5.835	6.000	6.142	6.264	6.372	6.467	6.551	6.626	6.694	6.756	6.811
4600	3.262	3.308	3.352	3.758	4.394	4.872	5.243	5.540	5.784	5.986	6.158	6.304	6.432	6.543	6.642	6.729	6.807	6.878	6.941	6.999
4700	3.317	3.364	3.411	3.830	4.490	4.985	5.370	5.678	5.930	6.140	6.317	6.470	6.601	6.717	6.819	6.909	6.990	7.063	7.129	7.189
4800	3.372	3.421	3.469	3.904	4.587	5.099	5.498	5.817	6.078	6.295	6.479	6.637	6.773	6.893	6.998	7.092	7.176	7.252	7.320	7.382
4900	3.428	3.478	3.528	3.978	4.685	5.215	5.628	5.957	6.227	6.452	6.643	6.806	6.947	7.071	7.180	7.277	7.364	7.442	7.513	7.577
5000	3.483	3.536	3.587	4.052	4.783	5.332	5.759	6.100	6.379	6.612	6.809	6.977	7.124	7.252	7.364	7.465	7.555	7.635	7.709	7.775
5100	3.539	3.593	3.646	4.127	4.883	5.450	5.891	6.244	6.532	6.773	6.976	7.151	7.302	7.434	7.551	7.655	7.748	7.831	7.907	7.975
5200	3.595	3.651	3.706	4.203	4.984	5.569	6.025	6.389	6.688	6.936	7.146	7.326	7.483	7.619	7.740	7.847	7.943	8.029	8.107	8.178
5300	3.652	3.709	3.766	4.279	5.085	5.690	6.160	6.537	6.844	7.101	7.318	7.504	7.665	7.807	7.931	8.042	8.141	8.230	8.311	8.384
5400	3.708	3.768	3.826	4.356	5.188	5.812	6.297	6.686	7.003	7.268	7.492	7.684	7.850	7.996	8.125	8.239	8.341	8.433	8.516	8.592
5500	3.765	3.827	3.887	4.433	5.291	5.935	6.436	6.836	7.164	7.437	7.668	7.866	8.038	8.188	8.320	8.438	8.544	8.638	8.724	8.802
5600	3.822	3.886	3.948	4.511	5.395	6.059	6.575	6.988	7.326	7.608	7.846	8.050	8.227	8.382	8.519	8.640	8.749	8.846	8.935	9.015
5700	3.880	3.945	4.009	4.589	5.501	6.185	6.717	7.142	7.490	7.780	8.026	8.236	8.419	8.578	8.719	8.844	8.956	9.057	9.148	9.231
5800	3.937	4.004	4.070	4.668	5.607	6.311	6.859	7.298	7.656	7.955	8.208	8.425	8.612	8.777	8.922	9.051	9.166	9.270	9.364	9.449
5900	3.995	4.064	4.132	4.747	5.714	6.439	7.003	7.455	7.824	8.132	8.392	8.615	8.808	8.978	9.127	9.260	9.378	9.485	9.582	9.670
6000	4.053	4.124	4.194	4.827	5.822	6.569	7.149	7.613	7.993	8.310	8.578	8.808	9.007	9.181	9.334	9.471	9.593	9.703	9.803	9.893
6100	4.11	4.16	4.26	4.91	5.93	6.70	7.30	7.77	8.16	8.49	8.77	9.00	9.21	9.39	9.54	9.68	9.81	9.92	10.03	10.12
6200	4.17	4.25	4.32	4.99	6.04	6.83	7.44	7.94	8.34	8.67	8.96	9.20	9.41	9.59	9.76	9.90	10.03	10.15	10.25	10.35
6300	4.23	4.31	4.38	5.07	6.15	6.96	7.59	8.10	8.51	8.86	9.15	9.40	9.61	9.80	9.97	10.12	10.25	10.37	10.48	10.58
6400	4.29	4.37	4.45	5.15	6.26	7.10	7.75	8.27	8.69	9.04	9.34	9.60	9.82	10.02	10.19	10.34	10.48	10.60	10.71	10.81
6500	4.35	4.43	4.51	5.24	6.38	7.23	7.90	8.43	8.87	9.23	9.54	9.80	10.03	10.23	10.41	10.56	10.70	10.83	10.94	11.05
6600	4.41	4.49	4.57	5.32	6.49	7.37	8.05	8.60	9.05	9.42	9.74	10.01	10.24	10.45	10.63	10.79	10.93	11.06	11.18	11.29
6700	4.47	4.55	4.64	5.40	6.61	7.51	8.21	8.77	9.23	9.61	9.94	10.21	10.45	10.67	10.85	11.02	11.16	11.30	11.42	11.53
6800	4.53	4.61	4.70	5.49	6.72	7.65	8.37	8.94	9.41	9.81	10.14	10.42	10.67	10.89	11.08	11.25	11.40	11.54	11.66	11.77
6900	4.59	4.68	4.77	5.57	6.84	7.79	8.53	9.12	9.60	10.00	10.34	10.64	10.89	11.11	11.31	11.48	11.64	11.78	11.90	12.02
7000	4.65	4.74	4.83	5.66	6.96	7.93	8.69	9.29	9.79	10.20	10.55	10.85	11.11	11.34	11.54	11.71	11.87	12.02	12.15	12.27
7100	4.71	4.80	4.90	5.74	7.07	8.07	8.85	9.47	9.98	10.40	10.76	11.07	11.33	11.56	11.77	11.95	12.12	12.26	12.40	12.52
7200	4.77	4.87	4.96	5.83	7.19	8.22	9.01	9.65	10.17	10.60	10.97	11.28	11.56	11.79	12.01	12.19	12.36	12.51	12.65	12.77

X

$V - V_K = 19$ SIGMA = 2465.0

H	3.4	3.2	3.0	2.8	2.6	2.4	2.2	2.0	1.8	1.6	1.4	1.2	1.0	0.8	0.6	0.4	0.2	0.10	0.09	0.08
7300	13.03	12.90	12.76	12.61	12.43	12.24	12.03	11.78	11.50	11.18	10.80	10.36	9.83	9.18	8.36	7.31	5.92	5.03	4.93	4.83
7400	13.29	13.16	13.01	12.85	12.68	12.48	12.26	12.01	11.73	11.40	11.01	10.56	10.01	9.34	8.51	7.43	6.00	5.09	4.99	4.89
7500	13.55	13.41	13.27	13.11	12.93	12.72	12.50	12.24	11.95	11.61	11.22	10.75	10.19	9.51	8.66	7.56	6.09	5.16	5.06	4.95
7600	13.81	13.67	13.52	13.36	13.18	12.97	12.74	12.48	12.18	11.83	11.43	10.95	10.38	9.68	8.80	7.68	6.18	5.23	5.12	5.02
7700	14.08	13.94	13.78	13.62	13.43	13.22	12.98	12.71	12.40	12.05	11.64	11.15	10.56	9.85	8.95	7.80	6.27	5.29	5.19	5.08
7800	14.35	14.20	14.05	13.87	13.68	13.47	13.22	12.95	12.64	12.27	11.85	11.35	10.75	10.02	9.11	7.93	6.36	5.36	5.25	5.14
7900	14.62	14.47	14.31	14.13	13.94	13.72	13.47	13.19	12.87	12.50	12.07	11.56	10.94	10.19	9.26	8.06	6.45	5.43	5.32	5.20
8000	14.89	14.74	14.58	14.40	14.20	13.97	13.72	13.43	13.10	12.72	12.28	11.76	11.13	10.37	9.41	8.18	6.54	5.50	5.38	5.27
8100	15.17	15.02	14.85	14.66	14.46	14.23	13.97	13.67	13.34	12.95	12.50	11.97	11.33	10.55	9.57	8.31	6.63	5.57	5.45	5.33
8200	15.45	15.29	15.12	14.93	14.72	14.49	14.22	13.92	13.58	13.18	12.72	12.18	11.52	10.72	9.72	8.44	6.73	5.63	5.51	5.39
8300	15.73	15.57	15.40	15.20	14.99	14.75	14.48	14.17	13.82	13.42	12.94	12.39	11.72	10.90	9.88	8.57	6.82	5.70	5.58	5.46
8400	16.01	15.85	15.67	15.47	15.26	15.01	14.73	14.42	14.06	13.65	13.17	12.60	11.92	11.08	10.04	8.70	6.91	5.77	5.65	5.52
8500	16.30	16.14	15.95	15.75	15.53	15.28	15.00	14.67	14.31	13.89	13.40	12.81	12.12	11.27	10.20	8.83	7.00	5.84	5.71	5.58
8600	16.59	16.42	16.23	16.03	15.80	15.54	15.26	14.93	14.56	14.13	13.62	13.03	12.32	11.45	10.36	8.96	7.10	5.91	5.78	5.65
8700	16.88	16.71	16.52	16.31	16.07	15.81	15.52	15.19	14.81	14.37	13.85	13.25	12.52	11.63	10.52	9.10	7.19	5.98	5.85	5.71
8800	17.18	17.00	16.81	16.59	16.35	16.09	15.79	15.45	15.06	14.61	14.09	13.47	12.73	11.82	10.69	9.23	7.29	6.05	5.92	5.78
8900	17.47	17.29	17.10	16.88	16.63	16.36	16.06	15.71	15.31	14.85	14.32	13.69	12.93	12.01	10.85	9.37	7.38	6.12	5.99	5.84
9000	17.77	17.59	17.39	17.16	16.92	16.64	16.33	15.97	15.57	15.10	14.56	13.91	13.14	12.20	11.02	9.50	7.48	6.19	6.05	5.91
9100	18.08	17.89	17.68	17.45	17.20	16.92	16.60	16.24	15.83	15.35	14.80	14.14	13.35	12.39	11.19	9.64	7.58	6.27	6.12	5.97
9200	18.38	18.19	17.98	17.75	17.49	17.20	16.87	16.51	16.09	15.60	15.04	14.37	13.56	12.58	11.36	9.78	7.68	6.34	6.19	6.04
9300	18.69	18.49	18.28	18.04	17.78	17.48	17.15	16.78	16.35	15.85	15.28	14.59	13.78	12.78	11.53	9.92	7.77	6.41	6.26	6.11
9400	19.00	18.80	18.58	18.34	18.07	17.77	17.43	17.05	16.61	16.11	15.52	14.83	13.99	12.97	11.70	10.06	7.87	6.48	6.33	6.17
9500	19.31	19.11	18.88	18.64	18.36	18.06	17.71	17.33	16.88	16.37	15.77	15.06	14.21	13.17	11.87	10.20	7.97	6.55	6.40	6.24
9600	19.62	19.42	19.19	18.94	18.66	18.35	18.00	17.60	17.15	16.62	16.01	15.29	14.43	13.37	12.04	10.34	8.07	6.63	6.47	6.31
9700	19.94	19.73	19.50	19.25	18.96	18.66	18.29	17.88	17.42	16.89	16.26	15.53	14.65	13.57	12.22	10.48	8.17	6.70	6.54	6.37
9800	20.26	20.05	19.81	19.55	19.26	18.94	18.58	18.16	17.69	17.15	16.51	15.77	14.87	13.77	12.39	10.63	8.27	6.77	6.61	6.44
9900	20.58	20.37	20.13	19.86	19.57	19.24	18.87	18.45	17.97	17.41	16.77	16.01	15.09	13.97	12.57	10.77	8.37	6.85	6.68	6.51
10000	20.91	20.69	20.44	20.17	19.87	19.54	19.16	18.73	18.24	17.68	17.02	16.25	15.32	14.17	12.75	10.92	8.48	6.92	6.75	6.58
10100	21.24	21.01	20.76	20.49	20.18	19.84	19.46	19.02	18.52	17.95	17.28	16.49	15.54	14.38	12.93	11.06	8.58	7.00	6.82	6.64
10200	21.57	21.34	21.08	20.80	20.49	20.15	19.75	19.31	18.81	18.22	17.54	16.73	15.77	14.59	13.11	11.21	8.68	7.07	6.89	6.71
10300	21.90	21.66	21.41	21.12	20.81	20.45	20.05	19.60	19.09	18.49	17.80	16.98	16.00	14.80	13.29	11.36	8.78	7.15	6.97	6.78
10400	22.23	22.00	21.73	21.44	21.12	20.76	20.36	19.90	19.37	18.77	18.06	17.23	16.23	15.01	13.48	11.51	8.89	7.22	7.04	6.85
10500	22.57	22.33	22.06	21.77	21.44	21.08	20.66	20.19	19.66	19.05	18.33	17.48	16.46	15.22	13.66	11.66	8.99	7.30	7.11	6.92
10600	22.91	22.67	22.39	22.09	21.76	21.39	20.97	20.49	19.95	19.32	18.60	17.73	16.70	15.43	13.85	11.81	9.10	7.37	7.18	6.99
10700	23.25	23.00	22.73	22.42	22.08	21.71	21.28	20.80	20.24	19.61	18.86	17.99	16.93	15.64	14.03	11.96	9.20	7.45	7.25	7.06
10800	23.60	23.35	23.06	22.75	22.41	22.02	21.59	21.10	20.54	19.89	19.13	18.24	17.17	15.86	14.22	12.12	9.31	7.52	7.33	7.13
10900	23.95	23.69	23.40	23.09	22.74	22.35	21.90	21.43	20.83	20.18	19.41	18.50	17.41	16.08	14.41	12.27	9.42	7.60	7.40	7.20
11000	24.30	24.03	23.74	23.42	23.07	22.67	22.22	21.71	21.13	20.46	19.68	18.76	17.65	16.30	14.60	12.43	9.53	7.68	7.48	7.27
11100	24.65	24.38	24.09	23.76	23.40	23.00	22.54	22.02	21.43	20.75	19.96	19.02	17.89	16.52	14.80	12.58	9.63	7.76	7.55	7.34
11200	25.01	24.73	24.43	24.10	23.73	23.32	22.86	22.35	21.74	21.04	20.24	19.28	18.14	16.74	14.99	12.74	9.74	7.83	7.62	7.41
11300	25.37	25.09	24.78	24.45	24.07	23.65	23.18	22.64	22.04	21.34	20.52	19.55	18.38	16.96	15.18	12.90	9.85	7.91	7.70	7.48
11400	25.73	25.44	25.13	24.79	24.41	23.99	23.51	22.97	22.35	21.63	20.80	19.81	18.63	17.19	15.38	13.06	9.96	7.99	7.77	7.55

x

SIGMA = 2465.0

V-Vc = 19

H \ x	3.4	3.2	3.0	2.8	2.6	2.4	2.2	2.0	1.8	1.6	1.4	1.2	1.0	0.8	0.6	0.4	0.2	0.10	0.09	0.08
11500	26.09	25.80	25.49	25.14	24.75	24.32	23.84	23.29	22.66	21.93	21.08	20.08	18.88	17.41	15.58	13.22	10.07	8.07	7.85	7.62
11600	26.45	26.16	25.84	25.49	25.10	24.66	24.17	23.61	22.97	22.23	21.37	20.35	19.13	17.64	15.78	13.38	10.18	8.15	7.92	7.69
11700	26.82	26.53	26.20	25.84	25.45	25.00	24.50	23.93	23.28	22.53	21.66	20.62	19.38	17.87	15.97	13.54	10.29	8.23	8.00	7.77
11800	27.19	26.89	26.56	26.20	25.79	25.34	24.83	24.26	23.60	22.83	21.95	20.90	19.64	18.10	16.18	13.70	10.40	8.31	8.07	7.84
11900	27.57	27.26	26.93	26.56	26.15	25.69	25.17	24.58	23.91	23.14	22.24	21.17	19.89	18.33	16.38	13.87	10.52	8.39	8.15	7.91
12000	27.94	27.63	27.29	26.92	26.50	26.03	25.51	24.91	24.23	23.45	22.53	21.45	20.15	18.56	16.58	14.03	10.63	8.47	8.23	7.98
12100	28.32	28.01	27.66	27.28	26.86	26.38	25.85	25.24	24.55	23.76	22.83	21.73	20.41	18.80	16.79	14.20	10.74	8.55	8.30	8.06
12200	28.70	28.38	28.03	27.65	27.21	26.73	26.19	25.58	24.88	24.07	23.13	22.01	20.67	19.04	16.99	14.36	10.86	8.63	8.38	8.13
12300	29.09	28.76	28.41	28.01	27.58	27.09	26.54	25.92	25.20	24.38	23.42	22.29	20.93	19.27	17.20	14.53	10.97	8.71	8.46	8.20
12400	29.47	29.14	28.78	28.38	27.94	27.44	26.89	26.25	25.53	24.70	23.73	22.58	21.20	19.51	17.41	14.70	11.09	8.79	8.54	8.28
12500	29.86	29.53	29.16	28.76	28.31	27.80	27.24	26.59	25.86	25.02	24.03	22.86	21.46	19.75	17.62	14.87	11.20	8.87	8.61	8.35
12600	30.25	29.91	29.54	29.13	28.67	28.16	27.59	26.94	26.19	25.34	24.33	23.15	21.73	20.00	17.83	15.04	11.32	8.95	8.69	8.43
12700	30.64	30.30	29.92	29.51	29.04	28.53	27.94	27.28	26.53	25.66	24.64	23.44	22.00	20.24	18.04	15.21	11.43	9.03	8.77	8.50
12800	31.04	30.69	30.31	29.89	29.42	28.89	28.30	27.63	26.86	25.98	24.95	23.73	22.27	20.48	18.25	15.38	11.55	9.12	8.85	8.57
12900	31.44	31.09	30.70	30.27	29.79	29.26	28.66	27.98	27.20	26.31	25.26	24.03	22.54	20.73	18.47	15.55	11.67	9.20	8.93	8.65
13000	31.84	31.48	31.09	30.65	30.17	29.63	29.02	28.33	27.54	26.64	25.57	24.32	22.82	20.98	18.68	15.73	11.79	9.28	9.01	8.72
13100	32.25	31.88	31.48	31.04	30.55	30.00	29.39	28.69	27.89	26.97	25.89	24.62	23.09	21.23	18.90	15.90	11.91	9.37	9.09	8.80
13200	32.65	32.28	31.88	31.43	30.93	30.38	29.75	29.04	28.23	27.30	26.21	24.92	23.37	21.48	19.12	16.08	12.03	9.45	9.16	8.88
13300	33.06	32.69	32.28	31.82	31.32	30.75	30.12	29.40	28.58	27.63	26.52	25.22	23.65	21.73	19.34	16.25	12.15	9.53	9.24	8.95
13400	33.47	33.09	32.68	32.22	31.70	31.13	30.49	29.76	28.93	27.97	26.85	25.52	23.93	21.99	19.56	16.43	12.27	9.62	9.32	9.03
13500	33.89	33.50	33.08	32.61	32.09	31.51	30.86	30.12	29.28	28.30	27.17	25.82	24.21	22.24	19.78	16.61	12.39	9.70	9.41	9.10
13600	34.30	33.91	33.49	33.01	32.49	31.96	31.24	30.49	29.63	28.64	27.49	26.13	24.50	22.50	20.00	16.79	12.51	9.79	9.49	9.18
13700	34.72	34.33	33.89	33.41	32.88	32.29	31.62	30.86	29.99	28.99	27.82	26.44	24.78	22.76	20.23	16.97	12.63	9.87	9.57	9.26
13800	35.14	34.74	34.30	33.82	33.28	32.67	31.99	31.23	30.35	29.33	28.15	26.75	25.07	23.02	20.45	17.15	12.76	9.96	9.65	9.33
13900	35.57	35.16	34.72	34.22	33.68	33.06	32.38	31.60	30.71	29.68	28.48	27.06	25.36	23.28	20.68	17.34	12.88	10.04	9.73	9.41
14000	35.99	35.58	35.13	34.63	34.08	33.46	32.76	31.97	31.07	30.02	28.81	27.37	25.65	23.54	20.91	17.52	13.00	10.13	9.81	9.49
14100	36.42	36.01	35.55	35.04	34.48	33.85	33.15	32.35	31.43	30.38	29.14	27.69	25.94	23.80	21.13	17.70	13.13	10.21	9.89	9.57
14200	36.86	36.43	35.97	35.46	34.89	34.25	33.54	32.72	31.80	30.73	29.48	28.00	26.23	24.07	21.37	17.89	13.25	10.30	9.98	9.65
14300	37.29	36.86	36.39	35.87	35.30	34.65	33.93	33.10	32.17	31.08	29.82	28.32	26.53	24.34	21.60	18.07	13.38	10.39	10.06	9.73
14400	37.73	37.29	36.82	36.29	35.71	35.05	34.32	33.49	32.54	31.44	30.16	28.64	26.83	24.61	21.83	18.26	13.50	10.48	10.14	9.81
14500	38.17	37.73	37.25	36.71	36.12	35.46	34.72	33.87	32.91	31.80	30.50	28.96	27.12	24.88	22.06	18.45	13.63	10.56	10.23	9.89
14600	38.61	38.16	37.68	37.14	36.54	35.87	35.11	34.26	33.28	32.16	30.84	29.28	27.42	25.15	22.30	18.64	13.76	10.65	10.31	9.96
14700	39.05	38.60	38.11	37.56	36.95	36.28	35.51	34.65	33.66	32.52	31.19	29.61	27.73	25.42	22.54	18.83	13.88	10.74	10.39	10.04
14800	39.50	39.04	38.54	37.99	37.38	36.69	35.92	35.04	34.04	32.88	31.53	29.94	28.03	25.69	22.77	19.02	14.01	10.83	10.48	10.12
14900	39.95	39.49	38.98	38.42	37.80	37.10	36.32	35.43	34.42	33.25	31.88	30.27	28.34	25.97	23.01	19.21	14.14	10.92	10.56	10.20
15000	40.40	39.93	39.42	38.85	38.22	37.52	36.73	35.83	34.80	33.62	32.24	30.60	28.64	26.25	23.25	19.40	14.27	11.00	10.65	10.28

SIGMA = 2225.0

$V-V_c = 20$

H x	3.4	3.2	3.0	2.8	2.6	2.4	2.2	2.0	1.8	1.6	1.4	1.2	1.0	0.8	0.6	0.4	0.2	0.10	0.09	0.08
0	1.000	1.000	1.000	1.000	1.000	1.000	1.000	1.000	1.000	1.000	1.000	1.000	1.000	1.000	1.000	1.000	1.000	1.000	1.000	1.000
20	1.016	1.016	1.016	1.016	1.016	1.015	1.015	1.015	1.015	1.015	1.014	1.014	1.014	1.013	1.012	1.012	1.011	1.010	1.010	1.010
40	1.032	1.032	1.032	1.031	1.031	1.031	1.031	1.030	1.030	1.029	1.029	1.028	1.027	1.026	1.025	1.023	1.021	1.020	1.020	1.019
60	1.048	1.048	1.048	1.047	1.047	1.047	1.046	1.045	1.045	1.044	1.043	1.042	1.041	1.039	1.037	1.035	1.032	1.029	1.029	1.029
80	1.065	1.064	1.064	1.063	1.063	1.062	1.062	1.060	1.060	1.059	1.058	1.056	1.055	1.053	1.050	1.047	1.042	1.039	1.039	1.039
100	1.081	1.081	1.080	1.080	1.079	1.078	1.077	1.076	1.075	1.074	1.072	1.071	1.068	1.066	1.063	1.058	1.053	1.049	1.049	1.048
120	1.098	1.097	1.097	1.096	1.095	1.094	1.093	1.092	1.090	1.089	1.087	1.085	1.082	1.079	1.075	1.070	1.063	1.059	1.059	1.058
140	1.115	1.114	1.113	1.112	1.111	1.110	1.109	1.108	1.106	1.104	1.102	1.099	1.096	1.093	1.088	1.082	1.074	1.069	1.069	1.068
160	1.131	1.131	1.130	1.129	1.128	1.126	1.125	1.123	1.121	1.119	1.117	1.114	1.110	1.106	1.101	1.094	1.085	1.079	1.079	1.078
180	1.148	1.148	1.146	1.145	1.144	1.143	1.141	1.139	1.137	1.135	1.132	1.129	1.125	1.120	1.114	1.106	1.095	1.089	1.088	1.087
200	1.166	1.165	1.163	1.162	1.161	1.159	1.157	1.155	1.153	1.150	1.147	1.143	1.139	1.133	1.127	1.118	1.106	1.099	1.098	1.097
220	1.183	1.182	1.180	1.179	1.177	1.176	1.174	1.171	1.169	1.166	1.162	1.158	1.153	1.147	1.140	1.130	1.117	1.109	1.108	1.107
240	1.200	1.199	1.197	1.196	1.194	1.192	1.190	1.188	1.185	1.181	1.178	1.173	1.168	1.161	1.153	1.142	1.128	1.119	1.118	1.117
260	1.218	1.216	1.215	1.213	1.211	1.209	1.207	1.204	1.201	1.197	1.193	1.188	1.182	1.175	1.166	1.154	1.139	1.129	1.128	1.127
280	1.235	1.234	1.232	1.230	1.228	1.226	1.223	1.220	1.217	1.213	1.208	1.203	1.197	1.189	1.179	1.166	1.149	1.139	1.138	1.136
300	1.253	1.251	1.250	1.248	1.245	1.243	1.240	1.237	1.233	1.229	1.224	1.218	1.211	1.203	1.192	1.179	1.160	1.149	1.147	1.146
320	1.271	1.269	1.267	1.265	1.263	1.260	1.257	1.253	1.250	1.245	1.240	1.234	1.226	1.217	1.206	1.191	1.171	1.159	1.157	1.156
340	1.289	1.287	1.285	1.283	1.280	1.277	1.274	1.270	1.266	1.261	1.256	1.249	1.241	1.231	1.219	1.203	1.182	1.169	1.167	1.166
360	1.307	1.305	1.303	1.300	1.298	1.294	1.291	1.287	1.283	1.277	1.271	1.264	1.256	1.245	1.232	1.216	1.193	1.179	1.177	1.176
380	1.325	1.323	1.321	1.318	1.315	1.312	1.308	1.304	1.299	1.294	1.287	1.280	1.271	1.260	1.246	1.228	1.204	1.189	1.187	1.186
400	1.344	1.341	1.339	1.336	1.333	1.329	1.326	1.321	1.316	1.310	1.303	1.295	1.286	1.274	1.259	1.240	1.215	1.199	1.197	1.195
420	1.362	1.360	1.357	1.354	1.351	1.347	1.343	1.338	1.333	1.327	1.320	1.311	1.301	1.288	1.273	1.253	1.226	1.209	1.207	1.205
440	1.381	1.378	1.375	1.372	1.369	1.365	1.361	1.356	1.350	1.344	1.336	1.327	1.316	1.303	1.287	1.265	1.237	1.219	1.217	1.215
460	1.400	1.397	1.394	1.391	1.387	1.383	1.378	1.373	1.367	1.360	1.352	1.343	1.331	1.318	1.300	1.278	1.248	1.229	1.227	1.225
480	1.418	1.416	1.412	1.409	1.405	1.401	1.396	1.391	1.384	1.377	1.369	1.359	1.347	1.332	1.314	1.291	1.259	1.240	1.237	1.235
500	1.437	1.434	1.431	1.428	1.423	1.419	1.414	1.408	1.402	1.394	1.385	1.375	1.362	1.347	1.328	1.303	1.271	1.250	1.247	1.245
520	1.457	1.453	1.450	1.446	1.442	1.437	1.432	1.426	1.419	1.411	1.402	1.391	1.378	1.362	1.342	1.316	1.282	1.260	1.258	1.255
540	1.476	1.472	1.469	1.465	1.461	1.456	1.450	1.444	1.437	1.428	1.419	1.407	1.393	1.377	1.356	1.329	1.293	1.270	1.268	1.265
560	1.495	1.492	1.488	1.484	1.479	1.474	1.468	1.462	1.454	1.446	1.435	1.424	1.409	1.392	1.370	1.342	1.304	1.280	1.278	1.275
580	1.515	1.511	1.507	1.503	1.498	1.493	1.487	1.480	1.472	1.463	1.452	1.440	1.425	1.407	1.384	1.355	1.315	1.291	1.288	1.285
600	1.534	1.531	1.526	1.522	1.517	1.511	1.505	1.498	1.490	1.480	1.469	1.456	1.441	1.422	1.398	1.367	1.327	1.301	1.298	1.295
620	1.554	1.550	1.546	1.541	1.536	1.530	1.524	1.516	1.508	1.498	1.486	1.473	1.457	1.437	1.412	1.380	1.338	1.311	1.308	1.305
640	1.574	1.570	1.565	1.561	1.555	1.549	1.542	1.535	1.526	1.516	1.504	1.490	1.473	1.452	1.427	1.393	1.349	1.321	1.318	1.315
660	1.594	1.590	1.585	1.580	1.574	1.568	1.561	1.553	1.544	1.533	1.521	1.506	1.489	1.468	1.441	1.407	1.361	1.332	1.328	1.325
680	1.614	1.610	1.605	1.600	1.594	1.587	1.580	1.572	1.562	1.551	1.538	1.523	1.505	1.483	1.455	1.420	1.372	1.342	1.339	1.335
700	1.634	1.630	1.625	1.619	1.613	1.607	1.599	1.590	1.580	1.569	1.556	1.540	1.521	1.498	1.470	1.433	1.384	1.352	1.349	1.345
720	1.655	1.650	1.645	1.639	1.633	1.626	1.618	1.609	1.599	1.587	1.573	1.557	1.538	1.514	1.484	1.446	1.395	1.363	1.359	1.355
740	1.675	1.670	1.665	1.659	1.653	1.645	1.637	1.628	1.617	1.605	1.591	1.574	1.554	1.530	1.499	1.459	1.406	1.373	1.369	1.365
760	1.696	1.691	1.685	1.679	1.673	1.665	1.657	1.647	1.636	1.624	1.609	1.592	1.571	1.545	1.513	1.473	1.418	1.383	1.379	1.376
780	1.716	1.711	1.706	1.699	1.693	1.685	1.676	1.666	1.655	1.642	1.627	1.609	1.587	1.561	1.528	1.486	1.429	1.394	1.390	1.386
800	1.737	1.732	1.726	1.720	1.713	1.705	1.696	1.685	1.674	1.660	1.645	1.626	1.604	1.577	1.543	1.499	1.441	1.404	1.400	1.396

V-V$_c$ = 20

SIGMA = 2225.0

H	0.08	0.09	0.10	0.2	0.4	0.6	0.8	1.0	1.2	1.4	1.6	1.8	2.0	2.2	2.4	2.6	2.8	3.0	3.2	3.4
820	1.406	1.410	1.414	1.453	1.513	1.558	1.593	1.621	1.644	1.663	1.679	1.693	1.705	1.715	1.725	1.733	1.740	1.747	1.753	1.758
840	1.416	1.420	1.425	1.464	1.526	1.573	1.609	1.638	1.661	1.681	1.698	1.712	1.724	1.735	1.745	1.753	1.761	1.768	1.774	1.779
860	1.426	1.431	1.435	1.476	1.540	1.587	1.625	1.654	1.679	1.699	1.716	1.731	1.744	1.755	1.765	1.774	1.781	1.788	1.795	1.801
880	1.436	1.441	1.446	1.487	1.553	1.602	1.641	1.671	1.697	1.717	1.735	1.750	1.763	1.775	1.785	1.794	1.802	1.809	1.816	1.822
900	1.447	1.451	1.456	1.499	1.567	1.618	1.657	1.689	1.714	1.736	1.754	1.770	1.783	1.795	1.806	1.815	1.823	1.831	1.837	1.843
920	1.457	1.462	1.467	1.511	1.580	1.633	1.673	1.706	1.732	1.754	1.773	1.789	1.803	1.815	1.826	1.836	1.844	1.852	1.859	1.865
940	1.467	1.472	1.477	1.523	1.594	1.648	1.690	1.723	1.750	1.773	1.792	1.809	1.823	1.836	1.847	1.856	1.865	1.873	1.880	1.887
960	1.477	1.482	1.488	1.534	1.608	1.663	1.706	1.740	1.768	1.792	1.812	1.829	1.843	1.856	1.867	1.878	1.887	1.895	1.902	1.909
980	1.487	1.493	1.498	1.546	1.622	1.678	1.722	1.758	1.787	1.811	1.831	1.848	1.863	1.877	1.888	1.899	1.908	1.916	1.924	1.931
1000	1.498	1.503	1.509	1.558	1.636	1.694	1.739	1.775	1.805	1.829	1.850	1.868	1.884	1.897	1.909	1.920	1.929	1.938	1.946	1.953
1020	1.508	1.514	1.519	1.570	1.649	1.709	1.756	1.793	1.823	1.848	1.870	1.888	1.904	1.918	1.930	1.941	1.951	1.960	1.968	1.975
1040	1.518	1.524	1.530	1.582	1.663	1.725	1.772	1.810	1.842	1.868	1.890	1.908	1.925	1.939	1.952	1.963	1.973	1.982	1.990	1.997
1060	1.529	1.534	1.540	1.594	1.677	1.740	1.789	1.828	1.860	1.887	1.909	1.929	1.945	1.960	1.973	1.984	1.995	2.004	2.012	2.020
1080	1.539	1.545	1.551	1.606	1.691	1.756	1.806	1.846	1.879	1.906	1.929	1.949	1.966	1.981	1.994	2.006	2.017	2.026	2.035	2.043
1100	1.549	1.555	1.562	1.618	1.705	1.771	1.823	1.864	1.897	1.925	1.949	1.969	1.987	2.002	2.016	2.028	2.039	2.048	2.057	2.065
1120	1.559	1.566	1.572	1.629	1.720	1.787	1.840	1.882	1.916	1.945	1.969	1.990	2.008	2.024	2.038	2.050	2.061	2.071	2.080	2.088
1140	1.570	1.576	1.583	1.642	1.734	1.803	1.857	1.900	1.935	1.964	1.989	2.010	2.029	2.045	2.059	2.072	2.083	2.094	2.103	2.111
1160	1.580	1.587	1.593	1.654	1.748	1.819	1.874	1.918	1.954	1.984	2.009	2.031	2.050	2.067	2.081	2.094	2.106	2.116	2.126	2.134
1180	1.590	1.597	1.604	1.666	1.762	1.835	1.891	1.936	1.973	2.004	2.030	2.052	2.071	2.088	2.103	2.116	2.128	2.139	2.149	2.157
1200	1.601	1.608	1.615	1.678	1.777	1.851	1.908	1.954	1.992	2.024	2.050	2.073	2.093	2.110	2.125	2.139	2.151	2.162	2.172	2.181
1220	1.611	1.618	1.625	1.690	1.791	1.867	1.926	1.973	2.011	2.044	2.071	2.094	2.114	2.132	2.148	2.161	2.174	2.185	2.195	2.204
1240	1.622	1.629	1.636	1.702	1.805	1.883	1.943	1.991	2.031	2.064	2.091	2.115	2.136	2.154	2.170	2.184	2.197	2.208	2.219	2.228
1260	1.632	1.640	1.647	1.714	1.820	1.899	1.961	2.010	2.050	2.084	2.112	2.136	2.158	2.176	2.192	2.207	2.220	2.232	2.242	2.252
1280	1.642	1.650	1.658	1.726	1.834	1.915	1.978	2.028	2.070	2.104	2.133	2.158	2.179	2.198	2.215	2.230	2.243	2.255	2.266	2.276
1300	1.653	1.661	1.668	1.739	1.849	1.931	1.996	2.047	2.089	2.124	2.154	2.179	2.201	2.221	2.238	2.253	2.266	2.279	2.290	2.300
1320	1.663	1.671	1.679	1.751	1.863	1.948	2.013	2.066	2.109	2.145	2.175	2.201	2.223	2.243	2.260	2.276	2.290	2.302	2.313	2.324
1340	1.674	1.682	1.690	1.763	1.878	1.964	2.031	2.085	2.129	2.165	2.196	2.223	2.246	2.266	2.283	2.299	2.313	2.326	2.337	2.348
1360	1.684	1.693	1.701	1.775	1.893	1.981	2.049	2.104	2.148	2.186	2.217	2.244	2.268	2.288	2.306	2.323	2.337	2.350	2.362	2.372
1380	1.695	1.703	1.712	1.788	1.907	1.997	2.067	2.123	2.168	2.206	2.239	2.266	2.290	2.311	2.330	2.346	2.361	2.374	2.386	2.397
1400	1.705	1.714	1.722	1.800	1.922	2.014	2.085	2.142	2.188	2.227	2.260	2.288	2.313	2.334	2.353	2.370	2.385	2.398	2.410	2.421
1420	1.716	1.725	1.733	1.812	1.937	2.030	2.103	2.161	2.208	2.248	2.282	2.310	2.335	2.357	2.376	2.393	2.409	2.422	2.435	2.446
1440	1.726	1.735	1.744	1.825	1.952	2.047	2.121	2.180	2.229	2.269	2.303	2.333	2.358	2.380	2.400	2.417	2.433	2.447	2.459	2.471
1460	1.737	1.746	1.755	1.837	1.967	2.064	2.139	2.200	2.249	2.290	2.325	2.355	2.381	2.403	2.423	2.441	2.457	2.471	2.484	2.496
1480	1.747	1.757	1.766	1.850	1.982	2.081	2.157	2.219	2.269	2.311	2.347	2.377	2.404	2.427	2.447	2.465	2.481	2.496	2.509	2.521
1500	1.758	1.767	1.777	1.862	1.997	2.097	2.176	2.238	2.290	2.333	2.369	2.400	2.427	2.450	2.471	2.489	2.506	2.521	2.534	2.546
1520	1.768	1.778	1.788	1.875	2.012	2.114	2.194	2.258	2.310	2.354	2.391	2.422	2.450	2.474	2.495	2.514	2.530	2.546	2.559	2.572
1540	1.779	1.789	1.799	1.887	2.027	2.131	2.213	2.278	2.331	2.375	2.413	2.445	2.473	2.497	2.519	2.538	2.555	2.571	2.584	2.597
1560	1.789	1.800	1.810	1.900	2.042	2.148	2.231	2.297	2.352	2.397	2.435	2.468	2.496	2.521	2.543	2.563	2.580	2.596	2.610	2.623
1580	1.800	1.810	1.821	1.913	2.057	2.166	2.250	2.317	2.372	2.418	2.457	2.491	2.520	2.545	2.567	2.587	2.605	2.621	2.635	2.648
1600	1.811	1.821	1.831	1.925	2.072	2.183	2.269	2.337	2.393	2.440	2.480	2.514	2.543	2.569	2.592	2.612	2.630	2.646	2.661	2.674
1620	1.821	1.832	1.842	1.938	2.088	2.200	2.287	2.357	2.414	2.462	2.502	2.537	2.567	2.593	2.616	2.637	2.655	2.672	2.687	2.700
1640	1.832	1.843	1.853	1.950	2.103	2.217	2.306	2.377	2.435	2.484	2.525	2.560	2.591	2.617	2.641	2.662	2.681	2.697	2.713	2.726

x

V–V$_c$ = 20   SIGMA = 2225.0

x	0.08	0.09	0.10	0.2	0.4	0.6	0.8	1.0	1.2	1.4	1.6	1.8	2.0	2.2	2.4	2.6	2.8	3.0	3.2	3.4
1660	1.843	1.854	1.864	1.963	2.118	2.235	2.325	2.397	2.457	2.506	2.548	2.584	2.615	2.642	2.666	2.687	2.706	2.723	2.739	2.753
1680	1.853	1.864	1.876	1.976	2.134	2.252	2.344	2.418	2.478	2.528	2.571	2.607	2.638	2.666	2.690	2.712	2.731	2.749	2.765	2.779
1700	1.864	1.875	1.887	1.989	2.149	2.269	2.363	2.438	2.499	2.550	2.593	2.630	2.663	2.690	2.715	2.737	2.757	2.775	2.791	2.806
1720	1.875	1.886	1.898	2.001	2.165	2.287	2.382	2.458	2.521	2.573	2.616	2.654	2.687	2.715	2.741	2.763	2.783	2.801	2.817	2.832
1740	1.885	1.897	1.909	2.014	2.180	2.305	2.401	2.479	2.542	2.595	2.640	2.678	2.711	2.740	2.766	2.788	2.809	2.827	2.844	2.859
1760	1.896	1.908	1.920	2.027	2.196	2.322	2.421	2.499	2.564	2.617	2.663	2.702	2.735	2.765	2.791	2.814	2.835	2.854	2.870	2.886
1780	1.907	1.919	1.931	2.040	2.211	2.340	2.440	2.520	2.585	2.640	2.686	2.726	2.760	2.790	2.816	2.840	2.861	2.880	2.897	2.913
1800	1.917	1.930	1.942	2.053	2.227	2.358	2.459	2.541	2.607	2.663	2.710	2.750	2.785	2.815	2.842	2.866	2.887	2.907	2.924	2.940
1820	1.928	1.941	1.953	2.066	2.243	2.376	2.479	2.562	2.629	2.685	2.733	2.774	2.809	2.840	2.868	2.892	2.914	2.933	2.951	2.967
1840	1.939	1.952	1.964	2.079	2.259	2.394	2.498	2.582	2.651	2.708	2.757	2.798	2.834	2.866	2.893	2.918	2.940	2.960	2.978	2.994
1860	1.950	1.963	1.975	2.092	2.274	2.411	2.518	2.603	2.673	2.731	2.780	2.823	2.859	2.891	2.919	2.944	2.967	2.987	3.005	3.022
1880	1.960	1.974	1.987	2.105	2.290	2.430	2.538	2.624	2.695	2.754	2.804	2.847	2.884	2.917	2.945	2.971	2.994	3.014	3.033	3.050
1900	1.971	1.985	1.998	2.118	2.306	2.448	2.558	2.645	2.717	2.777	2.828	2.872	2.909	2.942	2.971	2.997	3.020	3.041	3.060	3.077
1920	1.982	1.996	2.009	2.131	2.322	2.466	2.577	2.667	2.740	2.801	2.852	2.896	2.935	2.968	2.998	3.024	3.047	3.069	3.088	3.105
1940	1.993	2.007	2.020	2.144	2.338	2.484	2.597	2.688	2.762	2.824	2.876	2.921	2.960	2.994	3.024	3.051	3.075	3.096	3.115	3.133
1960	2.004	2.018	2.032	2.157	2.354	2.502	2.617	2.709	2.785	2.847	2.901	2.946	2.985	3.020	3.050	3.078	3.102	3.124	3.143	3.161
1980	2.014	2.029	2.043	2.170	2.370	2.521	2.637	2.731	2.807	2.871	2.925	2.971	3.011	3.046	3.077	3.105	3.129	3.151	3.171	3.189
2000	2.025	2.040	2.054	2.183	2.387	2.539	2.657	2.752	2.830	2.895	2.949	2.996	3.037	3.072	3.104	3.132	3.157	3.179	3.199	3.218
2020	2.036	2.051	2.065	2.197	2.403	2.557	2.678	2.774	2.853	2.918	2.974	3.021	3.063	3.099	3.131	3.159	3.184	3.207	3.228	3.246
2040	2.047	2.062	2.077	2.210	2.419	2.576	2.698	2.796	2.875	2.942	2.998	3.047	3.089	3.125	3.157	3.186	3.212	3.235	3.256	3.275
2060	2.058	2.073	2.088	2.223	2.435	2.594	2.718	2.817	2.898	2.966	3.023	3.072	3.115	3.152	3.184	3.214	3.240	3.263	3.284	3.304
2080	2.069	2.084	2.099	2.236	2.452	2.613	2.739	2.839	2.921	2.990	3.048	3.098	3.141	3.178	3.212	3.241	3.268	3.291	3.313	3.332
2100	2.080	2.095	2.111	2.250	2.468	2.632	2.759	2.861	2.945	3.014	3.073	3.123	3.167	3.205	3.239	3.269	3.296	3.320	3.342	3.361
2120	2.091	2.106	2.123	2.263	2.484	2.651	2.780	2.883	2.968	3.038	3.098	3.149	3.193	3.232	3.266	3.297	3.324	3.348	3.370	3.391
2140	2.102	2.118	2.133	2.276	2.501	2.669	2.800	2.905	2.991	3.062	3.123	3.175	3.220	3.259	3.294	3.325	3.352	3.377	3.399	3.420
2160	2.113	2.129	2.145	2.290	2.517	2.688	2.821	2.927	3.014	3.087	3.148	3.201	3.246	3.286	3.321	3.353	3.381	3.406	3.428	3.449
2180	2.123	2.140	2.156	2.303	2.534	2.707	2.842	2.950	3.038	3.111	3.173	3.227	3.273	3.313	3.349	3.381	3.409	3.435	3.458	3.479
2200	2.134	2.151	2.168	2.317	2.551	2.726	2.863	2.972	3.061	3.136	3.199	3.253	3.300	3.341	3.377	3.409	3.438	3.464	3.487	3.508
2220	2.145	2.162	2.179	2.330	2.567	2.745	2.884	2.994	3.085	3.160	3.224	3.279	3.327	3.368	3.405	3.437	3.466	3.493	3.516	3.538
2240	2.156	2.174	2.190	2.343	2.584	2.764	2.905	3.017	3.109	3.185	3.250	3.305	3.354	3.396	3.433	3.466	3.495	3.522	3.546	3.568
2260	2.167	2.185	2.202	2.357	2.601	2.784	2.926	3.039	3.133	3.210	3.276	3.332	3.381	3.423	3.461	3.494	3.524	3.551	3.576	3.598
2280	2.178	2.196	2.213	2.371	2.618	2.803	2.947	3.062	3.156	3.235	3.301	3.358	3.408	3.451	3.489	3.523	3.553	3.581	3.605	3.628
2300	2.189	2.207	2.225	2.384	2.634	2.822	2.968	3.085	3.180	3.260	3.327	3.385	3.435	3.479	3.518	3.552	3.583	3.610	3.635	3.658
2320	2.200	2.219	2.236	2.398	2.651	2.841	2.989	3.108	3.204	3.285	3.353	3.412	3.463	3.507	3.546	3.581	3.612	3.640	3.665	3.689
2340	2.212	2.230	2.248	2.411	2.668	2.861	3.011	3.131	3.229	3.310	3.380	3.439	3.490	3.535	3.575	3.610	3.642	3.670	3.696	3.719
2360	2.223	2.241	2.259	2.425	2.685	2.880	3.032	3.154	3.253	3.336	3.406	3.466	3.518	3.563	3.604	3.639	3.671	3.700	3.726	3.750
2380	2.234	2.252	2.271	2.439	2.702	2.900	3.054	3.177	3.277	3.361	3.432	3.493	3.546	3.592	3.632	3.669	3.701	3.730	3.756	3.780
2400	2.245	2.264	2.282	2.452	2.719	2.919	3.075	3.200	3.302	3.387	3.458	3.520	3.573	3.620	3.661	3.698	3.731	3.760	3.787	3.811
2420	2.256	2.275	2.294	2.466	2.736	2.939	3.097	3.223	3.326	3.412	3.485	3.547	3.601	3.649	3.690	3.728	3.761	3.791	3.818	3.842
2440	2.267	2.286	2.306	2.480	2.754	2.959	3.119	3.246	3.351	3.438	3.512	3.575	3.629	3.677	3.720	3.757	3.791	3.821	3.848	3.873
2460	2.278	2.298	2.317	2.494	2.771	2.979	3.140	3.270	3.375	3.464	3.538	3.602	3.658	3.706	3.749	3.787	3.821	3.852	3.879	3.905
2480	2.289	2.309	2.329	2.507	2.788	2.998	3.162	3.293	3.400	3.489	3.565	3.630	3.686	3.735	3.778	3.817	3.851	3.882	3.910	3.936

SIGMA = 2225.0

$V-V_c = 20$

H	3.4	3.2	3.0	2.8	2.6	2.4	2.2	2.0	1.8	1.6	1.4	1.2	1.0	0.8	0.6	0.4	0.2	0.10	0.09	0.08
2500	3.967	3.942	3.913	3.882	3.847	3.808	3.764	3.714	3.657	3.592	3.515	3.425	3.317	3.184	3.018	2.805	2.521	2.341	2.321	2.300
2520	3.999	3.973	3.944	3.912	3.877	3.838	3.793	3.743	3.685	3.619	3.542	3.450	3.340	3.206	3.038	2.823	2.535	2.352	2.332	2.311
2540	4.031	4.004	3.975	3.943	3.907	3.867	3.822	3.771	3.713	3.646	3.568	3.475	3.364	3.228	3.058	2.840	2.549	2.364	2.343	2.323
2560	4.063	4.036	4.006	3.974	3.938	3.897	3.852	3.800	3.741	3.673	3.594	3.500	3.388	3.250	3.078	2.858	2.563	2.376	2.355	2.334
2580	4.095	4.067	4.038	4.005	3.968	3.927	3.881	3.829	3.769	3.701	3.620	3.525	3.412	3.272	3.099	2.875	2.577	2.387	2.366	2.345
2600	4.127	4.099	4.069	4.036	3.999	3.957	3.911	3.858	3.798	3.728	3.647	3.551	3.436	3.295	3.119	2.893	2.591	2.399	2.378	2.356
2620	4.159	4.131	4.101	4.067	4.029	3.987	3.940	3.887	3.826	3.755	3.673	3.576	3.460	3.317	3.139	2.910	2.605	2.411	2.389	2.367
2640	4.191	4.163	4.132	4.098	4.060	4.018	3.970	3.916	3.854	3.783	3.700	3.602	3.484	3.340	3.159	2.928	2.619	2.422	2.401	2.379
2660	4.224	4.195	4.164	4.130	4.091	4.048	4.000	3.945	3.883	3.811	3.727	3.627	3.508	3.362	3.180	2.945	2.633	2.434	2.412	2.390
2680	4.256	4.228	4.196	4.161	4.122	4.079	4.030	3.975	3.911	3.839	3.753	3.653	3.532	3.385	3.200	2.963	2.647	2.446	2.424	2.401
2700	4.289	4.260	4.228	4.193	4.153	4.110	4.060	4.004	3.940	3.866	3.780	3.679	3.556	3.407	3.221	2.981	2.661	2.458	2.435	2.412
2720	4.322	4.292	4.260	4.224	4.185	4.140	4.090	4.034	3.969	3.894	3.807	3.704	3.581	3.430	3.241	2.999	2.675	2.469	2.447	2.424
2740	4.355	4.325	4.292	4.256	4.216	4.171	4.121	4.063	3.998	3.923	3.834	3.730	3.605	3.453	3.262	3.017	2.689	2.481	2.458	2.435
2760	4.388	4.358	4.325	4.288	4.248	4.202	4.151	4.093	4.027	3.951	3.862	3.756	3.630	3.476	3.283	3.034	2.704	2.493	2.470	2.446
2780	4.421	4.391	4.357	4.320	4.279	4.233	4.182	4.123	4.056	3.979	3.889	3.782	3.655	3.499	3.303	3.052	2.718	2.505	2.482	2.458
2800	4.455	4.424	4.390	4.353	4.311	4.265	4.212	4.153	4.085	4.007	3.916	3.809	3.679	3.522	3.324	3.070	2.732	2.517	2.493	2.469
2820	4.488	4.457	4.423	4.385	4.343	4.296	4.243	4.183	4.115	4.036	3.944	3.835	3.704	3.545	3.345	3.088	2.746	2.529	2.505	2.480
2840	4.522	4.490	4.456	4.417	4.375	4.327	4.274	4.213	4.144	4.064	3.971	3.861	3.729	3.568	3.366	3.107	2.761	2.541	2.516	2.492
2860	4.555	4.524	4.489	4.450	4.407	4.359	4.305	4.244	4.174	4.093	3.999	3.888	3.754	3.591	3.387	3.125	2.775	2.552	2.528	2.503
2880	4.589	4.557	4.522	4.483	4.439	4.391	4.336	4.274	4.204	4.122	4.027	3.914	3.779	3.614	3.408	3.143	2.789	2.564	2.540	2.514
2900	4.623	4.591	4.555	4.515	4.472	4.423	4.367	4.305	4.233	4.151	4.055	3.941	3.804	3.638	3.429	3.161	2.804	2.576	2.551	2.526
2920	4.657	4.624	4.588	4.548	4.504	4.454	4.399	4.335	4.263	4.180	4.083	3.968	3.830	3.661	3.450	3.179	2.818	2.588	2.563	2.537
2940	4.692	4.658	4.622	4.581	4.537	4.487	4.430	4.366	4.293	4.209	4.111	3.994	3.855	3.685	3.472	3.198	2.833	2.600	2.575	2.549
2960	4.726	4.692	4.655	4.615	4.569	4.519	4.462	4.397	4.323	4.238	4.139	4.021	3.880	3.708	3.493	3.216	2.847	2.612	2.586	2.560
2980	4.760	4.726	4.689	4.648	4.602	4.551	4.493	4.428	4.353	4.267	4.167	4.048	3.906	3.732	3.514	3.235	2.862	2.624	2.598	2.571
3000	4.795	4.761	4.723	4.681	4.635	4.583	4.525	4.459	4.384	4.297	4.195	4.075	3.931	3.756	3.536	3.253	2.876	2.636	2.610	2.583

X

$V-V_c = 20$ SIGMA = 2225.0

H \ X	3.4	3.2	3.0	2.8	2.6	2.4	2.2	2.0	1.8	1.6	1.4	1.2	1.0	0.8	0.6	0.4	0.2	0.10	0.09	0.08
3100	4.970	4.934	4.894	4.850	4.801	4.747	4.686	4.616	4.537	4.445	4.338	4.212	4.060	3.875	3.644	3.346	2.949	2.696	2.669	2.640
3200	5.148	5.110	5.068	5.022	4.971	4.913	4.849	4.776	4.692	4.596	4.484	4.351	4.192	3.997	3.753	3.440	3.023	2.757	2.728	2.698
3300	5.329	5.289	5.245	5.197	5.143	5.083	5.015	4.938	4.851	4.750	4.631	4.492	4.325	4.120	3.864	3.535	3.097	2.818	2.787	2.756
3400	5.513	5.471	5.425	5.375	5.318	5.255	5.184	5.104	5.012	4.905	4.782	4.635	4.460	4.245	3.977	3.632	3.172	2.879	2.847	2.814
3500	5.701	5.657	5.609	5.555	5.496	5.430	5.356	5.271	5.175	5.064	4.934	4.781	4.597	4.372	4.091	3.729	3.248	2.941	2.907	2.873
3600	5.891	5.845	5.795	5.739	5.677	5.608	5.530	5.442	5.341	5.225	5.089	4.928	4.736	4.501	4.206	3.828	3.324	3.003	2.968	2.932
3700	6.085	6.037	5.984	5.926	5.861	5.789	5.707	5.615	5.510	5.388	5.246	5.078	4.877	4.631	4.324	3.928	3.401	3.065	3.029	2.991
3800	6.281	6.231	6.176	6.116	6.048	5.972	5.887	5.791	5.681	5.554	5.406	5.230	5.020	4.763	4.442	4.029	3.479	3.128	3.090	3.050
3900	6.481	6.429	6.372	6.308	6.238	6.159	6.070	5.970	5.855	5.722	5.567	5.385	5.165	4.897	4.562	4.131	3.557	3.191	3.151	3.110
4000	6.684	6.630	6.570	6.504	6.430	6.348	6.256	6.151	6.031	5.893	5.732	5.541	5.313	5.033	4.684	4.235	3.636	3.255	3.213	3.170
4100	6.890	6.834	6.771	6.702	6.626	6.540	6.444	6.335	6.210	6.066	5.898	5.700	5.462	5.171	4.807	4.339	3.716	3.319	3.275	3.231
4200	7.100	7.041	6.976	6.904	6.824	6.735	6.635	6.522	6.392	6.242	6.067	5.861	5.613	5.310	4.932	4.445	3.796	3.383	3.338	3.291
4300	7.312	7.251	7.183	7.109	7.026	6.933	6.829	6.711	6.576	6.420	6.239	6.024	5.766	5.451	5.058	4.552	3.877	3.448	3.401	3.352
4400	7.527	7.464	7.394	7.316	7.230	7.134	7.026	6.903	6.763	6.601	6.412	6.189	5.922	5.594	5.186	4.660	3.959	3.513	3.464	3.414
4500	7.746	7.680	7.607	7.527	7.437	7.337	7.225	7.098	6.952	6.784	6.588	6.357	6.079	5.739	5.315	4.769	4.041	3.578	3.527	3.475
4600	7.968	7.899	7.824	7.740	7.647	7.544	7.427	7.295	7.144	6.970	6.767	6.526	6.238	5.886	5.446	4.879	4.124	3.644	3.591	3.537
4700	8.193	8.121	8.043	7.957	7.861	7.753	7.632	7.495	7.339	7.158	6.947	6.698	6.400	6.034	5.578	4.991	4.208	3.710	3.655	3.599
4800	8.421	8.347	8.266	8.176	8.077	7.965	7.840	7.698	7.536	7.349	7.131	6.873	6.563	6.185	5.712	5.103	4.293	3.777	3.720	3.662
4900	8.652	8.575	8.491	8.399	8.295	8.180	8.051	7.904	7.736	7.542	7.316	7.049	6.728	6.337	5.847	5.217	4.378	3.843	3.785	3.725
5000	8.886	8.807	8.720	8.624	8.517	8.398	8.264	8.112	7.938	7.738	7.504	7.227	6.896	6.490	5.984	5.332	4.463	3.911	3.850	3.788
5100	9.123	9.041	8.952	8.852	8.742	8.619	8.480	8.323	8.143	7.936	7.694	7.408	7.065	6.646	6.122	5.448	4.550	3.978	3.915	3.851
5200	9.364	9.279	9.186	9.084	8.970	8.842	8.699	8.536	8.351	8.136	7.887	7.591	7.237	6.803	6.262	5.565	4.637	4.046	3.981	3.915
5300	9.607	9.520	9.424	9.318	9.200	9.069	8.921	8.753	8.561	8.340	8.081	7.776	7.410	6.962	6.403	5.684	4.725	4.114	4.047	3.979
5400	9.854	9.764	9.665	9.555	9.434	9.298	9.145	8.972	8.774	8.545	8.279	7.964	7.586	7.123	6.546	5.803	4.813	4.183	4.114	4.043
5500	10.10	10.01	9.91	9.80	9.67	9.53	9.37	9.19	8.99	8.75	8.48	8.15	7.76	7.29	6.69	5.92	4.90	4.25	4.18	4.11
5600	10.36	10.26	10.16	10.04	9.91	9.76	9.60	9.42	9.21	8.96	8.68	8.34	7.94	7.45	6.84	6.05	4.99	4.32	4.25	4.17
5700	10.61	10.51	10.41	10.29	10.15	10.00	9.83	9.64	9.43	9.18	8.88	8.54	8.12	7.62	6.98	6.17	5.08	4.39	4.32	4.24
5800	10.87	10.77	10.66	10.53	10.40	10.24	10.07	9.87	9.65	9.39	9.09	8.74	8.31	7.79	7.13	6.29	5.17	4.46	4.38	4.30
5900	11.13	11.03	10.91	10.79	10.65	10.49	10.31	10.11	9.88	9.61	9.30	8.93	8.49	7.96	7.28	6.42	5.27	4.53	4.45	4.37
6000	11.40	11.29	11.17	11.04	10.90	10.73	10.55	10.34	10.10	9.83	9.51	9.13	8.68	8.13	7.43	6.54	5.36	4.60	4.52	4.44
6100	11.67	11.56	11.43	11.30	11.15	10.98	10.79	10.58	10.34	10.05	9.73	9.34	8.87	8.30	7.59	6.67	5.45	4.67	4.59	4.50
6200	11.94	11.83	11.70	11.56	11.41	11.23	11.04	10.82	10.57	10.28	9.94	9.54	9.06	8.48	7.74	6.80	5.55	4.75	4.66	4.57
6300	12.21	12.10	11.97	11.83	11.67	11.49	11.29	11.06	10.81	10.51	10.16	9.75	9.26	8.65	7.90	6.93	5.64	4.82	4.73	4.64
6400	12.49	12.37	12.24	12.09	11.93	11.75	11.54	11.31	11.06	10.74	10.38	9.96	9.45	8.83	8.06	7.06	5.73	4.89	4.80	4.70
6500	12.77	12.65	12.51	12.36	12.19	12.01	11.80	11.56	11.29	10.97	10.60	10.17	9.65	9.01	8.22	7.19	5.83	4.96	4.87	4.77
6600	13.06	12.93	12.79	12.64	12.46	12.27	12.05	11.81	11.53	11.21	10.83	10.38	9.85	9.20	8.38	7.33	5.93	5.04	4.94	4.84
6700	13.34	13.21	13.07	12.91	12.73	12.54	12.31	12.06	11.78	11.44	11.06	10.60	10.05	9.38	8.54	7.46	6.02	5.11	5.01	4.91
6800	13.64	13.50	13.35	13.19	13.01	12.81	12.58	12.32	12.03	11.68	11.29	10.82	10.25	9.57	8.70	7.60	6.12	5.18	5.08	4.97
6900	13.93	13.79	13.64	13.47	13.29	13.08	12.84	12.58	12.28	11.93	11.52	11.04	10.46	9.75	8.87	7.73	6.22	5.26	5.15	5.04
7000	14.23	14.08	13.93	13.76	13.57	13.35	13.11	12.84	12.53	12.17	11.75	11.26	10.67	9.94	9.04	7.87	6.32	5.33	5.22	5.11
7100	14.53	14.38	14.22	14.05	13.85	13.63	13.39	13.11	12.79	12.42	11.99	11.49	10.88	10.13	9.21	8.01	6.42	5.41	5.30	5.18
7200	14.83	14.68	14.52	14.34	14.14	13.91	13.66	13.37	13.05	12.67	12.23	11.71	11.09	10.33	9.38	8.15	6.52	5.48	5.37	5.25

SIGMA = 2225.0

$V-V_c$ - 20

x

H	0.08	0.09	0.10	0.2	0.4	0.6	0.8	1.0	1.2	1.4	1.6	1.8	2.0	2.2	2.4	2.6	2.8	3.0	3.2	3.4
7300	5.32	5.44	5.56	6.62	8.29	9.55	10.52	11.30	11.94	12.47	12.92	13.31	13.64	13.94	14.20	14.42	14.63	14.81	14.98	15.13
7400	5.39	5.51	5.63	6.72	8.44	9.72	10.72	11.52	12.17	12.72	13.18	13.57	13.92	14.22	14.48	14.72	14.93	15.12	15.29	15.44
7500	5.46	5.59	5.71	6.83	8.58	9.90	10.92	11.74	12.41	12.97	13.44	13.84	14.19	14.50	14.77	15.01	15.23	15.42	15.60	15.76
7600	5.53	5.66	5.79	6.93	8.73	10.07	11.12	11.96	12.64	13.21	13.70	14.11	14.47	14.79	15.06	15.31	15.53	15.73	15.91	16.07
7700	5.60	5.74	5.86	7.03	8.87	10.25	11.32	12.18	12.88	13.47	13.96	14.38	14.75	15.07	15.36	15.61	15.84	16.04	16.22	16.39
7800	5.68	5.81	5.94	7.14	9.02	10.43	11.53	12.40	13.12	13.72	14.23	14.66	15.04	15.36	15.66	15.91	16.14	16.35	16.54	16.71
7900	5.75	5.88	6.02	7.24	9.17	10.61	11.73	12.63	13.36	13.98	14.49	14.94	15.32	15.66	15.96	16.22	16.46	16.67	16.86	17.04
8000	5.82	5.96	6.10	7.35	9.32	10.79	11.94	12.86	13.61	14.23	14.76	15.22	15.61	15.96	16.26	16.53	16.77	16.99	17.18	17.36
8100	5.89	6.04	6.18	7.46	9.47	10.98	12.15	13.09	13.85	14.49	15.04	15.50	15.90	16.25	16.57	16.84	17.09	17.31	17.51	17.69
8200	5.96	6.11	6.26	7.56	9.62	11.16	12.36	13.32	14.10	14.76	15.31	15.79	16.20	16.56	16.87	17.16	17.41	17.64	17.84	18.03
8300	6.04	6.19	6.33	7.67	9.77	11.35	12.57	13.55	14.36	15.02	15.59	16.07	16.49	16.86	17.19	17.47	17.73	17.96	18.17	18.37
8400	6.11	6.26	6.41	7.78	9.93	11.54	12.79	13.79	14.61	15.29	15.87	16.36	16.79	17.17	17.50	17.80	18.06	18.30	18.51	18.71
8500	6.18	6.34	6.49	7.89	10.08	11.73	13.00	14.03	14.86	15.56	16.15	16.66	17.10	17.48	17.82	18.12	18.39	18.63	18.85	19.05
8600	6.26	6.42	6.57	8.00	10.24	11.92	13.22	14.27	15.12	15.83	16.44	16.95	17.40	17.79	18.14	18.45	18.72	18.97	19.19	19.40
8700	6.33	6.50	6.66	8.11	10.40	12.11	13.44	14.51	15.38	16.11	16.72	17.25	17.71	18.11	18.46	18.78	19.06	19.31	19.54	19.75
8800	6.41	6.57	6.74	8.22	10.55	12.30	13.67	14.75	15.64	16.39	17.02	17.55	18.02	18.43	18.79	19.11	19.40	19.65	19.89	20.10
8900	6.48	6.65	6.82	8.33	10.71	12.50	13.89	15.00	15.91	16.67	17.31	17.86	18.33	18.75	19.12	19.44	19.74	20.00	20.24	20.45
9000	6.56	6.73	6.90	8.44	10.88	12.70	14.11	15.25	16.18	16.95	17.60	18.16	18.65	19.07	19.45	19.76	20.08	20.35	20.59	20.81
9100	6.63	6.81	6.98	8.56	11.04	12.90	14.34	15.50	16.44	17.23	17.90	18.47	18.97	19.40	19.76	20.12	20.43	20.70	20.95	21.18
9200	6.71	6.89	7.06	8.67	11.20	13.10	14.57	15.75	16.72	17.52	18.20	18.78	19.29	19.73	20.12	20.47	20.78	21.06	21.31	21.54
9300	6.78	6.97	7.15	8.78	11.37	13.30	14.80	16.00	16.99	17.81	18.50	19.10	19.61	20.06	20.46	20.82	21.13	21.42	21.68	21.91
9400	6.86	7.05	7.23	8.90	11.53	13.50	15.03	16.26	17.26	18.10	18.81	19.41	19.94	20.40	20.81	21.17	21.49	21.78	22.04	22.28
9500	6.94	7.13	7.32	9.02	11.70	13.71	15.27	16.52	17.54	18.39	19.12	19.73	20.27	20.74	21.15	21.52	21.85	22.14	22.41	22.66
9600	7.01	7.21	7.40	9.14	11.87	13.91	15.51	16.78	17.82	18.69	19.43	20.06	20.60	21.08	21.50	21.88	22.21	22.51	22.79	23.03
9700	7.09	7.29	7.48	9.25	12.04	14.12	15.74	17.04	18.10	18.99	19.74	20.38	20.94	21.42	21.88	22.23	22.58	22.88	23.16	23.41
9800	7.17	7.37	7.57	9.37	12.21	14.33	15.98	17.31	18.39	19.29	20.05	20.71	21.27	21.77	22.21	22.60	22.96	23.26	23.54	23.80
9900	7.25	7.45	7.65	9.49	12.38	14.54	16.23	17.57	18.68	19.59	20.37	21.04	21.61	22.12	22.56	22.96	23.34	23.63	23.92	24.19
10000	7.32	7.53	7.74	9.61	12.55	14.75	16.47	17.84	18.96	19.90	20.69	21.37	21.96	22.47	22.93	23.33	23.72	24.01	24.31	24.58
10100	7.40	7.62	7.83	9.73	12.72	14.97	16.71	18.11	19.25	20.21	21.01	21.70	22.30	22.83	23.29	23.70	24.10	24.40	24.70	24.97
10200	7.48	7.70	7.91	9.85	12.90	15.18	16.96	18.38	19.55	20.52	21.34	22.04	22.65	23.18	23.65	24.07	24.48	24.78	25.09	25.37
10300	7.56	7.78	8.00	9.97	13.07	15.40	17.21	18.66	19.84	20.83	21.67	22.38	23.00	23.54	24.02	24.45	24.86	25.17	25.48	25.77
10400	7.64	7.86	8.09	10.09	13.25	15.62	17.46	18.94	20.14	21.15	22.00	22.72	23.36	23.91	24.40	24.83	25.25	25.57	25.88	26.17
10500	7.72	7.95	8.17	10.22	13.43	15.84	17.71	19.21	20.44	21.46	22.33	23.07	23.71	24.27	24.77	25.21	25.64	25.96	26.28	26.57
10600	7.80	8.03	8.26	10.34	13.61	16.06	17.97	19.49	20.74	21.78	22.66	23.42	24.07	24.64	25.15	25.60	26.03	26.36	26.69	26.98
10700	7.88	8.12	8.35	10.46	13.79	16.28	18.22	19.78	21.05	22.10	23.00	23.77	24.43	25.01	25.53	25.98	26.42	26.76	27.09	27.40
10800	7.96	8.20	8.44	10.59	13.97	16.51	18.48	20.06	21.35	22.43	23.34	24.12	24.80	25.39	25.91	26.36	26.79	27.16	27.50	27.81
10900	8.04	8.28	8.53	10.72	14.16	16.74	18.74	20.35	21.66	22.76	23.68	24.48	25.16	25.77	26.30	26.77	27.19	27.57	27.92	28.23
11000	8.12	8.37	8.62	10.84	14.34	16.96	19.00	20.64	21.97	23.09	24.03	24.83	25.53	26.15	26.66	27.17	27.57	27.98	28.33	28.65
11100	8.20	8.46	8.70	10.97	14.52	17.19	19.27	20.93	22.29	23.42	24.37	25.20	25.91	26.53	27.08	27.57	27.99	28.40	28.75	29.08
11200	8.28	8.54	8.79	11.10	14.71	17.42	19.53	21.22	22.60	23.75	24.72	25.56	26.28	26.91	27.47	27.97	28.41	28.81	29.17	29.50
11300	8.37	8.63	8.89	11.23	14.90	17.66	19.80	21.51	22.92	24.09	25.08	25.92	26.66	27.30	27.87	28.37	28.81	29.23	29.60	29.93
11400	8.45	8.71	8.98	11.35	15.09	17.89	20.07	21.81	23.24	24.43	25.43	26.29	27.04	27.69	28.27	28.78	29.24	29.65	30.03	30.37

SIGMA = 2225.0

$V-V_c$ = 20

H	3.4	3.2	3.0	2.8	2.6	2.4	2.2	2.0	1.8	1.6	1.4	1.2	1.0	0.8	0.6	0.4	0.2	0.10	0.09	0.08
11500	30.80	30.46	30.08	29.66	29.19	28.67	28.09	27.42	26.66	25.79	24.77	23.56	22.11	20.34	18.12	15.28	11.48	9.07	8.80	8.53
11600	31.25	30.89	30.51	30.08	29.61	29.08	28.48	27.81	27.04	26.15	25.11	23.88	22.41	20.61	18.36	15.47	11.61	9.16	8.89	8.61
11700	31.69	31.33	30.94	30.51	30.03	29.49	28.88	28.20	27.41	26.51	25.46	24.21	22.71	20.88	18.60	15.66	11.74	9.25	8.98	8.70
11800	32.13	31.77	31.38	30.94	30.45	29.90	29.29	28.59	27.79	26.88	25.80	24.54	23.02	21.16	18.84	15.85	11.87	9.34	9.06	8.78
11900	32.58	32.22	31.80	31.37	30.87	30.31	29.69	28.98	28.18	27.24	26.15	24.87	23.32	21.44	19.08	16.05	12.01	9.43	9.15	8.86
12000	33.04	32.66	32.25	31.80	31.30	30.73	30.10	29.38	28.56	27.61	26.51	25.20	23.63	21.72	19.32	16.24	12.14	9.53	9.24	8.95
12100	33.49	33.11	32.70	32.24	31.72	31.15	30.51	29.78	28.95	27.98	26.86	25.54	23.94	22.00	19.57	16.44	12.27	9.62	9.33	9.03
12200	33.95	33.57	33.14	32.68	32.16	31.58	30.92	30.18	29.34	28.36	27.22	25.87	24.26	22.28	19.81	16.64	12.41	9.71	9.42	9.12
12300	34.41	34.02	33.59	33.12	32.59	32.00	31.34	30.59	29.73	28.74	27.58	26.21	24.57	22.57	20.06	16.84	12.54	9.81	9.51	9.20
12400	34.88	34.48	34.05	33.56	33.03	32.43	31.76	30.99	30.12	29.12	27.94	26.55	24.89	22.85	20.31	17.04	12.68	9.90	9.60	9.29
12500	35.35	34.95	34.50	34.01	33.47	32.86	32.18	31.40	30.52	29.50	28.31	26.90	25.21	23.14	20.56	17.24	12.81	10.00	9.69	9.37
12600	35.82	35.41	34.96	34.47	33.91	33.30	32.60	31.82	30.92	29.88	28.67	27.24	25.53	23.43	20.81	17.44	12.95	10.09	9.78	9.46
12700	36.29	35.88	35.42	34.92	34.36	33.73	33.03	32.23	31.32	30.27	29.04	27.59	25.85	23.72	21.07	17.65	13.09	10.19	9.87	9.54
12800	36.77	36.35	35.89	35.38	34.81	34.17	33.46	32.65	31.73	30.66	29.41	27.94	26.18	24.02	21.32	17.85	13.23	10.28	9.96	9.63
12900	37.25	36.83	36.36	35.84	35.26	34.62	33.89	33.08	32.13	31.05	29.79	28.30	26.50	24.31	21.58	18.06	13.37	10.38	10.05	9.72
13000	37.74	37.30	36.83	36.30	35.72	35.06	34.33	33.50	32.54	31.45	30.16	28.65	26.83	24.61	21.84	18.27	13.51	10.48	10.14	9.80
13100	38.22	37.78	37.30	36.77	36.18	35.51	34.77	33.92	32.96	31.84	30.54	29.01	27.16	24.91	22.09	18.47	13.65	10.57	10.24	9.89
13200	38.71	38.27	37.78	37.24	36.64	35.96	35.21	34.35	33.37	32.24	30.92	29.37	27.50	25.21	22.36	18.68	13.79	10.67	10.33	9.98
13300	39.21	38.76	38.26	37.71	37.10	36.42	35.65	34.78	33.79	32.64	31.31	29.73	27.83	25.51	22.62	18.89	13.93	10.77	10.42	10.07
13400	39.70	39.25	38.74	38.19	37.57	36.88	36.10	35.22	34.21	33.05	31.69	30.09	28.17	25.82	22.88	19.11	14.07	10.87	10.51	10.16
13500	40.20	39.74	39.23	38.66	38.04	37.34	36.55	35.65	34.63	33.46	32.08	30.46	28.51	26.13	23.15	19.32	14.21	10.97	10.61	10.24
13600	40.71	40.23	39.72	39.15	38.51	37.80	37.00	36.09	35.06	33.87	32.47	30.83	28.85	26.43	23.41	19.53	14.36	11.06	10.70	10.33
13700	41.21	40.73	40.21	39.63	38.99	38.27	37.46	36.54	35.49	34.28	32.86	31.20	29.19	26.74	23.68	19.75	14.50	11.16	10.80	10.42
13800	41.72	41.24	40.70	40.12	39.46	38.73	37.91	36.98	35.92	34.69	33.26	31.57	29.54	27.06	23.95	19.97	14.65	11.26	10.89	10.51
13900	42.23	41.74	41.20	40.61	39.95	39.20	38.37	37.42	36.35	35.11	33.66	31.94	29.88	27.37	24.23	20.18	14.79	11.36	10.99	10.60
14000	42.75	42.25	41.70	41.10	40.43	39.68	38.84	37.88	36.79	35.53	34.06	32.32	30.23	27.68	24.50	20.40	14.94	11.46	11.08	10.69
14100	43.27	42.76	42.21	41.60	40.92	40.16	39.30	38.33	37.23	35.95	34.46	32.70	30.58	28.00	24.77	20.62	15.09	11.56	11.18	10.78
14200	43.79	43.28	42.72	42.10	41.41	40.64	39.77	38.79	37.67	36.37	34.86	33.08	30.94	28.32	25.05	20.84	15.23	11.66	11.27	10.87
14300	44.31	43.79	43.23	42.60	41.90	41.12	40.24	39.25	38.11	36.80	35.27	33.46	31.29	28.64	25.33	21.06	15.38	11.77	11.37	10.96
14400	44.84	44.32	43.74	43.10	42.40	41.61	40.72	39.71	38.56	37.23	35.68	33.85	31.65	28.96	25.61	21.29	15.53	11.87	11.46	11.05
14500	45.37	44.84	44.26	43.61	42.90	42.10	41.19	40.17	39.01	37.66	36.09	34.24	32.01	29.29	25.89	21.51	15.68	11.97	11.56	11.15
14600	45.90	45.37	44.78	44.12	43.40	42.59	41.67	40.64	39.46	38.10	36.51	34.63	32.37	29.61	26.17	21.74	15.83	12.07	11.66	11.24
14700	46.44	45.90	45.30	44.64	43.90	43.08	42.16	41.11	39.91	38.53	36.92	35.02	32.73	29.94	26.45	21.97	15.98	12.18	11.76	11.33
14800	46.98	46.43	45.82	45.15	44.41	43.58	42.64	41.58	40.37	38.97	37.34	35.41	33.10	30.27	26.74	22.19	16.13	12.28	11.85	11.42
14900	47.52	46.97	46.35	45.67	44.92	44.08	43.13	42.06	40.83	39.41	37.76	35.81	33.47	30.60	27.02	22.42	16.29	12.38	11.95	11.51
15000	48.07	47.51	46.88	46.20	45.43	44.58	43.62	42.54	41.29	39.86	38.19	36.21	33.84	30.94	27.31	22.65	16.44	12.49	12.05	11.61

x